CHEMISTRY OF
PYRROLES

CHEMISTRY OF
PYRROLES

Boris A. Trofimov
Al'bina I. Mikhaleva
Elena Yu Schmidt
Lyubov N. Sobenina

CRC Press
Taylor & Francis Group
Boca Raton London New York

CRC Press is an imprint of the
Taylor & Francis Group, an **informa** business

CRC Press
Taylor & Francis Group
6000 Broken Sound Parkway NW, Suite 300
Boca Raton, FL 33487-2742

First issued in paperback 2016

Version Date: 20140815

ISBN 13: 978-1-138-03402-0 (pbk)
ISBN 13: 978-1-4822-3242-4 (hbk)

This book contains information obtained from authentic and highly regarded sources. Reasonable efforts have been made to publish reliable data and information, but the author and publisher cannot assume responsibility for the validity of all materials or the consequences of their use. The authors and publishers have attempted to trace the copyright holders of all material reproduced in this publication and apologize to copyright holders if permission to publish in this form has not been obtained. If any copyright material has not been acknowledged please write and let us know so we may rectify in any future reprint.

Library of Congress Cataloging-in-Publication Data

Chemistry of pyrroles / Boris A. Trofimov, Al'bina I. Mikhaleva, Elena Yu Schmidt,
Lyubov N. Sobenina.
  pages cm
"A CRC title."
Includes bibliographical references and index.
ISBN 978-1-4822-3242-4 (hardcover : alk. paper) 1. Pyrroles. 2. Aromatic
compounds. 3. Heterocyclic chemistry. 4. Chemistry, Organic. I. Trofimov, B. A.
(Boris Aleksandrovich) II. Mikhaleva, A. I. (Al?bina Ivanovna) III. Schmidt, Elena Yu,
1980- IV. Sobenina, Lyubov N.

QD401.C4955 2015
547'.593--dc23   2014028236

Visit the Taylor & Francis Web site at
http://www.taylorandfrancis.com

and the CRC Press Web site at
http://www.crcpress.com

Contents

Preface

The book is devoted to the latest achievements in the chemistry of pyrroles, fundamental structural units of vitally important molecular systems (chlorophyll and hemoglobin), natural hormones and antibiotics, pigments and pheromones.

The core of the book is the discovery and development (by the authors and other research teams) of novel facile and highly effective method for the construction of the pyrrole ring from ketones (ketoximes) and acetylene in superbase catalytic systems (Trofimov reaction). Owing to this reaction, diverse pyrroles bearing aliphatic, cycloaliphatic, olefinic, aromatic, and heteroaromatic substituents; pyrroles fused with various cyclic and heterocyclic systems; as well as almost hitherto unknown N-vinylpyrroles became widely accessible. In the book, conditions of typical syntheses, limitations of their applicability, and possibility of vinyl chloride or dichloroethane application instead of acetylene are analyzed. Chemical engineering aspects of the first synthesis of tetrahydroindole and indole from commercially available oxime of cyclohexanone and acetylene are considered. New facets of pyrroles and N-vinyl pyrroles reactivity in the reactions with the participation of both the pyrrole ring and N-vinyl groups are discussed. About a thousand structures of novel pyrrole compounds and their yields and physical–chemical characteristics are given.

This new edition on pyrrole chemistry will attract the attention of synthetic chemists, photochemists and photophysicists, pharmacologists, biochemists, experts in the field of polymerization, and chemical engineers. The book will also be of interest to teachers, PhD students, and students of chemical specialties.

Introduction

The interest in pyrrole chemistry is progressing dramatically. More than 30 years ago, two fundamental monographs [1,2] covering various aspects of chemistry and physical chemistry of pyrroles were published. Since then, a flow of reviews and analytical publications related to synthesis, reactivity, and properties of pyrrole compounds have followed like an avalanche [3–39]. This is due to the ever-increasing knowledge of the essential role that pyrrole structures play in the chemistry of living organisms, drug design, and development of advanced materials. Correspondingly, a number of research papers, for example [40–51], dealing with the most diverse issues of synthetic, theoretical, and applied chemistry of pyrrole are snowballing.

Relatively simple pyrrole compounds are continued to be isolated from natural objects including antibiotics, pheromones, toxins, cell fission inhibitors, and immunomodulators [52,53]. Certainly, the steady attention to pyrroles is owing, first of all, to the fact that pyrrole moiety constitutes a core of numerous biologically important compounds such as chlorophyll, hemoglobin, vitamin B_{12}, and alkaloids, participating in the biotransformation of solar energy, oxygen transfer processes, and other life-sustaining reactions [54]. Marine organisms were found to comprise diverse polycyclic secondary metabolites bearing halopyrrole structural units [55,56]. Besides, a number of alkaloids incorporating alkyl- and aryl-substituted fragments were isolated from marine objects [57–59]. The halogenated pyrrole alkaloids were obtained from microorganisms, mushrooms, plants, and sea invertebrates [60]. A pyrrole compound used for the detection of histidine kinase [61] and possessing gram-positive antimicrobic activity against stable bacterial strains [62] was isolated from *Streptomyces rimosus* (Scheme I.1). Some polysubstituted pyrroles were found to be useful in the treatment of epidermoid carcinoma in humans [63,64]. 1,2-Diarylpyrroles turned out to be potent and selective inhibitors of cyclooxogenase-2 (COX-2) enzyme, which plays an important role in the development of inflammatory processes [65]. 1-Phenyl-3-(aminomethyl)pyrrole showed high affinity to the D2, D3, and D4 subtypes of a dopamine receptor [66]. Some aroyl(aminoacyl)pyrroles exhibited anticonvulsive activity [67].

Pyrroles are intensively employed in the synthesis of natural compound congeners [68] and as pharmacophores [69–72] and building blocks for drug design. For example, anticancer antibiotic CC-1065 (Scheme I.2) [73,74] incorporates pyrrole fragment in its structure. Lipitor (atorvastatin), one of the best-selling drug in pharmaceutical history used for lowering blood cholesterol, represents functionalized 2,3-diphenylpyrrole (Scheme I.2) [75,76]).

Over the last decade, such research areas as design of electroconductive polypyrroles [77,78], optoelectronic materials [79–81], and sensors [82] containing pyrrole structural units have been developing especially rapidly.

Note that the works cited here are just for illustrative purposes. In fact, the flow of publications devoted to these directions is tremendous and keeps increasing drastically.

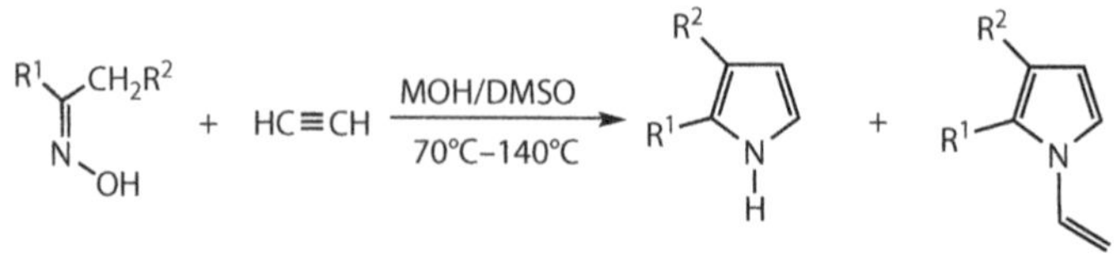

SCHEME I.1 Pyrroles from natural sources.

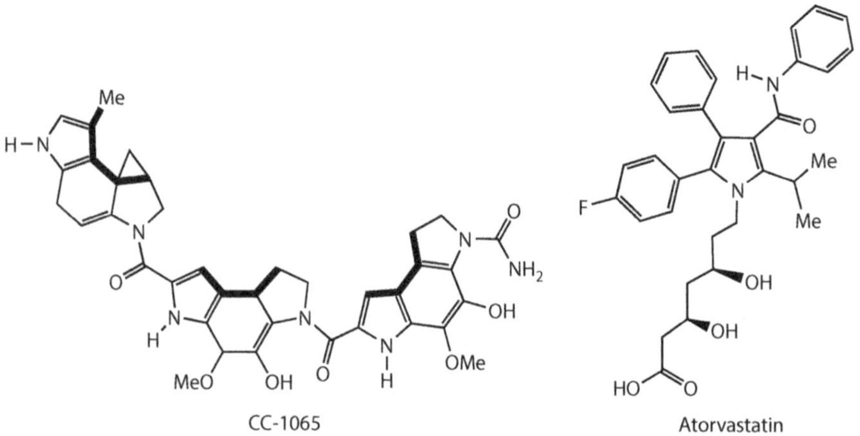

CC-1065   Atorvastatin

SCHEME I.2 Popular drugs having pyrrole scaffold.

The pyrrole core as the dominant subunit ensures play of colors both in animal and plant life [2]. The pyrrole scaffold is known to take a special place both in the chemical laboratory of the Lord and his art palette. Not only the miracle of respiration, solar energy transformation, and sophisticated and still incomprehensible regulation processes in higher organisms but also the simple green riot of forests, poetry of Indian summer, and the enchantment in a bunch of flowers have all originated from substances assembled mainly with pyrrole rings. At the dawn of its development, pyrrole chemistry, skipping many steps, impetuously intruded in the complicated pyrrole systems, porphyrin and phthalocyanine structures such as hemin, chlorophyll, bile pigments, hemoglobin, cytochromes, and vitamin B_{12}. However, the last decades have witnessed the phenomena that can be called "pyrrole renaissance" [2], that is, resurrection of the chemistry of pyrrole itself. "Pyrrole renaissance" was also instigated by the fact that along with continuous isolation and investigation of natural pyrroles [59,60,83], researchers focus their efforts on the preparation of synthetic analogs [84] and the development of expedient procedures for the construction of key "building blocks," carriers of the pyrrole nucleus [7,16,39,85–90].

However, as emphasized in the monograph [2], it is the synthesis of simple pyrroles, especially alkyl-substituted ones, that still presents a real challenge to chemists. For instance, from 39 reactions of the pyrrole ring construction, reported in the book [1], only few have real synthetic value. The majority of these reactions are multistage and laborious, the starting materials being hardly accessible.

N-Vinylpyrroles are versatile reactive carriers of the pyrrole fragment, suitable for the diverse purposes of organic synthesis and polymerization. For a long time, they were almost unknown and inaccessible compounds [1]. The exceptions were N-vinyl derivatives of indole [91–94] and carbazole [95,96], which, though containing the pyrrole nucleus, possess appreciably different properties than typical pyrroles. Genuine N-vinylpyrroles became available only several decades ago [7].

The situation changed suddenly and abruptly after the discovery and systematic development of the reaction between ketoximes (the simplest ketone derivatives) and acetylene in the superbase catalytic systems (KOH/DMSO type) yielding the pyrroles and N-vinylpyrroles in a one preparative stage (in a one-pot manner) [4–7,9–19,22,23,29,35]. This synthesis is included into the monographs [17,97], encyclopedias [98], and textbooks [99,100] and now is referred to as the Trofimov reaction [101–103]. Simple pyrroles and their N-vinyl derivatives cease to be expensive and exotic products and have turned into cheap and abundant compounds. Nowadays, they are extensively and systematically studied by many research teams as promising monomers, semiproducts for fine organic synthesis, and drug precursors. The available substituted pyrroles and N-vinylpyrroles became fertile ground for the verification and application of modern concepts of organic chemistry and reactivity [7].

Thus, a novel research direction has been originated (and now is rapidly progressing) at the interface of pyrrole and acetylene chemistry. This is the chemistry of alkyl-, cycloalkyl-, aryl-, and hetarylpyrroles as well as their N-vinyl derivatives synthesized from ketones (via ketoximes) and acetylene by facile, technologically feasible procedures.

The first advances in this area were summarized about 30 years ago in the book *N-Vinylpyrroles* [7], which focused mainly on synthesis and properties of N-vinylpyrroles, the least studied pyrrole compound at that time. Over the last decades, researches in the field of chemistry and physical chemistry of pyrroles and N-vinylpyrroles associated with the reaction of ketones (ketoximes) with acetylenes have progressed greatly. Indeed, the classical chemistry of pyrrole was supplemented with new pages. There was shaped its independent section representing the symbiosis of pyrrole and acetylene chemistry where possibilities of synthesis and reactivity of both pyrrole and acetylene compounds were highlighted in a new way. The results of these investigations presented in numerous publications require systematization, generalization, and in-depth analysis. The present book is devoted to these issues.

The authors believe that further development of the scientific basis of the chemistry of pyrrole, new monomers, and their functional derivatives using simple and the most available organic semiproducts, first of all acetylene, remains one of the strategic objectives of modern organic synthesis. The discovery of essentially new reactions and methods, which can be adopted by chemical engineering, is one of the ways to reach this goal. Now, various acetylene compounds are often employed as the starting materials when developing new methods for the synthesis of substituted pyrroles [45,104].

This book summarizes a part of the investigations dealing with the development of novel general approach to stimulate anionic transformations of acetylene. This approach, based on the application of the superbase media, reagents, and catalysts, allows new unexpected reactions to be implemented as well as known processes of nucleophilic addition to the triple bond to be accelerated significantly [6,10,22,32,34,35,105,106]. Since the new area of pyrrole chemistry, which the present monograph is dedicated to, is closely intertwined with new aspects of acetylene chemistry the latter should be briefly discussed here.

Long ago, acetylene was a key chemical feedstock [107–110]. Though in the 1960s it was forced out by petrochemical ethylene and propylene, some decades later, it again attracted the attention of synthetic chemists and engineers [5,6,111]. The reason is the upsurge in oil and gas prices in the world market [6,111]. It was supposed that due to the development of plasma, laser, and ultrasonic technologies, cheap acetylene can be obtained not only from fossil hydrocarbons but also from different types of coals and combustible shales [112]. At that time, R&D works aimed at the development of plasma technology of direct acetylene synthesis from coal [113] were launched. Also, fundamental and applied researches were carried out to employ liquid acetylene for considerable intensification of acetylene-based industrial syntheses [114].

The Lord himself widely uses acetylene and its close congeners in his chemical creativity, for instance, in biogenic synthesis of heterocycles [115,116]. It is common knowledge that acetylenic compounds are abundant in nature. Astronomers have discovered clouds of cyanoacetylene in interstellar space [117,118]; biologists have discussed for a long time the role of acetylene in prebiological syntheses and in the origination of living matter [117,119,120]. Mendeleyev has assumed that oil genesis is connected with acetylene, its derivatives, and metal carbides (hydrolysis of the carbides in Earth's depths and condensation of the formed acetylenes).

The acetylene molecule with its unique six-electron chemical bond, strength, high energy, and at the same time its vulnerability to diverse transformation hardly fits the Procrustean bed of modern theories of valency and reactivity. Being a steady challenge to theorists [121,122], it stimulates the development of fundamental works in the field of structure of matter and energy transformation.

It is known that the majority of chemical reactions involving acetylene do not require energy, but, on the contrary, release it. This can be considered as an important advantage with regard to the ever-growing deficiency of energy characterizing the development of our consumer society. There are predictions that in the future acetylene will be the only feedstock of industrial organic synthesis [98,123]. More than 30 years ago, it was estimated [124] that acetylene obtained from calcium carbide could be a real competitor to petrochemical ethylene and propylene. Such estimations were confirmed by the decline in consumption of these hydrocarbons. For example, in 1978, the production capacity of ethylene plants in the United States was only 50% [125].

In this line, further development of acetylene chemistry, the fundamentals of which were laid in the beginning of the twentieth century by the academician Favorsky [126], becomes an urgent challenge. Over the last decades, monographs dedicated to acetylene chemistry were published one after another [5,7,111,127–129], which evidenced the renaissance in this field of knowledge. The present book, which not only relates to pyrrole chemistry but pertains equally to acetylene chemistry, continues this tendency. The special section (1.1.1) of the book briefly covers a novel rapidly progressing direction in the acetylene chemistry based on extensive

application of superbasic media. The direct one-pot method for the synthesis of pyrroles from ketones (ketoximes) and acetylene is based on one of such reactions. The
first chapter deals with the reaction of ketoximes with acetylene in the superbase systems of alkali metal hydroxide–dimethylsulfoxide type. In particular, its preparative
potential, the influence of the reaction conditions and initial reactants structure on
the yields and composition of pyrroles and N-vinylpyrroles, side pathways, the extension of the reaction over aldoximes, technological regimes, and mechanistic aspects
are discussed. Chapter 2 is devoted to the main chemical properties of pyrroles and
N-vinylpyrroles synthesized from ketones and acetylenes of various structures.

The analysis of vast experimental material has revealed that the new reaction of
ketones (ketoximes) with acetylene has a general character and allows diverse 2-,
2,3-, 2,3,4- and 2,3,5-substituted pyrroles and their N-vinyl derivatives to be synthesized in high yields. The reaction tolerates almost all ketones (ketoximes) having at
least one methylene or methyl group in the α-position relative to the ketone (oxime)
function.

The generality of pyrrole synthesis from ketones (ketoximes) and acetylenes is
well illustrated by its successful application for the synthesis of such exotic, hitherto
unknown, or hardly available representatives of pyrroles such as adamantylpyrrole
[130], ferrocenylpyrrole [131], *para*-cyclophanylpyrrole [132], 1,4-bis-(N-vinylpyr-
rol-2-yl)benzene [133], and dipyrrolylpyridine [134] (Scheme I.3).

The first industrial technology for the synthesis of indole via 4,5,6,7-tetrahydro-
indole has been worked out on the basis of the reaction of large-scale cyclohexanone
with acetylene (Scheme I.4) [135–137]. The technology also enables preparation of
N-vinyl derivatives of 4,5,6,7-tetrahydroindole and indole.

Particularly, the book deals with regioselectivity of the reaction between asymmetric methyl alkyl ketoximes and acetylenes involving the construction of the pyrrole ring preferably via the methylene group of the alkyl radical. Also, the reasons of
the process selectivity violation at the elevated temperature are discussed. It is shown
that the latter phenomenon can be used for the preparation of not only 2,3-dialkyl
substituted but also 2-alkyl substituted N-vinylpyrroles.

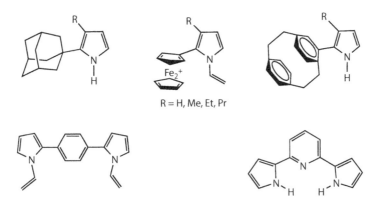

R = H, Me, Et, Pr

SCHEME I.3   Some exotic pyrroles synthesized from ketoximes and acetylene.

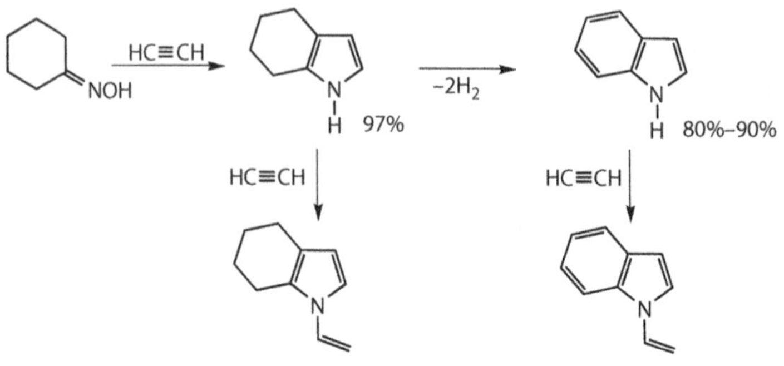

SCHEME I.4 Industrially feasible synthesis of indole from cyclohexanone oxime and acetylene.

The investigations of the reaction of ketoximes with acetylene have shown that the superbase system KOH/DMSO essentially facilitates vinylation of pyrroles with acetylene. This finding constitutes the basis of a new efficient method for vinylation of compounds having the N–H-bond. The process fundamentally differs from the known protocols since it is brought about under atmospheric pressure at moderate temperatures (80°C–100°C). The method is recommended for vinylation of any NH heterocycles (resistant to the action of alkalis) in simple reactors. Apart from the obvious promise for industry, the method is also indispensable for laboratories that do not have special operating building and equipment (autoclaves working with acetylene under pressure).

The informative tables included in this book contain structural formulas, yields, and classical physical and chemical characteristics (melting and boiling points) of all pyrroles and N-vinylpyrroles synthesized from ketoximes and acetylene. The same data are also given for selected O-vinyl oximes, key intermediates of new pyrrole synthesis, as well as for the functionalized compounds of the pyrrole series obtained in the course of pyrrole chemistry development. The tables provide references to original works, thus providing the reader a guide to a variety of the reactions and synthesized compounds discussed.

The authors had no intention to make the book an exhaustive bibliographic survey. Some early works related to N-vinyl derivative of carbazole and indole, which were summarized later in reviews and monographs, are not included in this book.

The stem of the book constitutes the works performed by the authors together with a huge team of experts from different research areas.

The authors are grateful to A.M. Vasiltsov, Dr.Sc.; O.V. Petrova, Ph.D; N.V. Zorina, Ph.D; I.V. Tatarinova, Ph.D; and I.G. Grushin for the help in the manuscript preparation.

Any suggestions and remarks concerning the book will be accepted with gratitude.

1 Synthesis of Pyrroles and N-Vinylpyrroles by the Reaction of Ketones (Ketoximes) with Acetylenes

Currently, the most intensively developed method for the synthesis of NH- and N-vinylpyrroles is based on heterocyclization of ketones (in the form of ketoximes) with acetylene in the superbase system alkali metal hydroxide–dimethyl sulfoxide (DMSO) (Scheme 1.1). This reaction was discovered about 30 years ago.

The main advantage of the reaction is that the pyrrole synthesis starts from available materials, that is, ketoximes, easily prepared from ketones (widespread class of organic compounds), and acetylenes.

The reaction allows not only diverse 2- and 2,3-substituted pyrroles but also their N-vinyl derivatives to be synthesized. Naturally, yields of the products depend on the structure of initial reactants and the reaction conditions, but in the case of simple ketoximes, they are quite constant and range 70%–95% under optimal conditions. Apart from alkyl- and aryl-substituted pyrroles, other hitherto hardly available or even unknown pyrroles became accessible.

The present chapter summarizes the results of systematic investigations of this reaction.

1.1 HETEROCYCLIZATION OF KETOXIMES WITH ACETYLENE

The reaction of ketoximes with acetylene proceeds smoothly at 70°C–140°C, usually at 80°C–100°C. Sometimes, heating of the reactants up to this temperature would suffice to initiate a mild exothermic process, which can be regulated by the addition of acetylene.

The synthesis is extremely feasible: acetylene under atmospheric pressure is passed through the heated stirring solution of the reactants and a catalyst in DMSO. The process takes 3–5 h to complete. Also, the reaction can be carried out in autoclave where the reaction time is reduced under pressure.

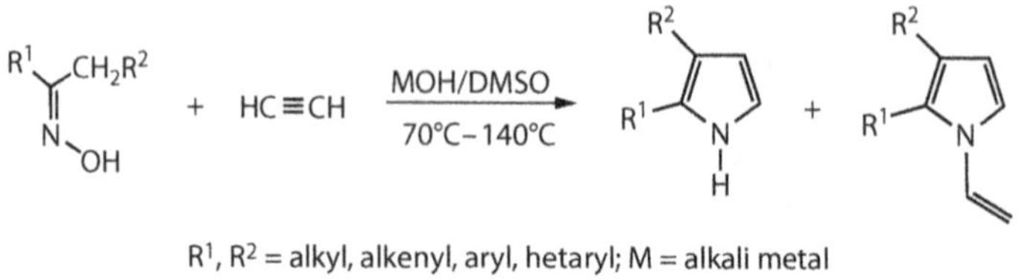

R^1, R^2 = alkyl, alkenyl, aryl, hetaryl; M = alkali metal

SCHEME 1.1   Synthesis of pyrroles and N-vinylpyrroles from ketoximes and acetylene.

1.1.1   SUPERBASE SYSTEM ALKALI METAL HYDROXIDE–DIMETHYL SULFOXIDE AS THE REACTION CATALYST

The reaction is catalyzed by a superbase pair alkali metal hydroxide/DMSO [6,10,22,32,34,105,106], although specially prepared alkali metal oximates are also active in the process.

Superbase systems are known to contain a strong base and a solvent or reactant capable of specifically binding the cation "baring" the conjugated anion [138]. Such systems can be prepared on the basis of linear or cyclic glycol ethers, microcyclic polyethers (crown ethers), highly polar non hydroxylic solvents (sulfoxides, e.g., DMSO), sulfones (sulfolane), amides (N-methylpyrrolidone, dimethylformamide, hexametapol), and phosphine oxides as well as from liquid ammonia, amines, etc. For example, basicity of sodium methylate in 95% DMSO is by seven orders higher than in pure methanol [139].

0.25 M NaOCH$_3$	
Solvent	Acidity, H_-
Methanol	12.2
95% DMSO + 5% methanol	19.4

The effects of basicity increase are especially pronounced in the region of extremely high concentrations of polar non hydroxylic solvent. So, tetramethylammonium hydroxide in 99.5% aqueous DMSO is by 14 orders more basic than in pure water [140].

As a first approximation, superbasicity of the KOH/DMSO system is due to loosening of ion pair of the base under the action of DMSO that leads to the formation of solvent-separated ion pair [141] and generation (in some cases) of highly basic and weakly solvated dimsyl anion (Scheme 1.2) [6,10,32,105]. However, other reasons also exist.

A more careful study of superbase systems should account for cooperative effects of dielectric permeability alteration, hydrogen bonding, activity of water, dispersive interactions, and changes in water structure and ion hydration degree [140,142,143].

Superbases are probably formed on phase interfaces in the conditions of phase-transfer catalysis [144–146], when highly concentrated or solid alkali (or other base) acts as a phase and anions are transferred by bulky organophilic cation incapable of forming the contact ion pair (Scheme 1.2).

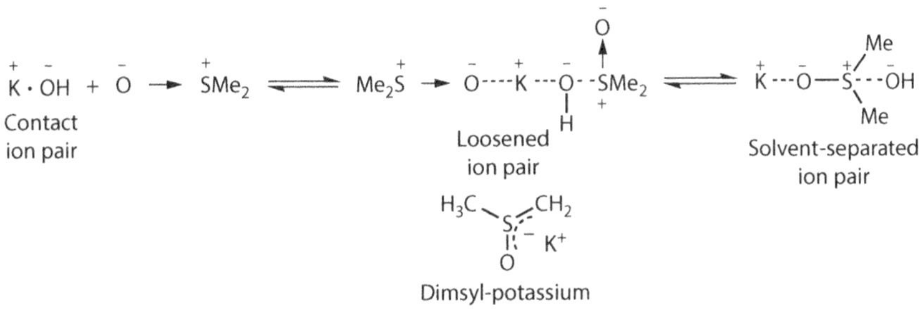

SCHEME 1.2 Superbasicity features of the KOH/DMSO system.

The significant acceleration of nucleophilic addition to the triple bond in media specifically solvating cations was found for the first time when studying vinylation of polyethylene glycols [5]. These reactions, owing to chelation of potassium cation with polyethylene glycol chains, are accompanied by side reactions of cleavage and disproportionation of polyethylene oxide fragments [147,148]. Also, the template cyclization occurs to give macrocyclic crown-like acetals and polyenes (acetylene oligomers), which along with linear oligomers autocatalytically accelerate the process.

Apparently, of the same nature is catalytic effect of side products of monobasic alcohols vinylation [109,149–151], low-molecular thermopolymers of alkylvinyl ethers [–CH$_2$–CH(OR)–]$_n$, and oligoacetylenes, which are also capable of chelating the cations (Scheme 1.3).

According to Miller [109], the black color of solutions observed during vinylation of alcohols with acetylene indicates the formation of complexes, though the latter have not been identified. Results of a kinetic study [152] speak in favor of such an assumption. These results cannot be interpreted if to believe that the addition of alkoxide ion to acetylene is a limiting stage of the process.

So-called complex bases of NaNH$_2$/t-BuONa type belong also to a class of the superbases [153]. It is shown that sodium alcoholate can be bounded with surface of sodium amide crystal and transfer the latter to solution in the form of complex aggregates (Scheme 1.4) [153].

In any case, synergism of the base action, the strongest activation of one base by another, is a real phenomenon that is widely and successfully employed in preparative organic chemistry [5,6,9,10,32,34,105].

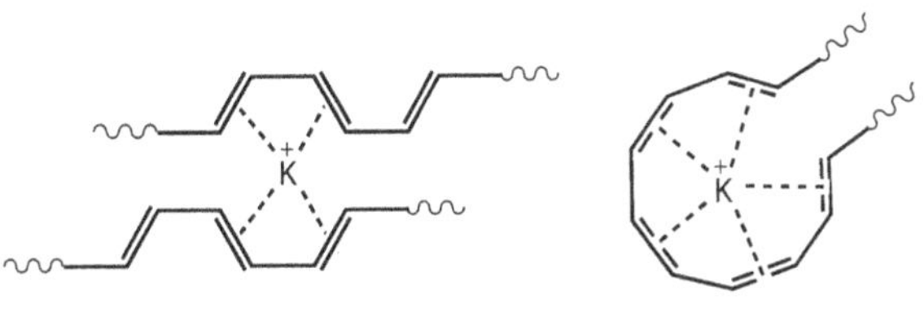

SCHEME 1.3 Tentative chelating of potassium cation by polyvinylene moieties.

$$\text{NaNH}_2 \text{ (solid)} + (\text{RONa})_m \text{(solvent)}_n \rightleftharpoons$$

$$\rightleftharpoons \{(\text{NaNH}_2)_l (\text{RONa})_p\} \text{(solvent)}_q$$

$$\begin{array}{c} \text{RO} \cdots \text{Na} \\ \vdots \qquad \vdots \\ \text{Na} \cdots \text{NH}_2 \end{array} \rightleftharpoons$$

SCHEME 1.4 Sodium amide/sodium alkoxide superbasic aggregates.

Apparently, there is a certain analogy between activation of $(A^-M^+)_n$ base via the formation of complexes with other base $(B^-M^+)_m$ and activation of $(A^-M^+)_n$ base through the solvation with polar non hydroxylic solvent.

In real systems, these two routes of activation should interact and are realized much more often than it seems at first glance. This fact is completely ignored when mechanisms of reactions with participation of bases are interpreted. For instance, synthesis of pyrroles in the system $KOH/R_2C=NOH/HC\equiv CH/DMSO$ may simultaneously involve several bases, namely, KOH, $R_2C=NOK$ (oximate), $HC\equiv CK$ (acetylenide), and $MeSOCH_2K$ (dimsyl potassium), which should give a plethora of complex bases in various combinations.

In DMSO-derived superbase systems, acetylene can be activated, at least, by three ways:

1. Specific interaction and complex formation with DMSO (Scheme 1.5)
2. Incorporation into inner solvate sphere of metal cation (Scheme 1.6)
3. Ionization and formation of acetylenides (Scheme 1.7)

The first ab initio calculations performed about 30 years ago have shown that there is a bonding interaction between acetylene and alkali metal cations and the complexes formed have nonclassical bridge structure where electronic density is transferred to metal [154–156].

$$HC\equiv C-H\cdots \overset{-}{O}-\overset{+}{S}Me_2 \qquad \begin{array}{c} HC\equiv CH \\ \downarrow \\ Me_2S-O \\ {}^{+}\quad {}^{-} \end{array} \qquad \begin{array}{c} HC\equiv C-H \\ \downarrow \\ Me_2S-O \\ {}^{+}\quad {}^{-} \end{array}$$

SCHEME 1.5 Complex formation of acetylene with DMSO.

$$\begin{array}{c} H \\ | \\ C \\ ||| \\ C \\ | \\ H \end{array} \longrightarrow \overset{+}{M}\cdots \overset{-}{O}-Me_2\overset{+}{S}$$

SCHEME 1.6 Insertion of acetylene into solvate sphere of metal cation.

$$HC\equiv CH + KOH \rightleftharpoons \overset{+}{K} \overset{-}{C}\equiv CH + H_2O$$

SCHEME 1.7 Reversible potassium acetylenide formation from KOH and acetylene.

**Energy of Complex Formation of
Acetylene with Alkali Metal
Cations, kcal/mol [154]**

M	$\underset{HC\equiv CH}{M^{\oplus}}$	$M-HC\overset{\oplus}{=}CH$
Li	39.3	22.0
Na	15.9	−10.0
K	14.4 [155,156]	—

**Electron Density Transfer of Acetylene
on Alkali Metal Cation [154]**

M	$\underset{HC\equiv CH}{M^{\oplus}}$	$M-HC\overset{\oplus}{=}CH$
Li	0.31	0.38
Na	0.08	0.10

As a first approximation, all this leads to the decrease in energy of low-lying molecular orbitals of acetylene in such a complex as compared to similar orbitals of free acetylene, and, consequently, the triple bond becomes more available for the nucleophilic attack. The nucleophiles themselves in the superbase systems turn to supernucleophiles due to the dramatic increase of their free energy [157,158].

Catalytic function of the KOH/DMSO system in heterocyclization of ketoximes with acetylene is clearly demonstrated in the example of application of mixed solvent DMSO–dioxane. The interaction of cyclohexanone oxime with acetylene [159] occurs when DMSO is added to dioxane solution already in the amount of 5%–10%. Varying DMSO concentration, one can accomplish the process selectively, that is, to obtain either 4,5,6,7-tetrahydroindole (at small concentration of DMSO) or N-vinyl-4,5,6,7-tetrahydroindole (in pure DMSO, Scheme 1.8).

Similar conclusions were drawn when studying the interaction of acetophenone oxime with acetylene furnishing 2-phenyl- and 2-phenyl-N-vinylpyrroles (Scheme 1.9) [160].

The decrease of DMSO concentration in mixtures with dioxane leads to drastic drop of 2-phenyl-N-vinylpyrrole yield. In this case, one can also selectively prepare either 2-phenylpyrrole or its N-vinyl derivatives by changing DMSO content in the reaction mixture.

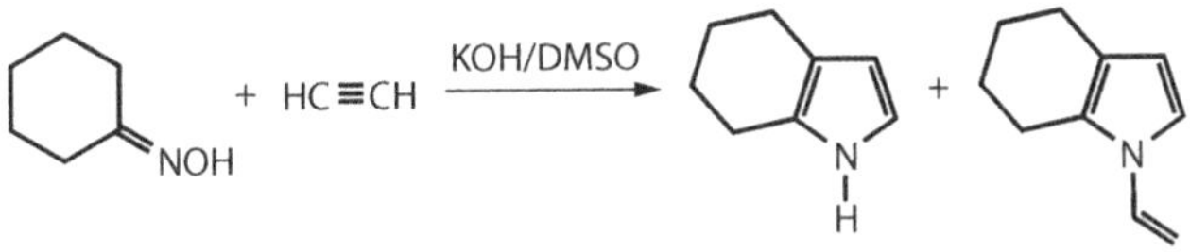

SCHEME 1.8 Synthesis of 4,5,6,7-tetrahydroindole and its N-vinyl derivative from cyclohexanone oxime and acetylene in the KOH/DMSO system.

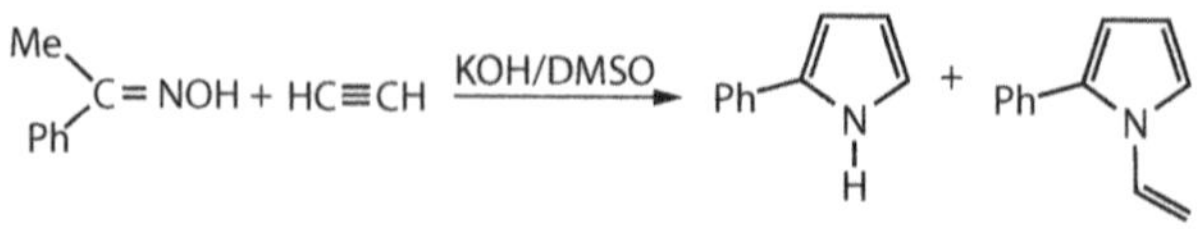

SCHEME 1.9 Synthesis of 2-phenyl- and 2-phenyl-N-vinylpyrroles from acetophenone oxime and acetylene in the KOH/DMSO system.

1.1.2 EFFECTS OF BASE NATURE AND CONCENTRATION

To synthesize pyrroles from ketoximes and acetylene, alkali metal hydroxides (LiOH, NaOH, KOH, CsOH, RbOH) are employed as strong bases, key component of the superbase system.

Dependence of the catalytic system activity upon cation nature has been studied on the example of the reactions between cyclohexanone and acetophenone oximes and acetylene [159,160]. The catalyst activity is shown to increase with growth of the cation atomic number: Li < Na < K < Rb < Cs [160,161].

The previous sequence, valid for many oximes of aliphatic and cycloaliphatic ketones, is not however absolute and can be varied depending on the reaction conditions and ketoxime type.

The reaction of acetophenone oxime with acetylene (100°C, 3 h) is well catalyzed by all alkali metal hydroxides (taken in 10%–30% from the oxime weight), but $Ca(OH)_2$ is inactive under these conditions [160]. Tetrabutylammonium hydroxide exerts weak catalytic action on the reaction only at harsher conditions (120°C). Potassium, zinc, and cadmium acetates as well as zinc, copper (I and II), and cobalt chlorides do not show catalytic activity in this reaction (starting acetophenone oxime is recovered almost completely [160]), though several cations of the aforementioned salts are known [95,162] to be catalysts of direct vinylation of NH-heterocycles with acetylene.

Alkali metal hydroxides differ not only in reactivity but also in selectivity of action. For example, LiOH selectively catalyzes the reaction of the pyrrole ring construction from alkyl aryl ketoximes [160,163–165], while it is almost inactive at the stage of vinylation of the pyrrole formed. At the same time, LiOH is ineffective for cycloaliphatic ketoximes at both stages [159]; the pyrrole ring formation is accelerated in this case by rubidium and tetrabutylammonium hydroxides [159,161].

A slight decrease of N-vinyl-2-phenylpyrrole yield in the reaction catalyzed by RbOH and CsOH is a result of resinification process [160], which probably can be suppressed by the decrease in temperature (during optimization of the conditions).

A much more efficient and facile tool influencing the process is the change of KOH concentration in the reaction medium. It is especially clearly shown on the example of cyclohexanone oxime transformation to 4,5,6,7-tetrahydroindole and N-vinyl-4,5,6,7-tetrahydroindole via the interaction with acetylene in the system KOH/DMSO [159]. Under quite mild conditions (100°C), the increase of KOH concentration (up to equimolar ratio with the starting oxime) leads to augmentation of N-vinyl-4,5,6,7-tetrahydroindole yield. In harsher conditions (120°C), alkali starts to accelerate the side processes. As a consequence, a reverse dependency of N-vinyl-4,5,6,7-tetrahydroindole yield upon base concentration is also possible [159].

Generally, the reaction rate increases with the growth of base concentration in the medium [159,160], and good isolated yields can be reached when base is used in superstoichiometric excess relative to ketoxime (about 10-fold) [4]. However, as a rule, optimum ratio of ketoxime and alkali is close to equimolar.

As for various catalytic activity of alkali metal hydroxides, it should be noted that this effect is not unique for this reaction and is observed in almost all base-catalytic processes involving alkalis, for example, vinylation reaction [5,6,109,152,166], nucleophilic substitution and elimination [153], Favorsky reaction [167], synthesis of divinyl sulfide from acetylene and alkali metal sulfides [168], and cyclization of cyanoacetylenic alcohols [169]. For instance, in vinylation of 2-ethoxyethanol with acetylene in the presence of different hydroxides, the following relative reaction rates are observed [109,152]:

LiOH	0.10
NaOH	0.76
KOH	1.00
RbOH	0.83
$[PhCH_2N^+Me_3]^-OH$	0.0

Such order of bases activity corresponds almost precisely to that observed in the synthesis of pyrroles from ketoximes and acetylene and, obviously, is caused by the same reasons. It is assumed [109,152] that the inability of trimethylbenzylam-monium hydroxide to catalyze vinylation reaction is due to the lack of coordination ability of this base. This fact as well as inhibition of the reaction with water, pyridine, phenanthroline, and diketones evidences [109,152] that the reaction proceeds via two mechanisms, that is, complex and ionic ones (in the latter case, participation of a complex ion as intermediate is not excluded).

According to the data of ab initio calculations [156], in complexes of acetylene with alkali metal cations (see also previous section), unoccupied $2s$ and $2p$ orbitals of Li^+ cation have the lowest energy values. Unoccupied $4s$ and $4p$ orbitals of K^+ cation are located higher, but remain (as well as at Li^+) below than the lowest unoccupied molecular orbital (LUMO) of acetylene. In the case of Na^+ cation, its unoccupied $3s$ and $3p$ atomic orbitals have positive energy values, and $3p$-AO is located higher than LUMO of acetylene molecule.

Energy values (in atomic units) of the lowest unoccupied orbitals of alkali metal cations in complexes with acetylene [150] are as follows:

Orbital	Li^+	Na^+	K^+
S	−0.1790	0.1269	−0.0596
P	−0.0958	0.4326	0.1233

Such relative disposition of the cation unoccupied orbitals indicates the decrease in orbital interaction energy in the following order: $C_2H_2Li^+ > C_2H_2K^+ > C_2H_2Na^+$.

Thus, the $C_2H_2Na^+$ and $C_2H_2K^+$ complexes have different nature (see the previous section). It is confirmed by the analysis of structure of bonding molecular orbitals of the complexes. The contribution of atomic orbitals of the cation in these molecular orbital (MO) decreases in the series $Li^+ > K^+ > Na^+$. Correspondingly, the contribution of electrostatic interaction to the energy of complex formation upon transition Na^+ to K^+ diminishes more drastically than upon transition from Li^+ to Na^+.

In solutions, the cation solvation energy lowers as follows: $Li^+ \gg Na^+ > K^+$, that is, their proneness to complex formation with acetylene should change even more in favor of potassium.

Certainly, semiquantitative data (values of pyrrole yields) do not allow to reveal completely the reasons of different catalytic activity of the bases in this reaction.

When discussing the mechanisms of nucleophilic reactions with participation of the C≡CH group, one should not underestimate the fact that acetylenes are quite strong C–H acids that, to some extent, should be ionized in the superbase media (Scheme 1.10).

Obviously, in acetylenides of various metals, π-system is polarized in a different manner depending upon electronegativity and structure of the cation electron shell. It is known [170] that cations with low-lying unoccupied d-orbitals, being in ionic couple with acetylenide ion, can interact with the latter not only electrostatically but also through the mechanism π-d bonding (back donation, Scheme 1.11).

The higher contribution of this interaction leads to higher "baring" of carbon nuclei (see structures **A** and **B**, Scheme 1.12), and, consequently, proneness of the triple bond to nucleophilic attack becomes also higher.

Hence, it naturally follows that polarization and deformation of acetylene π-electron shell depend on atomic number of a cation, which acetylene anion is bonded to. However, activation of acetylene in the reaction with ketoximes via acetylenides assumes the attack of nucleophile to the carbanion-like complex, and this is likely a weak spot of this hypothesis. Nevertheless, the electrophilic assistance of alkali metal cation (Na^+) in the course of nucleophilic addition to acetylene (Scheme 1.13) is confirmed by quantum-chemical calculations (ab initio, STO-3G) [171].

According to the frontier orbitals concept, in the absence of nucleophile (F^-), cation Na^+ is located symmetrically relative to acetylene. When both ions are

$$MOH + HC≡CH \underset{DMSO}{\overset{DMSO}{\rightleftharpoons}} \overset{+}{M}\ \overset{-}{C}≡CH + H_2O$$

SCHEME 1.10 Reversible acetylenide formation from acetylenes and alkali metals in DMSO.

$$M^+ \overset{\frown}{} C≡CH$$

SCHEME 1.11 Back donation in metal acetylenide.

$$
\begin{array}{cc}
\overset{+}{M} & \overset{+\delta'}{M} \\
\overset{..}{\underset{..}{C}} = C & \overset{-}{C} = C \\
\mathbf{A} & \underset{+\delta''}{H} \\
 & \mathbf{B}
\end{array}
$$

SCHEME 1.12 Tentative metal acetylenide structures.

$$H-C\equiv C-H \qquad\qquad H-C\equiv C-H$$
$$Na^+ \qquad\qquad\qquad Na^+$$
F⁻ (above second structure)

SCHEME 1.13 Electrophilic assistance of the metal cation to the nucleophilic attack at acetylene.

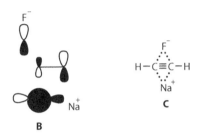

$$H-C\equiv C-H$$
$$Na^+$$
$$C$$

B

SCHEME 1.14 Computed structures of the F⁻/C₂H₂/Na⁺ complex.

present around the triple bond, the anion F⁻ is shifted by 0.12 Å toward the nearest carbon atom (in relation to the optimized distance without Na⁺), while Na⁺ cation is shifted by 0.52 Å to the other end of the triple bond (in relation to the optimized distance in the C₂H₂Na⁺ without anion). The optimum position of Na⁺ cation corresponds to scheme **B** (Scheme 1.14) [171]. The symmetric structure **C** is unstable [171].

Since in complexes of transition metals, acetylene is quite often stabilized in carbene (vinylidene) form, it is not excluded that in the presence of alkali metal cations, this form also becomes more stable and makes a certain contribution to activation of the triple bond.

1.1.3 EFFECT OF SOLVENT

The reaction of ketoximes with acetylene is effectively carried out only in high-polar non hydroxylic solvents, which form (as it was already stated) a superbase medium together with strong bases. Notably, among the solvents tested, DMSO was found to be best (Table 1.1) [159,160].

Thus, the solvent neither changes the solubility of acetylene in the reaction mixture (at 15 atm, DMSO dissolves acetylene only by 1.3–1.5 times better than hexametapol [172]) nor alters the polarity of medium (the dipole moments of these two solvents are also similar [173]). Apparently, the effect of solvent is caused by distinctions in specific solvation that activates reactants. In this case, this solvation does not essentially differ from catalysis.

Similar results have been obtained when studying the reaction of acetophenone oxime with acetylene [160]. So, 2-phenyl- and 2-phenyl-N-vinylpyrroles are formed only in DMSO; in sulfolane, these pyrroles are detected only as traces, while in dimethylformamide, hexametapole, dioxane, benzene, and methanol,

TABLE 1.1

Effect of Solvent on Yields of NH- and N-Vinyltetrahydroindoles[a]

Solvent	Total Yield of Products Mixture (%)	Composition of Mixture[b] (%)		
		Cyclohexanone oxime	Tetrahydroindole	N-vinyltetrahydroindole
DMSO	57	Traces	Traces	~100
Dibutylsulfoxide	40	Traces	Traces	~100
Hexametapol	30	20	50	30
Sulfolane	12	40	48	12
Dioxane	80	~100	Absent	Absent
Benzene	82	~100	Absent	Absent
Methanol	90	~100[c]	Absent	Absent

[a] 140°C, 1 h, 10% KOH from cyclohexanone oxime weight, initial acetylene pressure 16 atm.
[b] GLC data.
[c] Also, vinylmethyl ether and dimethylacetal are formed.

the reaction does not take place at all (100°C, 3 h, 30% KOH, excess C_2H_2 under the initial pressure of 12 atm).

As compared to DMSO, hexametapol and sulfolane form the systems, which catalyze the synthesis of NH- and N-vinylpyrroles from ketoximes and acetylene much less actively. At least, protocols involving these systems are underdeveloped and so far have no high preparative value. In such solvents as ether, alcohols, and hydrocarbons, the reaction does not proceed [4].

Interestingly, in aqueous medium, acetylene and ketoximes interact in an absolutely different fashion, delivering pyridines instead of pyrroles [4,174] (see Section 1.5.4). The reaction occurs also with acetylene obtained directly in the autoclave from calcium carbide [174].

Tertiary acetylenic alcohols (in up to 50% yield [175]) are formed from ketoximes and acetylene in aqueous DMSO. This is hitherto unknown analog of the Favorsky reaction (see Section 1.5.5).

It is pertinent to discuss the previously postulated equilibrium, which should be shifted toward DMSO (Scheme 1.15).

However, here, account should be taken of big hydration energy of KOH, its very high concentrations in DMSO applied in the pyrrole syntheses, and the ability of DMSO to form strong complexes with water [176,177]. All these factors lower the activity of water in the system and shift the equilibrium toward dimsyl anion.

$$\text{MeSMe} + \text{KOH} \rightleftharpoons \text{KCH}_2\text{SMe} + \text{H}_2\text{O}$$

$$\underset{O}{|} \qquad\qquad\qquad \underset{O}{|}$$

SCHEME 1.15   Reversible formation of dimsyl anion in KOH/DMSO system.

1.1.4 EFFECT OF PRESSURE

In autoclave, heterocyclization of ketoximes with acetylene into pyrroles and N-vinylpyrroles is usually performed under pressure (initial pressure at room temperature is 8–16 atm, predominantly 10–12 atm). The maximum pressure of acetylene in the reaction can reach 20–25 atm.

According to current legislation of Russian Federation and many other countries, the processes employing acetylene under pressure exceeding 1.5 atm are illegalized in industry (except for the compressed and dissolved acetylene in cylinders intended for gas welding and metal arc cutting). Therefore, pressure becomes a major challenge for engineering implementation of different acetylene reaction.

These restrictions were imposed after the experiments based on classical investigations into explosive properties of acetylene (performed at the dawn of acetylene industry) resulted in tragic events (see [109] and references therein). The technologies of acetylene processing under pressure proposed by Reppe [95,96,109] were developing in violation of the existing safety regulations. However, long-term exploitation of apparatus under pressure has proven their safety. To decrease the explosive properties of acetylene (phlegmatization), the latter was diluted with inert gases, usually with nitrogen or saturated hydrocarbons [178]. During the World War II in Germany, there were two plants for manufacturing vinylmethyl ether with a capacity of 5000 tons per year (working pressure in this process was 22 atm). In Russia, an efficient technology for the vinylation of lower alcohols with compressed undiluted acetylene has been developed on the basis of researches by Favorsky and Shostakovsky, carried out independently from Reppe's works. Vapors of such alcohols and their vinyl ethers, being in a gas phase, have proved to be reliable phlegmatizators [110].

The analysis of the experimental material testifies that the reaction rate, which can be roughly estimated by the amount of products formed for standard time, is approximately in direct dependence on the acetylene pressure. Since such a dependence is typical for many processes of "liquid–gas" type, the most important here is to answer the following question: how fast is the synthesis of pyrroles realized under the pressure about 1.5 atm and whether this is applicable for engineering.

The experiments related to the synthesis of 4,5,6,7-tetrahydroindole and N-vinyl-4,5,6,7-tetrahydroindole from cyclohexanone oxime and acetylene, performed in 5 and 25 L reactors under pressure 1.5 atm, give positive answer to the latter question. For instance, at 110°C and 0.4 mol/L KOH concentration, the productivity of 1 L of catalytic solution is 50–100 g of N-4,5,6,7-vinyltetrahydroindole per hour. It means that it is possible to produce up to 400 tons of N-4,5,6,7-vinyltetrahydroindole per year using small semi-industrial reactor of 1 m^3 volume that is quite acceptable for modern technology [179].

These results have confirmed high technological effectiveness of the reaction and laid the basis for the development of a novel feasible process for the synthesis of diverse pyrroles and their N-vinyl derivatives, which has been realized on a pilot plant.

Application of this pilot plant on industrial scale has shown that if to carry out the reaction in DMSO using KOH in amount of 1 mol per 1 mol of cyclohexanone oxime, N-4,5,6,7-vinyltetrahydroindole can be obtained in up to 83% yield and 99% purity under acetylene pressure from 0.3 to 1.5 atm without heat supply due to exothermicity of the process.

Further investigations have elaborated upon the conditions that ensure the synthesis of pyrroles and their N-vinyl derivatives under atmospheric pressure in the simplest reactor equipped with mixer and bubbler for acetylene supply (reaction temperature is 93°C–97°C). To reach this goal, the increase of alkali concentration in the reaction mixture (up to 50% from weight of the initial ketoxime) and approximately two-time-longer reaction time are required. The yield of N-vinylpyrroles remains rather high (~80%).

Recently, a novel convenient for industrial implementation technology for the synthesis of 4,5,6,7-tetrahydroindoles from cyclohexanone oxime and acetylene has been developed. The technology comprises the application of a nanocrystalline catalyst composed from sodium oximate and its complex with DMSO in 2:1 ratio [135,180]. The catalyst is synthesized by the interaction of cyclohexanone oxime with NaOH in DMSO at 100°C–130°C. To assure more complete shift of equilibrium toward formation of the complex, water is removed in the form of azeotrope with toluene. In reality, the catalytic system represents suspension, a solution of sodium cyclohexanone oximate in DMSO, the phase composition and activity of which is controlled by temperature. The crucial point is that the new catalyst has been elaborated on the basis of NaOH, a cheaper and less aggressive base, which is not applied in all variation of this reaction up to the moment. The technology is distinguished for high degree of cyclohexanone oxime conversion (70%–90%), high selectivity relative to the target product (95%–99%), enhanced safety (atmospheric or near to atmospheric acetylene pressure, elimination of diethyl ether, explosive reactant), and the improved ecological characteristics (it can be realized as low-waste process).

1.1.5 One-Pot Synthesis of Pyrroles from Ketones, Hydroxylamine, and Acetylene

A version of the pyrrole synthesis involving one-pot transformation of ketones into NH- and N-vinylpyrroles has been worked out. Preliminary isolation of oximes is always associated with consumption of reactants, solvents, and time. In some cases (e.g., for ketoximes having high solubility in water), this procedure presents certain preparative difficulties. To exclude the stage of ketoximes isolation, a possibility of ketone oximation in situ in DMSO and the subsequent application of the oxime solutions in the reaction with acetylene has been studied [181].

Various ketones were used as the starting compounds to determine the applicability limits of this version of the reaction (Scheme 1.16).

The process is carried out in the same conditions as a simple and safe protocol that can be realized in any laboratory: acetylene is fed to thermostatically controlled reactor under atmospheric pressure.

Oximation with the $NH_2OH \cdot HCl/NaHCO_3$ mixture (in equimolar ratio with ketone) proceeds at room temperature and is completed for 3–4 h. Before the addition of KOH, which catalyzes together with DMSO, the interaction of oximes with

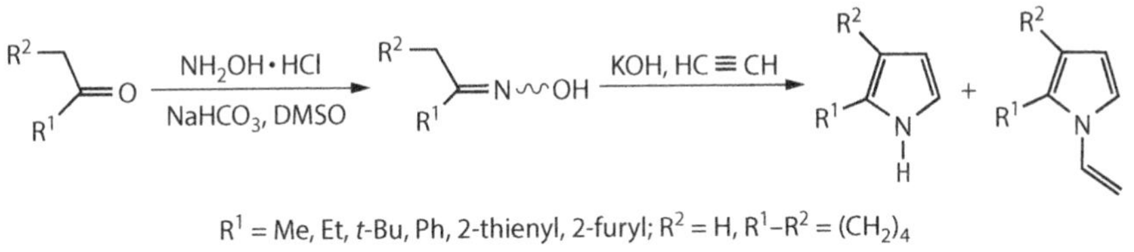

R^1 = Me, Et, t-Bu, Ph, 2-thienyl, 2-furyl; R^2 = H, R^1–R^2 = $(CH_2)_4$

SCHEME 1.16 One-pot synthesis of pyrroles and N-vinylpyrroles from ketones, hydroxylamine, and acetylene.

acetylene, the reaction mixture is heated to the reaction temperature (100°C) and is blown with acetylene to remove CO_2 from the reaction medium. The conditions found have allowed N-vinyl-4,5,6,7-tetrahydroindole (71%), 2-phenylpyrrole (60%), 2-thienylpyrrole (46%), and 4,5-dihydrobenz[g]indole (72%) to be selectively synthesized.

Among the advantages of this version of the reaction are its simplicity, feasibility (atmospheric pressure of acetylene), and possibility of application of more accessible (in comparison with oximes) ketones. The elimination of the oximation stage and the corresponding consumption of reactants, inevitable losses during oximes isolation lowering the total yields of the target pyrroles, increase the synthetic value of this method.

Another version of the reaction using excess acetylene under pressure has been developed to increase selectivity relative to N-vinylpyrroles [182]. The synthesis of N-vinylpyrroles is accomplished as follows: a heated (70°C) mixture of ketone, $NH_2OH \cdot HCl$ and $NaHCO_3$ in DMSO was blown with argon, and after adding KOH, treated with acetylene in autoclave at 100–120°C.

This variant of the reaction turns out to be efficient for aliphatic, cycloaliphatic, aromatic, and heteroaromatic as well as hydronaphthalinic ketones. The isolated yields of N-vinylpyrroles range 50%–80%.

4,5-Dihydrobenz[e]indole and its vinyl derivative have been synthesized via the three-component reaction of 2-tetralone, hydroxylamine, and acetylene under pressure in the MOH/DMSO systems (Scheme 1.17) [183].

N-vinyl-4,5-dihydrobenz[e]indole is prepared in 75% yield (KOH/DMSO, 110°C, 3 h, the initial acetylene pressure 14 atm). A mixture of vinylated and non vinylated products is obtained in the NaOH/DMSO system at a lower temperature (90°C–100°C).

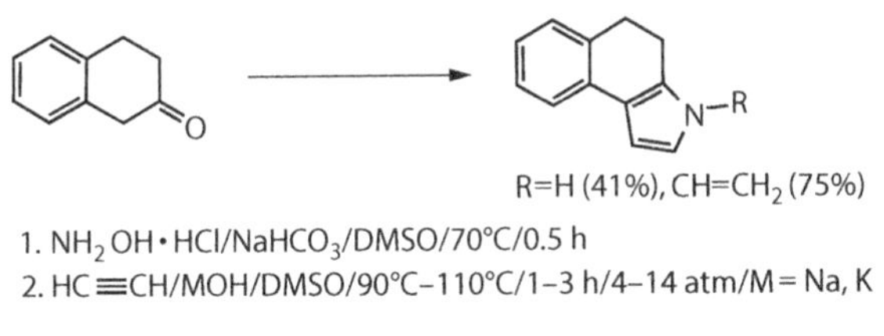

R=H (41%), CH=CH_2 (75%)

1. $NH_2OH \cdot HCl/NaHCO_3/DMSO/70°C/0.5$ h
2. $HC≡CH/MOH/DMSO/90°C–110°C/1–3$ h/4–14 atm/M= Na, K

SCHEME 1.17 The three-component reaction of 2-tetralone, hydroxylamine, and acetylene.

Despite all obvious benefits of this method, its realization requires usage of additional base (NaHCO$_3$) and preliminary elimination of carbon dioxide from the reaction mixture (remaining CO$_2$ decreases catalytic activity of the MOH/DMSO system).

This shortcoming has been circumvented in the developed one-pot synthesis of pyrroles from ketones and acetylene in the system NH$_2$OH·HCl/KOH/DMSO. In this protocol, KOH is used both to generate free hydroxylamine and to form superbase medium (a ratio of ketone–NH$_2$OH·HCl–KOH varies within 1:1–1.5:1.5–2.5). Realization of this synthesis under atmospheric pressure is demonstrated on the example of 4,5,6,7-tetrahydroindole and its N-vinyl derivative where selectivity of the reaction relative to both NH- and N-4,5,6,7-vinyltetrahydroindole has been attained (Scheme 1.18).

The reaction is carried out as follows: acetylene is passed through a mixture of cyclohexanone, NH$_2$OH·HCl, KOH, and DMSO heated to reaction temperature. The reaction time is 2.5–6 h (100°C) for 4,5,6,7-tetrahydroindole and 2.5–3.5 h (120°C) for its N-vinyl derivative. If 4,5,6,7-tetrahydroindole needs to be synthesized, its further vinylation can be slowed down, by increasing the content of water in DMSO up to 3%.

N-vinyl-4,5-dihydrobenz[g]indole has been prepared [184] from 1-tetralone and acetylene in the system NH$_2$OH·HCl/KOH/DMSO (molar ratio of tetralone–NH$_2$OH·HCl–KOH = 1:1:2.5) without application of the auxiliary base and removal of carbon dioxide from the reaction mixture (Scheme 1.19). The yield of chromatographically pure N-vinyl-4,5-dihydrobenz[g]indole is more than 70%.

Synthesis is realized in a one-pot manner as the typical multicomponent process involving a series of consecutive and parallel reactions, namely, the interaction of

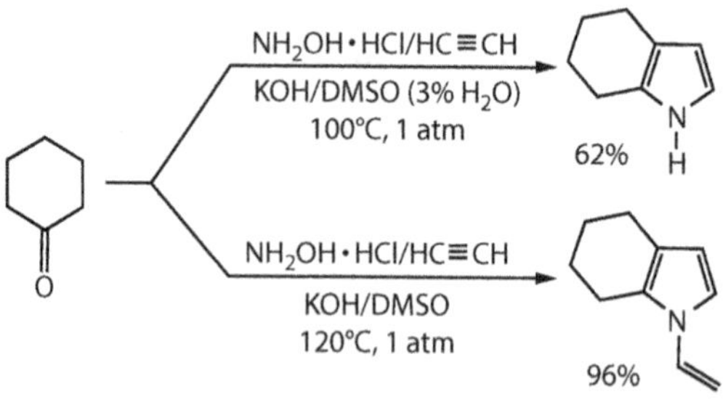

SCHEME 1.18 One-pot synthesis of N-4,5,6,7-vinyltetrahydroindole and its vinyl derivative from cyclohexanone, hydroxylamine, and acetylene in the KOH/DMSO system.

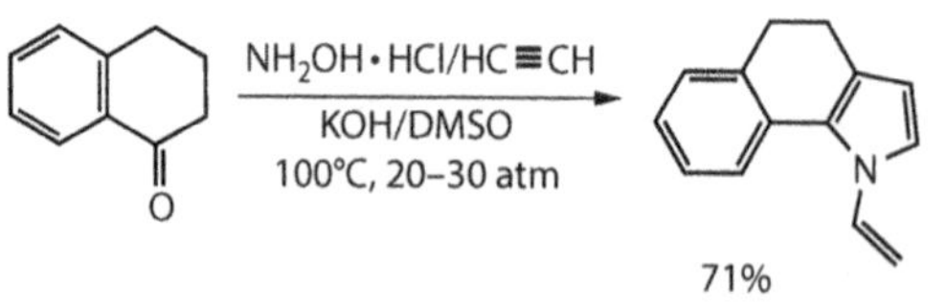

SCHEME 1.19 One-pot synthesis of N-vinyl-4,5-dihydrobenzo[g]indole from 1-tetralone and acetylene in the system NH$_2$OH·HCl/KOH/DMSO.

$NH_2OH \cdot HCl$ with KOH, ketone oximation, vinylation of the oxime formed with acetylene, transformation of O-vinyl oxime into intermediate NH-4,5-dihydrobenz[g]
indole, and further vinylation of the latter.

A method for the synthesis of pyrroles and N-vinylpyrroles directly from ketones
and acetylene in the system $NH_2OH \cdot HCl/KOH/DMSO$ offers a number of essential
advantages, that is, the total number of reactants decreases, the auxiliary reactor for
oximation in the presence of $NaHCO_3$ is removed from the process, degassing of
the reaction mixture (CO_2 elimination) is not required, and extraction of the target
products from aqueous DMSO is facilitated due to desalting with potassium chloride
that is formed at the oximation stage.

These advantages are apparently owing to the combination of usual direction of
the synthesis (through oxime vinylation) with two other parallel reactions, which
are not intrinsic for processes proceeding via ketoximes. One of these reactions is
triggered by the vinylation of hydroxylamine to generate O-vinylhydroxylamine **A**
(Scheme 1.20). The latter then oximizes a ketone to give O-vinyl oxime **B** that further rearranges to the pyrrole.

Another reaction comprises the interception of intermediate adduct of ketone and
hydroxylamine **C** by acetylene to afford O-vinyl derivative **D**, further dehydration
of which leads to O-vinyl oxime **B** (Scheme 1.21). Such interception can be more
preferable than vinylation of the corresponding ketoxime, since the hydroxyl group
is bonded to less electronegative nitrogen atom than nitrogen atom in ketoxime.

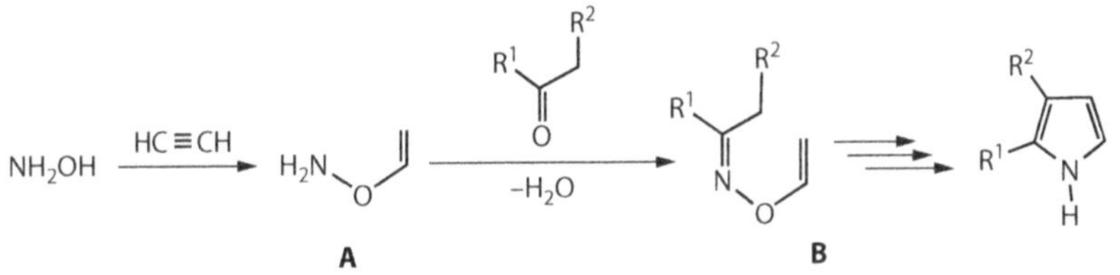

SCHEME 1.20 Tentative assistance of hydroxylamine to the one-pot pyrrole synthesis from
ketones, hydroxylamine, and acetylene.

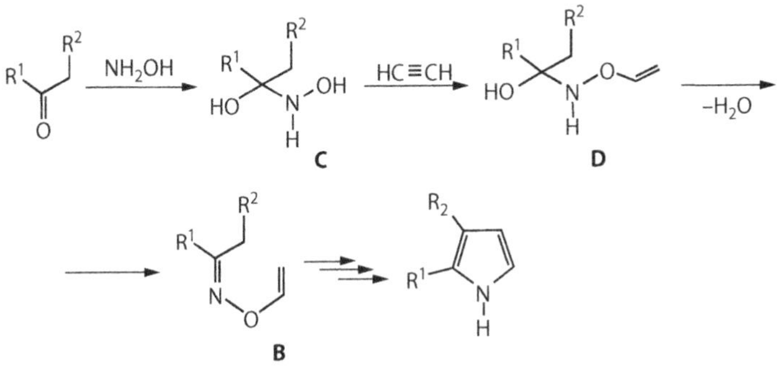

SCHEME 1.21 Alternative route to the pyrroles in the reaction of ketones, hydroxylamine,
and acetylene.

1.1.6 EFFECT OF KETOXIMES STRUCTURE ON YIELDS AND RATIO OF PYRROLES

The heterocyclization of ketoximes with acetylene leading to NH- and N-vinylpyrroles is viable for all ketoximes having at least one methylene or methyl group in α-position to the oxime moiety and containing no substituents sensitive to the action of bases [4,5,7].

Yields and physical–chemical constants of all NH- and N-vinylpyrroles synthesized are listed in Tables 1.2 and 1.3, respectively.

1.1.6.1 Dialkyl- and Alkylcycloalkylketoximes

The reaction of symmetric dialkylketone oximes with acetylene furnishes 2-alkyl- and 2,3-dialkylpyrroles and their N-vinyl derivatives in up to 95% yield [185–187]. Oximes of asymmetric dialkylketones, depending on the reaction conditions, give either 2,3-dialkyl-substituted pyrroles and their N-vinyl derivatives or their mixture with the corresponding 3-unsubstituted pyrroles (Scheme 1.22). The latter can be prepared in amounts suitable for preparative separation [188].

When the competition between methyl and methylene groups is possible, usually, the latter participate in the formation of the pyrrole ring [188].

Synthesis of N-vinylpyrroles from symmetric and asymmetric ketoximes is generally carried out in autoclave (10–20 atm, 120°C–140°C, 1–3 h) with a large excess acetylene, KOH–ketoxime ratio being 0.1–0.3:1 [185]. Preferable ketoxime–DMSO ratio is 1:8–16. NH-pyrroles are successfully obtained when acetylene is passed through the reaction mixture (atmospheric pressure, 93°C–100°C, 6–10 h, 40%–100% KOH from the ketoxime weight) in the presence of DMSO containing 4%–10% of water that suppresses vinylation reaction [186].

The interaction of alkyl cyclopropyl ketoximes with acetylene proceeds under acetylene atmospheric pressure (90°C–100°C) to deliver mixtures of 3-alkyl-2-cyclopropylpyrroles and their N-vinyl derivatives in 24%–64% yields (Scheme 1.23) [189].

Oxime of 1-acetyladamantane reacts with acetylene under pressure (13 atm) to give O-vinyl oxime (80% yield) [130], which upon heating (DMSO, 120°C, 1 h) rearranges to 2-(1-adamantyl)pyrrole (83% yield). 1-Acetyladamantane and adamantane are the side products of this reaction (Scheme 1.24).

Under harsher conditions, oxime of 1-acetyladamantane reacts with acetylene giving rise to 2-(1-adamantyl)-N-vinylpyrrole in a yield of 48% (Scheme 1.25), the minor product being 1-acetyladamantane (24%), which, if necessary, can be re-oximized and involved in the reaction with acetylene.

1.1.6.2 Oximes of Cyclic and Heterocyclic Ketones

N-vinyl-4,5,6,7-tetrahydroindole appears to be the first representative of N-vinylpyrroles synthesized from ketoximes and acetylene (Scheme 1.26) [190–195]. Later on, the conditions allowing to stop the reaction and the stages of 4,5,6,7-tetrahydroindole formation have been found [196]. The yields of NH- and N-vinyltetrahydroindoles reach 81% and 93%, respectively.

The fundamental peculiarities and experimental details of this synthesis are discussed in the work [161]. The formation of N-vinyl-4,5,6,7-tetrahydroindole from 4,5,6,7-tetrahydroindole is confirmed by easy vinylation of the latter under the

TABLE 1.2
Pyrroles Obtained from Ketoximes

No.	Structure	Yield (%)	Bp, °C/torr (Mp, °C)	d_4^{20}	n_D^{20}	References
1	2	3	4	5	6	7
1	Me	42	75/50	0.9246	1.5010	[186]
		80				[302]
		33				[298]
2	Me, Me	73	69–70/30	0.9322	1.4990	[201]
		85				[302]
		91				
3	Me, Et	72	60.5/4	0.9206	1.5010	[186]
		90				[302]
		30				[299]
4	Et, Me	73	71–72/13	0.9223	1.4990	[7]
5	Me, i-Pr	69	69–70/5	0.8240	1.4910	[7]
6	Et, n-Pr	73	85/3	0.8365	1.4950	[186]
		83				[302]
7	n-Pr, Me	83	71–72/5	0.9071	1.4935	[7]
8	i-Pr, Me	87	62–64/7	0.9000	1.4910	[7]

(Continued)

TABLE 1.2 (Continued)
Pyrroles Obtained from Ketoximes

No.	Structure	Yield (%)	Bp, °C/torr (Mp, °C)	d_4^{20}	n_D^{20}	References
1	2	3	4	5	6	7
9	*t*-Bu (pyrrole)	65	(32)			[7]
10	*n*-Pr, *n*-Bu (pyrrole)	72 72	90–91/~1	0.8800	1.4860	[186] [187]
11	*n*-Am, Me (pyrrole)	64	80/~1	0.9058	1.4903	[7]
12	Me (pyrrole)	30	90/8	0.8956	1.5158	[207]
13	Me, Me, Me (pyrrole)	38	98/2		1.5202	[209]
14	Me (pyrrole)	16				[209]
15	cyclopropyl (pyrrole)	8			1.5371	[189]

(Continued)

TABLE 1.2 (*Continued*)
Pyrroles Obtained from Ketoximes

No.	Structure	Yield (%)	Bp, °C/torr (Mp, °C)	d_4^{20}	n_D^{20}	References
1	2	3	4	5	6	7
16		24				[189]
17		3				[189]
18		25				[189]
19		83	(100–102)			[130]
20		5				[212]
21		97	(50)			[135] [180]
22		70	75–76/~1			[7]
23		79	(102)			[202]

(Continued)

TABLE 1.2 (*Continued*)
Pyrroles Obtained from Ketoximes

No.	Structure	Yield (%)	Bp, °C/torr (Mp, °C)	d_4^{20}	n_D^{20}	References
1	2	3	4	5	6	7
24		60	57–58/~0			[202]
25		24	(128–129)			[205]
26		65	(143–144)			[205]
27		62	(143)			[203]
28		13	(127–129)			[203]
29		35				[203]
30		16				[205]

(Continued)

TABLE 1.2 (*Continued*)
Pyrroles Obtained from Ketoximes

No.	Structure	Yield (%)	Bp, °C/torr (Mp, °C)	d_4^{20}	n_D^{20}	References
1	2	3	4	5	6	7
31		70	109			[199]
32		41	(121–122)			[183]
33		64 73	(129)			[160] [219]
34		38	(34)			[160]
35		21	(65)			[286]
36		28	(25–27)			[282]
37		62	122/~1			[164]
38		60	125/~1			[164]

(Continued)

TABLE 1.2 (Continued)
Pyrroles Obtained from Ketoximes

No.	Structure	Yield (%)	Bp, °C/torr (Mp, °C)	d_4^{20}	n_D^{20}	References
1	2	3	4	5	6	7
39	i-Pr, Ph	28	(67–68)			[7]
40	n-Bu, Ph	60	134/~1			[164]
41	n-Am, Ph	58	146/~1	1.0005	1.5751	[164]
42	n-$H_{13}C_6$, Ph	72	168/~1	0.9699	1.5548	[7]
43	Ph	42	(213–215)			[221]
44	Ph, Ph	72	(127)			[165]
45	Ph, Ph	15	(143)			[286]
46	Me	63	(149)			[7]
47	Et	63	(148–149)			[163]

(Continued)

TABLE 1.2 (*Continued*)
Pyrroles Obtained from Ketoximes

No.	Structure	Yield (%)	Bp, °C/torr (Mp, °C)	d_4^{20}	n_D^{20}	References
1	2	3	4	5	6	7
48		23	(74–76)			[222]
49		65	(140)			[163]
50		41	(49–52)			[221]
51		36	(158–160)			[7]
52		43	(149)			[7]
53		26	(61–63)			[221]
54		47	(129–130)			[221]
55		43	(152–154)			[221]
56		48	(180 разл.)			[221]

(Continued)

TABLE 1.2 (*Continued*)
Pyrroles Obtained from Ketoximes

No.	Structure	Yield (%)	Bp, °C/torr (Mp, °C)	d_4^{20}	n_D^{20}	References
1	2	3	4	5	6	7
57		34	(82–84)			[288]
58		74	(140–142)			[132]
59		24	(122–124)			[224]
60		38	(150–152)			[224]
61		41	(148–150)			[224]
62		22	(179–180)			[225]
63		64	(154–155)			[225]
64		24	(180–182)			[227]

(*Continued*)

TABLE 1.2 (*Continued*)
Pyrroles Obtained from Ketoximes

No.	Structure	Yield (%)	Bp, °C/torr (Mp, °C)	d_4^{20}	n_D^{20}	References
1	2	3	4	5	6	7
65		32	(202–204)			[227]
66		37	(128–130)			[227]
67	*n*-Pr	46				[227]
68	Me	11	(20–22)			[228]
69		22	(41)			[232]
70	Me	23	101/3	0.9674	1.5785	[233]
71	Et	39	159/10	1.1483	1.5107	[233]
72		61	(75)			[244]

(Continued)

TABLE 1.2 (Continued)
Pyrroles Obtained from Ketoximes

No.	Structure	Yield (%)	Bp, °C/torr (Mp, °C)	d_4^{20}	n_D^{20}	References
1	2	3	4	5	6	7
73	Me	57	112/1			[247]
74	Et	60	125–126/1	1.1093	1.6675	[247]
75	n-Pr	48	130/~2	1.1164	1.5880	[247]
76	n-Bu	60	145/~2	1.0956	1.5342	[247]
77	n-$H_{13}C_6$	53	176/1	1.0538	1.5773	[247]
78		36				[250]
79	Me	50				[250]
80	Et	55				[250]
81		24	(128–132)			[234]

(*Continued*)

TABLE 1.2 (*Continued*)
Pyrroles Obtained from Ketoximes

No.	Structure	Yield (%)	Bp, °C/torr (Mp, °C)	d_4^{20}	n_D^{20}	References
1	2	3	4	5	6	7
82		63	(78–80)			[248]
83		32	(88–89)			[251] [252]
84		29	(130–132)			[227]
85		58	(210)			[283]
86		16	(289–290)			[133]
87		6	(138–142)			[133]
88		8	(123–126)			[133]
89		4	(185–186)			[134]
90		3	(116–118)			[134]

TABLE 1.3
N-Vinylpyrroles Obtained from Ketoximes

No.	Structure	Yield (%)	Bp, °C/torr (Mp, °C)	d_4^{20}	n_D^{20}	References
1	2	3	4	5	6	7
1	Me	54	39–40/10	0.9308	1.5230	[185–187]
2	Me, Me	73	71/20	0.9297	1.5297	[185–187]
3	Me, Et	70	59–60/6	0.9158	1.5175	[185–187]
4	Et, Me	86	59–60/6	0.9112	1.5185	[185–187]
5	Me, i-Pr	83	61/~1	0.9020	1.5175	[185–187]
6	Et, n-Pr	73	86–87/3	0.9020	1.5160	[185–187]
7	n-Pr, Me	83	63/2	0. 9141	1.5159	[185–187]
8	i-Pr, Me	87	74/10	0.9016	1.5150	[185–187]

(Continued)

TABLE 1.3 (*Continued*)
N-Vinylpyrroles Obtained from Ketoximes

No.	Structure	Yield (%)	Bp, °C/torr (Mp, °C)	d_4^{20}	n_D^{20}	References
1	2	3	4	5	6	7
9	*t*-Bu	91	52/~1	0.9016	1.5133	[185–187]
10	*n*-Pr, *n*-Bu	95	86–87/2	0.8925	1.5025	[185–187]
11	*n*-Am, Me	76	86/2	0.8753	1.5078	[185–187]
12	Me, Me	80	98–99/5	0.9254	1.5717	[207,208]
13	Me, Me, Me	42	90/4		1.5703	[209]
14	Me	46	140/4		1.5542	[209]

(Continued)

TABLE 1.3 (*Continued*)
N-Vinylpyrroles Obtained from Ketoximes

No.	Structure	Yield (%)	Bp, °C/torr (Mp, °C)	d_4^{20}	n_D^{20}	References
1	2	3	4	5	6	7
15		45			1.5308	[209]
16		15		0.9792	1.5394	[189]
17		64				[189]
18		16			1.5964	[189]
19		34		1.0154	1.5913	[189]
20		44				[189]
21		48	(106–108)			[130]
22		65				[212]

(Continued)

TABLE 1.3 (*Continued*)
N-Vinylpyrroles Obtained from Ketoximes

No.	Structure	Yield (%)	Bp, °C/torr (Mp, °C)	d_4^{20}	n_D^{20}	References
1	2	3	4	5	6	7
23		25	(143–146)			[213]
24						[217]
25		8	(199–201)			[217]
26		4			1.6010	[271]
27		93	85–86/3	1.0010	1.5562	[190–195]
28		80	68–69/2	0.9801	1.5465	[7]
29		93	73–74/~1	0.9937	1.5560	[202]

(*Continued*)

TABLE 1.3 (*Continued*)
N-Vinylpyrroles Obtained from Ketoximes

No.	Structure	Yield (%)	Bp, °C/torr (Mp, °C)	d_4^{20}	n_D^{20}	References
1	2	3	4	5	6	7
30		100	65–66/~0	0.9809	1.5512	[202]
31		100	114–116/~0	–	1.5472	[202]
32		90	(25–27)	0.9805	1.5303	[205]
33		6				[203]
34		55				[203]
35		45				[203]
36		31				[203]

(*Continued*)

TABLE 1.3 (*Continued*)
N-Vinylpyrroles Obtained from Ketoximes

No.	Structure	Yield (%)	Bp, °C/torr (Mp, °C)	d_4^{20}	n_D^{20}	References
1	2	3	4	5	6	7
37		48	(106–108)			[130]
38		3				[270]
39		6	(188)			[271]
40		25 71	142–143/1	1.0705	1.6565	[184,199]
41		75				[183]
42		78	—			[445]
43		70	94/1	1.0443	1.6190	[160,220]
44		76	110/1–2	1.0509	1.5690	[160]

(Continued)

TABLE 1.3 (*Continued*)
N-Vinylpyrroles Obtained from Ketoximes

No.	Structure	Yield (%)	Bp, °C/torr (Mp, °C)	d_4^{20}	n_D^{20}	References
1	2	3	4	5	6	7
45	Ph, Me	74	124/1–2	1.0345	1.6009	[282]
46	Et, Ph	77	124/~1	1.0192	1.5910	[160]
47	n-Pr, Ph	80	132/1–2	0.9972	1.5831	[160]
48	i-Pr, Ph	70	108–112/1	1.0149	1.5932	[7]
49	n-Bu, Ph	75	126–128/~1	0.9878	1.5752	[160]
50	n-Am, Ph	79	145/1–2	0.9764	1.5670	[160]
51	$n\text{-}H_{13}C_6$, Ph	95	158/2–3	0.9700	1.5615	[7]
52	$n\text{-}H_{15}C_7$, Ph	80	160/1–2	0.9636	1.5551	[7]

(Continued)

TABLE 1.3 (*Continued*)
N-Vinylpyrroles Obtained from Ketoximes

No.	Structure	Yield (%)	Bp, °C/torr (Mp, °C)	d_4^{20}	n_D^{20}	References
1	2	3	4	5	6	7
53	n-H₁₇C₈ Ph	82	166/1–2	0.9575	1.5500	[7]
54	n-H₁₉C₉ Ph	82	170–171/~1	0.9474	1.5409	[7]
55	Ph Ph	73	(109–110)			[165]
56	Ph	68	(120–122)			[221]
57	Me	64	104/2	1.0250	1.6070	[7]
58	Me Me	51	104/1–2	1.0026	1.5830	[7]
59	Me Me Me Me	8			1.6815	[222]
60	Et	72	126–127/~1	1.0175	1.5979	[163]

(*Continued*)

TABLE 1.3 (*Continued*)
N-Vinylpyrroles Obtained from Ketoximes

No.	Structure	Yield (%)	Bp, °C/torr (Mp, °C)	d_4^{20}	n_D^{20}	References
1	2	3	4	5	6	7
61	Me / Et	80	129–130/2–3	1.0041		[221]
62	i-Pr	58	113/1–2	1.0050	1.5849	[7]
63	t-Bu	72	123/1–2	0.9989	1.5821	[7]
64	Cl	65	120/~1	1.1674	1.6290	[163]
65	i-Pr / Cl	49	125–126/~2	1.0759	1.5846	[221]
66	Br	22	123–124/~1	1.3860	1.6469	[7]
67	MeO	44	142–144/~1	1.0857	1.6075	[7]
68	i-Pr / MeO	48	140–142/~2–3	1.0274	1.5792	[221]
69	EtS	48	160–162/~3	1.0685	1.6338	[221]

(Continued)

TABLE 1.3 (*Continued*)
N-Vinylpyrroles Obtained from Ketoximes

No.	Structure	Yield (%)	Bp, °C/torr (Mp, °C)	d_4^{20}	n_D^{20}	References
1	2	3	4	5	6	7
70		55	188–190/~1	1.0953	1.6150	[221]
71		21	182			[286]
72		7	50–52			[7]
73		14				[132]
74		61	(48–50)			[226]
75		71	(50–51)			[226]
76		30	48–50			[227]
77		29				[227]

(Continued)

TABLE 1.3 (*Continued*)
N-Vinylpyrroles Obtained from Ketoximes

No.	Structure	Yield (%)	Bp, °C/torr (Mp, °C)	d_4^{20}	n_D^{20}	References
1	2	3	4	5	6	7
78		16	–			[227]
79		40	40–62/5	1.0513	1.5934	[228]
80		15			1.6024	[270]
81		7			1.5968	[271]
82		3				[270]
83		12			1.5780	[271]
84		3				[270]

(Continued)

TABLE 1.3 (*Continued*)
N-Vinylpyrroles Obtained from Ketoximes

No.	Structure	Yield (%)	Bp, °C/torr (Mp, °C)	d_4^{20}	n_D^{20}	References
1	2	3	4	5	6	7
85		50	75–76/3	1.0886	1.5977	[231]
86	Me	85	88/3	1.0803	1.5775	[233]
87	Et	65	96/~1	1.0492	1.5649	[233]
88	n-Pr	57	112–114/3	1.0834	1.5592	[233]
89	i-Pr	81	118/~1	0.9954	1.5545	[233]
90	n-Bu	64	115/5	1.0030	1.5557	[233]
91		50	110–111/~1	1.0526	1.6350	[244]
92	Me	67	115/~1	1.0812	1.6078	[245]
93	Et	70	118/~1	1.0450	1.5981	[245]

(*Continued*)

TABLE 1.3 (*Continued*)
N-Vinylpyrroles Obtained from Ketoximes

No.	Structure	Yield (%)	Bp, °C/torr (Mp, °C)	d_4^{20}	n_D^{20}	References
1	2	3	4	5	6	7
94	*n*-Pr	63	136/~1	1.0447	1.5823	[245]
95	*n*-Bu	62	146/~1	1.0206	1.5805	[245]
96	*n*-$H_{13}C_6$	60	157/~1	1.0241	1.5698	[245]
97		28				[250]
98	Me	42				[250]
99	Et	49				[250]
100		6			1.5165	[230]
101		46			1.6535	[234]

(*Continued*)

TABLE 1.3 (*Continued*)
N-Vinylpyrroles Obtained from Ketoximes

No.	Structure	Yield (%)	Bp, °C/torr (Mp, °C)	d_4^{20}	n_D^{20}	References
1	2	3	4	5	6	7
102		68	(68–70)			[248]
103		68	121–122/3			[251,252]
104		62	148–150/4			[252]
105		65	112–115/1			[252]
106		36			1.6084	[269]
107		2	–			[269]
108		30	(102–104)			[227]
109		7	(128–130)			[133]

(*Continued*)

TABLE 1.3 (*Continued*)
N-Vinylpyrroles Obtained from Ketoximes

No.	Structure	Yield (%)	Bp, °C/torr (Mp, °C)	d_4^{20}	n_D^{20}	References
1	2	3	4	5	6	7
110		30	(130–133)			[133]
111		25	(89–90)			[133]
112		6	(110–112)			[133]
113		14	(102–106)			[133]
114		5	(78–80)			[134]
115		16	(94–96)			[134]
116		13	82–84			[134]
117		15	(85–88)			[134]
118		12	87–90			[134]

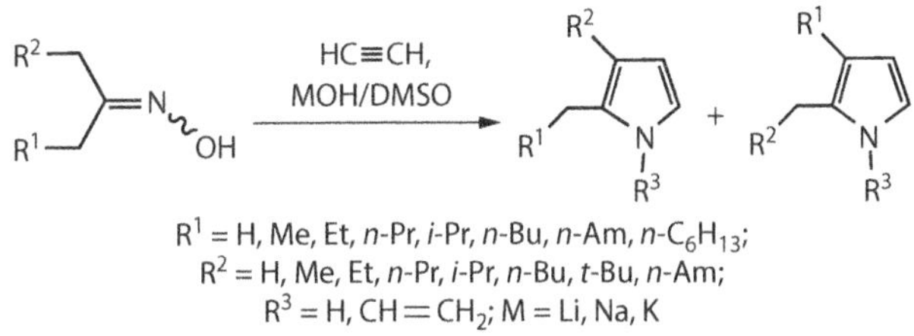
HC≡CH,
MOH/DMSO
R¹ = H, Me, Et, n-Pr, i-Pr, n-Bu, n-Am, n-C6H13;
R² = H, Me, Et, n-Pr, i-Pr, n-Bu, t-Bu, n-Am;
R³ = H, CH=CH2; M = Li, Na, K

SCHEME 1.22 Reaction of dialkyl ketoximes with acetylene.

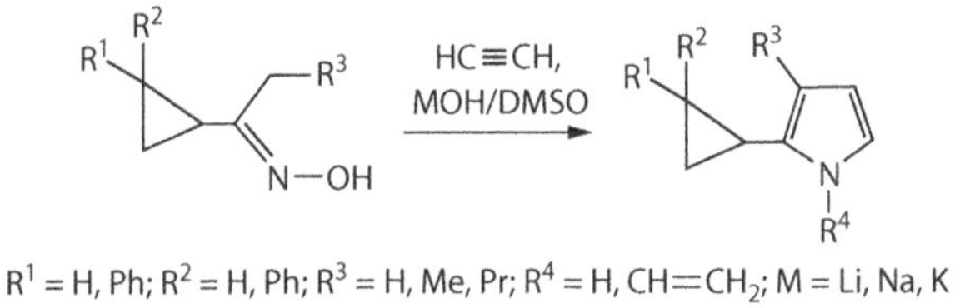
HC≡CH,
MOH/DMSO
R¹ = H, Ph; R² = H, Ph; R³ = H, Me, Pr; R⁴ = H, CH=CH2; M = Li, Na, K

SCHEME 1.23 Reaction of alkyl cyclopropyl ketoximes with acetylene.

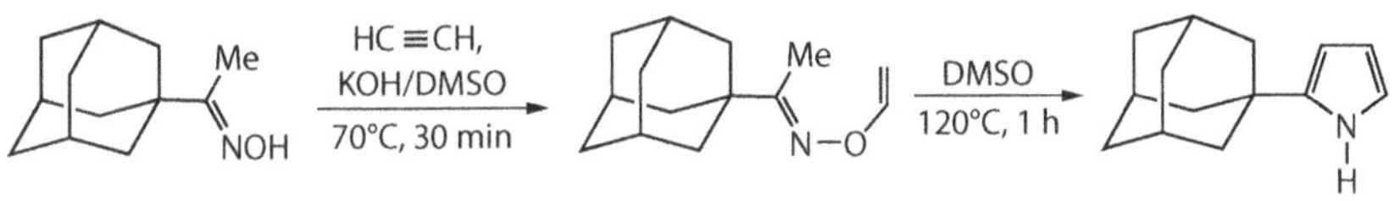
HC≡CH,
KOH/DMSO
70°C, 30 min
DMSO
120°C, 1 h

SCHEME 1.24 Two-step synthesis of 2-(1-adamantyl)pyrrole from 1-acetyladamantane oxime and acetylene.

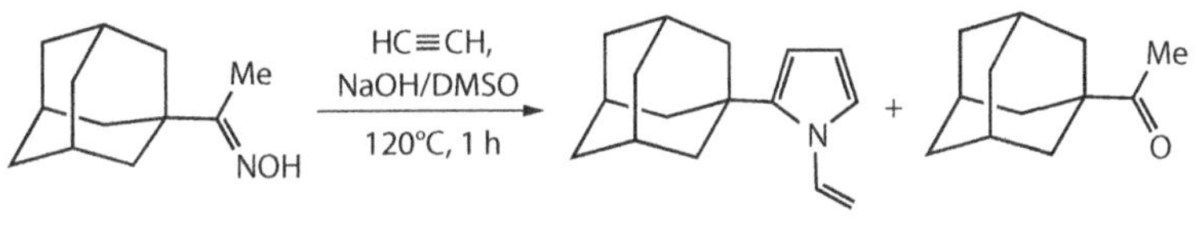
HC≡CH,
NaOH/DMSO
120°C, 1 h

SCHEME 1.25 Straightforward synthesis of 2-(1-adamantyl)-N-vinylpyrrole from 1-acetyladamantane oxime and acetylene.

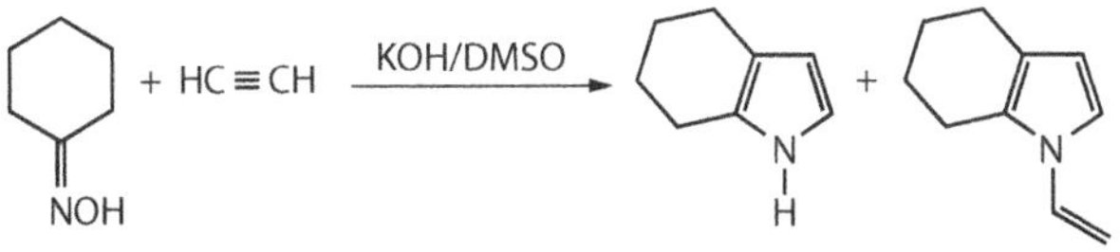
+ HC≡CH
KOH/DMSO

SCHEME 1.26 Synthesis of 4,5,6,7-tetrahydroindole and N-vinyl-4,5,6,7-tetrahydroindole from cyclohexanone oxime and acetylene.

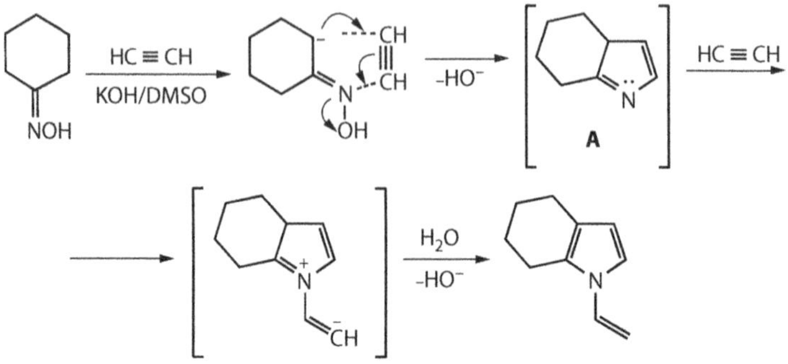

SCHEME 1.27 Alternative pathway to N-vinyl-4,5,6,7-tetrahydroindole.

conditions of this reaction (the yield is not less than 85%). However, the comparison of yields of both compounds under similar conditions allows to assume [161] that the preparation of N-vinyl-4,5,6,7-tetrahydroindole in a one step does not exclude its direct synthesis by the interaction of acetylene with any reactive intermediate (since the yield of N-vinyl-4,5,6,7-tetrahydroindole is always appreciably higher than that of 4,5,6,7-tetrahydroindole). According to the mechanism proposed in the first communications dedicated to this reaction, 3H-pyrrole A can serve as an intermediate of this type. Since the nitrogen atom of this pyrrole is not included in the aromatic system, it should possess an increased nucleophilicity. Therefore, it should react more actively with the second molecule of acetylene than the "pyrrole" nitrogen atom of 4,5,6,7-tetrahydroindole (Scheme 1.27).

No traces of O-vinylcyclohexanone oxime, the expected intermediate product, could be detected even under mild conditions and at low degrees of conversion of cyclohexanone oxime. Apparently, if O-vinyl oxime is the intermediate of this variant of the reaction, its rearrangement to 4,5,6,7-tetrahydroindole proceeds extremely quickly.

The synthesis of NH- and N-vinyl-4,5,6,7-tetrahydroindoles is implemented successfully at cyclohexanone oxime–acetylene molar ratio of 1:(2–5) at 90°C–140°C. At temperature below 90°C, the reaction proceeds slowly, and the rate of side processes increases above 140°C. The bases (alkali metal hydroxides and alcoholates) taken in 10%–50% amounts relative to cyclohexanone oxime serve as the reaction catalysts. The reaction is efficiently catalyzed by potassium, rubidium, and tetrabutylammonium hydroxides. It should be noted that the last two bases have a selective effect on the construction of the tetrahydroindole ring (synthesis of 4,5,6,7-tetrahydroindole) [196], while KOH, as it was already discussed earlier, is also active in the vinylation step.

For the reaction of cyclohexanone oxime with acetylene, it has been established that the most effective polar solvents in this reaction is DMSO, and the system DMSO–dioxane is the most efficient of the mixed solvents [159]. Application of the mixed solvent DMSO–dioxane allows to conduct the reaction selectively, that is, to obtain either 4,5,6,7-tetrahydroindole (DMSO concentration is about 5%–10%) or N-vinyl-4,5,6,7-tetrahydroindole (in pure DMSO). It is not expedient to use more than 25-fold excess of the solvent, while the lack of the latter requires an increase in the reaction temperature, which leads to resinification.

The kinetics of the reaction between cyclohexanone oxime and acetylene in the systems MOH/DMSO (M=Li, Na, K, Cs) under atmospheric pressure has been studied, and the quantitative data about the effect of alkali metal cation on the yields of 4,5,6,7-tetrahydroindole and its vinyl derivative as well as on the reaction selectivity have been obtained [197].

Early qualitative observations (see Section 1.1.2) that pyrrolization rate considerably increases on going from LiOH to NaOH and further to the KOH and the LiOH/CsF systems are confirmed. In the presence of LiOH and NaOH/CsF systems, the vinylation is suppressed and selectivity of 4,5,6,7-tetrahydroindole formation rises.

7-Methyl-4,5,6,7-tetrahydroindole (70% yield) and its N-vinyl derivative (80% yield) are prepared from 2-methylcyclohexanone oxime and acetylene (Scheme 1.28) [7].

4,4,6,6-Tetramethyl-4,5,6,7-tetrahydroindole is synthesized in 70% yield by the Trofimov reaction from the corresponding ketone (via oxime) and acetylene (Scheme 1.29) [198].

Hitherto unknown 4,5-dihydrobenz[g]indole is obtained from 1-tetralone oxime and acetylene in the system KOH/DMSO in one stage (Scheme 1.30) [199]. 4,5-Dihydrobenz[g]indoles and their N-vinyl derivatives substituted at the benzene ring are isolated from the reaction of the corresponding 1-tetralone oximes and acetylene in the systems MOH/DMSO (Scheme 1.30) [200].

From cycloheptanone [201], cyclooctanone [202], and cyclododecanone [202] oximes, the corresponding pyrroles and their N-vinyl derivatives are synthesized (90°C–140°C) in high yields (up to quantitative ones, Scheme 1.31).

Oximes of 1,2,5-trimethylpiperidin-4-one, 2-methyl-, and 2-(2-furyl)-4-ketodecahydroquinolines react with acetylene (85°C–95°C, atmospheric pressure) to afford substituted pyrrolo[3,2-c]piperidines and their N-vinyl derivatives in yields up to 69% (Scheme 1.32) [203].

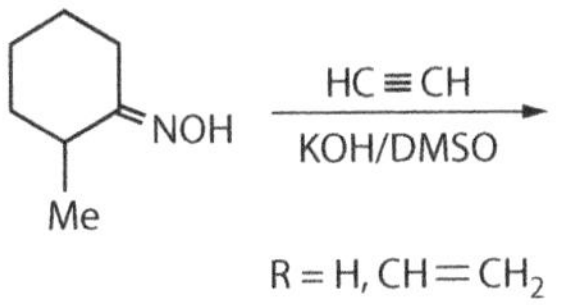

R = H, CH=CH$_2$

SCHEME 1.28 Synthesis of 7-methyl-4,5,6,7-tetrahydroindole and its N-vinyl derivative from 2-methylcyclohexanone oxime and acetylene.

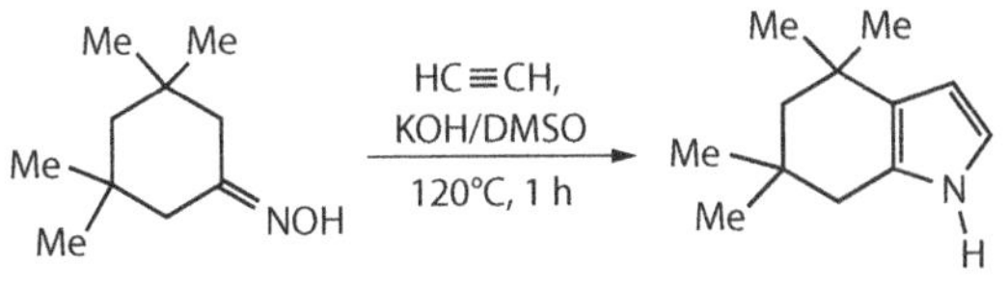

SCHEME 1.29 Synthesis of 4,4,6,6-tetramethyl-4,5,6,7-tetrahydroindole from oxime of 3,3,5,5 tetramethylcyclohexanone and acetylene.

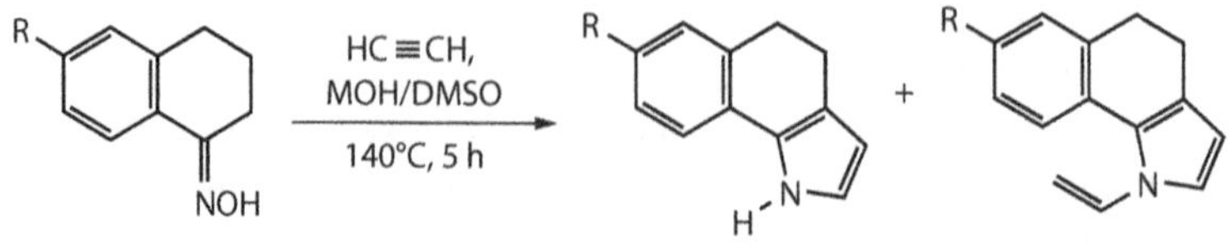

R = H (71%), OMe (65%); R = H (21%), OMe (85%)

SCHEME 1.30 Synthesis of 4,5-dihydrobenz[g]indoles from 1-tetralone oximes and acetylene.

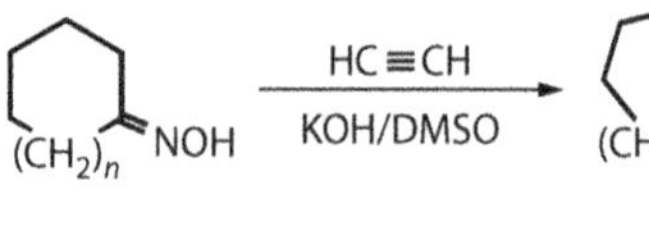

n = 2, 3, 7; R = H, CH=CH$_2$

SCHEME 1.31 Synthesis of pyrroles from oximes of macrocyclic ketones and acetylene.

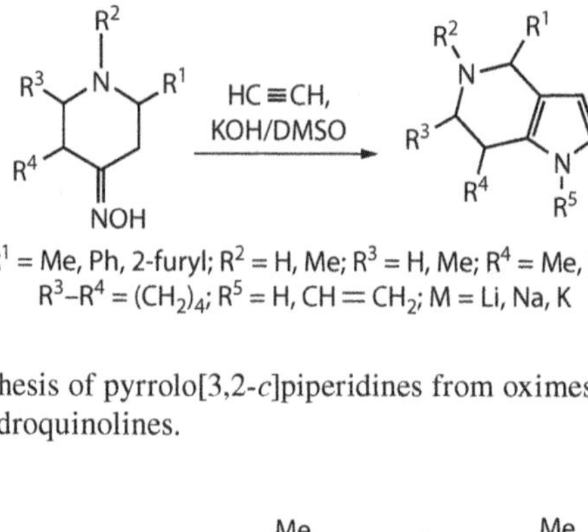

R^1 = Me, Ph, 2-furyl; R^2 = H, Me; R^3 = H, Me; R^4 = Me, Ph;
R^3–R^4 = (CH$_2$)$_4$; R^5 = H, CH=CH$_2$; M = Li, Na, K

SCHEME 1.32 Synthesis of pyrrolo[3,2-c]piperidines from oximes of substituted piperidinones and ketodecahydroquinolines.

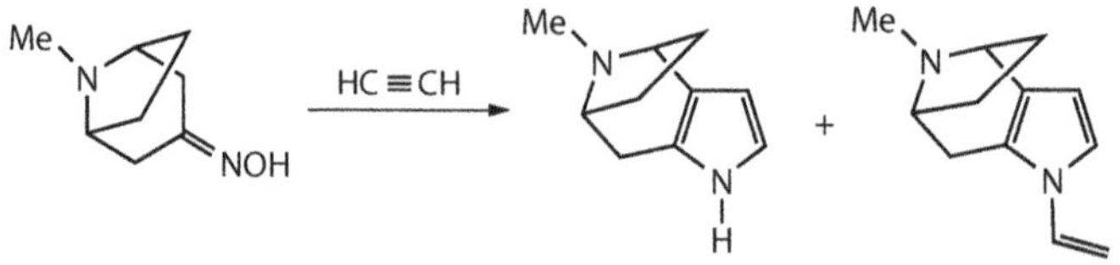

SCHEME 1.33 Synthesis of pyrrole and N-vinylpyrrole from tropinone oxime and acetylene.

Varlamov et al. [204] have reported the synthesis of pyrrole and N-vinylpyrrole from tropinone oxime and acetylene via the Trofimov reaction (Scheme 1.33), the yields being 49% and 41%, respectively.

The corresponding 4,4,6,6-tetramethyl-5-nitroxyl-4,5,6,7-tetrahydro-5-azaindole (X=Ö, R=H), 4,4,6,6-tetramethyl-4,5,6,7-tetrahydro-5-azaindole (X=H, R=H), and its N-vinyl derivative (X=H, R=CH=CH$_2$) are prepared (90°C–105°C, acetylene atmospheric pressure) in 24%–90% yields from nitroxyl-centered oxime, 2,2,6,6-tetramethyl-1-nitroxyl-4-hydroxyiminopiperidine (X=Ö), and also from

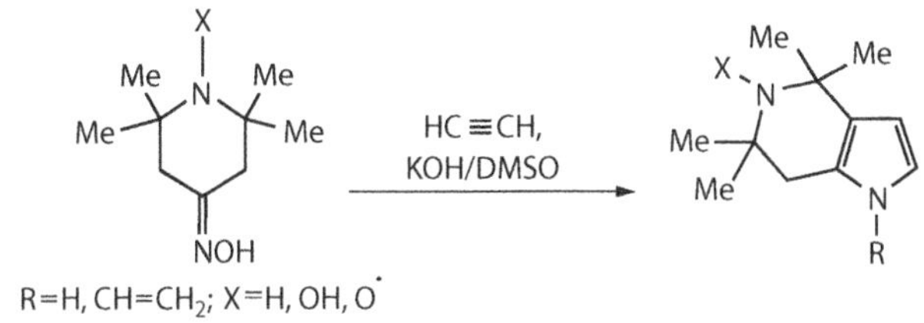

R=H, CH=CH$_2$; X=H, OH, O$^\cdot$

SCHEME 1.34 Synthesis of 4,5,6,7-tetrahydro-5-azaindoles and their nitroxyl and vinyl derivatives.

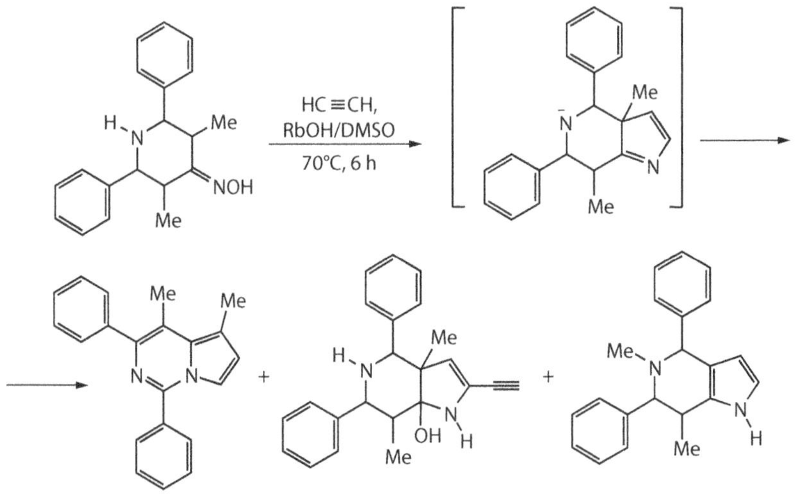

SCHEME 1.35 Reaction of oxime of 3,5-dimethyl-2,6-diphenylpiperid-4-one with acetylene in the RbOH/DMSO system.

2,2,6,6-tetramethyl-4-piperidone oxime (X=H) and its 1-hydroxyl derivative (X=OH, Scheme 1.34) [205].

The reaction of 3,5-dimethyl-2,6-diphenylpiperid-4-one oxime with acetylene in the system RbOH/DMSO proceeds in a peculiar fashion: 3a-Me group migrates toward anionic nitrogen atom to give 5,7-dimethyl-4,6-diphenyl-4,5,6,7-tetrahydropyrrolo[3,2-*c*]pyridine [206]. Tetrahydropyrrolo[1,2-*c*]pyrimidine is formed by decomposition of the intermediate 3*H*-pyrrole via the retro-Mannich reaction (Scheme 1.35).

1.1.6.3 Oximes of Terpenoid Ketones and Their Analogs

The first syntheses of pyrroles bearing unsaturated substituents from γ,δ-ethylenic ketones have been reported in works [207,208]. It is shown that oximes of these ketones easily react with acetylene to form the expected alkenylpyrroles in good or moderate yields.

One-pot conversion of 3-butenyl methyl ketoxime to 2-methyl-3-(2-propenyl) pyrrole and 2-methyl-3-(1-propenyl)-N-vinylpyrrole (Scheme 1.36) demonstrates

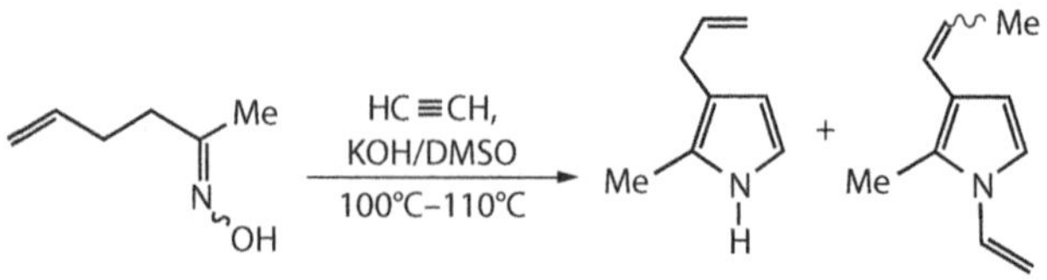

SCHEME 1.36 Synthesis of pyrroles from oximes of γ,δ-ethylenic ketones.

two important features of this reaction: it can be either selectively stopped at the stage of the pyrrole ring construction (without vinylation across the N–H bond and prototropic isomerization of the alkenyl radical) or continued until the formation of N-vinylpyrrole in which the double bond of the alkenyl radical is extremely shifted toward the pyrrole cycle.

Later on, the structural effects of the alkenyl group of terpenoid ketone (5-alken-2-ones) oximes on the formation of pyrroles and N-vinylpyrroles as well as on proto-tropic isomerization of the alkenyl substituent have been studied [209].

Under standard conditions (90°C–95°C, 3 h, initial acetylene pressure 10–12 atm), the pyrrolization of terpenoid ketoximes proceeds with the partici-pation of α-methylene group of the ketoxime alkenyl radicals (Scheme 1.37). 2-(3-Alkenyl)pyrroles, which could be formed due to involvement of the methyl moiety into heterocyclization, are not detected in the reaction mixtures. This fact correlates well with the findings [7,16,18] that upon condensation of alkyl methyl ketoximes with acetylene, the methyl group participates in the pyrrole ring assem-bly only at above 120°C. Prototropic isomerization of the alkenyl radical also does not occur.

The structure of the alkenyl radical dramatically influences the yields and ratio of NH- and N-vinylpyrroles in spite of the fact that the substituent structure changes on considerable distance from the reactive site and only in the alkyl surrounding of the double bond. This can be explained by a partial prototropic isomerization of the initial ketoximes affording oximes of α,β-ethylenic ketones (Scheme 1.38), which are not prone to conversion into pyrroles under the reaction conditions.

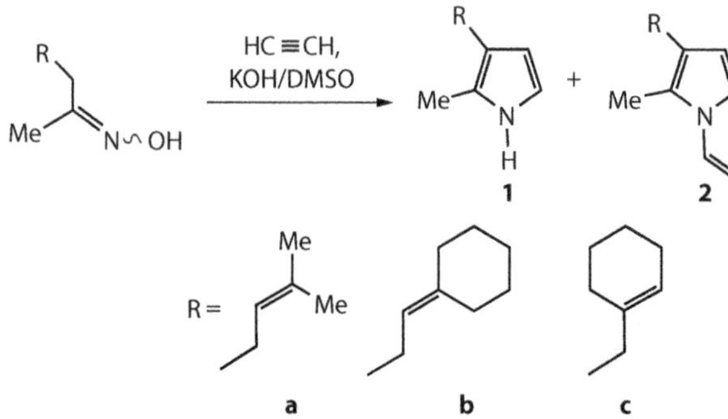

SCHEME 1.37 Synthesis of pyrroles from oximes of terpenoid ketones.

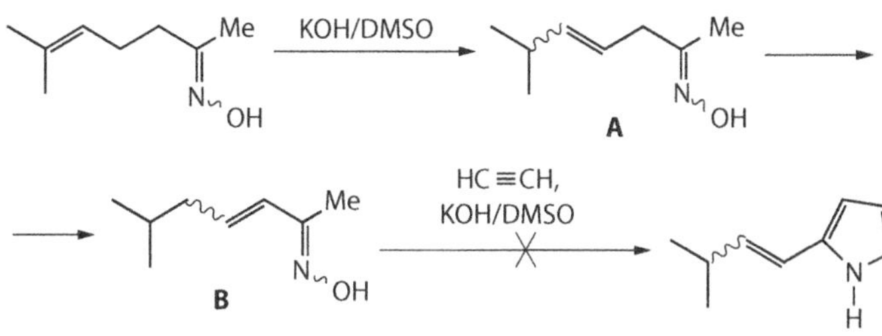

SCHEME 1.38 Prototropic isomerization of alkenyl ketoximes.

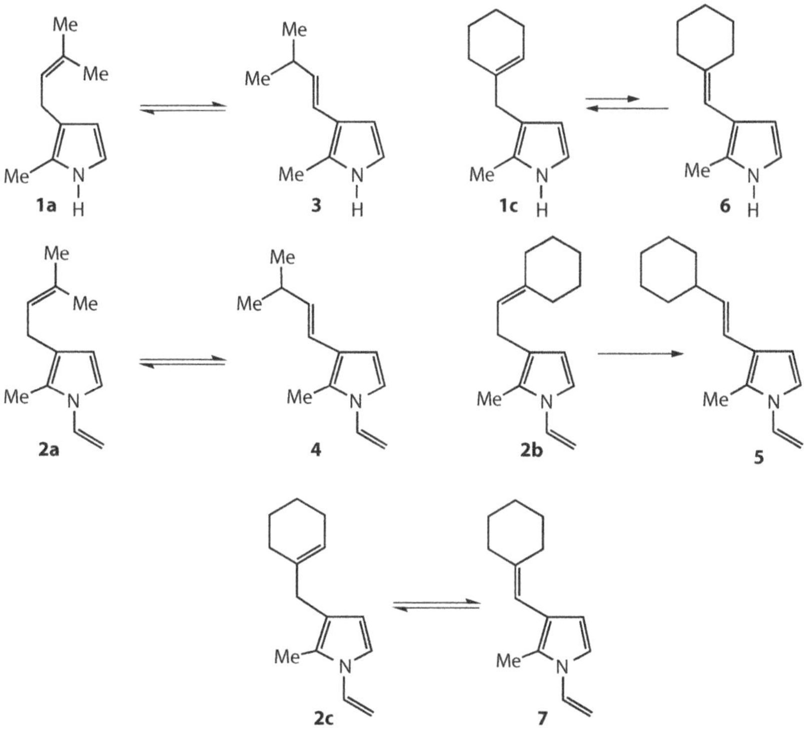

SCHEME 1.39 Prototropic isomerization of alkenyl substituents in pyrroles and N-vinylpyrroles in the KOH/DMSO and *t*-BuOK/DMSO systems.

The absence of pyrroles in the reaction mixture corresponding to the intermediate oximes of type **A** can be caused by rapid conversion of the latter into more stable conjugated oximes **B**.

The prototropic isomerization of alkenyl radicals of terpenoid structure in pyrroles and N-vinylpyrroles (Scheme 1.39) has been studied (Table 1.4) in the systems KOH/DMSO (120°C, 1 h, KOH–pyrrole molar ratio = 1:1, concentration of pyrrole in DMSO = 10%) and *t*-BuOK/DMSO (60°C–65°C, 1 h, concentration of pyrrole in DMSO 10%, *t*-BuOK 10%).

TABLE 1.4
Prototropic Isomerization of Alkenyl Substituents in Pyrroles 1, 2

Pyrrole	Isomerization Product	Conversion, % (Content of E-isomer, %)
3-Allylpyrrole	3-(1-Propenyl)pyrrole	93 (70) [208]
1a	3	30 (~100)
1c	6	~5
2a	4	75 (~100)
2b	5	100 (~100)
2c	7	50 (~100)

The both systems give the identical results testifying to the formation of equilibrium mixture of isomerization products (Table 1.4). Thermodynamic control is confirmed by repeated heating of the isomerization product (the same conditions), the structure of which does not change. The system t-BuOK/DMSO is more active owing to much better solubility of t-BuOK in DMSO in comparison with KOH, while the basicity of both systems is approximately the same (at equal concentration of the base in DMSO) [210]. Composition and structure of isomerizate components have been proved by ^{1}H NMR technique. The values of spin-spin coupling constants ($^3J_{\alpha\beta} = 15.4$–15.6 Hz) indicate that the alkenyl substituents are of E-configuration exclusively.

The alkenyl radicals in N-vinylpyrroles isomerize much deeper (Table 1.4) than in the corresponding NH-pyrroles. It evidences the existence of conjugation between 3-(1-alkene) and N-vinyl fragments, which is transmitted through the pyrrole ring and nitrogen atom. The energy of this conjugation, estimated as the difference of the reactions 1a ⇆ 3 and 2a ⇆ 4 free energies (Gibbs energy) is ~6 kJ/mol. The same value calculated for 1c ⇆ 6 and 2c ⇆ 7 reactions equals 10 kJ/mol.

Differences in isomerization degree (equilibrium constants) for pyrroles 1a, c and 2a–c are caused by distinctions in energy values of the ethylene fragments, that is, by their different stability. The cyclohexene fragment is the most stable (pyrroles 1c and 2c), and, in this case, the isomerization degree is the lowest. On the contrary, pyrrole 2b containing conjugated cyclohexylidene group (the least stable) isomerizes almost completely.

At the same time, under the aforementioned conditions of pyrrole synthesis, the isomerization of alkenyl radicals does not take place, and the reaction is regioselective that opens a shortcut to pyrroles bearing alkenyl fragments of terpenoid nature. Therefore, new opportunities for the direct synthesis of fragrant and biologically active agents emerge. For example, in the chain of transformations occurring during pyrrole 7 synthesis, all four compounds (ketone, oxime, pyrrole, and N-vinylpyrrole) appear to be the carriers of long-lasting odor. Besides, pyrroles and N-vinylpyrroles synthesized from terpenoid ketones can serve as valuable monomers, cross-linking agents, and starting materials for the preparation of functionally substituted pyrroles.

1.1.6.4 Oximes of Ketosteroids

A combination of the pyrrole ring and steroid scaffold in a one molecule is of primary interest for the design of novel pharmacological substances. A spectacular example of such a synergism is extremely high biological activity of batrachotoxin, a product of skin secretion from Colombian frog *Phyllobates aurotaenia*, the structure of which combines the steroid and pyrrole fragments [211]. The South American Indians still use this deadly toxin to prepare darts for their blowguns.

A possibility of successful application of the Trofimov reaction for the construction of the pyrrole ring bonded to the steroid skeleton has been exemplified by oximes of ketosteroids.

The study of the reaction between the oxime of Δ^5-pregnen-3β-ol-20-one and acetylene has allowed to find out the conditions (KOH/DMSO, 100°C, 5 h, initial acetylene pressure at room temperature 14 atm) ensuring not only formation of the pyrrole ring but also exhaustive vinylation of NH-pyrrole and hydroxyl functions (Scheme 1.40) [212].

Under these conditions, the initial ketoxime reacts with acetylene quantitatively to deliver O-vinyl oxime (10%), N-vinylpyrrole (63%), and vinyl ethers of pregnenolone as a mixture of epimers (in 25% total yield). The expected migration of the endocyclic double bond or change in the initial configuration of the steroid function is not observed. The major product, doubly vinylated pyrrole-steroid ensemble, contains steroid, vinyl ether, and N-vinylpyrrole fragments in a one molecule, which ensures that different types of chemical modifications are possible including addition and polymerization reactions.

When the reaction is carried out under milder conditions (75°C, 5 min.), along with N-vinylpyrrole, the formation of its non vinylated precursor is also observed.

The annulation of steroid skeleton with N-vinylpyrrole fragment is accomplished via the reaction of Δ^5-cholesten-3-one oxime with acetylene under pressure in the system KOH/DMSO (120°C, 30 min, Scheme 1.41) [213]. The pyrrolization

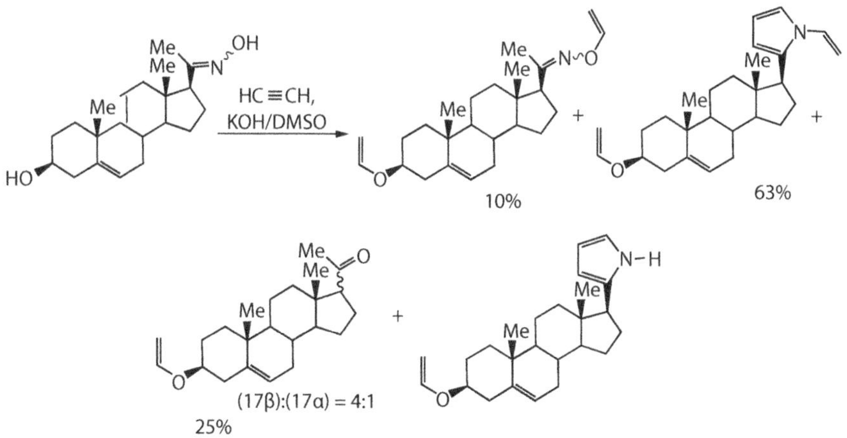

SCHEME 1.40   Reaction of oxime of Δ^5-pregnen-3β-ol-20-one with acetylene.

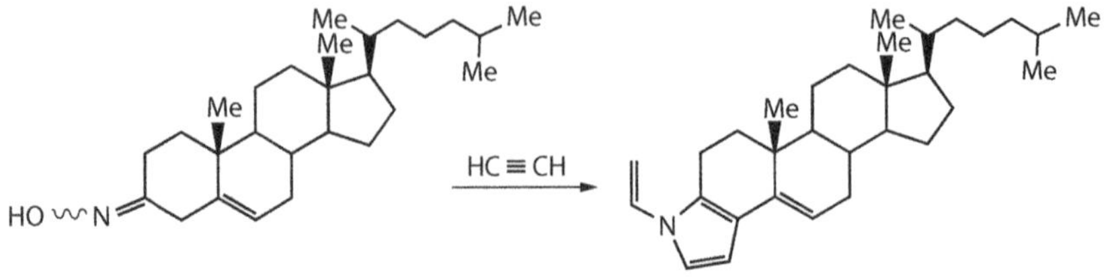

SCHEME 1.41 Synthesis of N-vinylpyrrole from Δ⁵-cholesten-3-one oxime and acetylene in the KOH/DMSO system.

proceeds regioselectively at the methylene group in position 4 to furnish N-vinyl-5-pyrrolo[3,4-*b*]cholestene in 25% isolated yield.

Decrease in the reaction temperature to 100°C and lower leads to deceleration of pyrrolization in comparison with rates of side processes, the most probable of which are prototropic isomerization of the $\Delta^5 \rightarrow \Delta^4$ bond in the initial ketoxime and partial deoximation. At 100°C (5 h), along with oxime of Δ^4-chloesten-3-one (isomer of Δ^5-cholesten-3-one), only traces of N-vinylpyrrolocholestene are detected in the reaction mixture (^{1}H NMR). When the temperature is raised above 120°C, a considerable amount of resin is formed that complicates isolation of the target pyrroles.

In parallel with the development of new methodology for the annulation of steroid skeleton with the pyrrole ring, it has been described a multistage assembly of the condensed system combining fragments of estrone and indole (yields at the last stage are up to 31%) [214]. Synthesis of N-phenylpyrrolo[2,3-*b*]cholestanes from cholestan-2,3-dione derivatives illustrates how laborious can be an alternative way of multistage annulation of steroids with the pyrrole fragments [215].

Comparison of reactivity of the endocyclic and exocyclic ketoxime functions in the position 3 of Δ^5-cholesten-3-one oxime and position 20 of pregnenolone oxime shows that in both cases, quite harsh reaction conditions are required. For Δ^5-cholesten-3-one oxime (120°C, 30 min, 25% yield), drastic decrease in the oxime group reactivity in comparison with that of cyclohexanone oxime may be due to the presence of the double 5,6-carbon–carbon bond and annulation with cyclohexane cycle, which apparently poses conformational hindrances to the reaction. On the other hand, the regiospecific pyrrolization involving methylene group in position 4, being of allylic character, testifies to higher reactivity of the latter as compared to the methylene group in position 2. In the case of pregnenolone oxime, low activity of the oximated acetyl fragment in comparison with that of in alkyl methyl ketoximes can be attributed to the steric hindrances imposed by the steroid skeleton.

Most recently, annulation of the steroid skeleton with the pyrrole nucleus using novel methodology has been successfully expanded over the reaction of 28-triphenylmethyloxy-lup-20(29)-en-3-one oxime with acetylene in the system KOH/DMSO at atmospheric pressure [216]. Pyrrole and N-vinylpyrrole, annulated with triterpenoid fragment, are synthesized in 60% and 15% yields, respectively (Scheme 1.42).

The progesterone molecule combines almost exactly the structural fragments of pregnenolone and Δ^5-cholesten-3-one (except for the position of the C=C bond). However, the results of interaction between progesterone dioxime and acetylene in

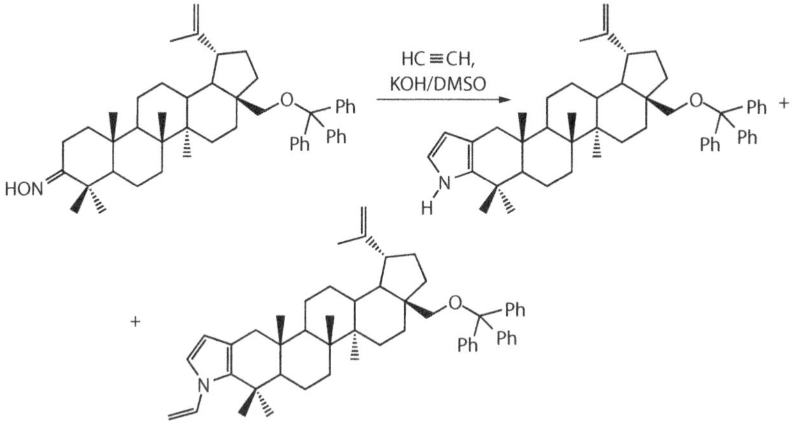

SCHEME 1.42 Reaction of 28-triphenylmethyloxy-lup-20(29)-en-3-one oxime with acetylene.

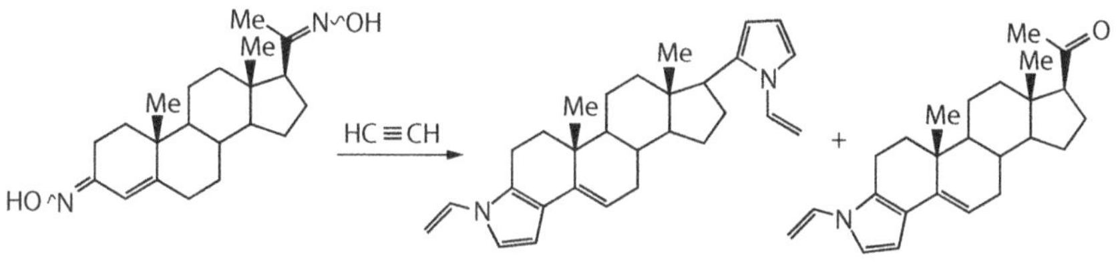

SCHEME 1.43 Synthesis of pyrroles from progesterone dioxime and acetylene in the system KOH/DMSO.

the system KOH/DMSO under acetylene pressure (120°C, 1 h, Scheme 1.43) [217] turn out to be quite different from those obtained when studying the reaction of pregnenolone and Δ^5-cholesten-3-one oximes with acetylene. If the oxime function of the side chain undergoes a partial deoximation as it is observed in pregnenolone oxime, the formation of the pyrrole ring with the participation of oxime group of the steroid skeleton proceeds regiospecifically and also is accompanied by prototropic migration of the double bond into Δ^5-position (Scheme 1.43).

Such reaction course is in agreement with inability of oximes of α,β-ethylenic ketones to react with acetylene under these conditions due to alternative transformations with the participation of electrophilic double bond. In this case, the reaction is probably preceded by migration of the double bond to generate oxime of nonconjugated β,γ-ethylenic ketone containing the oxime moiety capable to pyrrolization (Scheme 1.44). However, since this process is thermodynamically unfavorable (owing to violation of conjugation), yields of pyrroles are small.

Such a migration of the double bond is known for steroids and is observed, for example, during the transformation of Δ^4-cholesten-3-one to Δ^5-cholesten-3-one ethylene ketal [218].

At 120°C (1 h), the total yield of pyrroles is ~7%. When the temperature reaches 140°C for the same time, strong resinification of the reaction mixture occurs, while at

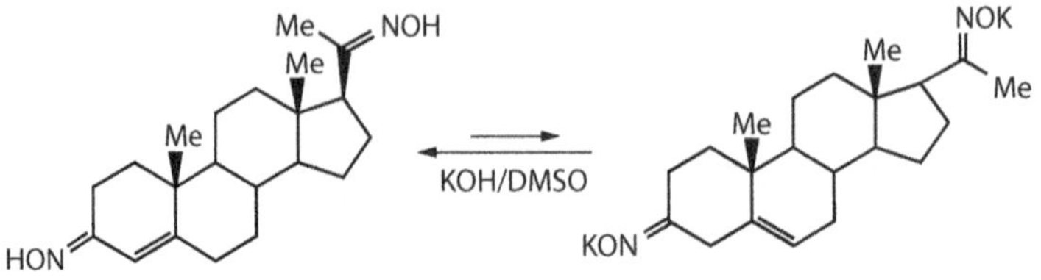

SCHEME 1.44 Premigration of the double bond in progesterone dioxime as prerequisite of the pyrrole synthesis.

100°C, progesterone dioxime is almost inert. Even in rather soft conditions (120°C, 1 h), a large amount of polymeric products is formed in the reaction mixture.

To avoid the undesirable deoximation, the reaction (120°C, 1 h) with dicesium salt of progesterone dioxime, prepared from dilithium derivative and cesium fluoride (CsF) has been carried out [217]. In this case, only dipyrrole derivative of progesterone (8% yield) is formed, whereas deoximation product is not detected. This can be explained by lower (trace) content of water necessary for deoximation in this system.

Thus, a general methodology providing for both combination and annulation of the steroid skeleton with the pyrrole ring in one preparative stage has been developed. The methodology is based on the reaction of ketosteroid oximes with acetylene in the presence of the superbase catalytic systems.

1.1.6.5 Oximes of Alkyl Aryl Ketones

A number of alkyl phenyl ketoximes is successfully involved in the reaction with acetylene. Thus, a series of hardly available 3-alkyl(phenyl)-2-phenylpyrroles [163–165,219] and hitherto unknown 3-alkyl(phenyl)-2-phenyl-N-vinylpyrroles have been synthesized (Scheme 1.45) [163,165,220].

The reaction readily proceeds upon heating of alkyl aryl ketoximes (100°C, 3 h) with acetylene under pressure in DMSO in the presence of 20%–30% (from ketoxime weight) lithium hydroxide (in the case of NH-pyrrole synthesis) or potassium hydroxide (when N-vinylpyrroles are synthesized). If acetylene lacks, the major reaction product is the corresponding 3-alkyl(phenyl)-2-phenylpyrrole [163]. However, the best yields of NH-pyrroles are attained using LiOH [164,165]. This catalyst, which accelerates the stage of the pyrrole ring formation, is almost inactive at the vinylation stage. In certain cases (especially in the presence of LiOH), the synthesis of unsubstituted (at nitrogen atom) pyrroles is accompanied by partial regeneration

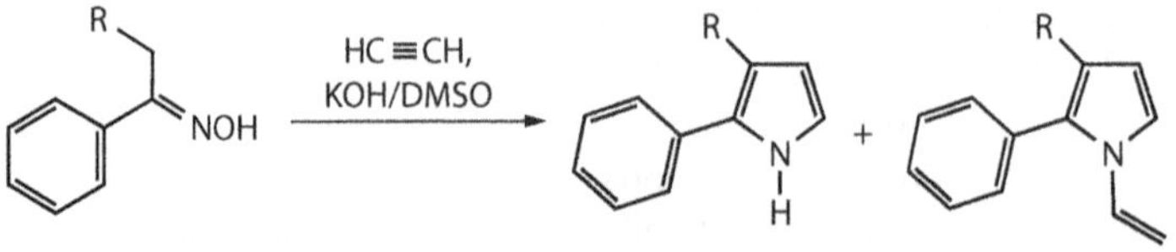

R = H, Me, Et, n-Pr, n-Bu, n-Am, n-C$_6$H$_{13}$, n-C$_8$H$_{17}$, n-C$_9$H$_{19}$, Ph

SCHEME 1.45 Synthesis of 2-phenyl-3-alkyl(phenyl)pyrroles from alkyl phenyl ketoximes and acetylene.

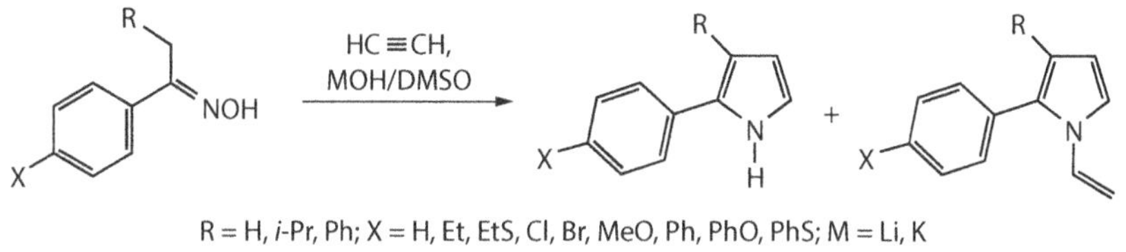

R = H, i-Pr, Ph; X = H, Et, EtS, Cl, Br, MeO, Ph, PhO, PhS; M = Li, K

SCHEME 1.46 Synthesis of 2-aryl-3-alkyl(phenyl)pyrroles from alkyl aryl ketoximes and acetylene.

of alkyl phenyl ketones [164]. When acetylene excess is employed, the reaction proceeds further to give 3-alkyl(phenyl)-2-phenyl-N-vinylpyrroles.

The structure of the starting alkylphenyl ketoximes does not affect substantially on N-vinylpyrrole yields, though the tendency to the yields increase with growth of carbon atoms in the alkyl radical is noticeable.

This new synthesis of 3-alkyl(phenyl)-2-phenylpyrroles has been extended over alkyl aryl ketoximes (Scheme 1.46) [163,221]. As a consequence, earlier unknown or difficult to produce 2-arylpyrroles can be synthesized using this methodology.

The reaction smoothly proceeds also at 100°C in the presence of 30% (from ketoxime weight) KOH in DMSO in autoclave under initial acetylene pressure of 8–16 atm. The maximum pressure reached at the reaction temperature is 20–25 atm. Then, intensive consumption of acetylene begins, and pressure quickly decreases. As it was already noted, initially, N-unsubstituted pyrroles are formed, which further are vinylated in the presence of acetylene excess. If it is necessary to obtain the corresponding NH-pyrrole, the synthesis is carried out with calculated amount of acetylene or with its lack. LiOH appears to be a selective catalyst of the pyrrole ring construction, the application of which does not require strict dosing of acetylene.

The yield of 2-aryl-N-vinylpyrroles essentially depends on the substituent in the benzene nucleus. Sharp drop of 4-bromophenyl-N-vinylpyrrole (from 4-bromoacetophenone oxime) is logical: it is caused by bromine saponification with strong base. From 4-nitroacetonenone oxime, the pyrrole is not formed under the conditions studied due to instability of the substituent in superbase medium. When the temperature decreases to 80°C, the reaction of 4-phenoxyacetophenone oxime and acetylene leads to the corresponding pyrrole in 43% yield [221].

The reaction of ethyl mesityl ketoxime with acetylene under atmospheric pressure affords 2-mesityl-3-methylpyrrole (23%), 2-mesityl-3-methyl-N-vinylpyrrole (8%), Z-(5%), and E-(2%) isomers of O-vinyl mesityl ketoximes (Scheme 1.47) [222].

The interaction of 5-acetyl[2,2]paracyclophane oxime with acetylene (KOH/DMSO or CsOH/DMSO, 100°C–105°C, 4–5 h) gives a mixture of 2-([2,2] para-cyclophan-5-yl)pyrrole and its N-vinyl derivative in low yields (19% and 14%, respectively) [132,223]. 2-[2,2] Paracyclophanylpyrrole (74% yield) is synthesized by thermal rearrangement of acetylparacyclophane O-vinyl oxime, a key intermediate of this reaction (Scheme 1.48) [132]. O-vinyl oxime (78% yield), in its turn, is prepared by vinylation of Cs salt of 5-acetyl-[2,2]paracyclophane oxime with acetylene under pressure in two-phase system DMSO/n-pentane [132].

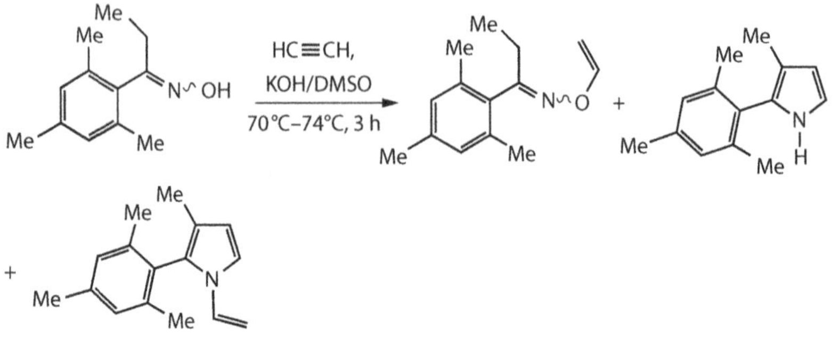

SCHEME 1.47   Reaction of ethyl mesityl ketoxime with acetylene.

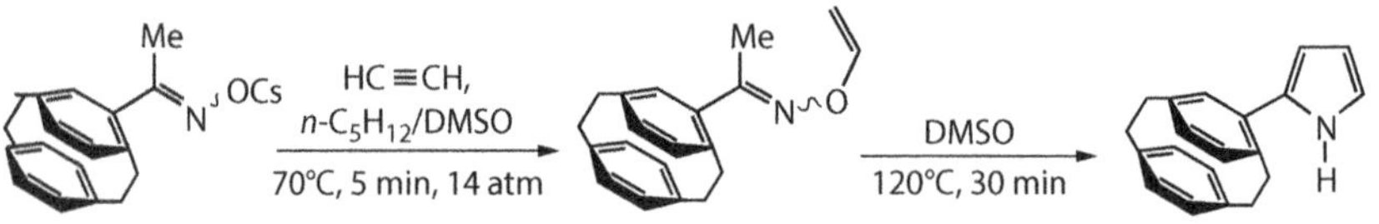

SCHEME 1.48   Two-step synthesis of 2-([2,2] paracyclophan-5-yl)pyrrole from 5-acetyl[2,2] paracyclophane oxime and acetylene via O-vinyl ketoximes.

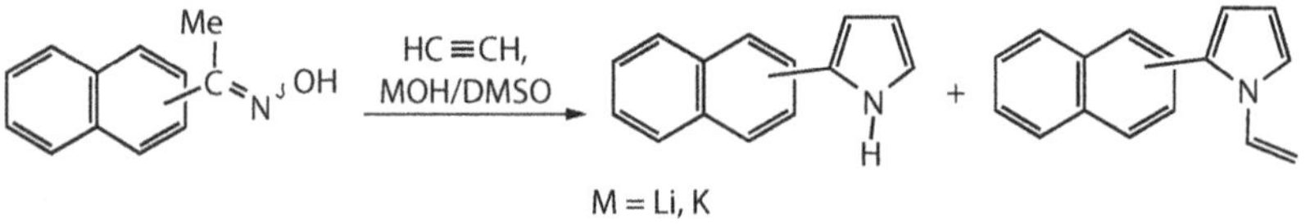

SCHEME 1.49   Synthesis of 2-naphthylpyrroles from oximes of acetylnaphthalenes.

Later on [224], 2-([2,2]paracyclophanyl)-3-alkylpyrroles were obtained in up to 41% directly from oximes of the corresponding ketones and acetylene in the presence of catalytic system LiOH/CsF/DMSO under acetylene pressure at 70°C (10 min).

Synthesis of pyrroles from ketoximes and acetylene has been successfully extended over 1- and 2-acetylnaphthalenes (their oximes, Scheme 1.49) [225].

In the conditions developed for alkyl aryl ketoximes (100°C, 3 h, 30% KOH from ketoxime weight, acetylene pressure 12–16 atm), the reaction with acetylnaphthalene oximes is accompanied by considerable resinification, and the yields of pyrrole are low. The best results are attained at 90°C: a mixture of 2-(1-naphthyl)pyrrole and 2-(1-naphthyl)-N-vinylpyrrole in 15% and 48% yields, respectively, is formed from 1-acetylnaphthalene oxime (2 h, KOH) [225].

Like in other cases, the nature of alkali metal cation dramatically effects on the yield and ratio of the pyrroles. For instance, in the presence of KOH (70°C, 3 h), the yield of 2-(2-naphthyl)pyrrole is 33%, while LiOH at the same temperature does not almost catalyze the process.

Naphthylpyrroles can be synthesized in good yields in a flask under atmospheric pressure by passing acetylene through the heated (90°C–100°C) stirred reaction

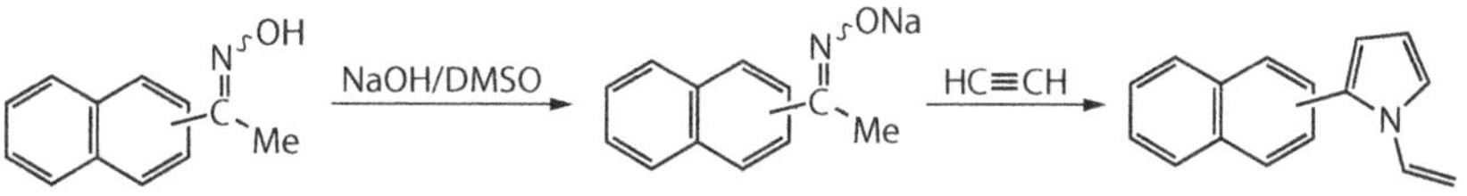

SCHEME 1.50 Selective synthesis of 2-naphthyl- N-vinylpyrroles from pre-prepared sodium oximates of acetylnaphthalenes.

mixture. For 4.5 h, the conversion of 2-acetylnaphthalene oxime is ~70%, and the yield of 2-(2-naphthyl)pyrrole reaches 64%. At 110°C–120°C (13 h), the reaction proceeds further (though resinification processes become more intensive): the reaction mixture consists of 2-(2-naphthyl)pyrrole and 2-(2-napthyl)-N-vinylpyrrole in 1:2 ratio. From this mixture, pure products in 18% and 31% yields are isolated using column chromatography on Al_2O_3.

Recently, a novel modification of the reaction has been developed to selectively synthesize 2-(1-naphthyl)-N-vinyl- and 2-(2-naphthyl)-N-vinylpyrroles (Scheme 1.50). According to the protocol, oximates are preformed from the corresponding oximes and sodium hydroxide in DMSO followed by their treatment with acetylene under pressure (120°C, 1 h) [226]. The yields of N-vinylpyrroles are 61% and 64%, respectively [226].

The reaction turns to be effective for oximes of acyl derivative of polycondensed aromatic compounds including anthracene, phenanthrene, and pyrene as well as for oximes of acylferrocenes (Scheme 1.51) [227].

Oximes of 1-pyrenyl pentanone and ferrocenyl butanone react with acetylene under atmospheric pressure (KOH/DMSO, 120°C, 5 h) to deliver the corresponding pyrroles (in 46% and 29% yields, respectively) and N-vinylpyrroles (in 16% and 30% yields, respectively), Scheme 1.51.

Previously, pyrroles bearing the ferrocene nucleus were synthesized through the reaction between acetyl- and benzylferrocene oximes and acetylene in the superbase catalytic system KOH/DMSO (95°C, 3.5 h) [131] at atmospheric pressure, the yields of pyrroles being 40% and 25%, respectively (Scheme 1.52).

Fundamentally, novel variants of the pyrrole synthesis have been elaborated to produce pyrroles with polycondensed aromatic substituents [227]. The methods are based on application of the combined superbases of the CsF/LiOH/DMSO type and deep drying of the reaction media to decrease solvatation of the reacting

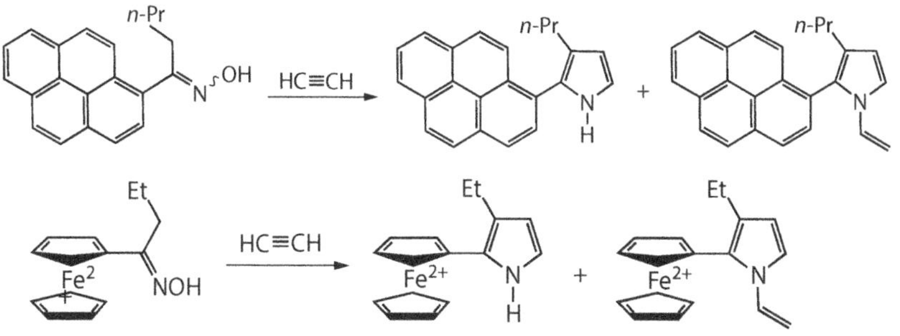

SCHEME 1.51 Synthesis of 2-pyrenyl- and 2-ferrocenylpyrroles.

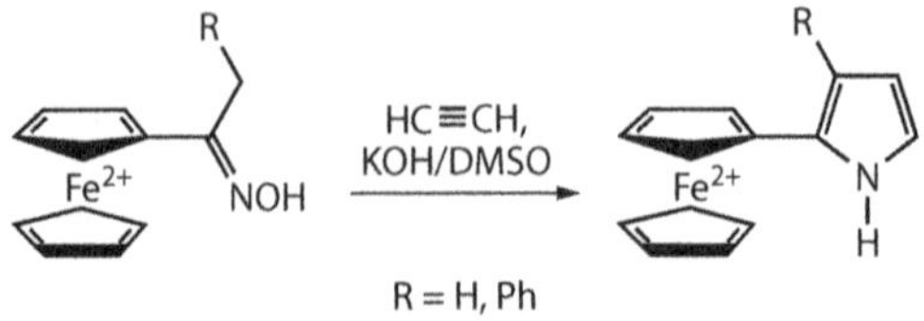

SCHEME 1.52   Synthesis of 2-ferrocenyl- and 2-ferrocenyl-3-phenylpyrroles.

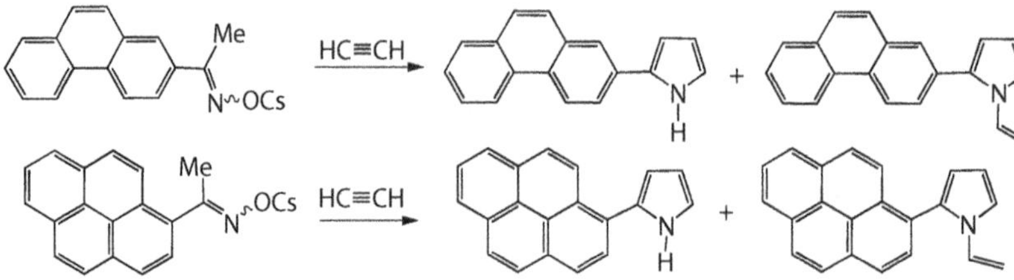

SCHEME 1.53   Synthesis of pyrroles bearing polycondensed aromatic substituents via Cs oximates.

oximate anions. For example, advances in the synthesis of 2-(2-phenanthrenyl)- and 2-(1-pyrenyl)pyrroles have been achieved due to employment of CsOH as catalyst precursor (it is generated in situ from CsF and LiOH) and owing to introduction of cesium salt oximes into the reaction with acetylene. Such a modification allows to obtain anhydrous reaction mixture that hinders deoximation. Besides, in this case, the reaction mixture has the increased basicity. Cesium salts of these oximes add to acetylene under elevated pressure in DMSO at 100°C for 1 h to give pyrroles with 32% and 37% yield, respectively, as well as their N-vinyl derivatives (in 30% and 29% yields, respectively) (Scheme 1.53) [227].

Cesium derivative of 2-acetylanthracene oxime upon short contact with acetylene (5 min) under mild conditions (80°C) selectively forms O-vinyl oxime of 2-acetylan-thracene (44% yield) [227]. The subsequent 3,3-sigmatropic rearrangement of this O-vinyl oxime tautomer leads to 2-(2-anthryl)pyrrole in 24% yield (Scheme 1.54). The rearrangement is realized in DMSO in the absence of a base (120°C, 30 min).

1.1.6.6   Oximes of Alkyl Hetaryl Pyrroles

The reaction of 2-acetyl-N-methylpyrrole oxime with acetylene [228] at atmospheric (80°C, 7 h) or elevated pressure (90°C, 3 h), catalyzed by the system KOH/DMSO, affords N-methyl-2-(pyrrol-2-yl)pyrrole (11%), its N-vinyl derivative (40%), and the intermediate methyl-(N-methylpyrrol-2-yl)-O-vinyl oxime (Scheme 1.55).

Yurovskaya et al. [229] have reported on previously unknown 2-(3-indolyl)pyr-roles that are synthesized in two stages from 3-acylindole oximes and acetylene in the presence of KOH/DMSO system (Scheme 1.56). Initially, the reaction of these oximes with acetylene under atmospheric pressure or at 10–12 atm (40°C–70°C, 3–18 h) affords the corresponding O-vinyl oximes. The latter, in the presence of a superbase catalyst and in the absence of acetylene, are converted at 100°C–105°C (8–10 h) into 2-(3-indolyl)pyrroles. The yields of indolylpyrroles range 29%–36%.

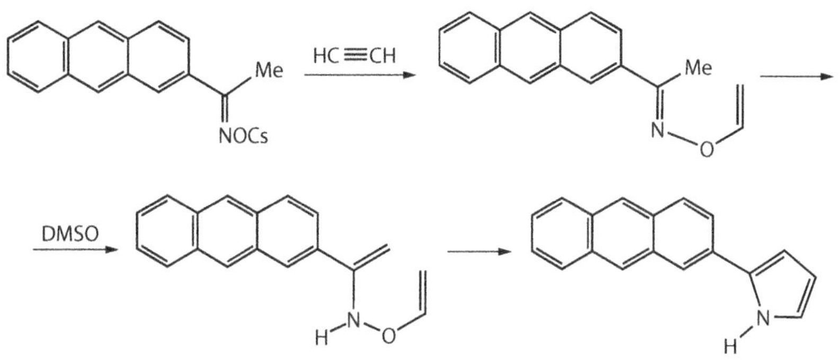

SCHEME 1.54 Two-step synthesis of 2-(2-anthryl)pyrrole from 2-acetylanthracene oxime and acetylene via O-vinyl oxime.

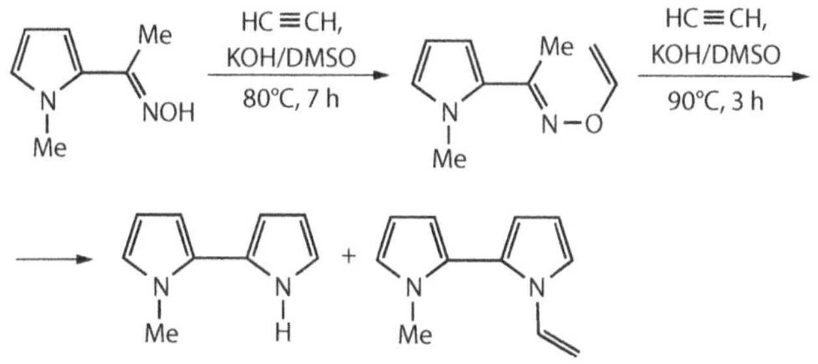

SCHEME 1.55 Reaction of 2-acetyl-N-methylpyrrole oxime with acetylene.

R¹ = H, Me; R² = H, Me, *i*-Pr

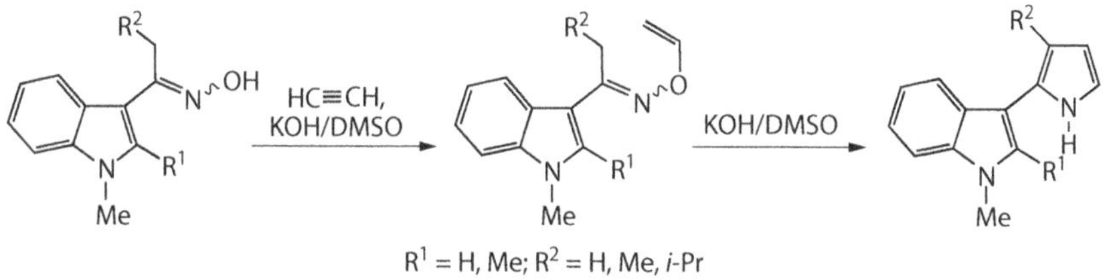

SCHEME 1.56 Two-step synthesis of 2-(3-indolyl)pyrroles from 3-acylindole oximes and acetylene via O-vinyl oxime.

Later [230], it has been revealed that the reaction of 3-acetylindole oxime with acetylene in the KOH/DMSO system leads, along with the expected indolylpyrrole, to δ-carboline (pyridoindole) in 40% yield (Scheme 1.57) (see Section 1.6).

The first data on behavior of furyl ketoximes in the Trofimov reaction have been published in works [231,232] where it was reported the synthesis of 3-alkyl-2-(2-furyl)-N-vinylpyrroles [231] and peculiarities of the spin-spin coupling in their ¹H NMR spectra [232].

Experimental details of the reaction between oximes of alkyl furyl ketones and acetylene (Scheme 1.58) have been described for the first time in [233].

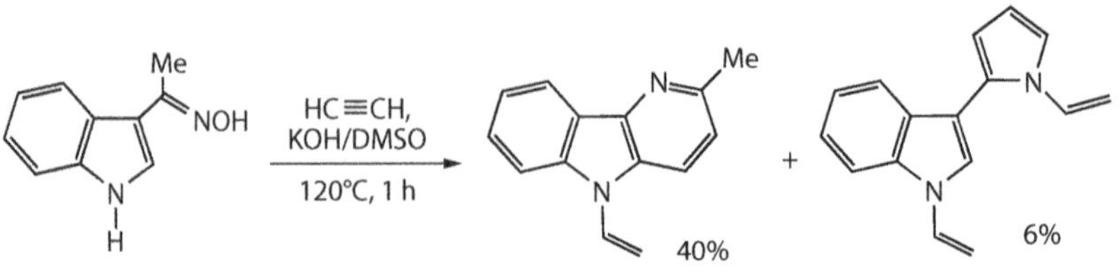

SCHEME 1.57 Reaction of 3-acetylindole oxime with acetylene to give indolylpyrrole and δ-carboline.

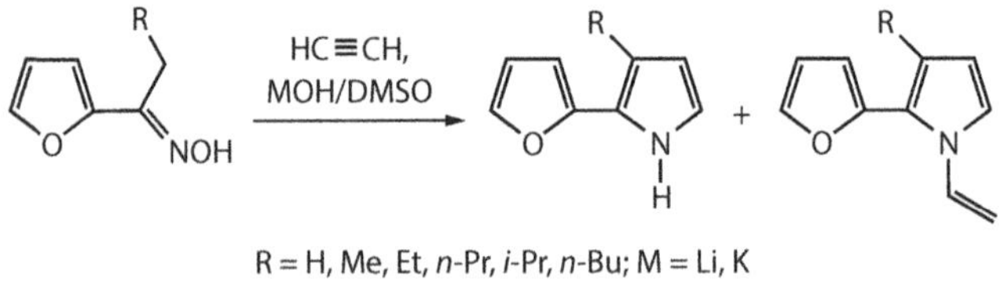

R = H, Me, Et, n-Pr, i-Pr, n-Bu; M = Li, K

SCHEME 1.58 Synthesis of 3-alkyl-2-(2-furyl)pyrroles and their N-vinyl derivatives from alkyl furyl ketoximes and acetylene.

The best conditions for the synthesis of 2-(2-furyl)-N-vinylpyrroles are as follows: reaction temperature of 110°C–130°C, acetylene excess, process duration of 3 h, and ketoxime–KOH ratio of 1:1. The yields of 2-(2-furyl)-N-vinylpyrroles achieved under these conditions range 50%–85%. At 130°C–140°C, alkyl furyl ketoximes easily form N-vinylpyrroles also at atmospheric pressure though the reaction time increases to 6–8 h.

The possibility of transformation of 2-acetylcoumarone oxime, a close structural analog of 2-acetylfuran oxime, to earlier unknown 2-pyrrolylbenzo[b]furan and its N-vinyl derivative has been demonstrated [234]. Upon short contact of this oxime with acetylene under pressure (70°C, 5 min) in the system KOH/DMSO, the expected pyrrole is formed in 24% yield, while the intermediate O-vinyl oxime is obtained with 22% yield (Scheme 1.59).

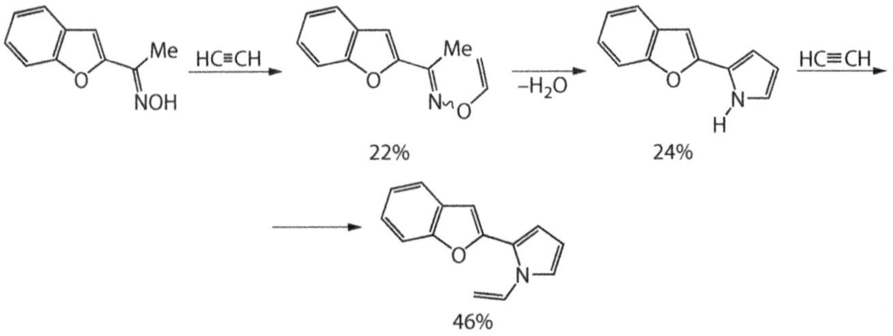

SCHEME 1.59 Reaction of 2-acetylcoumarone oxime with acetylene.

The ready pyrrolization of 2-acetylcoumarone O-vinyl oxime in comparison with the furan analog is evidently caused [231–233] by specific effect of the benzene ring annulated with the furan cycle, that is, by the formation of more extended conjugation system (and hence thermodynamically more stable) due to the incorporation of the pyrrole ring into it.

At a higher temperature (100°C) and increasing the reaction time (1 h), the corresponding N-vinylpyrrole is generated (46% yield) as the sole product that is characteristic also for alkyl furyl ketoximes forming 2-furyl-N-vinylpyrroles under similar conditions [233].

Methods for the synthesis of 2-(2-thienyl)pyrroles are known to be few in number. As a rule, they are multistage and require hardly accessible starting compounds, and the yields of the target pyrroles are far from to be high [235–238]. Some protocols comprise the introduction of the thienyl substituent into already-formed pyrrole ring [239]; others employ 1,4-dicarbonyl compounds as the initial reactant [240–242]. Also, multistage synthesis of N-aryl-2-(2-thienyl)pyrroles from thiophene has been disclosed [243]. At the first stage of the reaction, thiophene is acylated by succinic anhydride to afford acetylthiophene, which then reacts with arylamines in the presence of 1,3-dicyclohexylcarbodiimide leading to amides. The latter, upon contact with Lawesson's reagent, give mixtures of N-aryl-2-(2-thienyl)pyrroles and 5-arylamino-2,2-bithiophenes, the yields of the target pyrroles not exceeding 58%.

For the first time, the reaction of ketoximes with acetylene has been applied for the preparation of pyrroles bonded with the thiophene ring in the work [244], which describes synthesis of 2-(2-thienyl)pyrrole (in 60% yield) from 2-acetylthiophene (via its oxime) in the system KOH/DMSO (100°C–140°C). In the presence of acetylene excess under the same conditions, 2-(2-thienyl)-N-vinylpyrroles (50% yield) is formed.

3-Alkyl-2-(2-thienyl)-N-vinylpyrroles (70% yield) have been synthesized from alkyl thienyl ketoximes and acetylene (Scheme 1.60) [245–247]. The reaction smoothly proceeds at 100°C–130°C and can be realized both at elevated and atmospheric pressure of acetylene. 3-Alkyl-2-(2-thienyl)pyrroles can be obtained in up to 53% yield using the superbase pair LiOH/DMSO either upon addition of water in amount of 5%–15% from DMSO weight or by strict dosing of acetylene. Length of carbon chain of the alkyl radical has no noticeable effect on ratio and yields of the pyrroles [245].

2-(Benz[*b*]thiophen-3-yl)-N-vinylpyrrole (68% yield) has been selectively synthesized from 3-acetylbenz[*b*]thiophene oxime and acetylene (Scheme 1.61) under atmospheric pressure in the system KOH/DMSO (120°C, 5 h) [248].

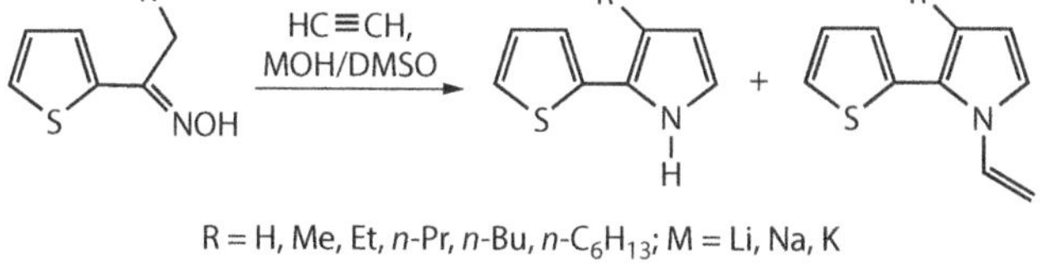

R = H, Me, Et, *n*-Pr, *n*-Bu, *n*-C$_6$H$_{13}$; M = Li, Na, K

SCHEME 1.60 Synthesis of 3-alkyl-2-(2-thienyl)pyrroles and their N-vinyl derivatives from alkyl thienyl ketoximes and acetylene.

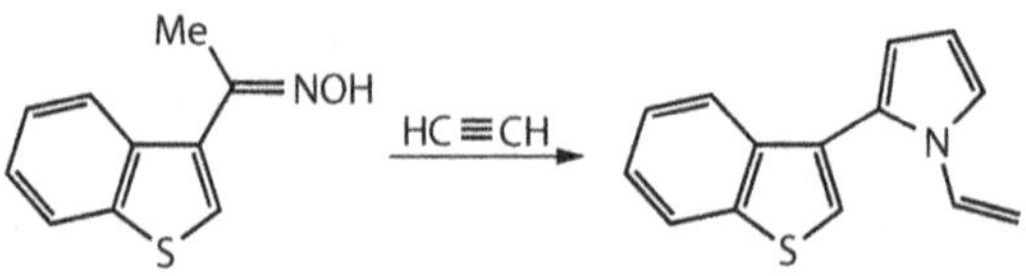

SCHEME   1.61   Synthesis   of   2-(benzo[b]thiophen-3-yl)-N-vinylpyrrole   from 3-acetylbenzo[b]thiophene oxime and acetylene.

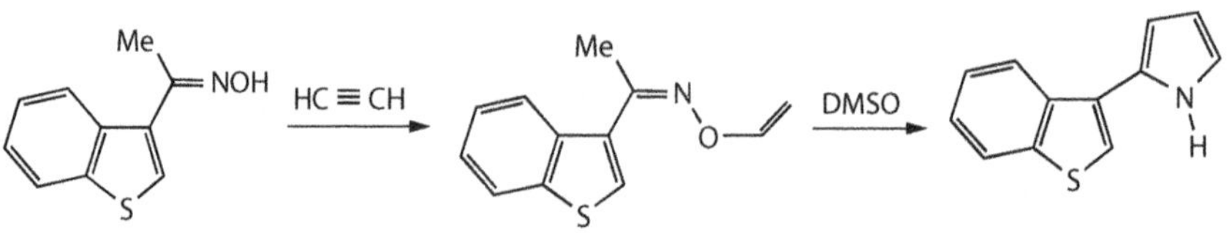

SCHEME 1.62   Two-step synthesis of 2-(benzo[b]thiophen-3-yl)pyrrole via O-vinyl oxime of 3-acetylbenzo[b]thiophene.

The reaction of 3-acetylbenz[b]thiophene oxime with acetylene under pressure at 90°C–120°C (1–1.5 h) results in strong resinification of the reaction mixture, in which 3-acetylbenz[b]thiophene (~10%), its oxime (~5%), and trace N-vinylpyrroles are identified using gas-liquid chromatography (GLC) and ^{1}H NMR techniques [248]. When the reaction temperature decreases to 80°C (KOH/DMSO, initial acetylene pressure 14 atm, 30 min), 3-acetylbenz[b]thiophene oxime reacts with acetylene to deliver O-vinyl oxime of 3-acetylbenz[b]thiophene (55% yield). The isolated O-vinyl oxime further undergoes the previously described rearrangement into pyrrole without a base in DMSO (120°C, 30 min) to give 2-(benz[b]thiophen-3-yl)pyrrole in 51% yield (Scheme 1.62) [248].

The synthesis of 2-(2-selenophenyl)pyrrole (10% yield) and 2-(2-selenophenyl)-N-vinylpyrrole (2% yield) from 2-acetylselenohene oxime and acetylene at 95°C–97°C has been reported for the first time in [249]. Seventeen years later, this reaction was employed to prepare 3-alkyl-2-(2-selenophenyl)pyrroles and their N-vinyl derivative from oximes of acylselenophenes (Scheme 1.63) [250]. The process is carried out in the system MOH/DMSO (M = Na, K) under acetylene pressure at 100°C (30–60 min).

After systematic optimization of this reaction, the conditions allowing selective obtaining either 2-(2-selenophenyl)pyrroles (NaOH, 60 min) or their N-vinyl

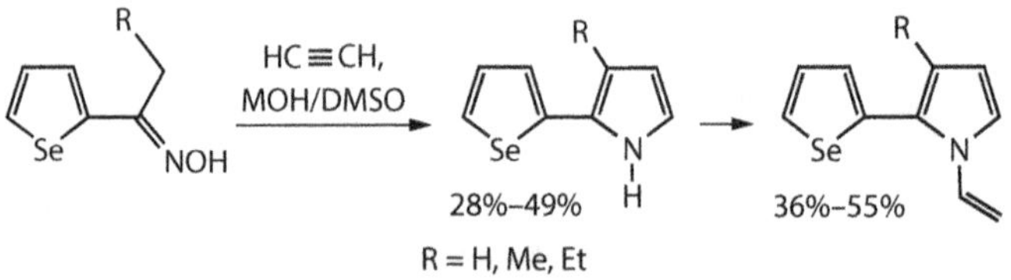

28%–49%   36%–55%

R = H, Me, Et

SCHEME 1.63   Synthesis of 2-(2-selenophenyl)pyrrole and their N-vinyl derivatives from 2-acetylselenohene oxime and acetylene.

derivatives (KOH, 30 min) have been found [250]. The yields reach 49% and 55% for pyrroles and N-vinylpyrroles, respectively.

In the system LiOH/DMSO under the same conditions, the reaction of acetylsele-nophene oxime with acetylene does not take place. However, if the process is carried out for a longer time (3 h), O-vinyl oxime is formed and isolated as the only reaction product (44% yield) (Scheme 1.64). Rearrangement of O-vinyl oxime (140°C, 30 min, DMSO) gives the pyrrole in 87% yield.

It is obvious that the synthesis of O-vinyl oximes of acylated selenophenes in the system LiOH/DMSO followed by their rearrangement can serve as an alternative methodology for selective preparation of 2-(2-selenophenyl)pyrroles.

One-pot transformation of 2-, 3-, and 4-acetylpyridine oximes into pyrroles and N-vinylpyrroles has been implemented [251,252]. Upon heating (KOH/DMSO, 105°C–110°C, 3 h), 2-acetylpyridine oxime and acetylene under pressure give a mixture of 2-(2-pyridyl)pyrrole (32%) and its N-vinyl derivative (36%). At 120°C, only 2-(2-pyridyl)-N-vinylpyrrole is formed in 68% yield (Scheme 1.65) [251].

2-(3-Pyridyl)-N-vinylpyrrole (62% yield) and 2-(4-pyridyl)-N-vinylpyrrole (65% yield) have been selectively synthesized from oximes of 3-acetyl- and 4-acetylpyridine, respectively, at 120°C (Scheme 1.65) [252].

The analysis of the reasons of lower reactivity of acetylpyridine oximes has shown that, according to the IR and NMR data, the starting oximes are pure E-isomers [252], while ketoximes of Z-configuration are the most reactive in the studied reaction [7]. Since acetylpyridine oximes exist at room temperature as E-isomers exclusively, their decreased proneness to formation of pyrroles in the reactions with acetylene becomes apprehensible. With the temperature increase and under the effect of the

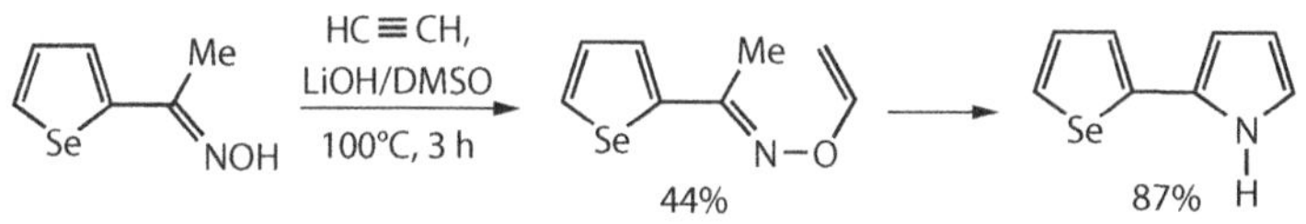

SCHEME 1.64 Two-step synthesis of 2-(2-selenophenyl)pyrroles via O-vinyl oxime.

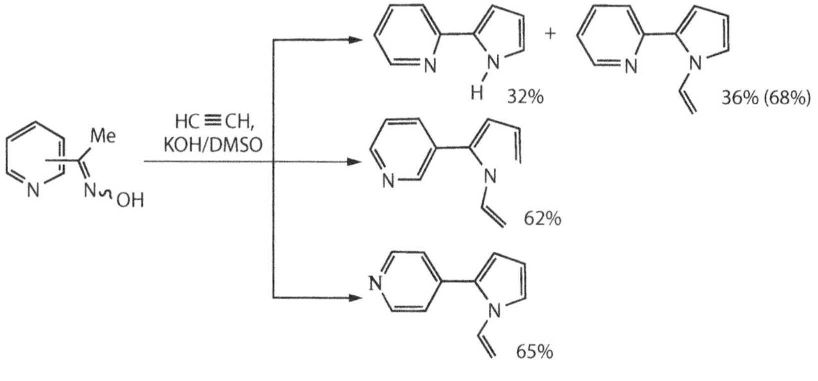

SCHEME 1.65 Synthesis of 2-pyridylpyrroles from acetylpyridine oximes and acetylene.

catalyst (KOH), oximes undergo configurational transformations that lead, finally, to the formation of pyrroles. However, this process proceeds not so easy as usually because $E{\rightarrow}Z$-isomerization of the initial ketoxime likely represents a rate-limiting stage.

1.1.6.7 Functionally Substituted Ketoximes

The extension of the reaction of ketoximes with acetylene over oximes of hydroxy, alkoxy, and vinyloxy alkyl ketones (Scheme 1.66) could lead to the synthesis of hardly accessible functionally substituted pyrroles. However, it turns out that oximes of such ketones react with acetylene in the system KOH/DMSO in a peculiar manner [253].

As a rule, such reaction involves splitting of the oxygen-containing radical to give a mixture of simpler pyrroles in low yields. Having no great synthetic value, these results, nevertheless, shed additional light on some peculiarities of this reaction as a whole.

The simplest representative of a series of studied ketoximes, 1-hydroxy-2-propanoe oxime, is unstable and completely resinifies under usual conditions of the reaction with acetylene.

1-Ethoxy-2-propanone oxime is more stable, and its interaction with excess acetylene affords an unexpected product, 2-methyl-N-vinylpyrrole in up to 15% yield (Scheme 1.67), which often exceeds the yield of the anticipated 2-ethoxymethyl-N-vinylpyrrole.

The results obtained support the conclusions [188,254] that carbanions bearing charge at α-carbon relative to the oxime moiety play an important role at the first stages of the pyrrole ring construction. From these positions, one can explain a regiospecificity of the reaction between 1-ethoxy-2-propanone oxime and acetylene proceeding with the formation of 2-ethoxymethyl-N-vinylpyrrole instead of 3-ethoxy-2methyl-N-vinylpyrrole. The intermediate carbanion **A** (Scheme 1.68), tentative precursor of 3-ethoxy-2-methylpyrrole, should be destabilized owing to repulsion of the negative charge from lone electron pairs of the oxygen atom. Therefore, the reaction occurs through more stable carbanion **B** delivering 2-ethoxymethylpyrrole.

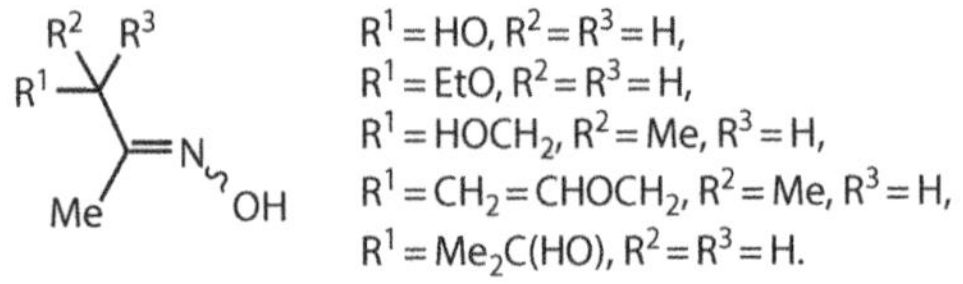

$R^1 = HO, R^2 = R^3 = H,$
$R^1 = EtO, R^2 = R^3 = H,$
$R^1 = HOCH_2, R^2 = Me, R^3 = H,$
$R^1 = CH_2 = CHOCH_2, R^2 = Me, R^3 = H,$
$R^1 = Me_2C(HO), R^2 = R^3 = H.$

SCHEME 1.66 Functionally substituted ketoximes studied in the reactions with acetylene in superbase system KOH/DMSO.

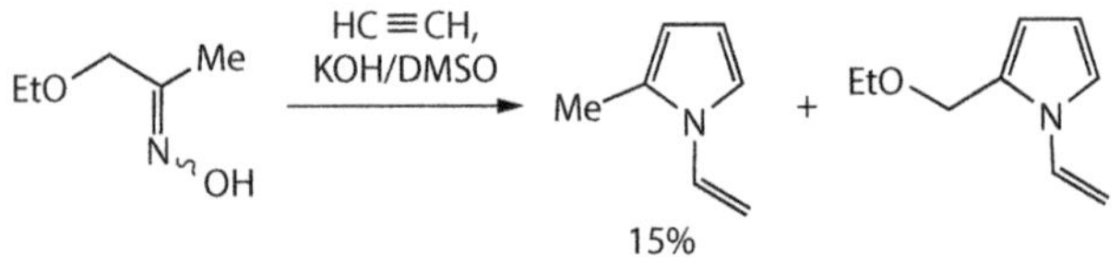

SCHEME 1.67 Reaction of 1-ethoxy-2-propanone oxime with acetylene.

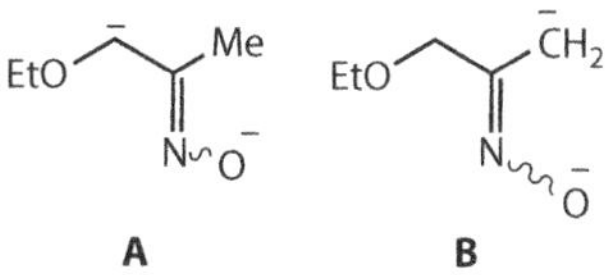

SCHEME 1.68 Generation of alternative carbanions from 1-ethoxy-2-propanone oxime under the action of KOH/DMSO system.

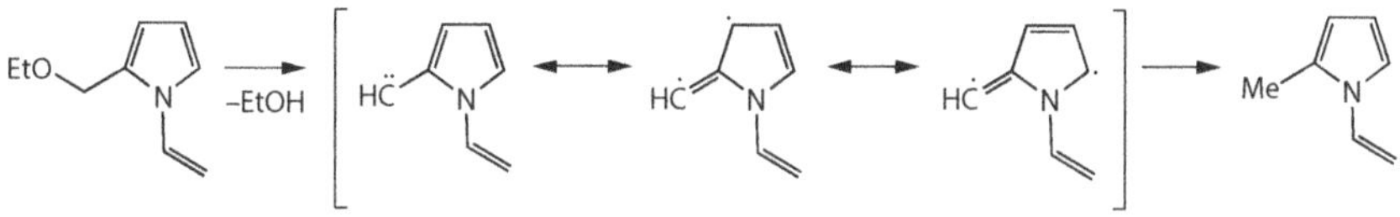

SCHEME 1.69 Tentative scheme of 2-methyl-N-vinylpyrrole formation.

It is more difficult to rationalize the formation of 2-methyl-N-vinylpyrrole in this reaction. However, bearing in mind that this pyrrole emerges in the reaction mixture as 2-ethoxymethyl-N-vinylpyrrole disappears, the authors [253] leave room for carbenoid splitting of the ethoxymethyl group followed by the reduction of carbene (Scheme 1.69). This deduction is confirmed by identification of ethanol among the reaction products.

Similar α-elimination is not excluded for 1-ethoxy-2-propanone oxime itself, which is converted to acetoxime that according to the usual scheme can lead to 2-methyl-N-vinylpyrrole.

The reaction of 1-hydroxy-2-methyl-3-butanone oximes with excess acetylene at 100°C furnishes two abnormal products (Scheme 1.70), 2,3-dimethyl-N-vinylpyrrole (this was also reported in [255]) and 2-(2-propenyl)-N-vinylpyrrole in 20% and 8% yields, respectively.

2-Methyl-1-vinyloxy-3-butanone oxime reacts with acetylene to give also 2-(2-propenyl)-N-vinylpyrrole (Scheme 1.71), but 2,3-dimethyl-N-vinylpyrrole is not formed in this case. During fractionation and purification, 2-(2-propenyl)-N-vinylpyrrole readily polymerizes, which, apparently, is one of the reasons of its low yield.

2-hydroxy-2-methyl-4-pentanone oxime interacts with acetylene giving rise to 2-methyl-N-vinylpyrrole in about 20% yield (Scheme 1.72).

The formation of 2,3-dimethyl-N-vinylpyrrole from 1-hydroxy-2-methyl-3-butanone oxime and 2-methyl-N-vinylpyrrole from 2-hydroxy-2-methyl-4-pentanone

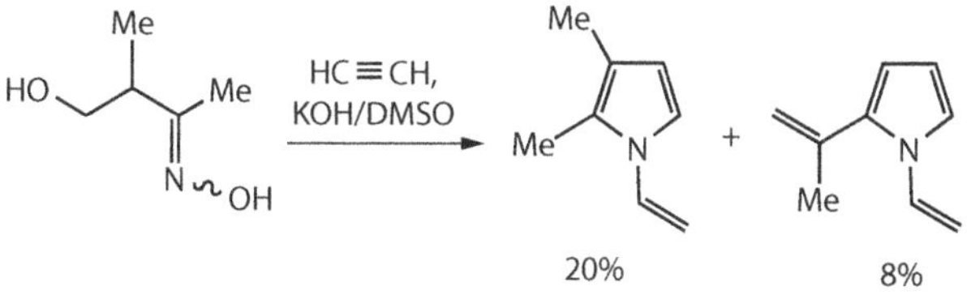

SCHEME 1.70 Reaction of 1-hydroxy-2-methyl-3-butanone oximes with acetylene.

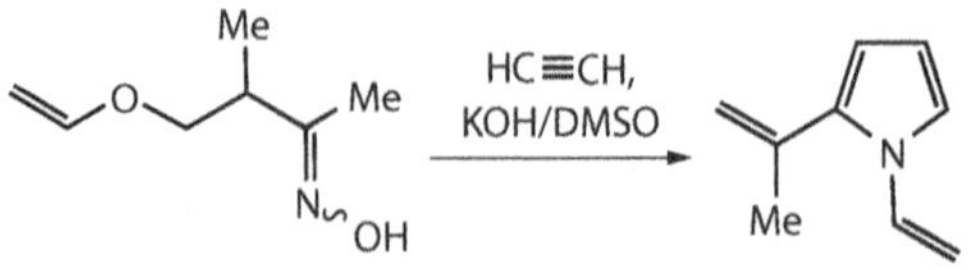

SCHEME 1.71   Reaction of 2-methyl-1-vinyloxy-3-butanone oxime with acetylene in KOH/DMSO system.

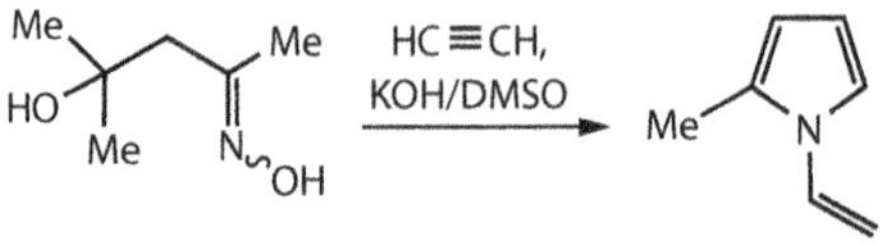

SCHEME 1.72   Reaction of 2-hydroxy-2-methyl-4-pentanone oxime with acetylene in KOH/DMSO system.

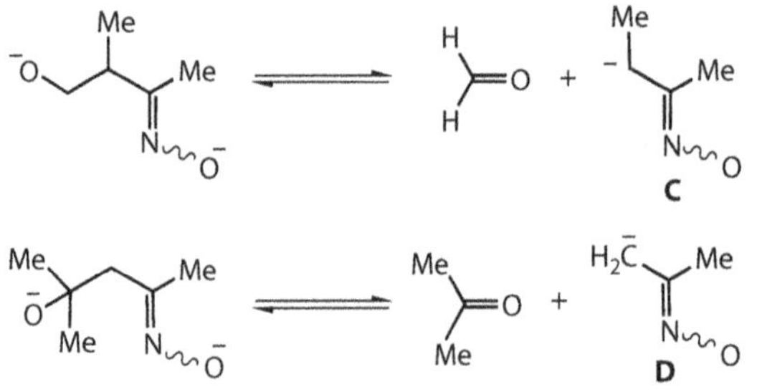

SCHEME 1.73   Tentative scheme of decomposition of 1-hydroxy-2-methyl-3-butanone oxime and 2-hydroxy-2-methyl-4-pentanone oxime.

oxime is likely caused by decomposition of these oximes in the form of dianions according to the scheme resembling *retro*-aldol condensation (Scheme 1.73).

Carbanions **C** and **D** react further with acetylene to yield the corresponding pyrroles.

The formation of 2-(2-propenyl)-N-vinylpyrrole from 2-methyl-1-hydroxy-3-butanone oxime and 2-methyl-1-vinyloxy-3-butanone oxime is apparently due to β-elimination of water or vinyl alcohol from the corresponding oximes or intermediate 2-hydroxy(vinyloxy)propyl-N-vinylpyrroles (Scheme 1.74).

1.1.6.8   Oximes of Diketones: Synthesis of Dipyrroles

Dipyrroles, especially those bounded through positions 2 and 3, are abundant in nature [256–262] and possess unique biological properties. Unlike dipyrroles bridged across the nitrogen atoms, in such ensembles, mutual influence of the functional groups, electron excitement, and polarization are transferred especially easily. Therefore, they are often incorporated in electroconductive polymers and various polyconjugated systems with high spectral and electrochemical response to the

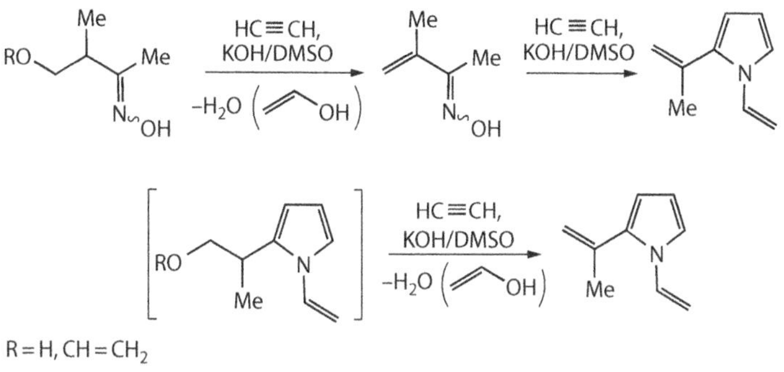

SCHEME 1.74 Plausible scheme of 2-(2-propenyl)-N-vinylpyrrole formation from 1-hydroxy-2-methyl-3-butanone oxime and 2-methyl-1-vinyloxy-3-butanone oxime.

change of external conditions (pH of medium, presence of certain ions and molecules, electromagnetic radiation, etc.) [263–268].

A direct approach to the synthesis of dipyrroles, including those separated by aromatic and heteroaromatic counterparts, is the reaction of diketone oximes with acetylene in the presence of the superbase catalytic systems.

1.1.6.9 Dioximes of 1,2-Diketones

The reaction of dimethylglyoxime, a simplest representative of 1,2-diketone dioximes, with acetylene in the system KOH/DMSO (100°C–140°C, initial acetylene pressure 14 atm), leads to the expected N,N'-divinyl-2,2'-dipyrrole (15% yield), among other products isolated from the reaction mixture being O-vinyl methylglyoxime and 2-acetyl-N-vinylpyrrole (Scheme 1.75) [269]. Also, 2-(2-pyridyl)-N-vinylpyrrole has been unexpectedly identified (see Section 1.5.4).

Later, it has been shown [270] that in the KOH/DMSO system with rise of temperature from 100°C to 140°C, there is an increase in the relative amount of 2-acetyl-N-vinylpyrrole, while the ratio of the dipyrrole and methylpyridylpyrrole remains practically constant. The growth of KOH content in the system intensifies the deoximation process leading to acetylpyrrole and subsequent ethynylation of iminoaldehyde formed as a result of [3,3]-sigmatropic rearrangement that affords pyridylpyrrole.

The application of NaOH as base (120°C, 1 h) results in formation of polymeric products. This is probably due to a lower basicity of NaOH and its more clearly

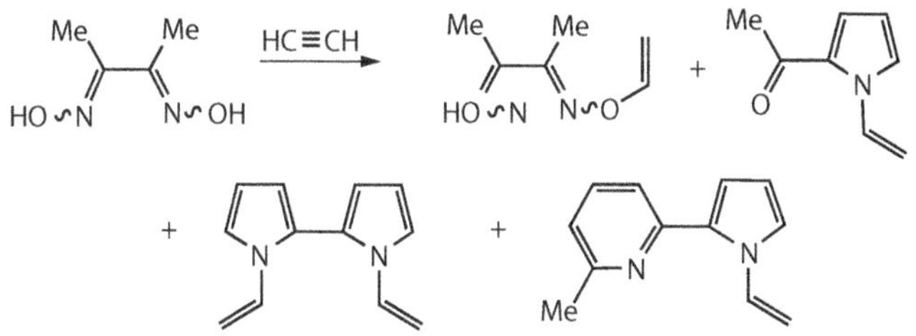

SCHEME 1.75 The reaction of dimethylglyoxime with acetylene in the system KOH/DMSO.

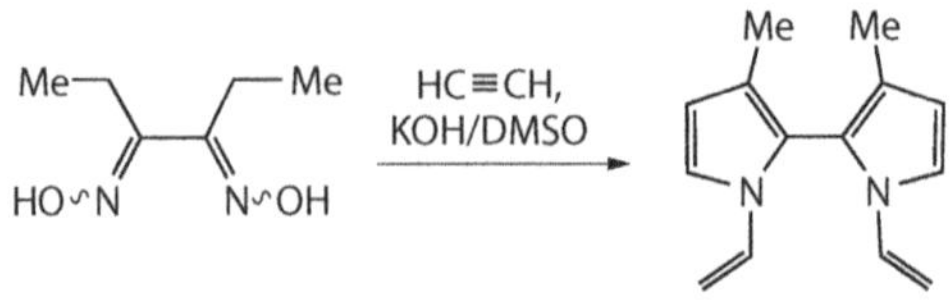

SCHEME 1.76 Formation of divinyldipyrrole from 3,4-hexanedione dioxime and acetylene.

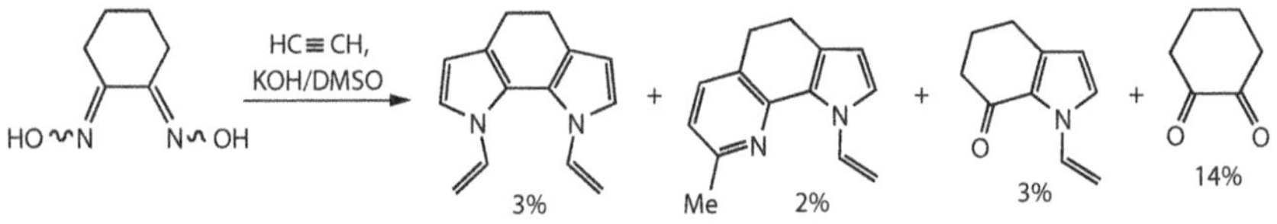

SCHEME 1.77 Reaction of 1,2-cyclohexanedione dioxime with acetylene.

defined ability, compared with KOH, to promote the deoximation processes producing dicarbonyl compounds susceptible to self-condensation in alkaline media.

When 3,4-hexanedione dioxime is involved into the reaction with acetylene (KOH/DMSO, 100°C, 1 h), the expected divinyldipyrrole is formed with unoptimized yield of 3% (Scheme 1.76) [270].

1,2-Cyclohexanedione dioxime under these conditions (120°C, 1 h) gives, together with 2,2'-dipyrrole (3%), 2-pyridyl (2%), 2-acyl (3%) pyrrole derivatives as well as the product from total deoximation of the initial dioxime, 1,2-cyclohexanedione (14%) (Scheme 1.77) [270].

Probably, the six-membered cyclic system containing two conjugated oxime groups is prone to a greater degree to the side processes leading to the formation of the pyridine ring and deoximation.

1.1.6.10 Dioximes of 1,3-Diketones

To obtain dipyrroles bridged across the positions 2 and 3, the reaction of 1,3-diketone dioximes, acetylacetone, benzoylacetone, 5-ethylnonane-4,6-dione, and 5,5-dimethylcyclohexane-1,3-dione (dimedone) with acetylene in the superbase system KOH/DMSO has been studied [271].

The selectivity of the interaction of acetylacetone dioxime with acetylene depends on the reaction temperature. At 100°C (KOH/DMSO, 1 h), O-vinyl oxime is formed exclusively (11% yield), while at 120°C (KOH/DMSO, 1 h), the major product is 2,3'-divinyldipyrrole (7% yield) (Scheme 1.78).

Noteworthy is the unusually high stability of the O-vinyl oxime group. The formation of the second pyrrole ring with its participation is observed only when the reaction temperature increases up to 120°C.

Benzoylacetone dioxime shows a lower selectivity in the reaction with acetylene. The reaction proceeds in the system KOH/DMSO at 100°C to afford, already after 5 min, a mixture containing N-vinylpyrroles (4% and 13% yields) and isoxazoles (18% total yield) (Scheme 1.79).

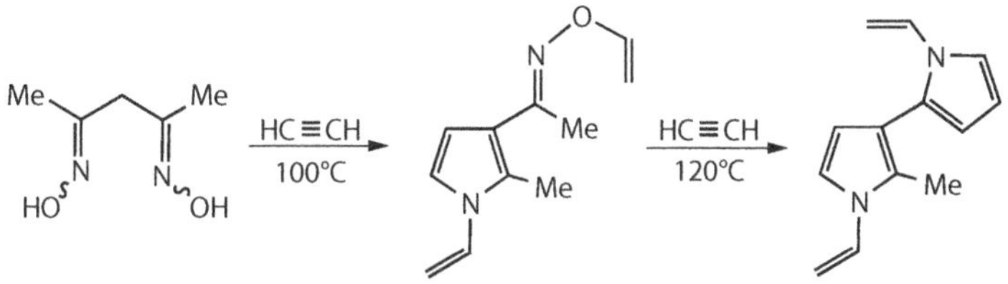

SCHEME 1.78 Synthesis of 2,3′-divinyldipyrrole from acetylacetone dioxime and acetylene.

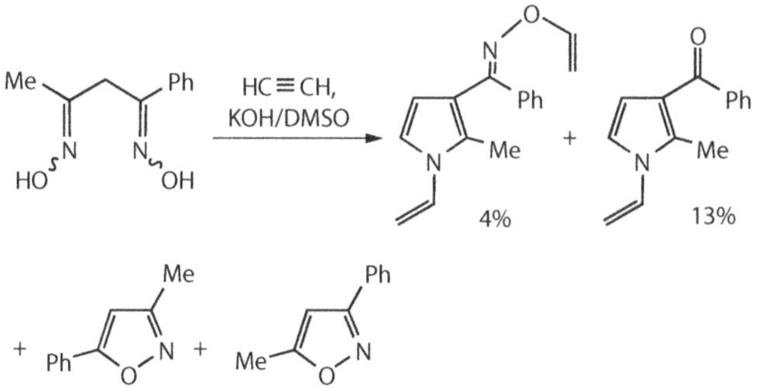

SCHEME 1.79 Reaction of benzoylacetone dioxime with acetylene.

Attention is attracted by the regioselectivity of benzoylacetone dioxime pyrroliza-tion with the participation of the oxime function of the acetyl fragment. In view of the relative ease of E,Z-isomerization of ketoximes under the reaction conditions [16,18], it may be assumed that this regioselectivity is not attributed to the configura-tion of the oxime functions of benzoylacetone dioxime, which is a mixture of E,Z- and E,E-isomers in 1:1 ratio (Scheme 1.80).

Apparently, higher electron saturation and lower steric hindrance of the acetyl oxime group enable its easier (as compared to the benzoyl oxime fragment) nucleo-philic addition to acetylene.

From 5-ethylnonane-4,6-dione oxime and acetylene, only 3,5-di-n-propyl-4-ethyloxazole (21%), formed owing to the intramolecular cyclization of dioxime, has been isolated (Scheme 1.81). It is possible that the steric effect of the propyl groups hinders addition of the oxime functions to acetylene.

The interaction of 5,5-dimethylcyclohexane-1,3-dione dioxime with acetylene in the KOH/DMSO system also does not result in the formation of pyrroles or O-vinyl

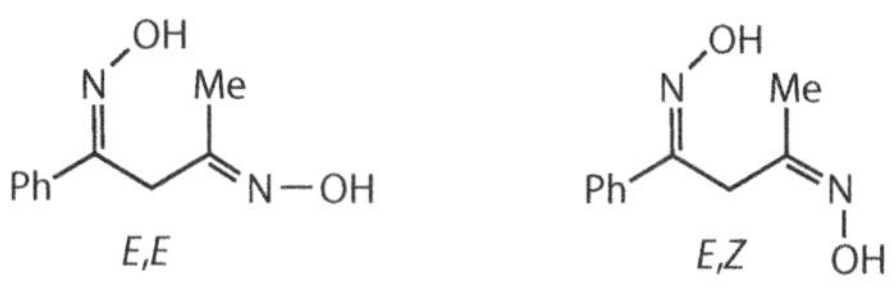

SCHEME 1.80 E,Z-isomerization of benzoylacetone dioxime.

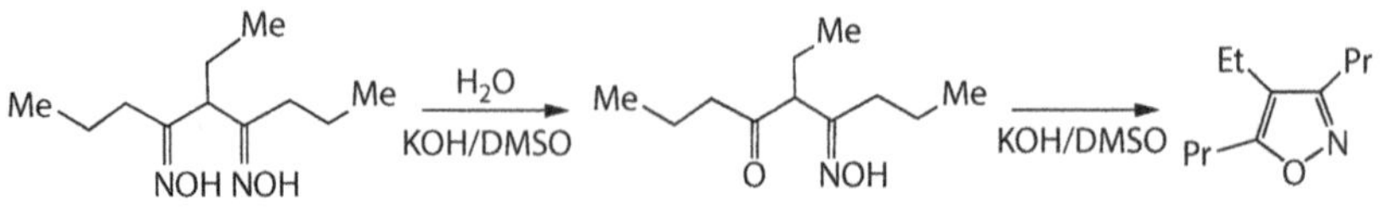

SCHEME 1.81 Formation of 3,5-di-*n*-propyl-4-ethyloxazole from 5-ethylnonane-4,6-dione dioxime.

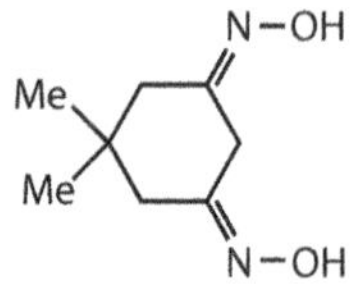

SCHEME 1.82 *Z,Z*-configuration of 5,5-dimethylcyclohexane-1,3-dione dioxime.

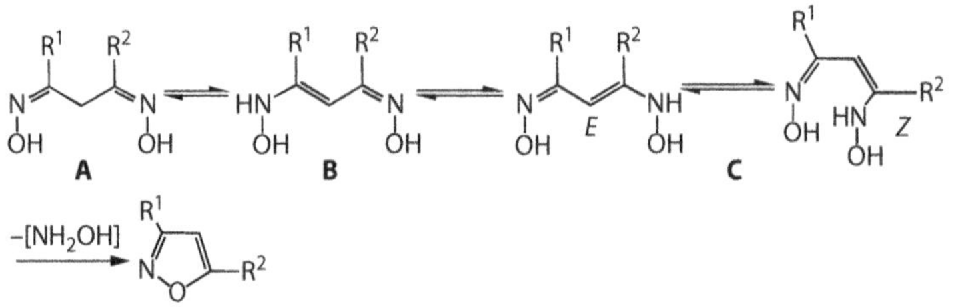

SCHEME 1.83 Imine-enamine tautomerism of 1,3-diketone dioximes.

oximes. The reaction mixture is strongly resinified that hinders its extraction. Probably, cyclohexane-1,3-dione backbone creates conformational encumbrances, and resinous compounds are generated due to the condensation of deoximation products and opening of the cyclohexane ring under the action of the KOH/DMSO system. It is also possible that this behavior of the dioxime is caused by its *Z,Z*-configuration (Scheme 1.82), leading to a steric interaction of the oxime functions, thus reducing their reactivity.

Undoubtedly, the main and the most general reason of complicated character of the reactions between 1,3-diketone dioximes and acetylene is the imine–enamine tautomerism (Scheme 1.83).

Population of oxime–hydroxylamine tautomers **B** and **C** can be high, since unlike the initial dioximes **A**, they are conjugated systems and the energy gain of their formation should be considerable. Besides, highly basic medium is favorable for pro-totropic migration of the double bonds and "enolization." The essential contribution of tautomers **B** and **C** in overall reaction is supported by the formation of isoxazoles, the predictable intramolecular cyclization of the tautomer in their *Z*-configurations. It is obvious that tautomers **B** and **C** as very reactive species (combination of the oxime and hydroxylamine functions with azadiene system), in the system KOH/DMSO/acetylene, can tolerate various reactions, namely, vinylation, cyclization, elimination, polymerization, and polycondensation.

1.1.6.11   Dioximes of 1,4-Diketones

The reaction of hexane-2,5-dione- and cyclohexane-1,4-dione dioximes (representatives of open-chained and cyclic compounds of this series) with acetylene has been studied [272].

In the case of hexane-2,5-dione dioxime (KOH/DMSO, 100°C, 1 h, initial acetylene pressure 14 atm), three dipyrroles **8–10** were expected to be formed via the rearrangements involving both methylene and methyl groups in the intermediates (Scheme 1.84).

However, N,N′-divinyl-2,2′-dimethyl[3,3′]dipyrrole **8** was isolated as the major product with not-optimized yield of 12%. Probably, dipyrroles **9** and **10** are the minor components of the reaction mixtures because they would be assembled with the participation of the methyl groups of **A** and **B** that are less reactive in comparison with the methylene functions [7]. The thermodynamic effect on distribution of the products also can be considerable owing to high conjugation of pyrrole **8**.

Nonconjugated dipyrrolic cyclophane (4,8-dihydropyrrolo[2,3-*f*]indole) **11** is formed from cyclohexane-1,4-dione dioxime in the same conditions (Scheme 1.85). This dipyrrole was isolated in 6% yield instead of the anticipated conjugated 3,3′-dipyrrolic system **12** [272].

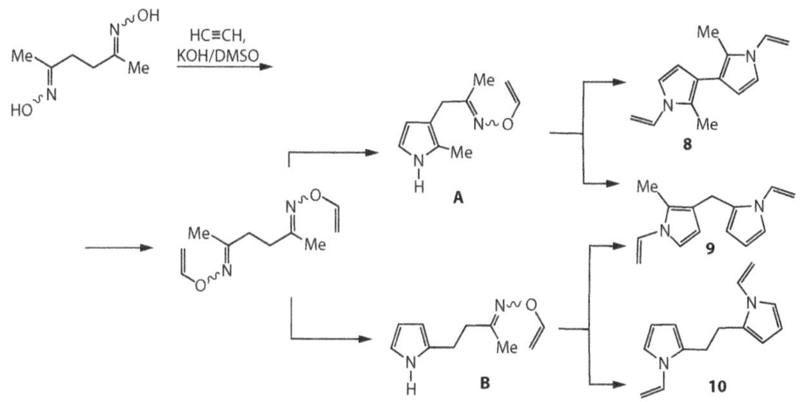

SCHEME 1.84   Plausible products of the interaction between hexane-2,5-dione dioxime and acetylene.

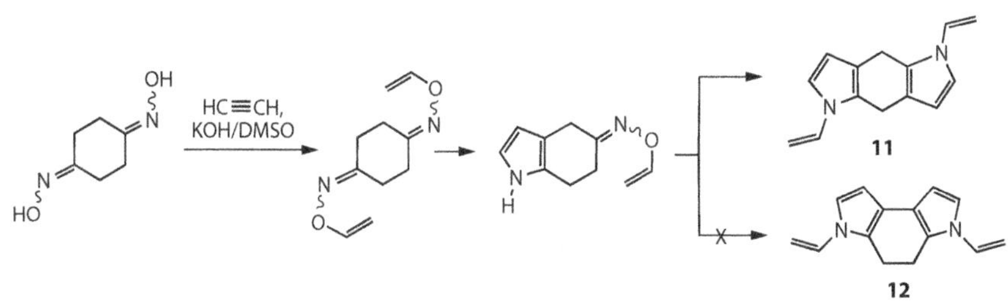

SCHEME 1.85   Alternative pathways of the reaction between cyclohexane-1,4-dione dioxime and acetylene.

Quantum-chemical calculations [273] performed in B3LYP/6-311++G (d,p) basis set shows high stability ($\Delta E = 1.33$ kcal/mol) of isomer **11** in comparison with isomer **12**.

The reason of lower thermodynamic stability of structure **12** as compared to compound **11** can be repulsion of hydrogen atoms in two adjacent methylene groups of the cyclohexane ring distorting the plane of whole tricyclic skeleton [data of B3LYP/6-311++G(d,p)], thus diminishing conjugation of the pyrrole rings. The calculations evidence that isomer **11** has plain structure with the vinyl groups deviating from the plane by less than 1°.

Thus, the reaction of 1,4-diketone dioximes with acetylene in the system KOH/DMSO opens a straightforward route to 3,3-dipyrroles and 4,8-dihydropyrrolo[2,3-*f*]indoles—promising pharmacophores, monomers, and building blocks for the design of biologically important molecules.

1.1.6.12   Dioximes Separated by Conjugated Systems

The assessment of application scopes of novel pyrrole synthesis has revealed [274] that it can be extended to such essentially new group of oximes as aromatic dioximes. For example, dipyrroles separated by the phenylene fragment, dipyrrolylbenzenes, can be prepared in a one-stage fashion from 1,4-diacetylbenzene dioxime.

After optimization of heterocyclization conditions of 1,4-diacetylbenzene dioxime [133], it has been found that the latter reacts with acetylene under pressure (110°C, 1 h, initial acetylene pressure 15 atm) in the system KOH/DMSO to afford dipyrrolylbenzene and its monovinyl derivative in 15% and 27% isolated yields, respectively (Scheme 1.86).

When the reaction is carried out in the system CsOH/DMSO (100°C) for 2 h, 1,4-bis-[2-(N-vinyl)pyrrolyl]benzene (6%) along with dipyrrole (16%) and its monovinyl derivative (7%) have been isolated from the reaction mixture (Scheme 1.87).

In the case of LiOH/DMSO system (140°C, 3 h, initial acetylene pressure 15 atm), monovinyldipyrrole is obtained as the sole product (Scheme 1.88).

The behavior of 4,4′-diacetyldiphenyl dioxime under conditions typical for the previous reaction proved to be unexpected: its treatment with acetylene in the systems MOH/DMSO (M = Li, K, Cs) at 100°C–150°C both at atmospheric pressure and under acetylene pressure does not lead to either the expected *O*-vinyl oximes or pyrroles [275]. Instead of them, either the initial dioxime is mainly recovered or the monooxime or complex mixtures of products are formed.

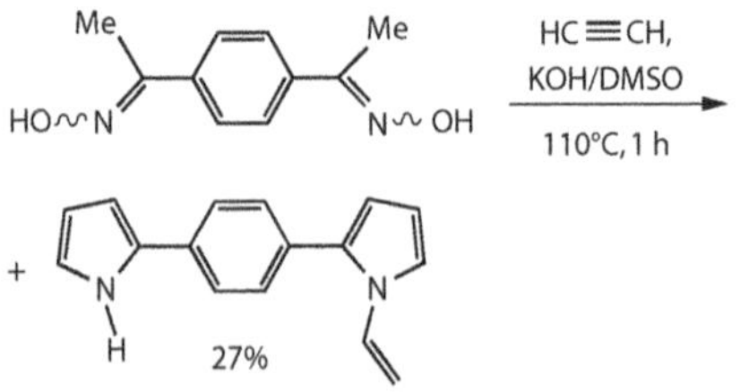

SCHEME 1.86   Synthesis of dipyrrolylbenzenes from 1,4-diacetylbenzene dioxime and acetylene in the system KOH/DMSO.

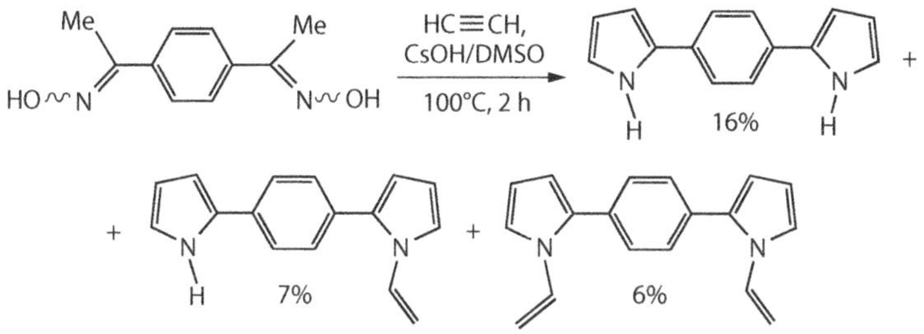

SCHEME 1.87 Synthesis of dipyrrolylbenzenes from 1,4-diacetylbenzene dioxime and acetylene in the system CsOH/DMSO.

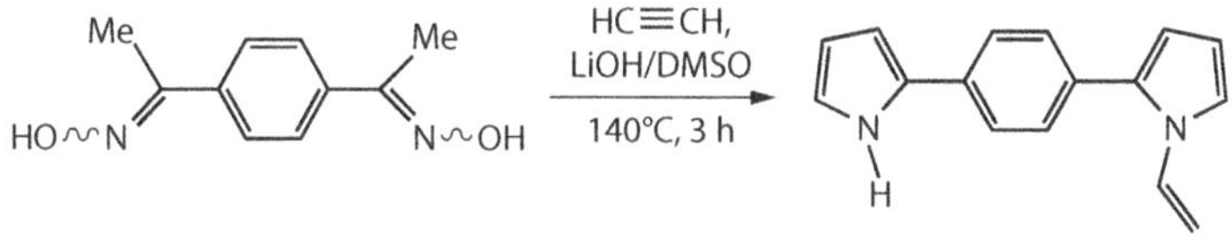

SCHEME 1.88 Selective synthesis of mono-N-vinyl derivative of 1,4-dipyrrolebenzene in the LiOH/DMSO system.

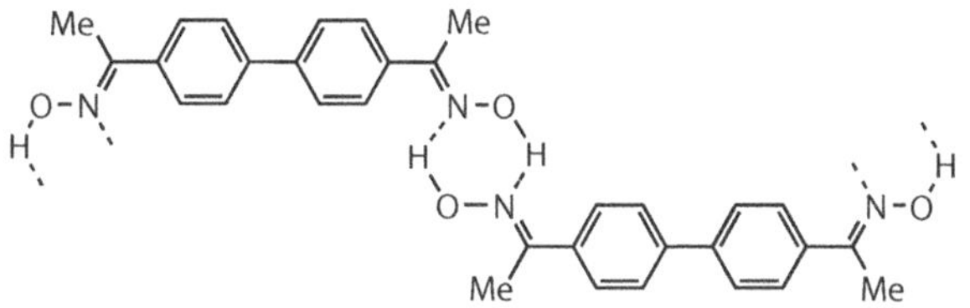

SCHEME 1.89 Intermolecular H-association of 4,4′-diacetyldiphenyl dioxime in CCl$_4$ solution.

According to IR spectra, 4,4′-diacetyldiphenyl dioxime is strongly associated even in very diluted solutions (4×10^{-4} mol/L, CCl$_4$). The spectra show the bands of intermolecular associates (3133, 3212, and 3291 cm^{-1}) with six-membered H-bonded structures (Scheme 1.89).

The x-ray diffraction study [275] has revealed additional features of the structure of this dioxime. The latter is found to be crystallized from DMSO as a coordination polymer with alternating molecules of dioxime and DMSO linked by strong bifurcated hydrogen bonds (Scheme 1.90). This is likely the reason of reduced reactivity of the dioxime towards alkali metal hydroxides and acetylene.

Interestingly, the interaction of 4,4′-diacetyldiphenyl oxide dioxime with acetylene in the system KOH/DMSO at 70°C (30 min) results in the only reaction product, di(O-vinyl)dioxime (15% yield), while under harsher conditions (120°C, 1 h) 4,4′-bis[2-(N-vinyl)pyrrolyl]diphenyl oxide is formed exclusively in 25% yield (Scheme 1.91) [133].

Divinylpyrrole is selectively synthesized in 14% yield from 4,4′-diacetyldiphenyl sulfide dioxime in the system KOH/DMSO under acetylene pressure (110°C, 1 h) (Scheme 1.92) [133].

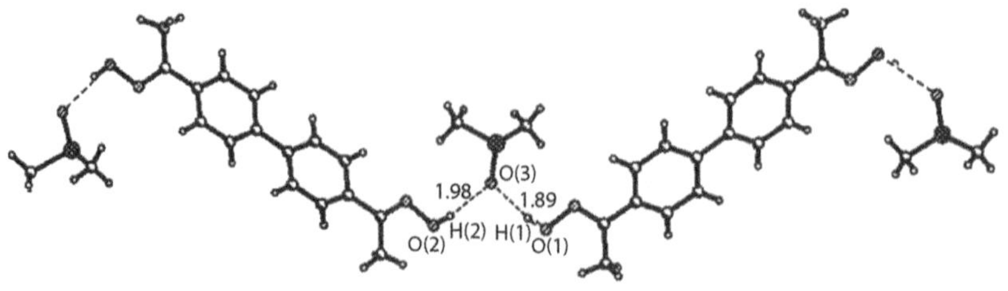

SCHEME 1.90 X-ray structure of the coordination polymer formed by 4,4'-diacetyldiphenyl dioxime and DMSO molecules.

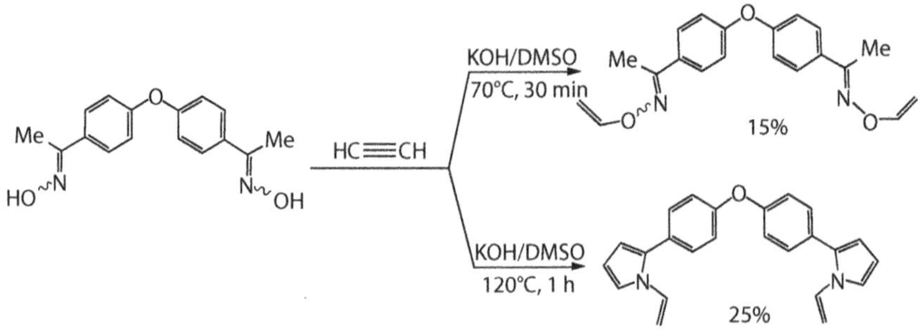

SCHEME 1.91 Synthesis of di(O-vinyl)dioxime and divinyldipyrroles from 4,4'-diacetyldiphenyl oxide dioxime with acetylene.

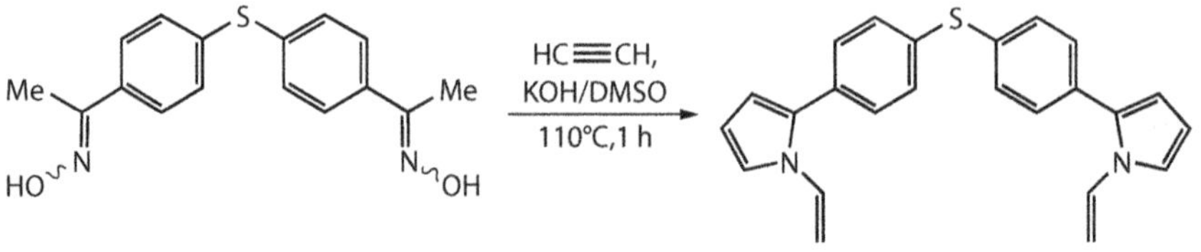

SCHEME 1.92 Selective synthesis of divinylpyrrole from 4,4'-diacetyldiphenyl sulfide dioxime and acetylene in the system KOH/DMSO.

When acetylene is employed at atmospheric pressure (KOH/DMSO, 120°C, 5 h), deoximation of the dioxime and resinification processes, due to autocondensation of the dicarbonyl compounds under the action of KOH/DMSO system, become prevailing. In these cases, dipyrroles are detected in trace amounts only.

Lower selectivity of the reaction is observed for the system LiOH/DMSO. In this reaction (130°C, 3 h), along with the expected dipyrrole, its mono- and divinyl derivatives are formed in 8%, 13%, and 6% isolated yields, respectively (Scheme 1.93) [133].

Upon short contact of 4,4'-diacetyldiphenyl oxide dioxime with acetylene under pressure (90°C, 5 min), di(O-vinyl) dioxime, monopyrrole derivative, and dipyrrole are formed in 12%, 6%, and 7%, isolated yields, correspondingly (Scheme 1.94) [133].

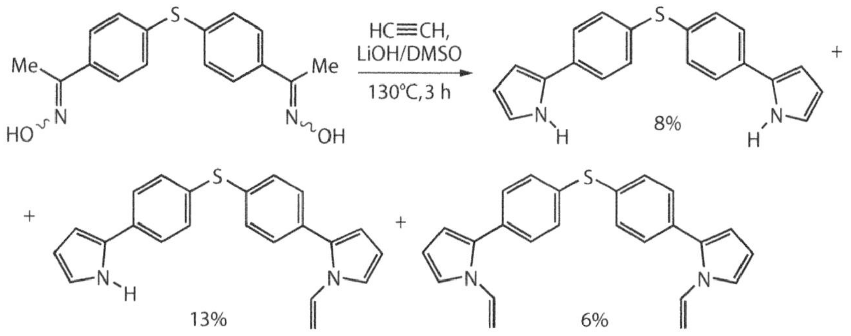

SCHEME 1.93 Synthesis of dipyrroles from 4,4′-diacetyldiphenyl sulfide dioxime and acetylene in the system LiOH/DMSO.

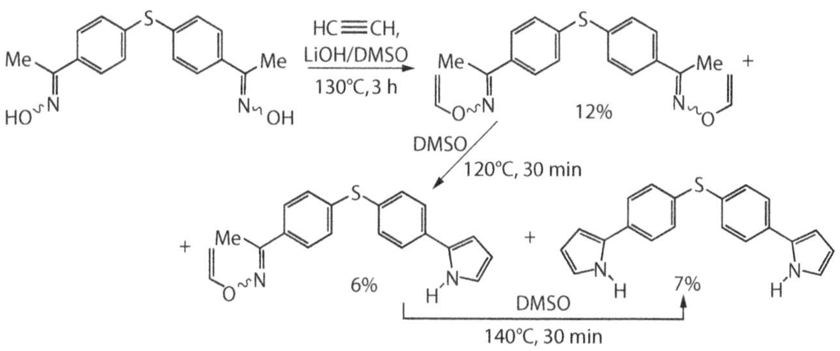

SCHEME 1.94 Synthesis of O-vinyl oximes and pyrroles from 4,4′-diacetyldiphenyl sulfide dioxime and acetylene in the system LiOH/DMSO. Thermal (noncatalytic) rearrangement of O-vinyl oximes to pyrroles.

Thermal non-catalytic rearrangement of di(O-vinyl) dioxime in DMSO (120°C, 30 min) furnishes monopyrrole, which further is transformed into dipyrrole only at a higher temperature (140°C, 30 min).

Apparently, the intermediate carbanion **A**, in the case of O-vinyl oxime with the pyrrole moiety, is additionally quenched by acidic pyrrole proton, which is absent in di(O-vinyl) dioxime (Scheme 1.95). As a consequence, the formation of next intermediate, divinylhydroxylamine **B**, and the construction of the pyrrole ring slow down.

Dipyrroles separated by pyridine spacers have been synthesized by the reaction of 2,6-diacetylpyridine and 3,5-diacetyl-2,6-dimethylpyridine dioximes with acetylene [134].

In the LiOH/DMSO system (140°C, 1 h), only 2,6-di(pyrrol-2-yl)pyridine is formed from 2,6-diacetylpyridine dioxime (Scheme 1.96). The product is first isolated from the reaction mixture as a 1:1 complex with DMSO (17% yield).

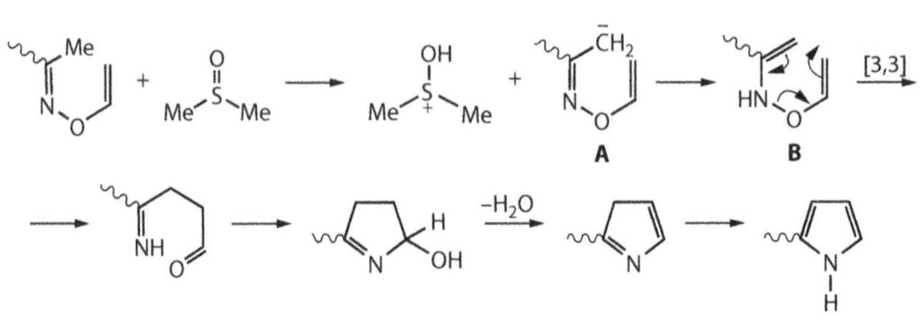

SCHEME 1.95 Tentative participation of DMSO molecule in the rearrangement of O-vinyl oximes to pyrroles.

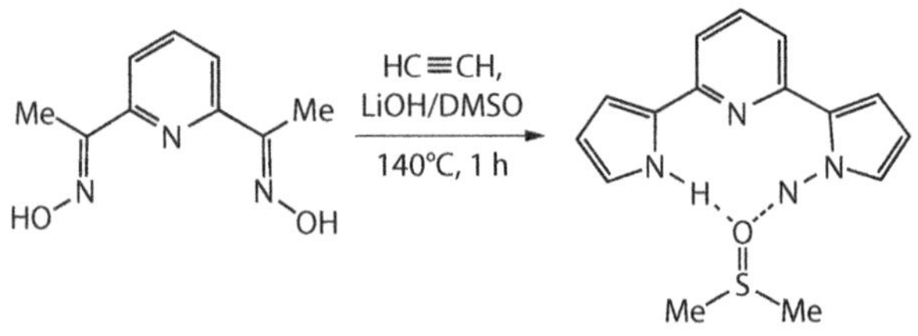

SCHEME 1.96 Formation of a complex of 2,6-di(pyrrol-2-yl)pyridine with DMSO from 2,6-diacetylpyridine dioxime and acetylene in the LiOH/DMSO system.

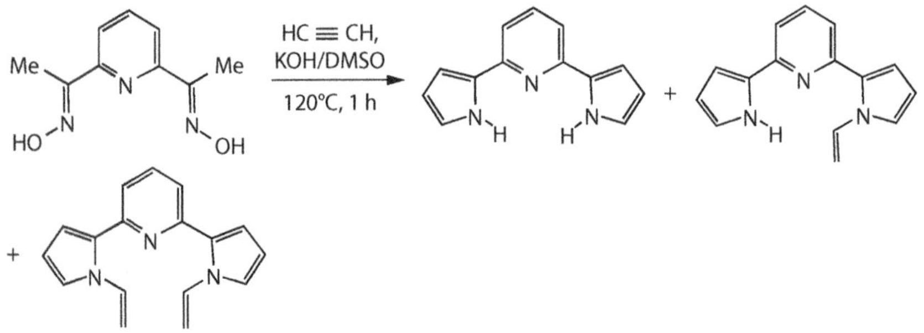

SCHEME 1.97 Synthesis of 2,6-dipyrrolylpyridines from 2,6-diacetylpyridine dioxime and acetylene in the KOH/DMSO system.

In the KOH/DMSO system, a mixture of three dipyrroles in 25% total yield is obtained, that is, the formation of mono- and divinyl derivatives together with non-vinylated dipyrrolylpyridine takes place (Scheme 1.97).

All three products have been isolated as individual compounds by column chromatography (Al_2O_3). Obviously, the yields and ratio of the dipyrrolylpyridines can be controlled by varying the reaction conditions.

When 2,6-diacetylpyridine dioxime is treated with acetylene in the system KOH/DMSO at a lower temperature (80°C) and for a shorter time (5 min), the dipyrrole precursors, the O-vinyl-oximes, are formed (Scheme 1.98) [134].

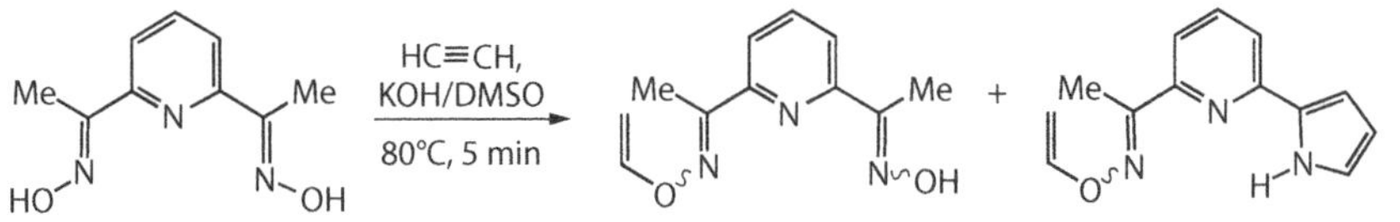

SCHEME 1.98 Reaction of 2,6-diacetylpyridine dioxime and acetylene in the KOH/DMSO system at a short contact of the reactants.

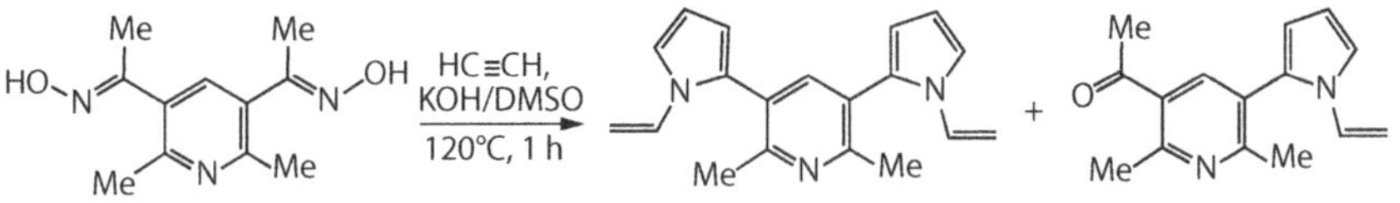

SCHEME 1.99 Reaction of 3,5-diacetyl-2,6-dimethylpyridine dioxime with acetylene in the KOH/DMSO system.

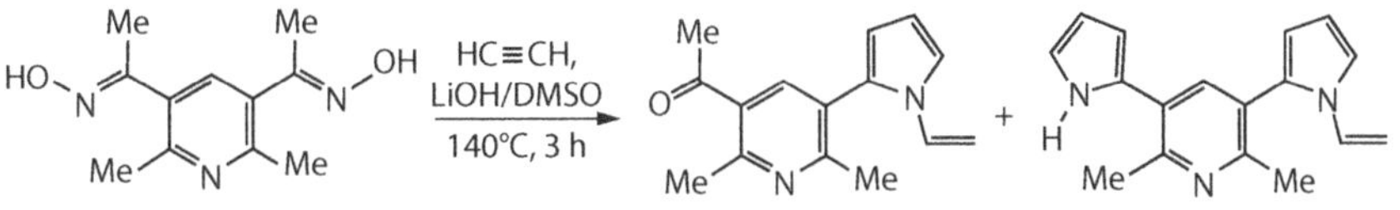

SCHEME 1.100 Reaction of 3,5-diacetyl-2,6-dimethylpyridine dioxime with acetylene in the LiOH/DMSO system.

3,5-Diacetyl-2,6-dimethylpyridine dioxime reacts with acetylene under pressure in the KOH/DMSO system (120°C, 1 h) to deliver a mixture of 3,5-di(N-vinyldipyrrolyl)pyridine (12% yield) and 3-acetyl-5-(N-vinylpyrrolyl)pyridine (13% yield) (Scheme 1.99) [134].

In the case of LiOH/DMSO system (140°C, 3 h), 3-acetyl-5-(N-vinylpyrrolyl) pyridine and monovinyl derivative are isolated in 9% and 15% yields, respectively (Scheme 1.100).

1.2 REGIOSPECIFICITY OF THE REACTION

As mentioned earlier, a main prerequisite for the formation of the pyrrole ring is that ketoxime should contain the methylene or methyl group in the α-position to the oxime function. In the case of unsymmetrical ketoximes where the competition between the methyl and methylene groups is possible, at moderate temperatures (to 120°C), the pyrrole ring is formed regioselectively, that is, involving the methylene group only (Scheme 1.101) [7,188]. However, already at 140°C, the relative content of the second isomeric pyrrole (construction owing to the methyl group) in the product mixture can reach 50%.

Since ketoximes exist predominantly in Z-configuration relative to the least bulky substituent [276,277], it becomes clear that, similar to the Beckman rearrangement, the reaction proceeds with participation of the radical, which is in E-position towards

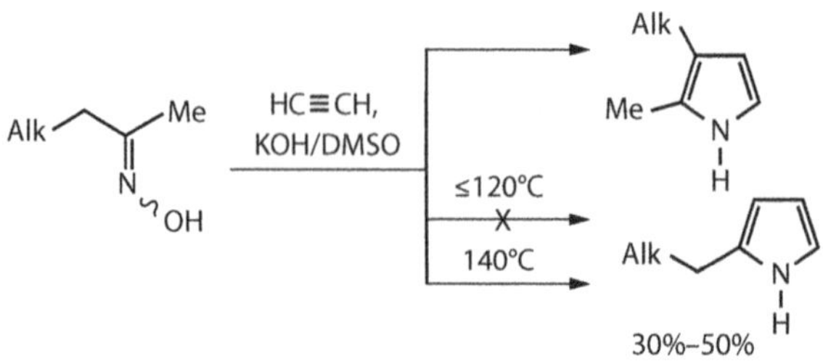

SCHEME 1.101 Competition between the methyl and methylene groups in the reaction of methyl alkyl ketoximes with acetylenes.

the hydroxyl group. It is assumed [7] that with the increase of the reaction temperature, the configurational equilibrium is obviously shifted to the E-isomer thus breaking regioselectivity of the reaction.

Unsymmetrical dialkyl ketoximes in heterocyclization reaction with acetylene can give two isomeric pyrroles or N-vinylpyrroles in the case of acetylene excess (Scheme 1.102).

The following questions are important for an understanding of the reaction mechanism and for better preparative use of the reaction: Which of the two groups of the ketoxime, the methyl (when $R^2=H$) or methylene group, primarily participates in the construction of the pyrrole ring? Are there differences in the reactivity of the two α-methylene moieties ($R^1 \neq R^2$, $R^2=Alk$) related to different alkyl groups with normal structures? Do the reaction conditions affect the ratios of the structural isomers? The answers to these questions have been found during a study of the interaction between a number of unsymmetrical ketoximes with acetylene in the presence of the KOH/DMSO system under conditions that lead to N-vinylpyrroles [188].

At 120°C, all methyl alkyl ketoximes react with acetylene exclusively at the methylene group of the alkyl radical irrespective of a structure of its other part (*iso*- or normal, the exception makes methyl-*iso*-propyl ketoxime that reacts exclusively at the methyl group, forming the only one isomer). With temperature increase, the regioselectivity of reaction is broken: at 140°C, pyrroles are already formed involving the methyl group in amount of 20%–50% from the mass of the mixture, though the total yield drops.

Thus, by selection of the conditions, one can change the direction of the reaction and synthesize not only 2,3-dialkyl-substituted but also 3-unsubstituted pyrroles

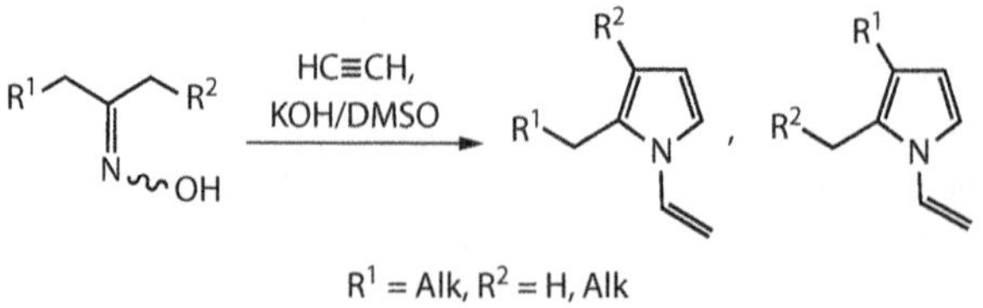

$R^1 = Alk, R^2 = H, Alk$

SCHEME 1.102 Formation of two isomeric pyrroles from unsymmetrical dialkyl ketoximes and acetylene.

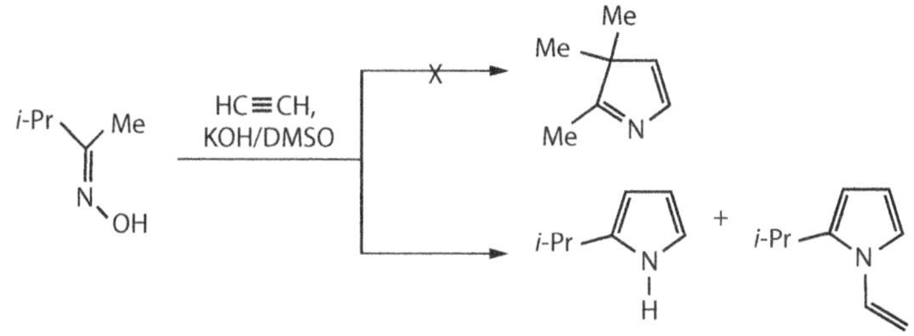

SCHEME 1.103   Formation of 2-*iso*-propylpyrrole from methyl-*iso*-propyl ketoxime with acetylene.

bearing various alkyl groups in the position 2. The two methylene groups of the two different alkyl radicals of normal structures (e.g., ethyl butyl ketoxime) have close reactivities: isomeric pyrroles are formed in approximately equal amounts even at 120°C.

In the case of methyl-*iso*-propyl ketoxime (competition between the Me and CH groups), only 2-*iso*-propylpyrrole and its N-vinyl derivative have been isolated in 10% and 15% yields, respectively (Scheme 1.103). In these conditions, 3*H*-pyrrole is likely unstable.

Strong bases usually easily abstract a proton closest to the C=N bond, for example, in *o*-methyldibenzyl ketoxime [278], aldimines [279], and hydrazones [280]. It is supposed [188] that since the less sterically hindered and hence more populated iso-mer of methyl alkyl ketoximes should have Z-orientation of the methyl group relative to the hydroxyl moiety, Z-1,3-dianions formed by the splitting of a proton from the methyl fragment should be unstable because of the repulsion of closely located nega-tive charges. However, it was reported [281] that Z-dianion, at least with Li cations, turned out to be significantly more stable than *E*-dianion, though attempts to prepare disodium and dipotassium salts of ketoximes failed [281].

Another factor that affects the structural directivity of the reaction may be the electron effect of radical [188].

Later, it has been proven that the initial stage of the process is a 1,3-prototropic shift in O-vinyl oximes, primary intermediates of the reaction. Therefore, the early interpretation of regiodirectivity of the pyrrole synthesis, based on the idea about different reactivity of *E*- and Z-configurations of the starting ketoximes, needs cor-rection. It is obvious that 1,3-prototropic shift in O-vinyl oximes will proceed easier with the participation of the methylene group and not the methyl moiety, because in the former case the disubstituted ethylene fragment of N,O-divinyl hydroxylamine **A** is known to be more thermodynamically stable in comparison with unsubstituted compound **B** (terminal vinyl group) (Scheme 1.104).

Thus, the visible regioselectivity of the reaction (preferable participation of the Z-configuration in the building up of the pyrrole ring) is defined by a ratio of the *E*- and Z-isomers of ketoximes only formally. Since 1,3-prototropic shift in O-vinyl oximes proceeds mainly with participation of the methylene (not methyl) group, that is, with participation of bulkier radical located in the Z-position to the hydroxyl

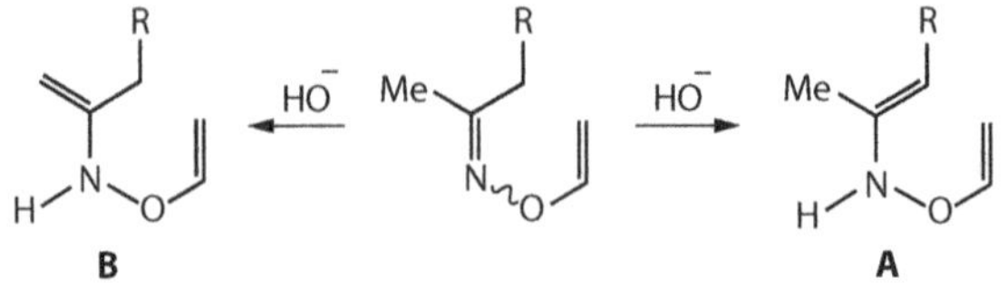

SCHEME 1.104 Prototropic rearrangement of alkyl methyl O-vinyl ketoximes to N,O-divinyl hydroxylamines.

group in the starting oxime, a quite reasonable at first sight conclusion about higher reactivity of the Z-configuration has been made.

If two different methylene groups compete for creation of the pyrrole ring (as it takes place in, e.g., benzyl ethyl ketoxime), the competition is won by group that is in E-position relative to hydroxyl, that is, by methylene function of the benzyl radical. In this case, conjugation with the benzene ring (formation of the styrene fragment) contributes additionally to stabilization of the ethylene counterpart of hydroxylamine O-vinyl ether formed.

Here, almost total selectivity, observed at below 80°C, is violated with the temperature increase (Scheme 1.105) [4,7,282].

For the further study of the regiospecificity of the reaction, methyl and ethyl benzyl ketoximes have been selected [282]. The former is obtained in the form of the pure E-isomer, while the latter represents a mixture of E- and Z-isomers in ~45:55 ratio. It is established that at room temperature solvents that differ fundamentally with respect to their solvation properties have virtually no effect on the configuration ratios. When benzyl methyl ketoxime is heated in nitrobenzene for 1 h, it is partially converted to the Z-isomer, the content of which increases with the temperature (¹H NMR data) from ~1% at 25°C to 36% at 150°C.

In the system KOH/DMSO, benzyl methyl ketoxime gives a mixture of E- and Z-isomers (74:26) even in the case of short heating (~80°C, 10 min), a further increase in the temperature (150°C) has almost no effect on the isomers ratio, which later remains the same at room temperature. Consequently, the percentage of Z-isomer in benzyl methyl ketoxime, which ranges 26%–36% (depending on the medium),

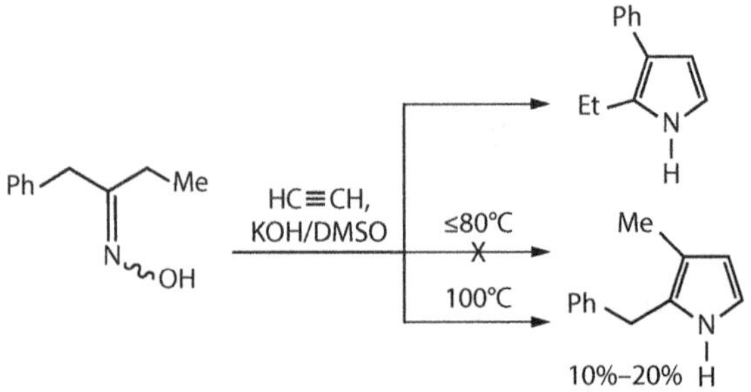

SCHEME 1.105 Thermally controlled regiodirection of reaction benzyl ethyl ketoxime with acetylene.

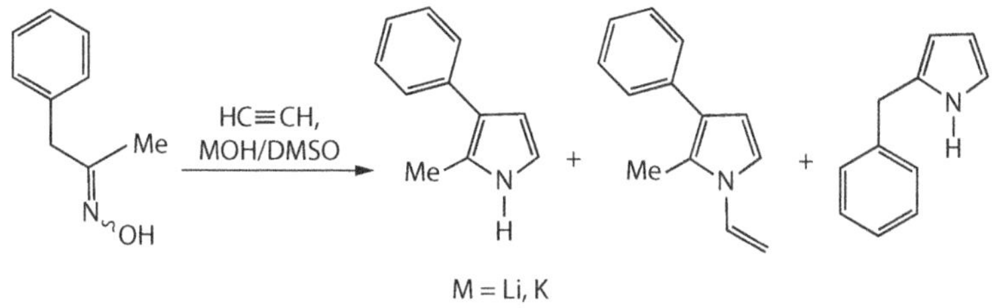

SCHEME 1.106 Reaction of benzyl methyl ketoxime with acetylene.

is evidently close to the equilibrium value. Unlike benzyl methyl ketoxime, the ratio of the isomers in a starting sample of benzyl ethyl ketoxime apparently already corresponds to the equilibrium one, since it remains almost unchanged in the case of heating in nitrobenzene (150°C) or KOH/DMSO (170°C).

The reaction of benzyl methyl ketoxime with excess acetylene (100°C, KOH) furnishes 2-methyl-3-phenyl-N-vinylpyrrole only (Scheme 1.106). The regioselectivity of the reaction is not violated even under harsher conditions (120°C) and when the reaction is carried out in the system LiOH/DMSO (120°C). Under these conditions, 2-methyl-3-phenyl-N-vinylpyrrole and its nonvinylated precursor are formed in approximately equal amounts (Scheme 1.106). Structural isomer, 2-benzylpyrrole, has been isolated only at increased concentrations of the reactants and base at 120°C. Thus, the methyl group also begins to participate in the construction of the pyrrole ring under these conditions (Scheme 1.106).

The results obtained evidence that benzyl methyl ketoxime reacts with acetylene mainly via the methylene group of the benzyl radical, which exists predominantly in the *E*-position relative to the hydroxyl group.

Benzyl ethyl ketoxime reacts with excess acetylene in the system KOH/DMSO at 100°C to yield two isomeric N-vinylpyrroles (Scheme 1.107) [282]; as the temperature is raised, the content of N-vinylpyrrole, which is formed through the benzyl group, decreases appreciably (from 85% at 100°C to 45% at 150°C).

At 60°C–80°C, the major reaction product is nitrogen-unsubstituted pyrrole. For example, a mixture of isomeric pyrroles in a ratio of 5.7:1 is obtained at 80°C. Thus, benzyl ethyl ketoxime reacts with acetylene predominantly through the methylene group of the benzyl radical.

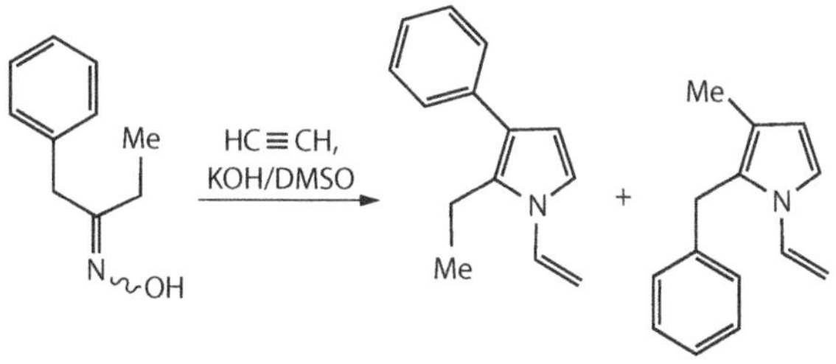

SCHEME 1.107 Reaction of benzyl methyl ketoxime with excess acetylene.

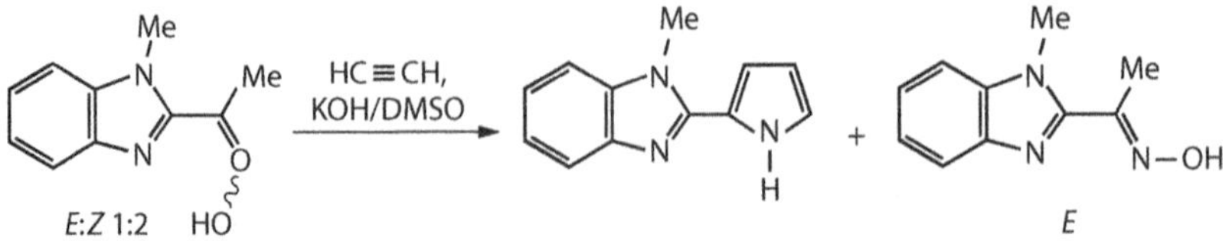

SCHEME 1.108 Low reactivity of *E*-isomer of 1-methyl-2-acetylbenzimidazole oxime in the reaction with and acetylene.

An interesting result is obtained when pyrrole is synthesized from 1-methyl-2-acetylbenzimidazole oxime and acetylene in the system KOH/DMSO [283]. Since under the reaction conditions the interconversion of *E*- and *Z*-isomers of this oxime meets with difficulties, only *Z*-isomer participates in the formation of the pyrrole ring. Apart from the pyrrole, pure *E*-isomers are isolated from the reaction mixture, which under the same condition do not react with acetylene (Scheme 1.108) [283].

In the light of the data on the key role of O-vinyl oximes as primary intermediates in the construction of the pyrrole ring, to explain this phenomenon, one should admit a dramatic difference in vinylation rates of ketoxime *E*- and *Z*-isomers.

1.3 SUBSTITUTED ACETYLENES IN THE REACTION WITH KETOXIMES

The reaction with ketoximes tolerates also several substituted acetylenes, stable in the presence of strong bases.

1.3.1 METHYLACETYLENE

The reaction of ketoximes with simplest homolog of acetylene, methylacetylene, its isomer (allene), and their mixtures has been studied [284]. The reaction proceeds in the systems MOH/DMSO (M = K, Cs) or KOBu-*t*/DMSO both at atmospheric and elevated pressure to furnish 5-methyl- and 4-methylpyrroles (Scheme 1.109).

The reaction temperature and time as well as structure of the starting propyne–allene mixture effect slightly on the ratio of pyrroles formed. Only a weak dependence of the regioisomers ratio upon the catalyst nature and propyne pressure (concentration) is observed. The reaction outcome strongly depends on ketoxime

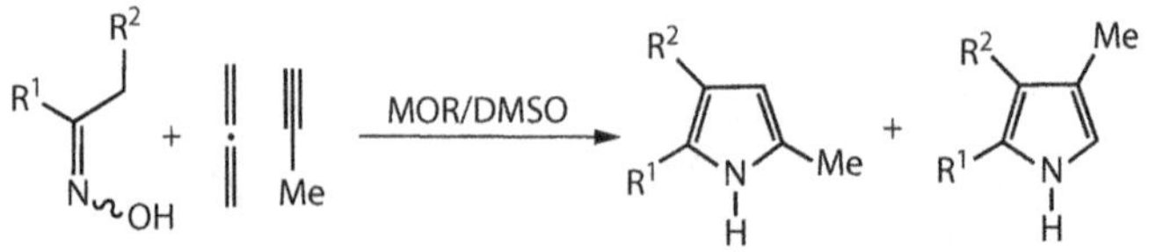

R^1 = Me, *i*-Pr, *t*-Bu, Ph, 2-furyl, 2-tienyl; R^2 = H, Me, R^1–R^2 = $(CH_2)_4$; R = H, *t*-Bu; M = K, Cs

SCHEME 1.109 Synthesis of 5-methyl- and 4-methylpyrroles from ketoximes and methyl acetylene-allene mixture.

structure and is defined by not only electron but also steric factors. With branching of the radical at the oxime function, the content of 4-methyl-substituted pyrrole in the mixture increases. It is clearly seen if to compare the results obtained for acetoxime, methyl isopropyl ketoxime and pinacoline oxime. Bulkier oximate anion, likely due to the steric reasons, preferably attacks the C^1 atom of methylacetylene. At the same time, cyclohexanone and 2-acetylfuran oximes react with the propyne in a regiose-lective manner to afford 5-methylpyrrole only. With oximes of ethyl methyl ketone, acetophenone, and 2-acetylthiophene, the yields of 5-methyl-substituted pyrroles range 90%–92%.

Alkylacetylenes are known to add nucleophiles to the central carbon atom [285] that explains preferential formation of 5-methyl-substituted pyrroles in the case of propyne.

Owing to the steric factors, ambident oximate anions bearing bulky substituents or bulky cations (Cs$^+$) have to be added to terminal carbon atom of propyne. As a consequence, 4-methylpyrroles are formed (Scheme 1.110).

Irrespective of whether the pure propyne or allene or their mixtures are employed, the reaction leads to the same results that is not surprising, since the propyne⇌allene equilibrium (~4:1) is reached very quickly in the presence of the superbases.

The reaction is expectedly accompanied by the addition of NH-pyrroles to pro-pyne or allene to deliver N-2-propenylpyrroles (Scheme 1.111).

The anticipated fact also is that propyne and allene are less reactive than acety-lene in the reactions of both formation and vinylation of the pyrrole [284].

The reaction of ethyl methyl ketoxime with propyne–allene mixture, despite the increased temperature (115°C–125°C), remains regioselective (only the methylene

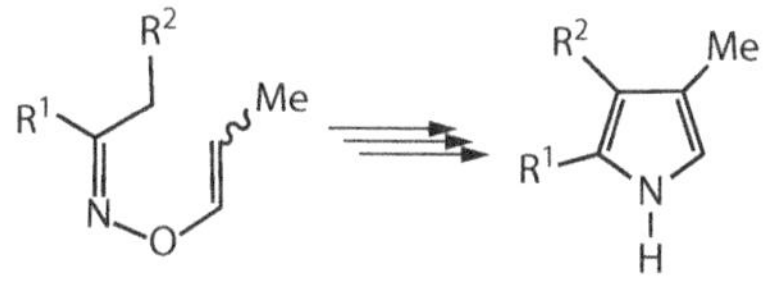

SCHEME 1.110   Formation of 4-methylpyrroles via O-1-propenyl ketoximes.

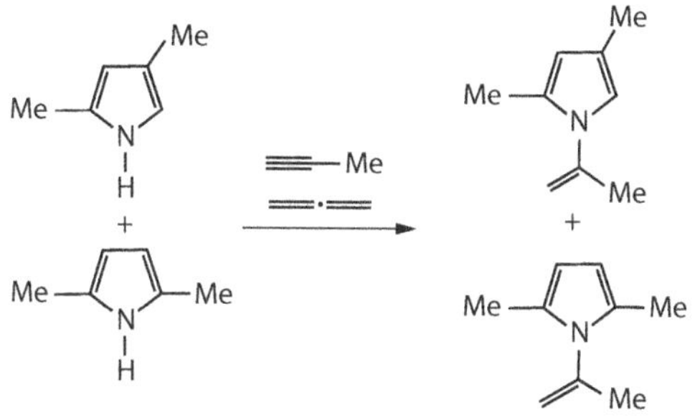

SCHEME 1.111   Vinylation of 2,4- and 2,5-dimethylpyrroles with methylacetylene-allene mixture in superbase media.

group participates in construction of the pyrrole ring). At the same time, with acetylene under the same conditions, the regioselectivity of the process is broken: not only methylene but also methyl group of the ketoxime, though to a small extent, is involved into the reaction [7].

Synthesis of pyrroles from ketones (through ketoximes) and propyne–allene mixture, a large-scale side product of hydrocarbons pyrolysis, essentially expands the preparative possibilities of the Trofimov reaction. 5-Methyl-substituted pyrroles, having diverse substituents in the positions 2 and 3, become readily available for the first time.

1.3.2 PHENYLACETYLENE

5-Phenylsubstituted pyrroles are formed exclusively from acetone or acetophenone oximes and phenylacetylene in the system KOH/DMSO (120°C–170°C, 5–7 h), no matter what the ratio of ketoxime–phenylacetylene is (1:1 or 1:2). The best yields of these pyrroles are 21% and 17%, respectively (Scheme 1.112) [286].

Later, Yurovskaya et al. have reported [287] that when the reaction of acetophenone oxime with phenylacetylene is conducted without a solvent, using potassium oximate as the base under rigorous conditions (140°C, 6 h), approximately equal amounts (5%–6%) of 2,4- and 2,5-diphenylsubstituted pyrroles are isolated.

A mixture of 2,3,4-triphenyl-, 2,3,5-triphenyl-, and 2,3,4-triphenyl-1-[(Z)-2-phenylvinyl]-pyrroles (their content in the crude product is 37%, 24%, and 30%, respectively) is formed from benzyl phenyl ketoxime and phenylacetylene in the system LiOH/DMSO (140°C, 6 h) (Scheme 1.113) [288].

The formation of two isomeric triphenylpyrroles may be rationalized by cyclization (via [3,3]-sigmatropic rearrangement) of isomeric O-vinyl oximes, the primary adducts of benzyl phenyl oximes to α- and β-carbon atoms of phenylacetylene (Scheme 1.114).

It is disclosed [289] that cyclohexanone oxime reacts with phenylacetylene at 80°C (KOH/DMSO) to give Z-[N-(β-phenylvinyl)]-3-phenyl-4,5,6,7-tetrahydroindole (42% yield) (Scheme 1.115).

1.3.3 ACYLACETYLENES

Recently [290], the results on regio- and stereoselective addition of ketoximes to acylacetylenes in the presence of 10 mol% PPh_3 have been published (Scheme 1.116).

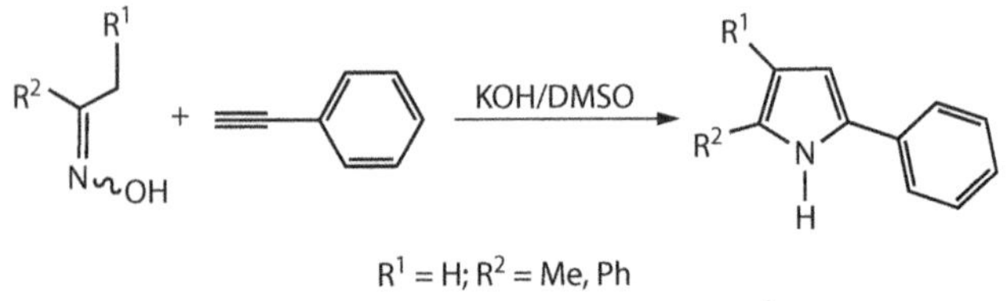

$R^1 = H; R^2 = Me, Ph$

SCHEME 1.112 Synthesis of 2,3-disubstituted-5-phenylpyrroles from ketoximes and phenylacetylene in the KOH/DMSO system.

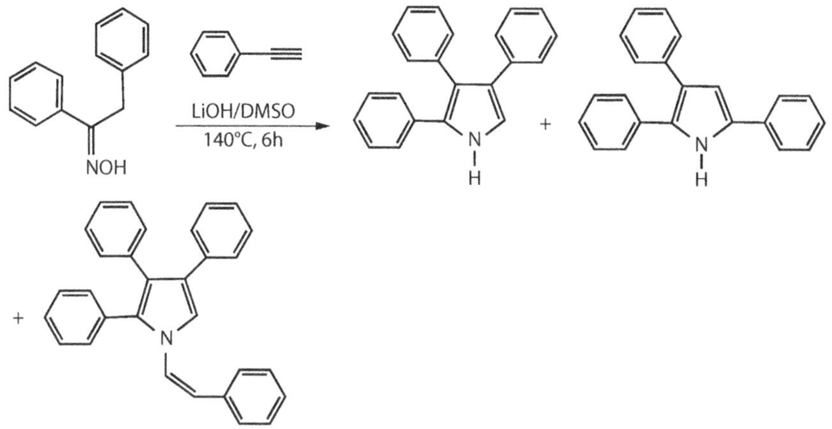

SCHEME 1.113 Synthesis of 2,3,4-triphenyl-, 2,3,5-triphenyl-, and 2,3,4-triphenyl-1-[(Z)-2-phenylvinyl]-pyrroles from benzyl phenyl ketoxime and phenylacetylene in the LiOH/DMSO system.

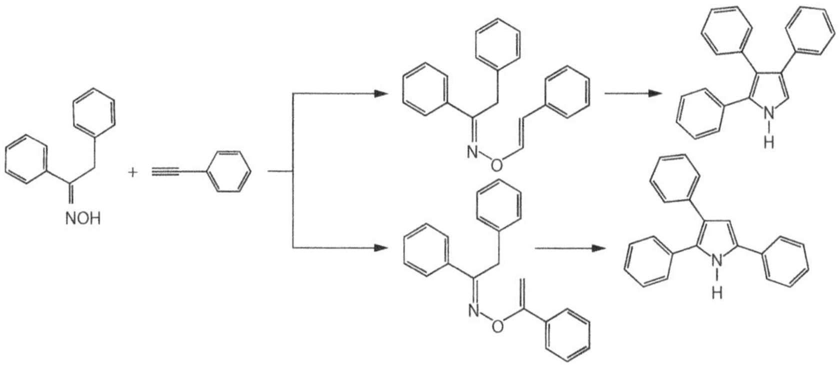

SCHEME 1.114 Formation of 2,3,4-triphenyl-, 2,3,5-triphenylpyrroles from benzyl phenyl ketoxime and phenylacetylene via O-vinyl oximes.

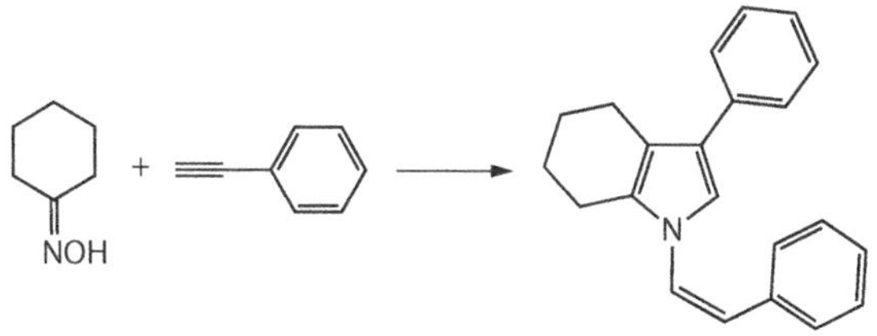

SCHEME 1.115 Synthesis of Z-[N-(β-phenylvinyl)]-3-phenyl-4,5,6,7-tetrahydroindole from cyclohexanone oxime and phenylacetylene.

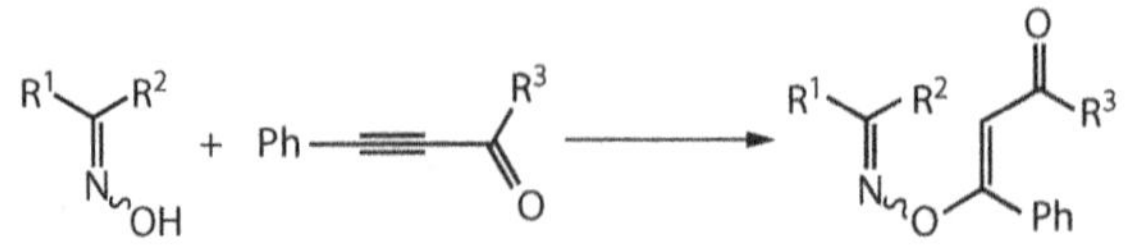

R^1 = Me, R^2 = Ph, 2-naphthyl, 2-thienyl; R^1–R^2 = (CH$_2$)$_5$; R^3 = Ph, 2-furyl, 2-thienyl

SCHEME 1.116 Regioselective and stereoselective addition of ketoximes to acylacetylenes in the presence of PPh$_3$.

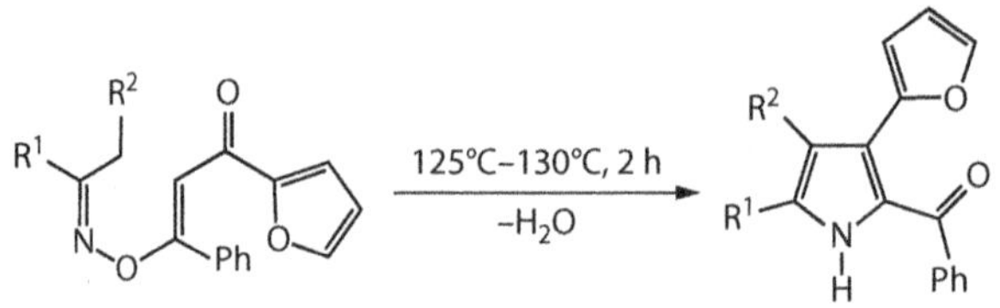

R^1 = Ph, 2-thienyl, R^1–R^2 = (CH$_2$)$_4$

SCHEME 1.117 Rearrangement of O-2-(acyl)vinyl ketoximes to 2-benzoylpyrroles.

A wide series of (*E*)-O-2-(acyl)vinyl ketoximes in 85% yields was synthesized (room temperature, 7 h).

(*E*)-O-2-(Acyl)vinyl ketoximes rearrange upon boiling (125°C–150°C) to acyl-substituted pyrroles of unusual structure [291]. The positions of the acyl and aryl(hetaryl) substituents in the pyrrole ring depends upon the reaction conditions.

When heated (125°C–130°C) without solvent, O-2-(acyl)vinyl ketoximes give 2-benzoylpyrroles in up to 14% yields (Scheme 1.117).

At a higher temperature (145°C–150°C, without solvent), a mixture of 2-acyl and 3-acyl-5-unsubstituted pyrroles in a ratio of ~1:1 is formed (Scheme 1.118).

Upon heating in *p*-xylene (138°C), another rearrangement takes place: a mixture of the isomeric 2-acylpyrroles in a ratio of ~1:2 has been isolated (Scheme 1.119) [291].

The formation of 2-acylpyrroles is likely preceded by the rearrangement of O-2-(acyl)vinyl ketoximes into diketones, intramolecular condensation of which (involving any carbonyl groups) leads to the pyrroles (Scheme 1.120).

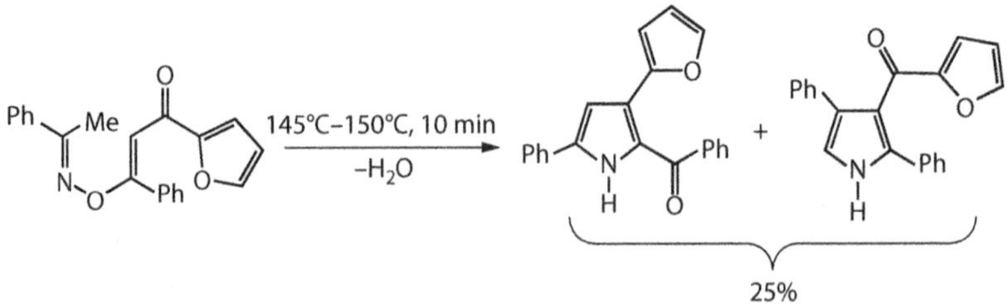

SCHEME 1.118 Partical redirection of O-2-(acyl)vinyl ketoximes rearrangement at higher temperature.

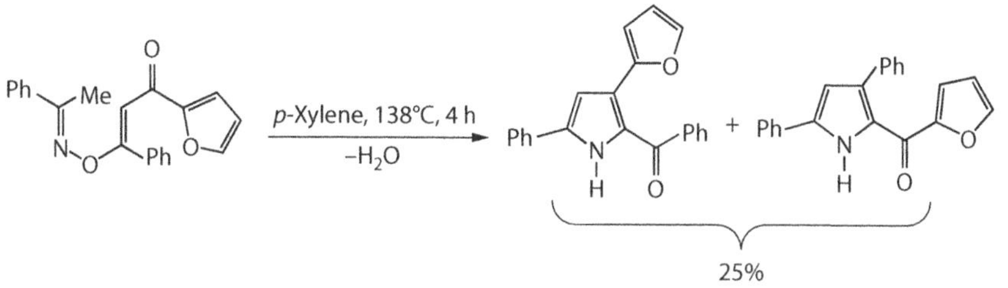

SCHEME 1.119 Alternative rearrangement of O-2-(acyl)vinyl ketoximes upon boiling in *p*-xylene.

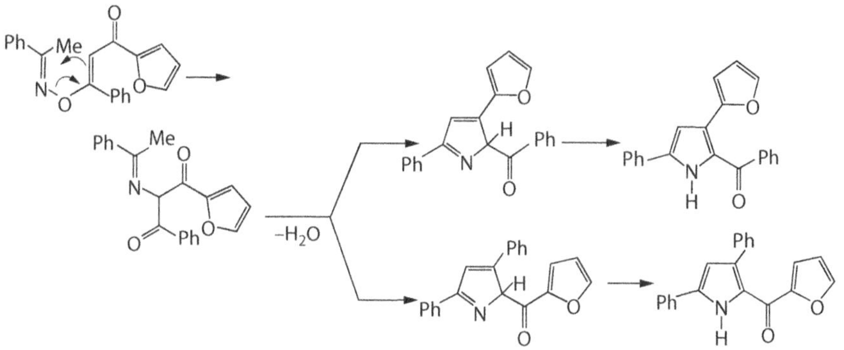

SCHEME 1.120 Plausible scheme of 2-acylpyrroles formation from O-2-(acyl)vinyl ketoximes.

The assembly of 3-acyl-5-unsubstituted pyrrole can be rationalized by the elimination of enol (or tautomeric 1,3-diketone) from the oxime to give azirine (Scheme 1.121). These two intermediates interact affording aziridinyl-1,3-diketone. The latter further rearranges into hydroxypyrroline, which dehydrates and aromatizes to deliver the target pyrrole, α-unsubstituted 3-furoylpyrrole.

In parallel, the one-pot nucleophilic catalysis method for the synthesis of highly substituted pyrroles from the reaction of oximes with ethyl propiolate and dimethyl acetylenedicarboxylate has been developed (Scheme 1.122) [42].

1.3.4 OTHER ACETYLENES

The extension of the reaction with ketoximes to vinylacetylene could lead to the preparation of 2- or 3-vinylpyrroles. However, the products, obtained upon heating (60°C–80°C) ketoximes and vinylacetylene, represent a mixture of acetylenic and allenic ethers (in up to 50% total yield) (Scheme 1.123) [292].

The formation of the expected ethynylpyrroles is not also observed in the reaction with diacetylene, though the corresponding Z-O-ethynyl vinyl ketoximes are isolated (Scheme 1.124) [293,294]. Their best yields (36%–41%) are attained when water content in the reaction mixture is 10%–30% and KOH concentration is 1%–2%.

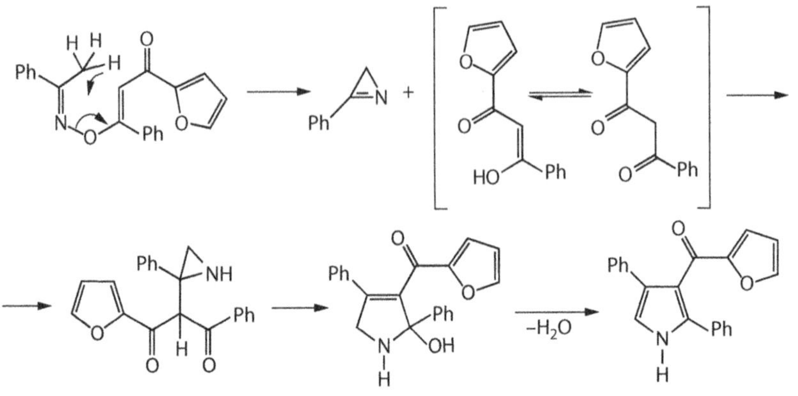

SCHEME 1.121   Assembly of 3-acyl-5-unsubstituted pyrrole from O-2-(acyl)vinyl ketoxime.

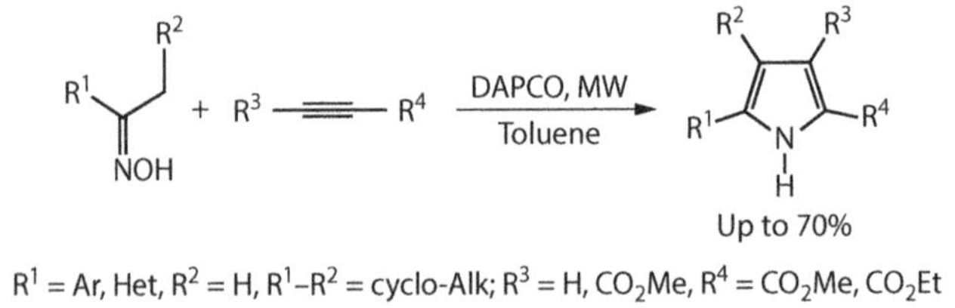

R^1 = Ar, Het, R^2 = H, R^1–R^2 = cyclo-Alk; R^3 = H, CO_2Me, R^4 = CO_2Me, CO_2Et

SCHEME 1.122   Synthesis of functionalized pyrroles from ketoximes and ethyl propiolate or dimethylacetylenedicarboxylate.

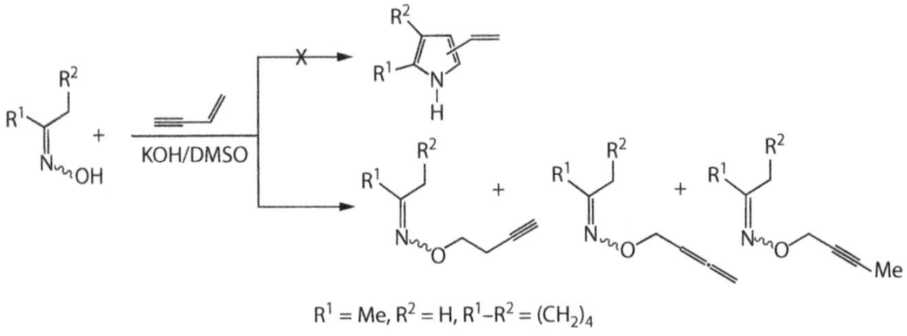

R^1 = Me, R^2 = H, R^1–R^2 = $(CH_2)_4$

SCHEME 1.123   Reaction of ketoximes with vinylacetylene in the KOH/DMSO system.

An attempt to synthesize phenylthiosubstituted pyrrole from phenylthioacetylene and cyclohexanone oxime results in Z-1,2-di(phenylthio)ethylene in 40% yield as the major product (Scheme 1.125) [295].

Cyanoacetylene adds ketoximes in a solution of ethanol and N-methylmorpholine at 0–20 °C to deliver the corresponding O-vinyl oximes (Scheme 1.126) [296].

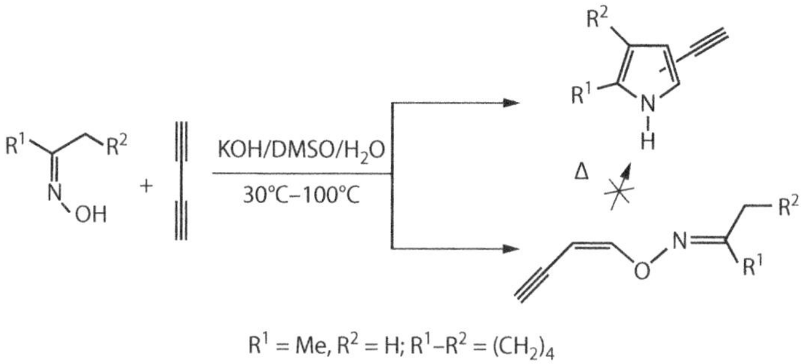

$R^1 = Me, R^2 = H; R^1-R^2 = (CH_2)_4$

SCHEME 1.124   Reaction of ketoximes with diacetylene in the KOH/DMSO/H$_2$O system.

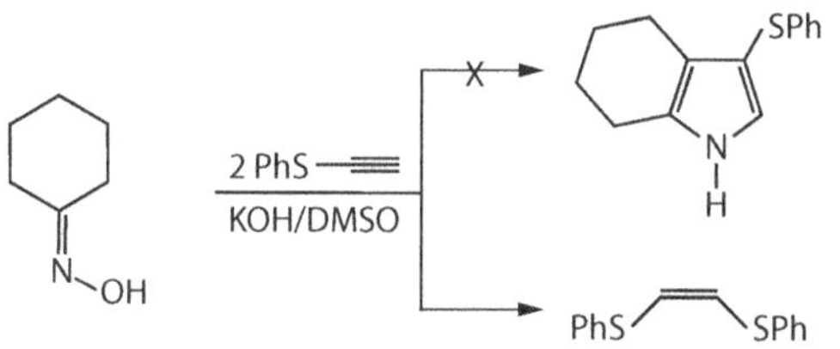

SCHEME 1.125   Reaction of cyclohexanone oxime with phenylthioacetylene in the KOH/DMSO system.

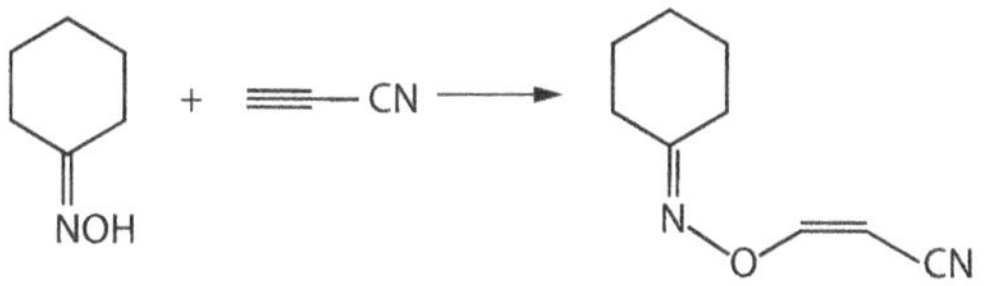

SCHEME 1.126   Reaction of cyclohexanone oxime with cyanoacetylene in the KOH/DMSO system.

For example, from cyanoacetylene and cyclohexanone oxime, the expected O-vinyl oxime is obtained in 96.5% yield. The data on possibility of transformation of such O-vinyl oximes to pyrroles are absent.

It has been revealed that the reaction of ketoximes with propargylic ethers affords, instead of pyrroles or their predecessors O-vinyl oximes, O-(3-alkoxy-2-E-propenyl) ketoximes (Scheme 1.127) [297].

Apparently, in the course of the reaction, the intermediate O-vinyl oximes undergo a fast stereoselective prototropic isomerization involving migration of the double bond from iminoxy to alkoxy group. These findings are fundamental as an example of abnormal addition of nucleophiles to propargylic ethers and also as the evidence of a stronger conjugation of the double bond with alkoxy than with iminoxy group (this is a driving force of isomerization).

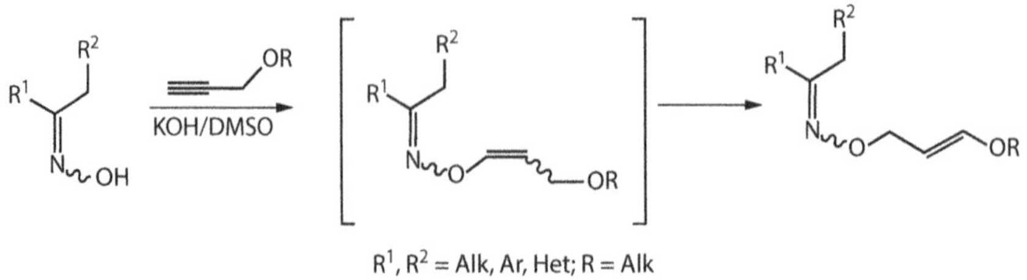

R¹, R² = Alk, Ar, Het; R = Alk

SCHEME 1.127 Reaction of ketoximes with propargylic ethers in the KOH/DMSO system.

1.4 VINYL HALIDES AND DIHALOETHANES AS SYNTHETIC EQUIVALENTS OF ACETYLENE

The interaction of ketoximes with vinyl halides at 80°C–130°C in the presence of alkali metal hydroxides and DMSO affords pyrroles and N-vinylpyrroles (Scheme 1.128) [298].

Varying DMSO concentration, reactants ratio, and reaction conditions, one can carry out the process selectively, that is, to prepare either pyrroles in a yield of about 40% or their N-vinyl derivatives (35% yield).

Along with vinyl halides as synthetic equivalents of acetylene, dihaloethanes can also be employed in the synthesis of pyrroles and N-vinylpyrroles from ketoximes (Scheme 1.129) [299]. The reaction of ketoximes with dihaloethanes in the presence of excess of alkali metal hydroxide in DMSO gives pyrroles and N-vinylpyrroles in ~30% yields. In further experiments, higher yields have been achieved [7].

Like in the case with vinyl halides, the reaction, depending on the reactants ratio and conditions, can be directed towards predominant formation of non-vinylated or vinylated pyrroles [299].

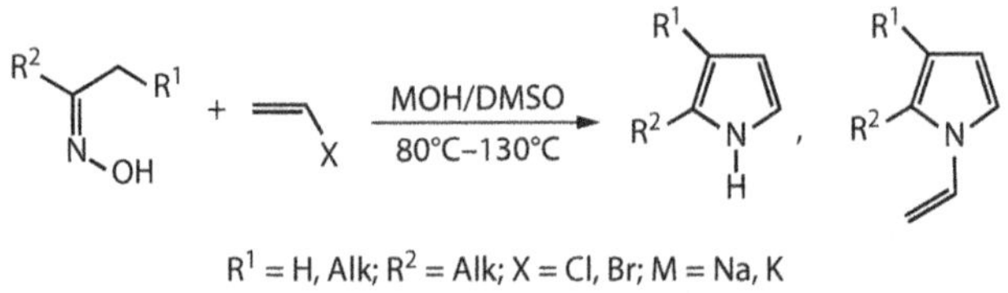

R¹ = H, Alk; R² = Alk; X = Cl, Br; M = Na, K

SCHEME 1.128 Reaction of ketoximes with vinyl halides in MOH/DMSO systems.

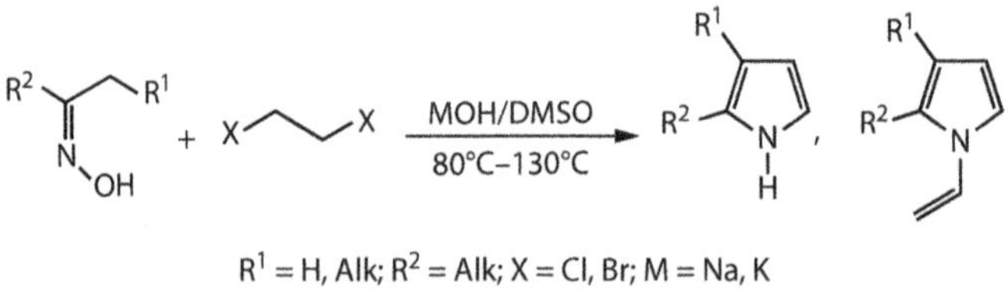

R¹ = H, Alk; R² = Alk; X = Cl, Br; M = Na, K

SCHEME 1.129 Reaction of ketoximes with dihaloethanes in MOH/DMSO systems.

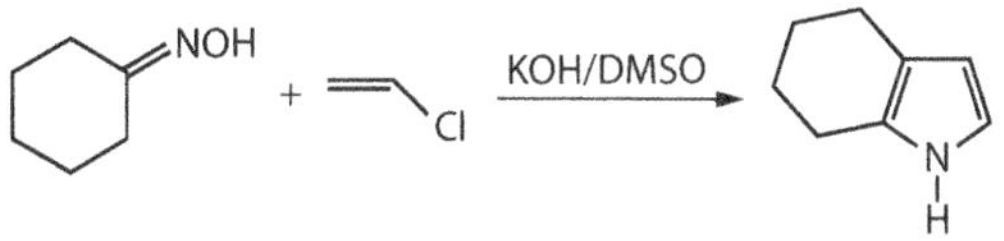
NOH
+
Cl
KOH/DMSO
N
H

SCHEME 1.130 Synthesis of 4,5,6,7-tetrahydroindole from cyclohexanone oxime and vinyl chloride.

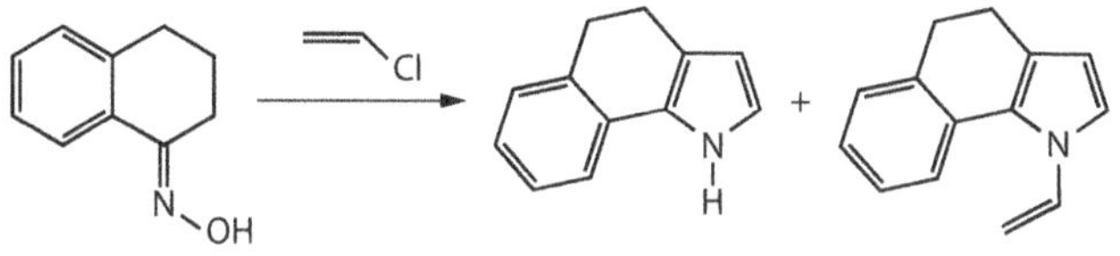
Cl
N
OH
N
H
+
N

SCHEME 1.131 Synthesis of 4,5-dihydrobenzo[*g*]indole and its N-vinyl derivative from α-tetralone oxime and vinyl chloride in the KOH/DMSO system.

A method for the preparation of 4,5,6,7-tetrahydroindole via condensation of cyclohexanone oxime with vinyl chloride has been developed (Scheme 1.130) [300]. The reaction proceeds in the presence of KOH in DMSO at 90°C–130°C and under atmospheric pressure. The yield of indole is 83%, oxime conversion being up to 75%.

The method is realized in simplest apparatus equipped with a stirrer. Besides, no special safety precautions are required, since vinyl chloride is not so explosive compound as acetylene. This method for the synthesis of 4,5,6,7-tetrahydroindole is facile, straightforward, safe, and based on application of cheap and available chemicals.

Using this methodology, 4,5-dihydrobenz[*g*]indole and its N-vinyl derivative (50% total yield) have been synthesized from α-tetralone oxime (KOH/DMSO, 140°C–150°C, 3–4 h) (Scheme 1.131) [301].

The synthesis of pyrroles from ketoximes and 1,2-dichloro- or 1,2-dibromoethanes is accomplished in the system KOH/DMSO (90°C–120° C) (Scheme 1.132) [302].

Ketoxime, 1,2-dihaloethane, and KOH should be used in 1:2–3:7–10 molar ratio, preferable ratio of ketoxime–DMSO being 1:10.

Similar to the method based on application of vinyl chloride, this protocol is simple, efficient, safe, and straightforward. Its important advantage is the usage of cheap and accessible starting materials. As it is stated previously, the interaction of ketoximes with dihaloethanes may lead to N-vinylpyrroles. In some cases,

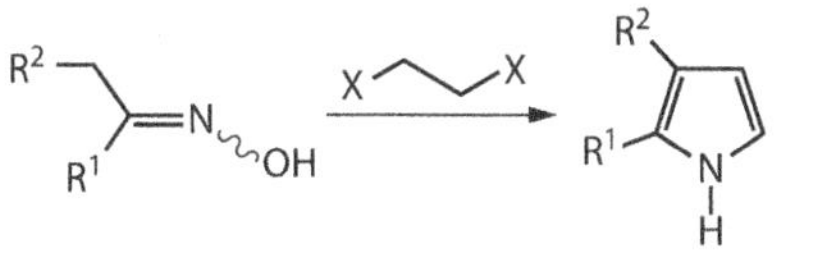
R2
R1
N
OH
X
X
R2
R1
N
H
R1 = Me, Et, n-Pr, n-Bu; R2 = H, Me, Et, n-Pr; X = Cl, Br

SCHEME 1.132 Synthesis of pyrroles from ketoximes and 1,2-dichloro- or 1,2-dibromoethanes in the KOH/DMSO system.

the reaction of nucleophilic substitution of halogen with oximate anions takes place to yield ketoxime glycol diethers.

The synthesis of 4,5,6,7-tetrahydroindoles from cyclohexanone oxime and 1,2-dihaloethanes has been disclosed [303]. The best overall yields (52%–61%) of NH- and N-vinyl-4,5,6,7-tetrahydroindoles are reached when molar ratio of cyclohexanone oxime–dichloroethane–KOH–DMSO is 1:1–2:7:10. For the successful synthesis of 4,5,6,7-tetrahydroindoles, it is important to add the alkali and dihaloethane to the solution of the ketoxime in DMSO in portions. Otherwise, the reaction of diether formation becomes appreciable. At the sacrifice of decreasing the yield to ~30%, one can attain 94%–95% selectivity relative to the major product, 4,5,6,7-tetrahydroindole. Like in the reaction with free acetylene, this is achieved mainly due to the addition of small amounts of water (10%–20%) to the reaction mixture. In this case, the water can be conveniently fed into the mixture by dissolving alkali in it, which simultaneously also facilitates the dispensing of both components. Somewhat poorer results are obtained with 1,2-dibromoethane under comparable conditions [303].

This version of the reaction, despite the lower yields of pyrroles, can be acceptable for laboratories possessing no free acetylene and having no experience in acetylene handling.

1.5   INTERMEDIATE STAGES AND SIDE REACTIONS

Systematic investigations into heterocyclization of ketoximes with acetylene to pyrroles and N-vinylpyrroles have shown that in certain conditions, O-vinyl oximes, 4H-2-hydroxy-2,3-dihydropyrroles, 3H-pyrroles, α-acetylenic alcohols, pyridines, and some other minor products are formed.

1.5.1   FORMATION OF O-VINYL OXIMES

The data on interaction of oximes with acetylenes are quite discrepant [4,7]. For example, according to the work [304], ketoximes are added to dimethyl acetylenedicarboxylate across the hydroxyl group to form O-vinyl derivatives. At the same time, it is known that such reactions lead to N-vinylnitrones [305–308] or N-vinyl zwitterion intermediates [305,309,310], which further, attaching the second molecule of acetylene, are converted to oxazole [306–308] or pyridine [305,309] derivatives.

Direct vinylation of ketoximes, namely, acetoxime, with acetylene has been implemented for the first time in more than 30 years [311]. The reaction is carried out under acetylene atmospheric pressure, the yields of O-vinyl oxime being 10% (Scheme 1.133).

Later, one has managed to improve these results by performing the reaction in autoclave (72% yield) and applying fractional vinylation [312]. However, this protocol turns out to be hardly reproducible.

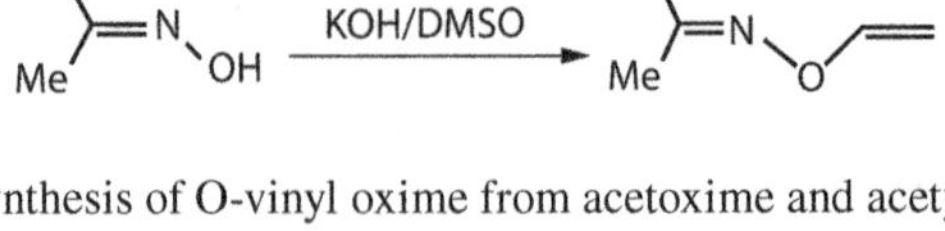

SCHEME 1.133   Synthesis of O-vinyl oxime from acetoxime and acetylene.

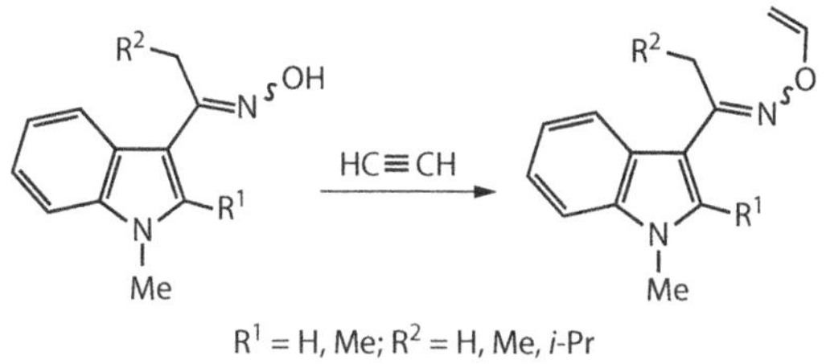
HC≡CH
OH
N
R1
R2
Me
O
R1 = H, Me; R2 = H, Me, i-Pr

SCHEME 1.134 Synthesis of O-vinyl oximes of 3-acylindoles by vinylation of 3-acylindole oximes with acetylene.

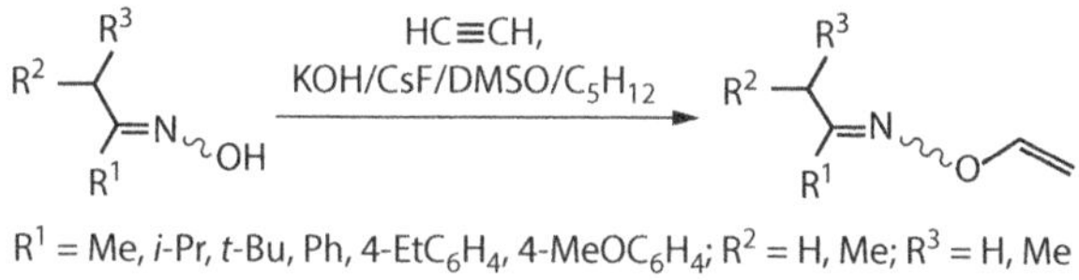
HC≡CH,
KOH/CsF/DMSO/C5H12
R1 = Me, i-Pr, t-Bu, Ph, 4-EtC6H4, 4-MeOC6H4; R2 = H, Me; R3 = H, Me

SCHEME 1.135 Vinylation of oximes with acetylene in the KOH/CsF/DMSO/pentane system.

O-vinyl oximes (in up to 54% yields) have been isolated by Yurovskaya et al. [229] also from the products of the reaction between 3-acylindole oximes with acetylene (Scheme 1.134) under mild conditions (KOH/DMSO, 40°C–70°C, acetylene atmospheric pressure).

Afterwards [313], systematic study of the reaction of ketoximes with acetylene have allowed developing the method for selective synthesis of O-vinyl oximes based on the application of two-phase catalytic system KOH/CsF/DMSO/pentane (60°C–80°C). The method enables to synthesize O-vinyl oximes in up to 90% yield (Scheme 1.135), to reduce the reaction time to minimum (~5–6 min), and to suppress pyrrolization of O-vinyl oximes and side reactions of acetylene with water leading to hardly separable vinyloxybutadienes [312], that is, to control the process.

Diaryl ketoximes are successfully vinylated with acetylene under pressure in the system KOH/DMSO (60°C–80°C, 5–7 min) to selectively afford the corresponding O-vinyldiaryl ketoximes in up to 90% yields (Scheme 1.136) [314].

Yields and physical–chemical characteristics of all O-vinyl oximes synthesized are given in Table 1.5.

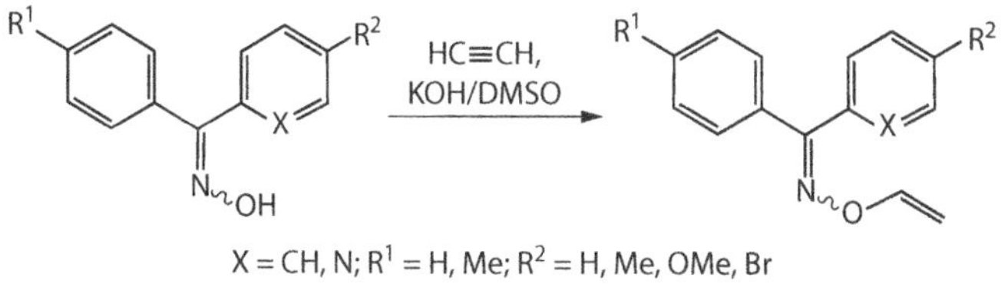
HC≡CH,
KOH/DMSO
X = CH, N; R1 = H, Me; R2 = H, Me, OMe, Br

SCHEME 1.136 Vinylation of diaryl ketoximes with acetylene in the KOH/DMSO system.

TABLE 1.5

O-Vinyl Oximes Obtained from Ketoximes

No.	Structure	Yield (%)	Bp, °C/torr (Mp, °C)	d_4^{20}	n_D^{20}	References
1	2	3	4	5	6	7
1		72	100–101	0.8625	1.4350	[313]
2		51				[313]
3		88				[313]
4		60	54–55/26		1.5542	[313]
5		10	(63)			[269]
6		51				[313]
7		41				[313]
8		52	(53–54)			[313]
9		20	109/3	1.0879	1.5774	[313]
10		90			1.6050	[314]
11		83			1.6000	[314]

(Continued)

TABLE 1.5 (*Continued*)
O-Vinyl Oximes Obtained from Ketoximes

No.	Structure	Yield (%)	Bp, °C/torr (Mp, °C)	d_4^{20}	n_D^{20}	References
1	2	3	4	5	6	7
12		82	(56)			[314]
13		87			1.6074	[314]
14		80	(84)			[314]
15		47			1.6070	[314]
16		46			1.5682	[396]
17		80			1.5806	[397]
18		21				[397]
19		11			1.5668	[271]

(*Continued*)

TABLE 1.5 (*Continued*)
O-Vinyl Oximes Obtained from Ketoximes

No.	Structure	Yield (%)	Bp, °C/torr (Mp, °C)	d_4^{20}	n_D^{20}	References
1	2	3	4	5	6	7
20		4			1.6010	[271]
21		38	45–53/1		1.5315	[313]
22		45	82/3	1.0879	1.5774	[313]
23		22			1.6025	[234]
24		55			1.6166	[248]
25		78	(86–88)			[132]
26		7				[132]
27		10				[212]

(*Continued*)

TABLE 1.5 (*Continued*)
O-Vinyl Oximes Obtained from Ketoximes

No.	Structure	Yield (%)	Bp, °C/torr (Mp, °C)	d_4^{20}	n_D^{20}	References
1	2	3	4	5	6	7
28		50	(124–126)			[270]
29		12			1.6031	[270]
30		15	(32–34)			[133]
31		12	(60–62)			[133]
32		6	(138–142)			[133]
33		44	(118–120)			[227]
34		8	(100–103)			[227]
35		8	(86–88)			[134]
36		3	(116–118)			[134]

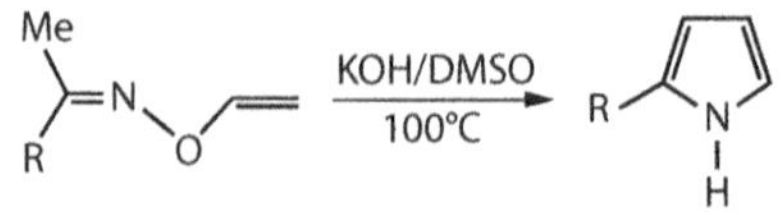

R = Ph, 4-ClC$_6$H$_4$, 2-furyl, 2-thienyl

SCHEME 1.137   Transformation of O-vinylaryl(hetaryl) ketoximes to pyrroles in the KOH/DMSO system.

It has been mentioned [311] that non-catalytic thermolysis of O-vinyl derivatives of aliphatic ketoximes does not result in pyrroles. At the same time, O-vinylaryl(hetaryl) ketoximes upon heating (~100°C, 1.5 h) in the system KOH/DMSO are transformed into the corresponding pyrroles (Scheme 1.137) [315].

Under similar conditions, O-vinyl acetoxime mainly undergoes resinification (yield of 2-methylpyrrole is ~5%). Thermolysis of pure O-vinyl acetophenone oxime gives 2-phenylpyrrole in trace amounts only [315].

Thus, superbase-catalyzed rearrangement of aromatic and heteroaromatic O-vinyl oximes can constitute one of the possible routes of the formation of pyrroles when the latter are synthesized from ketoximes and acetylene. Nevertheless, a reversibility of oxime vinylation reaction cannot be ruled out completely. In principle, regeneration of ketoxime and acetylene can lead to the formation of pyrroles through other scheme [7].

Later, it has been revealed that O-vinyl oximes are capable of rearranging to pyrroles in DMSO in the absence of alkali metal hydroxide [130,132,227,248,250]. At the moment, this approach (with preliminary isolation of O-vinyl oximes and followed by their rearrangement to pyrroles) is used as a selective method (without N-vinylpyrrole admixtures) for the synthesis of substituted pyrroles.

O-vinyl-α-(aminocarbonyl)acetamidoximes when refluxing in mesitylene for 15 min rearrange to 2-aminopyrroles and 2-pyrrolinones (Scheme 1.138) [41].

The attempt to synthesize pyrroles via prototropic isomerization of O-allyl ketoximes to O-(O-1-propenyl) ketoximes, followed by their [3,3]-sigmatropic rearrangement, failed [316,317]. Instead, in the case of O-allyl cyclohexanone oxime (KOH/DMSO or Bu^{t}OK/DMSO, 50°C–140°C, 1–6 h), a mixture of 2-methylene-1-cyclohexanol, 5,6,7,8-tetrahydroquinoline, and cyclohexanone in

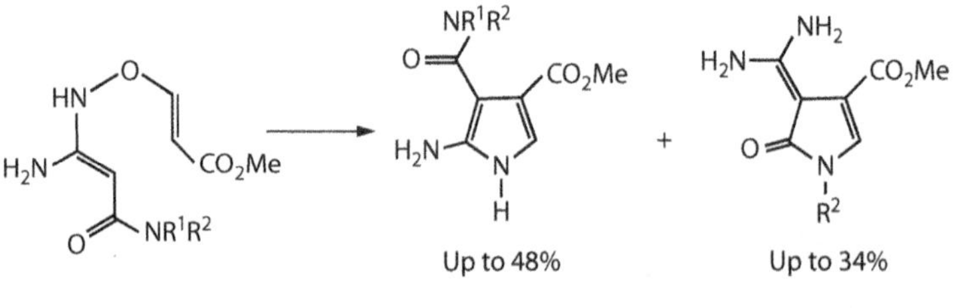

Up to 48%   Up to 34%

SCHEME 1.138   Transformation of O-vinyl-α-(aminocarbonyl)acetamidoximes to 2-aminopyrroles and 2-pyrrolinones.

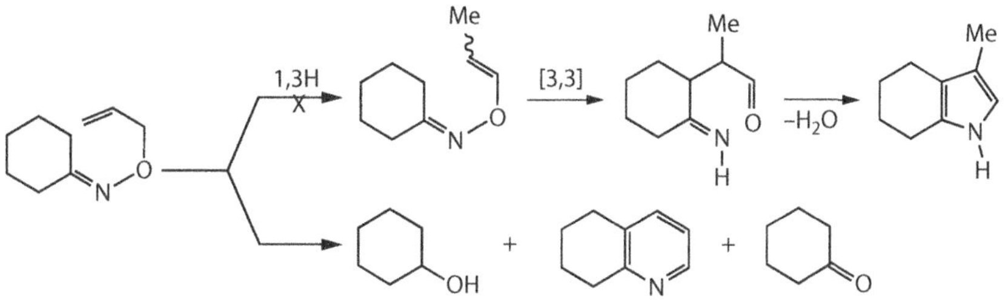

SCHEME 1.139   Transformation of O-allyl cyclohexanone oxime in superbase media.

50–70:20–30:5–10 ratio was obtained. The expected tetrahydroindole was not detected among the reaction products (Scheme 1.139).

Interestingly, the non-catalytic thermolysis of O-allyl oxime leads mainly to quinoline (when conducted on air) or isoxazolidines composed of two molecules of the starting O-allyl ketoximes (under argon) as depicted later on example of O-allyl cyclohexanone oxime (Scheme 1.140) [318].

The formation of isoxazolidines was explained by the [2,3]-sigmatropic rearrangement of the initial O-allyl ketoximes [318]. The nitrones primarily formed add as 1,3-dipoles to the second molecule of O-allyl ketoxime.

Similarly, O-allyl aldoximes upon thermolysis (180°C, 10 h, O-dichlorobenzene) gave a mixture of isomeric nitrones, the products of [2,3]-sigmatropic rearrangement, in 50%–80% yields [319]. Later, it was shown (ESR technique) that this reaction proceeds, at least partially, via the hemolytic mechanism [320]. Noteworthy, that in no case any pyrroles were discernible.

However, despite the aforementioned disappointing results, it was discovered that O-allyl ketoximes in the presence of Ir/Ag catalyst ([(cod)IrCl]$_2$, AgOTf, NaBH$_4$) readily isomerize to 1-propenyl ketoximes, which under mild heating with the molecular sieves rearranges to iminoaldehydes and further to 2,4- and 2,3,4-substituted pyrroles (Scheme 1.141) [43].

This catalytic isomerization of O-allyl ketoximes makes 1-propenyl ketoximes readily accessible that essentially extends the scope of the ketoxime-based pyrrole synthesis.

The regioselective synthesis of 2,3,4- or 2,3,5-trisubstituted pyrroles has been developed basing on [3,3]- and [1,3]-sigmatropic rearrangements of 1-propenyl oximes, respectively (Scheme 1.142) [45]. The regioselectivity is controlled either by

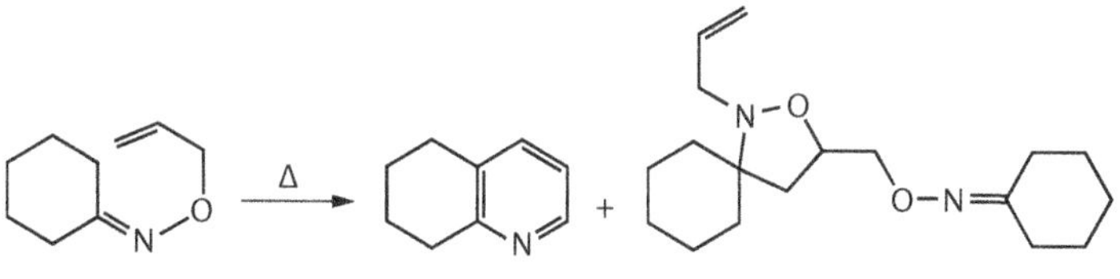

SCHEME 1.140   Thermolysis of O-allyl cyclohexanone oxime.

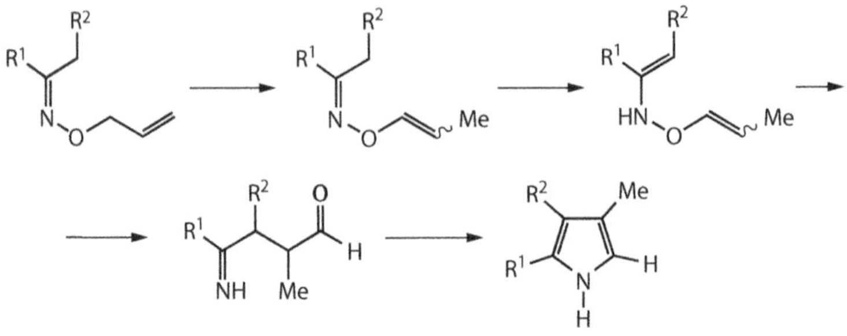

SCHEME 1.141 Rearrangements of O-allyl ketoximes in the presence of Ir/Ag catalyst and molecular sieves.

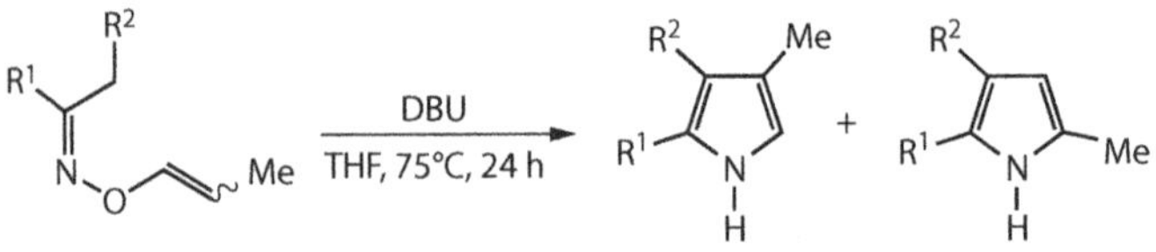

SCHEME 1.142 Rearrangements of O-1-propenyl oximes to 2,3,4- or 2,3,5-trisubstituted pyrroles in the presence of DBU.

the structure of the oxime substituent or through the addition of diazabicycloundecene (DBU).

When formation of N-O-dvinyl hydroxylamine is favored, its [3,3]-rearrangement leads to 2,3,4-trisubstituted pyrroles; while 1,3-prototropic shift to the O-1-propenyl oximes is hindered, [1,3]-sigmatropic rearrangement occurs to give 2,3,5-trisubstituted pyrrole (after cyclization of intermediate amino aldehyde) Scheme 1.143.

Recently, a multifaceted catalysis approach to substituted pyrroles from ketoximes and alkynes has been reported [46,49,51]. O-vinyl ketoximes rearrange in the presence of gold/silver catalyst (Ph₃PAuCl/AgOTf) in an efficient and regiocontrolled manner (Scheme 1.144).

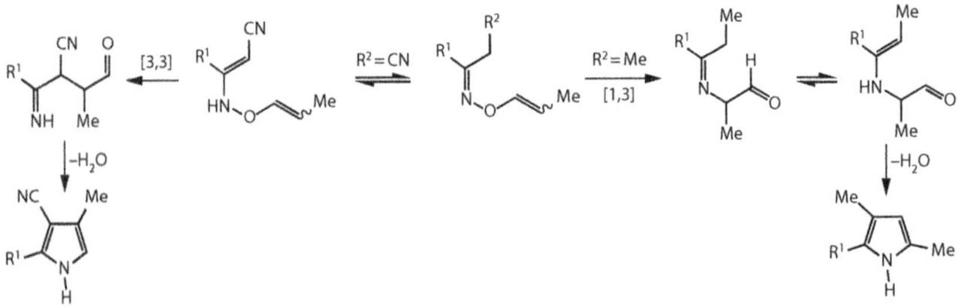

SCHEME 1.143 Alternative formation 2,3,4- and 2,3,5-trisubstituted pyrroles via rearrangements of O-1-propenyl oximes.

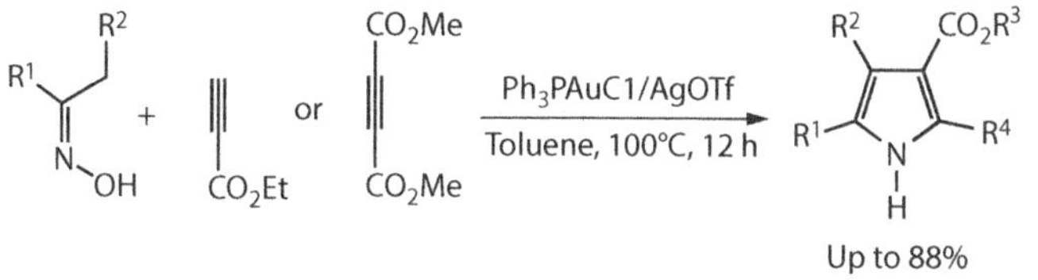

SCHEME 1.144 Synthesis of functionalized pyrroles from ketoximes and ethyl propiolate or dimethyl acetylenedicarboxylate in the presence of gold/silver catalyst.

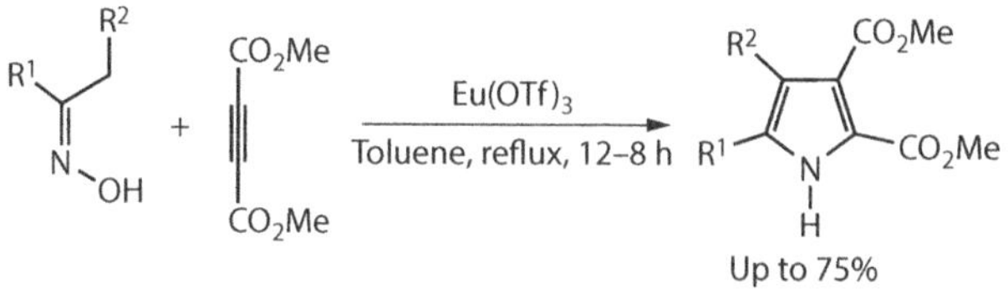

SCHEME 1.145 Eu-catalyzed synthesis of tetra-substituted pyrroles from ketoximes and dimethyl acetylenedicarboxylate.

Further on, it has been shown that the pyrroles can be synthesized directly from ketoximes and activated alkynes in this multifaceted catalysis process.

A cationic Au(I) species was found to catalyze different steps of reaction thus acting as a multifaceted catalysts [51].

The Eu-catalyzed synthesis of tetrasubstituted pyrroles from ketoximes and dimethyl acetylenedicarboxylate was also disclosed, the yields reaching 75% (Scheme 1.145) [50].

α-Tetralone oxime is added to methylpropyolate or dimethyl acetylenedicarboxylate in the Et_3N/DMSO system to yield O-vinyl oximes that rearrange on heating to dihydrobenz[g]indoles [321]. Thermal rearrangements of O-vinyl ketoximes of dihydrobenz[b]furan-4(5H)-one and 6,7-dihydroroindol-4(5h)-one leads to furo[g]- and pyrrolo[g]indoles [40].

1.5.2 FORMATION OF 4*H*-2-HYDROXY-2,3-DIHYDROPYRROLES

The most probable way of the pyrrole ring assembly from ketoximes and acetylene in the system KOH/DMSO is a heteroatomic version of Claisen rearrangement of the intermediate O-vinyl oxime, preliminarily isomerized to O,N-divinylhydroxylamine [4,7,16,18]. As shown in the previous section, O-vinyl oximes are isolable intermediates of the pyrrole synthesis capable of transformation to pyrroles under the action of KOH/DMSO and also in DMSO itself. However, other tentative intermediate stages of the rearrangement remained disputable for a long.

The first evidence that the reaction could be stopped successfully at the stage of formation of the intermediate 4*H*-2-hydroxy-2,3-dihydropyrroles has appeared in the work [322]. These exotic pyrrole derivatives are prepared from oximes of *iso*-propyl- and *iso*-butylphenylketones (Scheme 1.146).

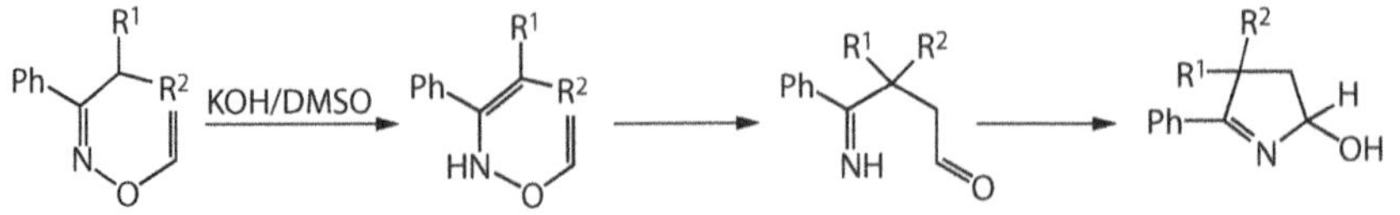

$R^1 = R^2 = Me\ (21\%);\ R^1 = H,\ R^2 = i\text{-}Pr\ (26\%)$

SCHEME 1.146 Formation of 4H-2-hydroxy-2,3-dihydropyrroles via rearrangements of O-vinyl oximes.

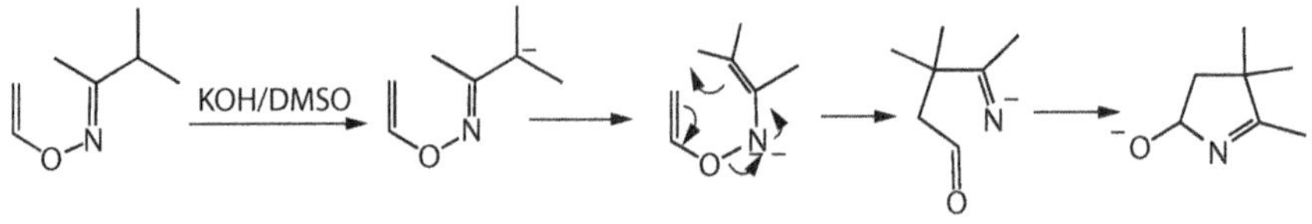

SCHEME 1.147 Anionic intermediates in rearrangements of O-vinyl oximes to 4H-2-hydroxy-2,3-dihydropyrroles.

Heating of O-vinyl oximes in the absence of the system KOH/DMSO does not lead to 4H-2-hydroxy-2,3-dihydropyrroles (GLC) [315]. Obviously, the first stage of this rearrangement (1,3-prototropic shift) can have the anionic nature (Scheme 1.147).

Two alkyl radicals in the position 4 make aromatization of the intermediates impossible and stabilize 4H-2-hydroxy-2,3-dihydropyrroles, whereas in the presence of hydrogen atom in this position, the rearrangement to the corresponding pyrrole readily occurs.

1.5.3 FORMATION OF 3H-PYRROLES

From ketoximes, containing no methylene or methyl group in α-position to the oxime function, such as *iso*-propylaryl(hetaryl) ketoximes and acetylene (70°C–90°C, 4–5 h, acetylene atmospheric pressure), one can synthesize hardly accessible 3H-pyrroles (in up to 53% yield) [323,324]. The latter are formed via 1,3-prototropic and [3,3]-sigmatropic rearrangement of O-vinyl oximes followed by cyclization of the intermediates iminoaldehyde (Scheme 1.148).

Therefore, three intermediates of the pyrrole synthesis from ketoximes and acetylene, namely, O-vinyl oximes, hydroxypyrrolines, and 3H-pyrroles, have been isolated and characterized. In certain cases, they are quite stable. Besides, very recently [45], it has been shown that iminoaldehyde is detected (NMR) in catalytic ([(cod)IrCl], AgOTf, NaBH$_4$, THF) version of the O-vinyl oxime rearrangement to pyrroles. This is another evidence in favor of the aforementioned mechanism of pyrroles formation.

Thus, the reaction of ketoximes with acetylene provides not only a facile route to pyrroles but also a shortcut to three almost unexplored classes of organic compounds, among which very reactive 3H-pyrroles, deprived of aromatic conjugation, attract particular interest. The prospects of their application in Diels–Alder reaction are especially alluring.

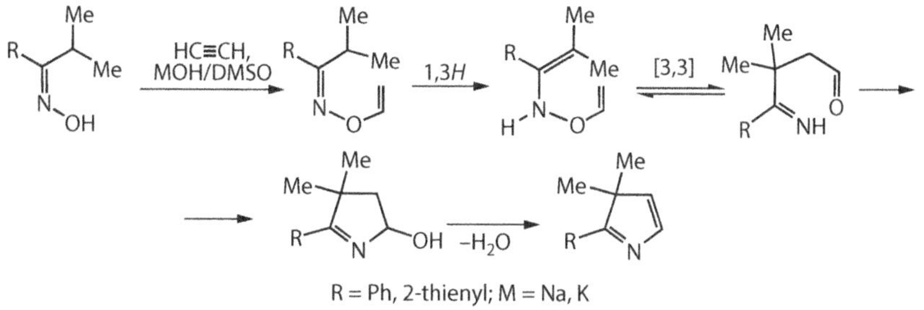

SCHEME 1.148 Synthesis of 3*H*-pyrroles from acetylene and ketoximes with just one C–H bond adjacent to the oxime function.

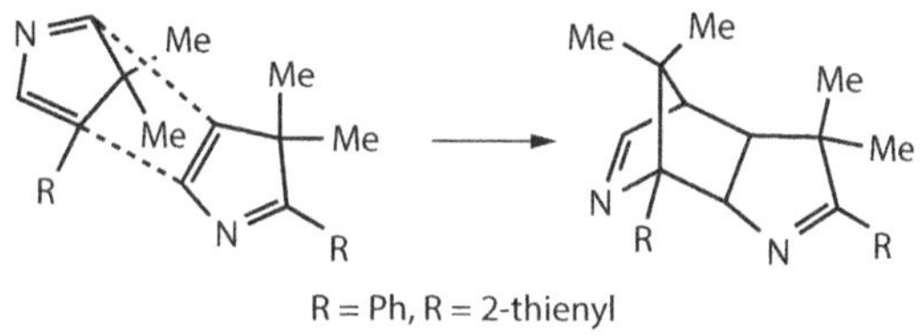

R = Ph, R = 2-thienyl

SCHEME 1.149 Auto Diels–Alder condensation of 3*H*-pyrroles.

On the example of 3,3-dimethyl-2-phenyl- and 3,3-dimethyl-2-(2-thienyl)-3*H*-pyrroles, a peculiar Diels-Alder condensation has been observed (Scheme 1.149). This is dimerization of 3*H*-pyrroles where one molecule of 3*H*-pyrrole acts as azadiene component and the second one represents dienophile [325].

This new reaction ensures a simple approach (probably diastereoselective) to hitherto unknown bridged tricyclic systems with partially hydrogenated pyrrolopyridine nucleus.

1.5.4 FORMATION OF PYRIDINES

An attempt to reproduce synthesis of O-vinyl oximes according to the protocol [326] from oximes and acetylene (generated in situ from calcium carbide) in aqueous medium has led to pyridines instead of the expected products [174]. For example, 2,4,6-trimethylpyridine in ~10% yield is prepared from acetoxime (Scheme 1.150).

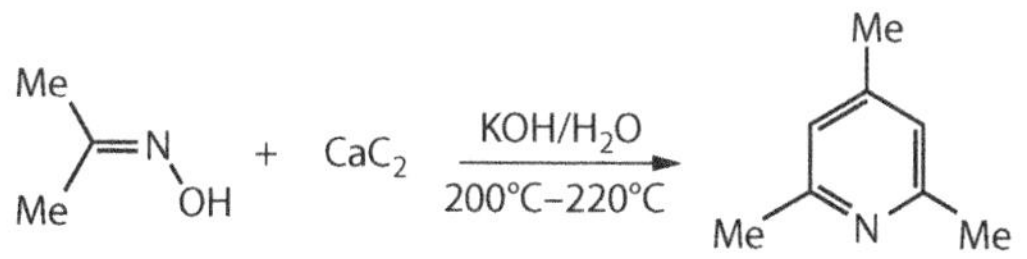

SCHEME 1.150 Formation of 2,4,6-trimethylpyridine from acetoxime and calcium carbide in aqueous KOH.

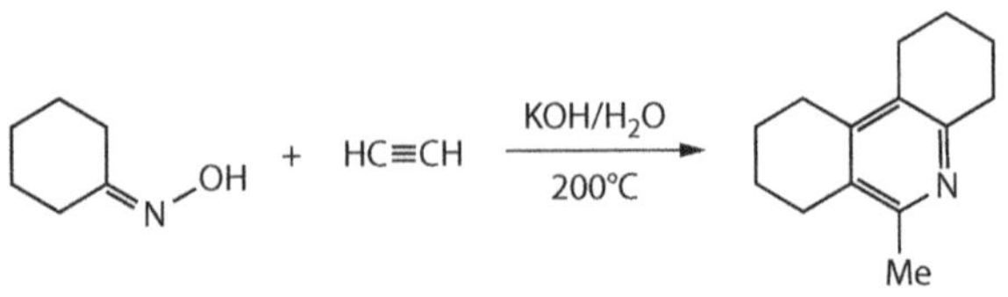

SCHEME 1.151 Formation of 6-methyl-1,2,3,4,7,8,9,10-octahydrophenanthridine from cyclohexanone oxime and acetylene in aqueous KOH.

Reaction is carried out at 200°C–220°C in autoclave for 8 h, the maximum pressure reached during the synthesis being 27 atm. When calcium carbide is used, the yields of pyridines do not exceed 10%, while with pure acetylene, their values attain 30%.

Cyclohexanone oxime in these conditions forms 6-methyl-1,2,3,4,7,8,9,10-octahydrophenanthridine (Scheme 1.151) [174].

Apparently, for ketoximes and acetylene in aqueous-alkaline medium, the following condensation processes leading to pyridines are possible:

A. Dimerization of two molecules of ketoxime with abstraction of hydroxylamine (analog of crotonic condensation) to afford oxime of α,β-unsaturated ketoxime (Scheme 1.152)
B. Hydration of acetylene to acetaldehyde (Scheme 1.153)
C. Condensation of acetaldehyde with oxime of α,β-unsaturated ketoxime (Scheme 1.154)

After optimization, this reaction could become an alternative route to the synthesis of substituted pyridines.

The reaction of acetophenone oxime with phenylacetylene in the LiOH/DMSO system (1:1 molar ratio, 130°C, 5 h) delivers 2,4,6-triphenylpyridine (2.5% yield), along with the expected 2,5-diphenylpyrrole and acetophenone as by-product (Scheme 1.155) [327].

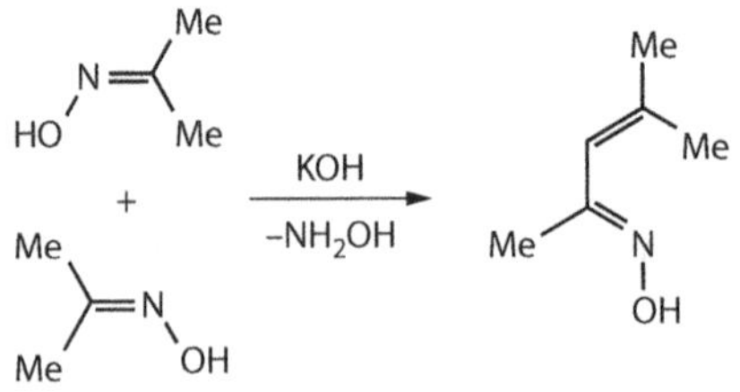

SCHEME 1.152 Possible formation of α,β-unsaturated ketoxime from acetoxime.

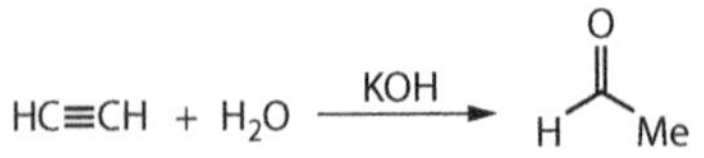

SCHEME 1.153 KOH-catalyzed hydration of acetylene to acetaldehyde.

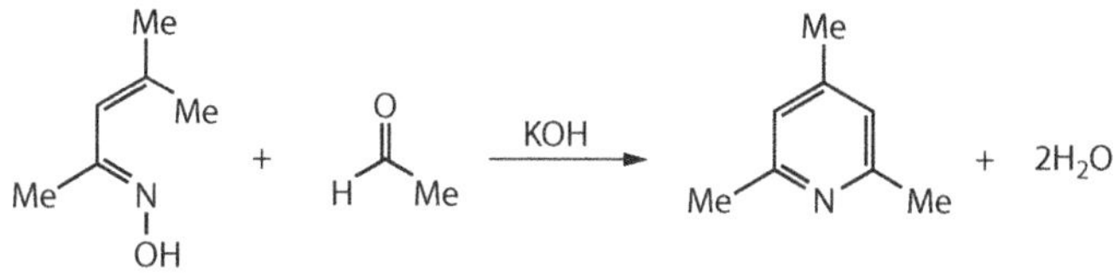

SCHEME 1.154 Plausible condensation of acetaldehyde with α,β-unsaturated ketoxime.

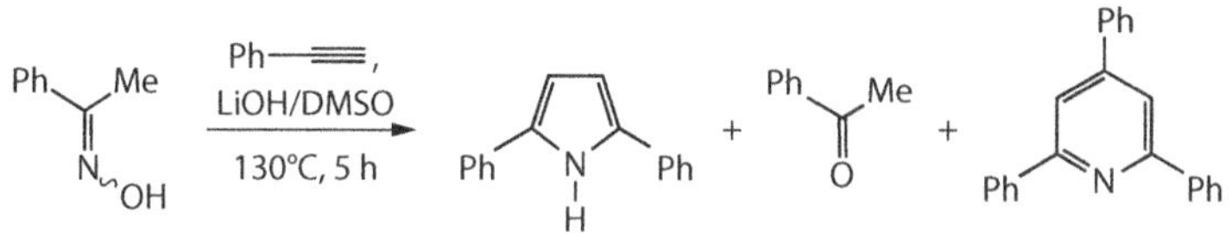

SCHEME 1.155 Side formation of 2,4,6-triphenylpyridine in the synthesis of 2,5-diphenyl-pyrrole from acetophenone oxime and phenylacetylene in the system LiOH/DMSO.

Apparently, the intermediate O-vinyl oxime rearranges to nitrone **A** [328], which is further fused with acetophenone to generate triene nitrone **B** (Scheme 1.156). Electrocyclization of the latter gives dihydropyridine-N-oxide **C**, which rearranges in N-hydroxydihydropyridine **D**. 1,4-Abstraction of methanol from pyridine **D** leads to 2,4,6-triphenylpyridine via the intermediate 1,4-biradical **E** (analog of 1,4-benzoid Bergman biradical [329,330]).

The closing of the pyridinic cycle is observed in the reaction of dimethylglyoxime with acetylene (Scheme 1.157) [269]. Among the expected reaction products such as O-vinyl oxime, pyrrole, and dipyrrole, N-vinyl-2-[2′-(6′-methylpyridyl)]pyrrole has also been isolated from the reaction mixture, prepared under usual conditions (KOH/DMSO, 100°C–140°C).

The content of pyridylpyrrole in a mixture of products depends on the reaction conditions, in the best cases reaching 36%. It is assumed [269] that O-vinyl oximes

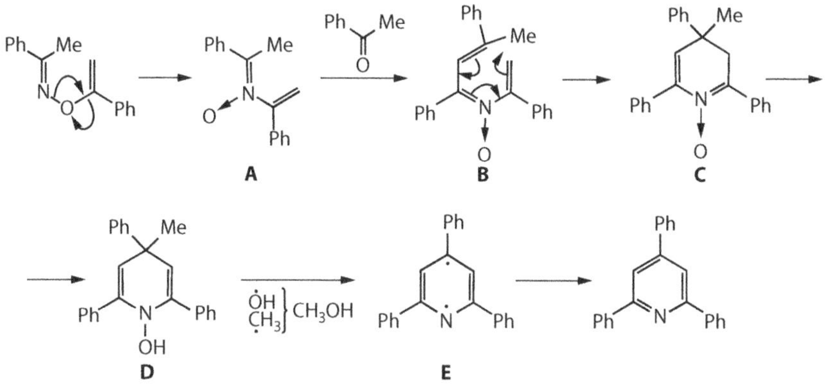

SCHEME 1.156 Plausible route to 2,4,6-triphenylpyridine from O-vinylacetophenone oxime.

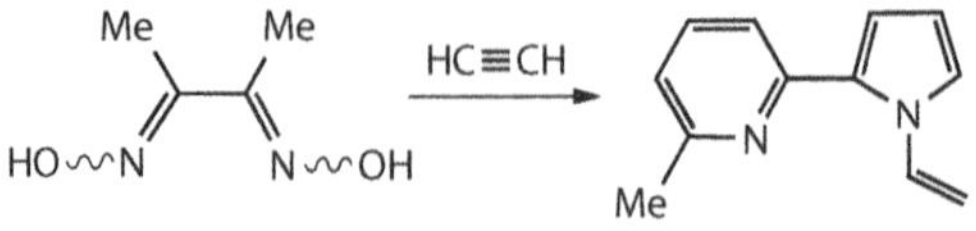

SCHEME 1.157 Formation of N-vinyl-2-[2′-(6′-methylpyridyl)]pyrrole from dimethylgly-oxime and acetylene.

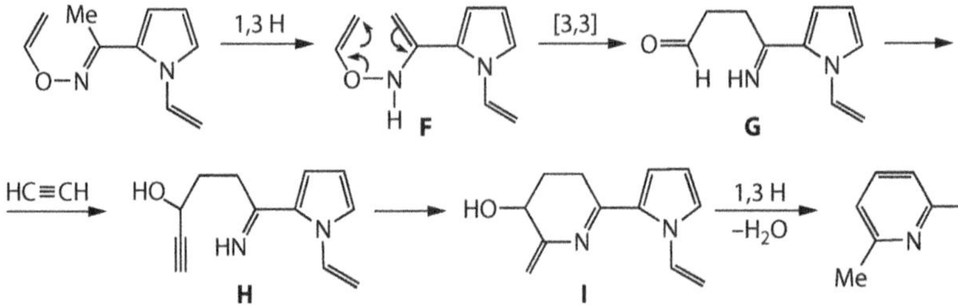

SCHEME 1.158 Transformations of intermediate O-vinyl oxime formed in the reaction of dimethylglyoxime with acetylene.

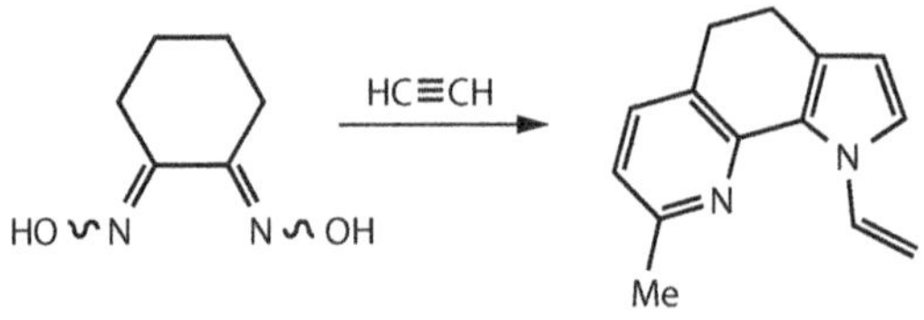

SCHEME 1.159 Side formation of pyrrolequinoline from 1,2-cyclohexanedione dioxime with acetylene.

undergo 1,3-prototropic shift under the action of superbase to afford vinylhydroxyl-amine **F** (Scheme 1.158). The latter rearranges to iminoaldehyde **G** ([3,3]-sigmatropic rearrangement) and intercepted by acetylene to form acetylenic alcohol **H** (Favorsky reaction). The subsequent cyclization to hydroxymethylenetetrahydropyridine **I** and its aromatization deliver pyridylpyrrole.

   The interaction of 1,2-cyclohexanedione dioxime with acetylene under the same conditions gives a mixture of products, from which pyrrolequinoline is isolated as side-product (2% yield) (Scheme 1.159) [270].

   The formation of the pyridine ring, observed in this cases, implies the interception of iminoaldehyde (intermediate **G**) by phenylacetylene and represents an additional experimental support of the ketoxime–acetylene mechanism of pyrrole synthesis that is presently accepted. On the other hand, this new direction of the reaction of ketoximes with acetylene, despite moderate yields of pyridylpyrroles, can have preparative value as straightforward one-pot synthesis of the unnatural nicotine-like alkaloid from simple available starting reactants (1,2-dioximes and acetylene).

1.5.5 FORMATION OF ACETYLENIC ALCOHOLS

It has been noted that the synthesis of pyrroles from ketoximes and acetylene in some case leads to tertiary α-acetylenic alcohols [4,5,7]. It indicates a possibility of oxime involvement into the Favorsky alkynol synthesis. The conditions of acetylenic alcohols formation from ketoximes and acetylene have been investigated on the example of the reaction between cyclohexanone oxime and acetylene (Scheme 1.160) [175].

The reaction is carried out in dioxane, hexametapol, and sulfolane (100°C–140°C, the initial acetylene pressure 12 atm). Catalyst and solvent have the most essential impact on the yield of carbynol. The efficiency of the studied catalysts drops in the following order: KOH ≥ RbOH > [(C$_4$H$_9$)$_4$N$^+$]$^-$OH > NaOH. LiOH and RbCl do not catalyze the reaction. Aqueous DMSO is the easiest medium to prepare acetylenic alcohol; dioxane takes the second place in terms of efficiency. The addition of DMSO to dioxane does not almost improve the carbynol yield. In hexametapol and sulfolane, 1-ethynyl-1-cyclohexanol is formed in trace amounts only.

The formation of ammonium is always observed in the course of the reaction. Though it is a common knowledge [331] that some ketoximes, upon heating with metal hydroxides, partially regenerate ketones and release ammonium, nevertheless, cyclohexanone likely is not an intermediate product of this reaction. This follows from a special experiment with cyclohexanone, acetylene and ammonium, the reaction of which results in a complex mixture of products containing no 1-ethynyl-1-cyclohexanol [175].

Almost entire inertness of cyclohexanone oxime in the experiments with NaOH and tetrabutylammonium hydroxide testify against the intermediate formation of cyclohexanone.

In optimum (or close to them) conditions of pyrroles and N-vinylpyrroles synthesis from ketoximes and acetylene in the system KOH/DMSO, acetylenic alcohols are not almost formed.

1.5.6 SIDE PRODUCTS FORMED IN TRACE AMOUNTS

While performing the synthesis of 2,5-diphenylpyrrole from acetophenone oxime and acetylene in the system NaOH/DMSO (140°C, 4 h), p-terphenyl has been isolated from the reaction mixture in trace amounts (1.5%) (Scheme 1.161) [332].

p-Terphenyl is not formed from acetophenone oxime or phenylacetylene taken separately under the same conditions. At the same time, it is isolated (1.7%) from the reaction of acetophenone with NaOH in DMSO (140°C, 4 h).

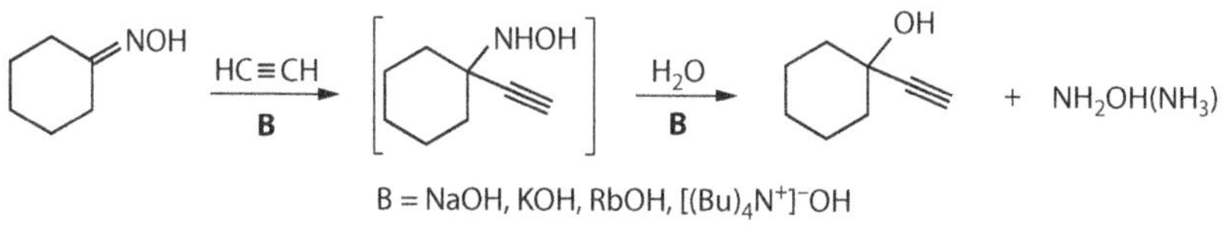

B = NaOH, KOH, RbOH, [(Bu)$_4$N$^+$]$^-$OH

SCHEME 1.160 Formation of 1-ethynyl-1-cyclohexanol in the reaction of cyclohexanone oxime and acetylene in superbase media.

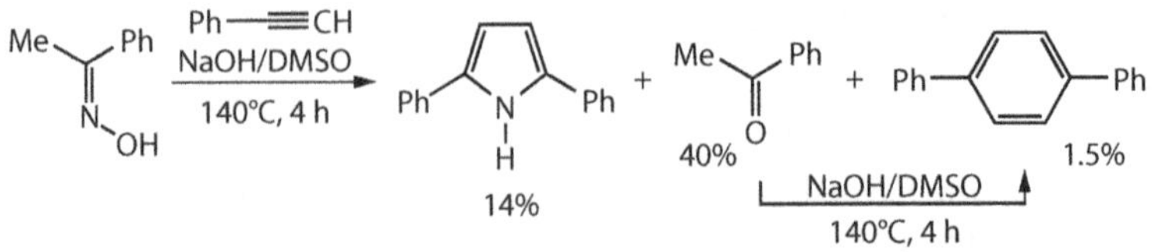

SCHEME 1.161 Trace formation of *p*-terphenyl in the reaction of acetophenone oxime with phenylacetylene in the system NaOH/DMSO.

Evidently, acetophenone in the system NaOH/DMSO undergoes hydroxymethylation [333–339] to produce substituted allyl alcohol **A** (Scheme 1.162), which further prototropically isomerizes to aldehyde **B**. Dimerization of aldehyde **B** via aldol–crotone mechanism leads to *p*-terphenyl.

Earlier, the adduct of DMSO with acetophenone **C** has been isolated and characterized as hydroxy derivative (after neutralization of the reaction mixture) [340]. In more active superbase system KOH/DMSO (140°C, 4 h), the yield of *p*-terphenyl increases to 14.6%.

Among the products of 2,5-diphenylpyrrole synthesis from acetophenone oxime and phenylacetylene in the presence of less basic (therefore, less prone to generating the dimsyl anions in sufficient concentration) system LiOH/DMSO, *p*-terphenyl is not found. In this case (130° C, 5 h), along with 2,5-diphenylpyrrole and acetophenone, 1,3,5-triphenylbenzene is isolated (6% yield) (Scheme 1.163) [341].

The assumption that 1,3,5-triphenylbenzene is formed via trimerization of phenylacetylene has not been confirmed. Also, it cannot be formed as a result of autocondensation of oxime or ketone. Hence it follows that 1,3,5-triphenylbenzene is a minor product of the synthesis of pyrrole from acetophenone oxime and aphenylcetylene.

Presumably [341], O-vinyl oxime undergoes rearrangement into nitrone **D** (as reported in [328] for cyclohexanone oxime O-allyl ether) (Scheme 1.164). The

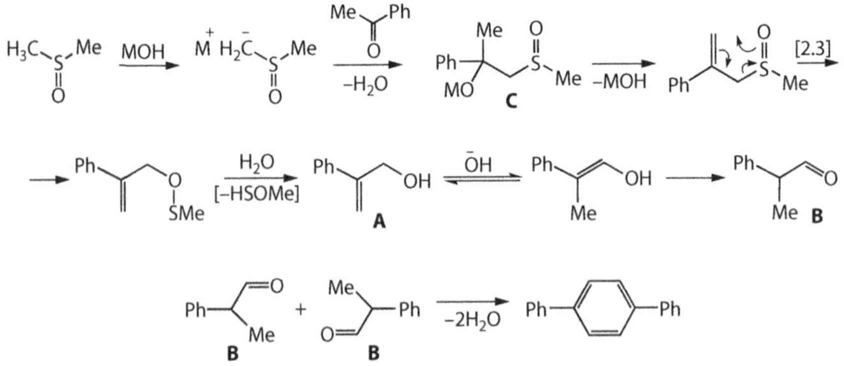

SCHEME 1.162 Plausible scheme of *p*-terphenyl formation from acetophenone in the MOH/DMSO systems.

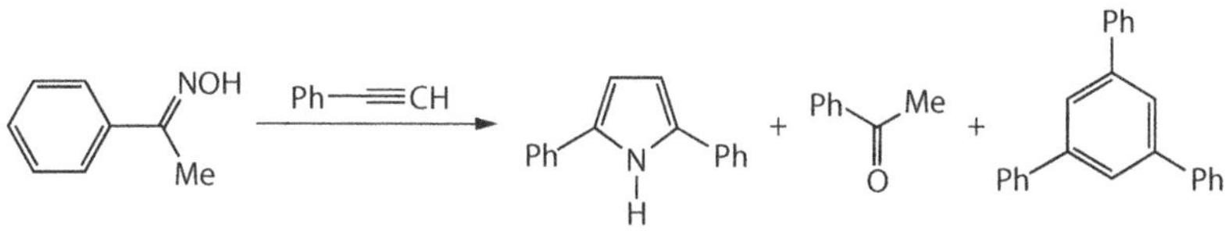

SCHEME 1.163 Trace formation of 1,3,5-triphenylbenzene in the reaction of acetophenone oxime with phenylacetylene in the LiOH/DMSO system.

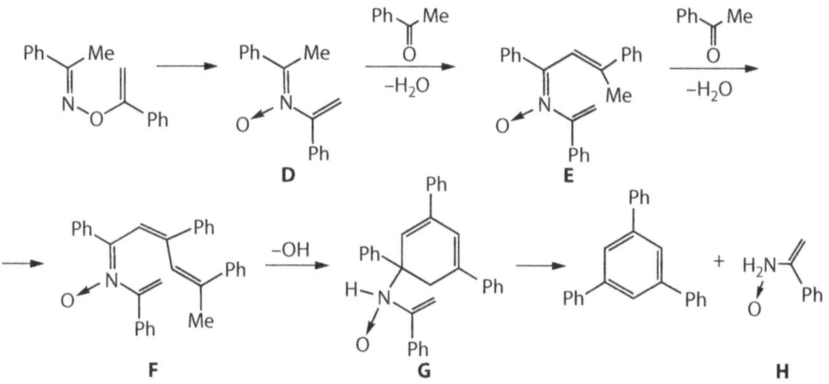

SCHEME 1.164 Plausible scheme of 1,3,5-triphenylbenzene formation via O-vinylacetophenone oxime in the LiOH/DMSO systems.

latter due to enhanced CH acidity of the methyl group (owing to strong electron-withdrawing effect of the nitrone function) is condensed with acetophenone. The azatriene **E** thus formed is condensed with the second acetophenone molecule to give azatetraene **F**, and cyclization of the latter, followed by aromatization of dihydrobenzene **G** via elimination of vinylnitrone **H** yields 1,3,5-triphenylbenzene.

The considered reactions enrich the ideas about the processes accompanying synthesis of pyrroles from ketoximes and acetylenes and, when optimized, can gain preparative value as simple approaches to *p*-terphenyls and 1,3,5-triarylbenzenes.

In the preparation of 2,3-diphenyl-N-vinlpyrrole (17%) from benzyl phenyl ketoxime and acetylene in the system KOH/DMSO at atmospheric pressure, the expected 2,3-diphenyl-N-vinylpyrrole along with N-benzylbenzamide (12%), benzoic acid (55%), and 2,3,6-triphenylpyridine (1.5%) has been identified among the reaction products (Scheme 1.165) [342].

The most unexpected by-product of the reaction in this case is 2,3,6-triphenylpyridine. Probably, benzyl phenyl ketone (the usual product of partial deoximation of oximes in the conditions for synthesis of pyrroles) condenses with acetaldehyde (product of acetylene hydration in superbase media [105], Equation 1, Scheme 1.166) to form an equilibrium mixture of ethylenic ketones **I** and **J** (Equation 2). The ketone **J** reacts with benzylamine to close 2,3,6-triphenyltetrahydropyridine cycle **K**. The latter aromatizes to the pyridine (Equation 3).

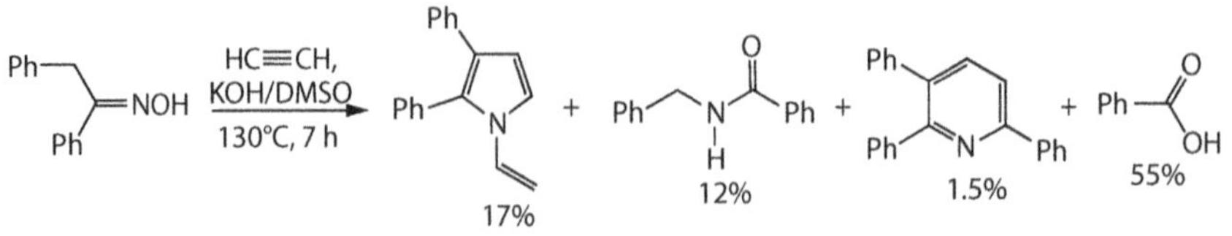

SCHEME 1.165 Side products formed in the synthesis of 2,3-diphenyl-N-vinylpyrrole from benzyl phenyl ketoxime and acetylene in the KOH/DMSO system.

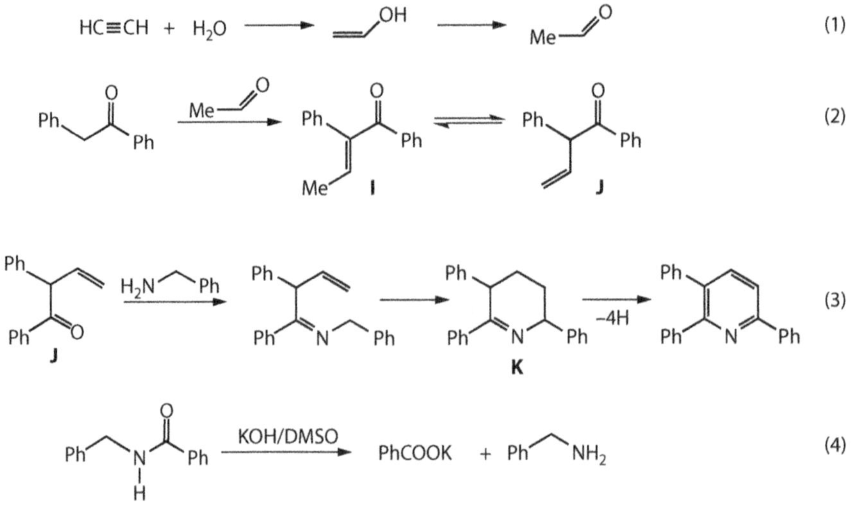

SCHEME 1.166 Plausible side reactions in the synthesis of 2,3-diphenyl-N-vinylpyrrole.

Benzylamine is the anticipated product of alkaline hydrolysis of N-benzylbenzamide, which is confirmed by the presence of benzoic acid in the reaction mixture (Equation 4). This scheme is proved by the separation of 2,3,6-triphenylpyridine from the reaction products of benzyl phenyl ketone with acetylene and benzylamine (130°C, 7 h).

A special experiment shows [343] that N-benzylbenzamide is not a product of oxime rearrangement (upon heating oxime under the same conditions without acetylene, it is nearly completely recovered). Obviously, the key role in the formation of amide from benzyl phenyl ketone belongs to acetylene.

Apparently, O-vinyl oxime is deprotonated under the action of the superbase forming carbanion **L** that further substitutes intramolecularly the vinyloxy anion to close the azirine cycle ring **M** (Scheme 1.167). Its hydration into hydroxyaziridine **N** followed by a rearrangement results in N-benzylbenzamide.

In essence, the stage of azirine formation resembles the azirine synthesis from oximes or their ethers by the Hoch–Campbell reaction [344–348].

2,3-Diphenyl-2*H*-azirine on heating is known to form 2,4,5-triphenylimidazole, along with other products [349]. The careful examination of the reaction mixture

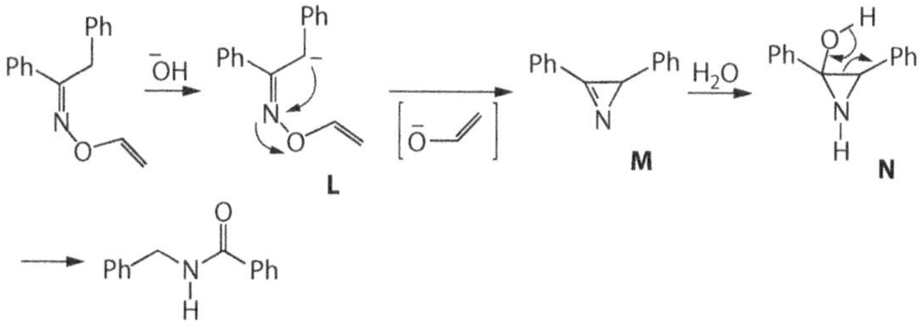

SCHEME 1.167 Plausible scheme of N-benzylbenzamide formation via O-vinyl benzyl phenyl oxime.

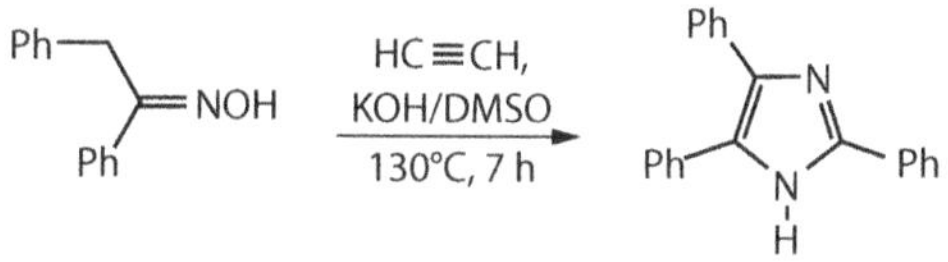

SCHEME 1.168 Trace formation of 2,4,5-triphenylimidazole from benzyl phenyl oxime and acetylene in the KOH/DMSO system.

obtained in the synthesis of pyrrole 2,3-diphenylpyrrole from benzyl phenyl oxime and acetylene [350] has allowed one to isolate 2,4,5-triphenylimidazole in trace amounts (1%) (Scheme 1.168). This finding proves the previously postulated formation of azirine **M** in the superbase system KOH/DMSO.

The ease of azirine **M** formation in this case seems to result from the enhanced CH acidity of the methylene group of benzyl phenyl oxime located between the oxime function and the benzene ring: stable carbanion **L** is formed, which simultaneously belongs to benzylic and α-carbonyl type.

Probably, the systematic elaboration of the reaction of ketoximes with acetylene in the superbase systems will open a fundamentally new route to 2H-azirines, which, if necessary, can be used as intermediates without preliminary isolation.

The reaction of oximes and acetylene [248,351] in the system KOH/DMSO or potassium oximate/DMSO affords the target pyrroles and N-vinylpyrroles together with 1-Z-[(2-methylthio)vinyl]pyrroles in trace amounts (0.05%–1.0%) (Scheme 1.169).

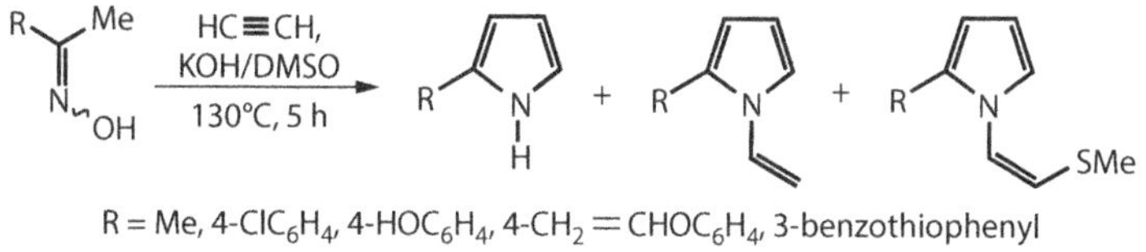

R = Me, 4-ClC$_6$H$_4$, 4-HOC$_6$H$_4$, 4-CH$_2$=CHOC$_6$H$_4$, 3-benzothiophenyl

SCHEME 1.169 Trace formation of 1-Z-[(2-methylthio)vinyl]pyrroles in the synthesis of pyrroles and N-vinylpyrroles in the system KOH/DMSO.

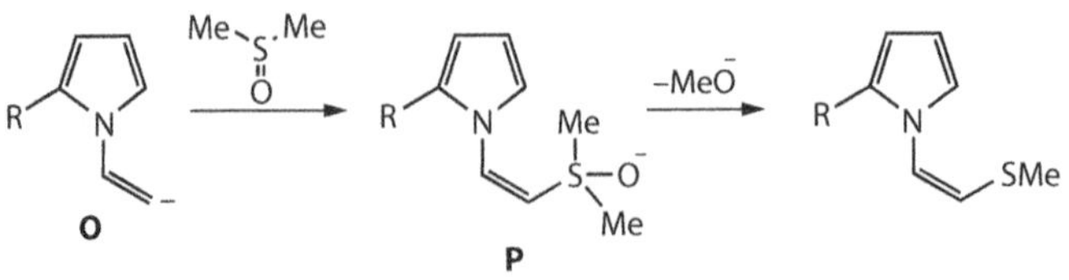

SCHEME 1.170 Plausible scheme of methylthiovinylpyrrole formation from N-vinylpyrroles and DMSO.

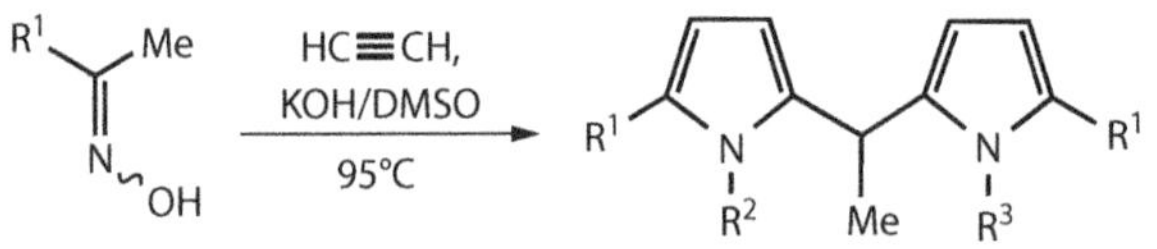

R^1 = Me, R^2 = R^3 = H (0.09%); R^1 = cyclo-C_3H_5, R^2 = R^3 = H (8.0%); R^1 = Me, R^2 = H, R^3 = CH=CH$_2$ (0.08%)

SCHEME 1.171 Trace formation of 1,1-(di-2-pyrrolyl)ethanes from ketoximes and acetylene.

This result can point to the interaction of intermediate vinyl carbanion **O** with DMSO leading to anion **P**, which further is transformed to methylthiovinylpyrrole (Scheme 1.170).

Thus, the stereoselective methylthiovinylation of pyrroles, formed from ketoximes and acetylene in the KOH/DMSO system, represents a rare example of interaction of the vinyl carbanion with sulfoxide function. Despite the negligible yield of methylthiovinylpyrroles, the fact of their formation sheds additional light on peculiarities of the reaction of ketoximes with acetylene in the systems MOH/DMSO.

In the synthesis of pyrroles from acetoxime or methyl cyclopropyl ketoximes and acetylene (KOH/DMSO, 95°C) leading to common products, pyrroles and N-vinylpyrroles, the minor reaction is realized to furnish 1,1-(di-2-pyrrolyl)ethanes (0.08%–8.0% yields) (Scheme 1.171) [352].

The formation of dipyrrolylethanes also takes place when preformed potassium oximate is employed instead of ketoximes and potassium hydroxide.

Probably, acetaldehyde, forming due to acetylene hydration, is condensed with pyrrole (though such condensation in the basic medium is unusual).

The product mixture obtained by the reaction of acetophenone oxime with acetylene (KOH/DMSO, 100°C–110°C) contains traces of 5-ethyl-2-phenylpyrrole (0.03%) along with the major products, 2-phenyl- and 2-phenyl-N-vinylpyrroles (Scheme 1.172) [353].

In the presence of aluminum oxide, the yield of abnormal pyrrole increases to 1.6%.

It is supposed [353] that the observed ethylation of the pyrrole ring results from reductive decomposition of one of the minor by-products, 1,1-di(2-phenyl-5-pyrrolyl) ethane (see earlier).

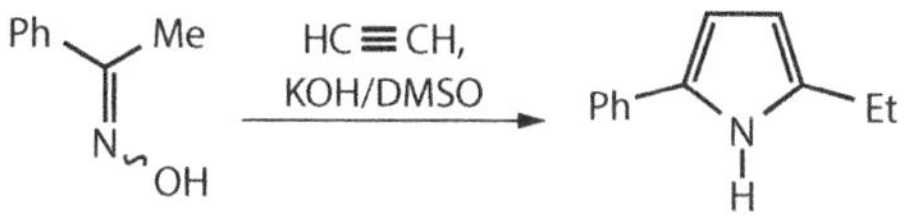

SCHEME 1.172 Trace formation of 5-ethyl-2-phenylpyrrole from acetophenone oxime and acetylene.

1.6 δ-CARBOLINES FROM 3-ACYLINDOLES AND ACETYLENE

The reaction of 3-acetylindole oxime and acetylene (KOH/DMSO, 120°C, 1 h, acetylene pressure 12–14 atm) affords mainly 2-methyl-5-vinyl-pyrido[3,2-*b*]indole (4-methyl-N-vinyl-δ-carboline, 40% yield) along with the expected N-vinyl-3-(N'-vinyl-2'-pyrrolyl)indole, the latter being a minor product in this case (only 6% yield) (Scheme 1.173) [230].

Interestingly that from N-alkyl-substituted oximes of 3-acylindoles, carbolines are not apparently formed in the similar conditions: only the expected O-vinyl oximes and 3-pyrrolylindoles are isolated from the reaction mixture [229].

Now, the mechanism of this promising reaction remains obscure. One of the plausible schemes of carboline formation [230] may involve homolysis of the N–O bond in O-vinyl oxime (Scheme 1.174) to give N-centered radical **A** that intramolecularly attacks the indole ring affording azirine radical **B**. The latter rearranges into the C-centered radical **C** that is added to acetylene, and the azadiene radical thus formed finalizes the assembly of δ-carboline nucleus.

The primary radical **A** seems to be less stable than its isomer **C** and can also be intercepted by acetylene to furnish γ-carboline (Scheme 1.175).

Obviously, this reaction (formation of γ-carboline) proceeds in insignificant degree. Though the expected doubling of signals in the ¹H NMR spectra of δ-carboline, which could be assigned to impurities of isomeric γ-carboline, is really observed, the intensity of these signals is low.

Certainly, the aforementioned mechanism of δ-carboline formation needs experimental support. For example, radical nature of the reaction assumes a possibility of its inhibition in the presence of typical inhibitors of radical processes. In such a case, intermediate radicals could be detected by ESR techniques. Also, the rearrangement of O-vinyl oxime to δ-carboline without formation of radicals and ions must not be ruled out.

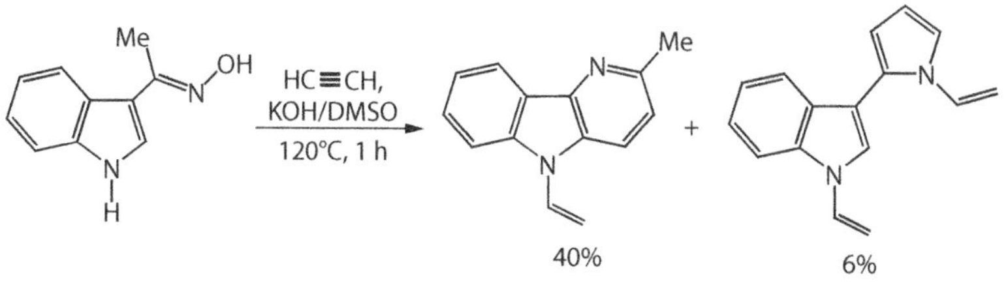

40%   6%

SCHEME 1.173 δ-Carboline from 3-acetylindole oxime and acetylene.

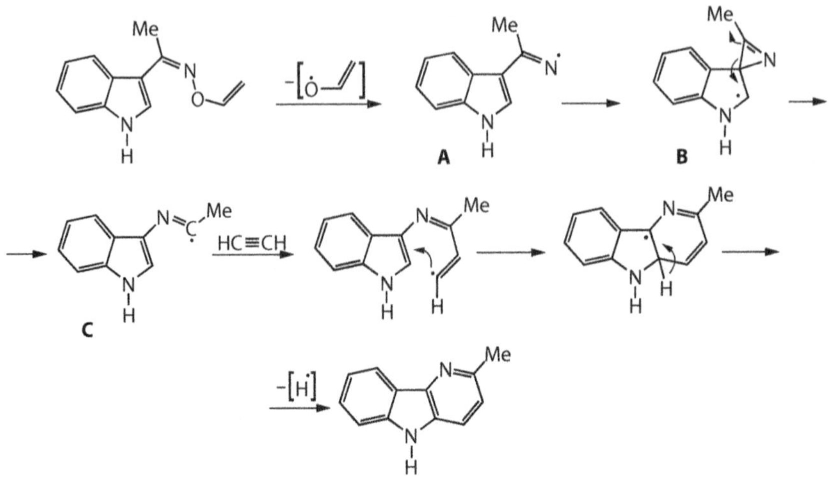

SCHEME 1.174 Plausible radical transformations of 3-acetylindole O-vinyl oxime to δ-carboline.

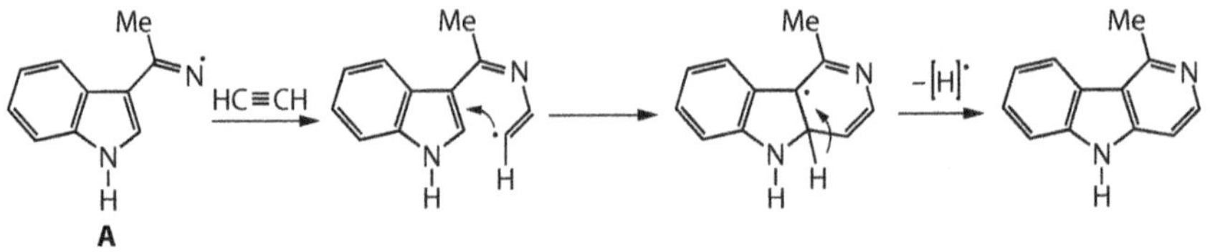

SCHEME 1.175 Expected formation of γ-carboline via primary radical **A**.

Currently, Scheme 1.176 is supposed to be more probable.

O-vinyl oxime rearranges in azirinoindole aldehyde **D**, which in its zwitterion form **E** is transformed further into δ-hydroxydihydrocarboline **F**. The latter aromatizes due to dehydration to deliver 4-methyl-δ-carboline.

Vinyl derivative of δ-carboline can be considered as N-protected δ-carboline since the N-vinyl group is easily deprotected to liberate NH-function. Besides, possible addition reactions to the N-vinyl moiety offer wide possibilities for the synthesis of novel N-substituted δ-carboline ensembles.

δ-Carbolines are abundant in nature (alkaloids cryptoquinoline [354] and cryptolepine [355]) possessing antitumor [356], antimalarial [357], antimicrobic [358], and antifungal activities [358]. Known methods for the synthesis of δ-carbolines are multistage, require hardly accessible chemicals, and give the target products, mainly functional representatives of carbolines, in low or moderate yields. Simple alkyl-substituted δ-carbolines, much less poorly studied, are synthesized from 1-acetylindolyl-3-ones [359] or functionalized pyridines [360].

The first example of one-pot assembly of 4-methyl-N-vinyl-δ-carboline from available 3-acetylindole and acetylene in the system KOH/DMSO assumes possible appearance of the new general reaction leading to δ-carboline scaffold.

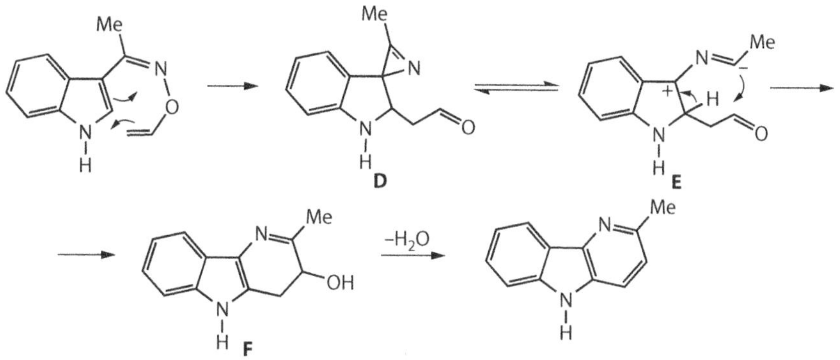

SCHEME 1.176 Alternative zwitterionic transformation of 3-acetylindole O-vinyl oxime to δ-carboline.

1.7 REACTION OF KETOXIMES WITH ACETYLENE IN THE PRESENCE OF KETONES: ONE-POT ASSEMBLY OF 4-METHYLENE-3-OXA-1-AZABICYCLO[3.1.0]HEXANES

The reaction of alkyl aryl(hetaryl) ketoximes with acetylene in the system KOH/ DMSO (80°C, 5–60 min), when conducted in the presence of a third component, aliphatic ketone, furnishes, along with the anticipated products (O-vinyl oximes, pyrroles, and N-vinylpyrroles), the unexpected complex bicyclic systems, 4-methylene-3-oxa-1-azabicyclo[3.1.0]hexanes (Scheme 1.177) [361]. In the superbase system LiOH/CsF/DMSO, the reaction proceeds selectively, that is, bicyclohexanes are formed in up to 75% yield without O-vinyl oximes or pyrroles.

Like the pyrrole synthesis, the assembly of 4-methylene-3-oxa-1-azabicyclo[3.1.0]hexanes is likely triggered by the formation of O-vinyl oxime (detected by GLC and NMR). The deprotonation of O-vinyl oxime in α-position relative to the oxime function and the further intermolecular nucleophilic substitution of the vinyloxy group can lead to azirine **A** (Scheme 1.178). The latter reacts with acetylene (in the form of carbanion) to give acetylenic ethynyl aziridine **B** (the nitrogen analog of the Favorsky reaction), which is added to the third

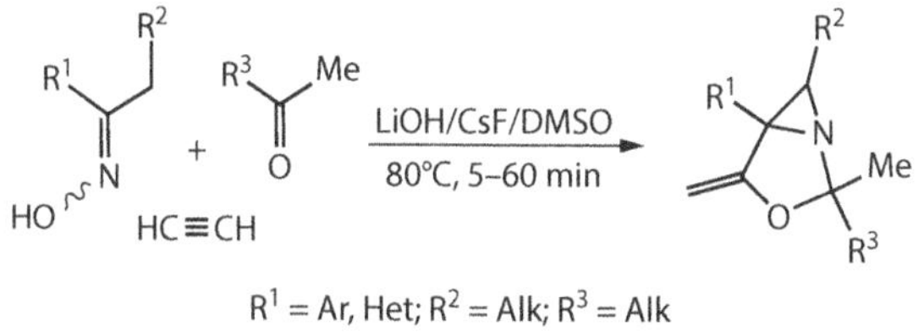

$R^1 = Ar, Het; R^2 = Alk; R^3 = Alk$

SCHEME 1.177 Assembly of 4-methylene-3-oxa-1-azabicyclo[3.1.0]hexanes from alkylaryl(hetaryl) ketoximes, ketones, and acetylene in LiOH/CsF/DMSO system.

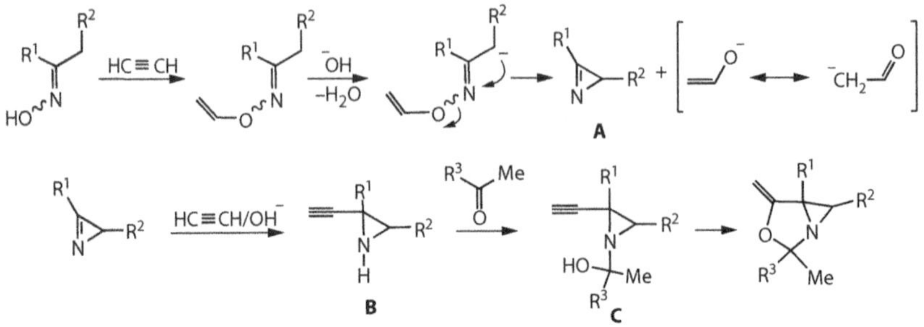

SCHEME 1.178 Probable mechanism of 4-methylene-3-oxa-1-azabicyclo[3.1.0]hexanes assembly from alkylaryl(hetaryl) ketoximes, ketones and acetylene.

component, aliphatic ketone. Hemiaminal **C** is intramolecularly vinylated to close the bicycle with exocyclic methylene group.

In support of this mechanism is the fact that dialkyl ketoximes practically do not give 4-methylene-3-oxa-1-azabicyclo[3.1.0]hexanes under the studied conditions. For example, in the case of acetoxime, only negligible amounts (7%) of the corresponding bicyclohexane are detected in the reaction mixture (^{1}H NMR). This corresponds to the weaker CH acidity of the dialkyl ketoxime and smaller positive charge at the oxime nitrogen atom owing to the electron-donating effect of both alkyl substituents.

3-Oxa-1-azabicyclo[3.1.0]hexanes are almost unexplored. To the best of our knowledge, only two publications deal with such structures that are ascribed to dimers of aziridine aldehydes [362,363]. However, similar compounds, namely, 3-oxa-1-azabicyclo[3.1.0]hexan-2-ones, are structural congeners of many natural and synthetic molecules possessing significant biological activity [364]. Among them are immunomodulators [365], antimicrobial agents [366], drugs against glaucoma [367], and Alzheimer's disease [368].

Actually, 4-methylene-3-oxa-1-azabicyclo[3.1.0]hexanes represent new heterocyclic systems, in which exocyclic double bond is conjugated with both oxygen atom and the aziridine cycle. The combination of three pharmacologically and synthetically important fragments (aziridine, 1,3-oxazolidine, and vinyl ether) in a one molecule can impart essentially new, nontypical for each functionality separately, properties to these systems.

1.8   TRANSFORMATIONS OF ALDOXIMES IN THE SYSTEMS MOH/DMSO AND MOH/DMSO/ACETYLENE

It is a common knowledge that nitriles are synthesized in high yields from aldoximes under mild conditions in the presence of acids or their derivatives, diorganylphosphites [369]; chlorothionoformates [370]; phosphonitrilic chloride [371]; diphosphorus tetraiodide [372]; ortho esters with $MeSO_3H$, SO_2, or HCl [373]; sulfur, selenium, and sulfur chlorides [374]; a trimethylamine/SO_2 complex [375]; selenium dioxide [376]; benzene sulfochloride [377]; trifluoroacetic anhydride in pyridine

[378]; and trifluoromethanesulfonic acid anhydride with triethylamine [379]. Some aldoximes can be dehydrated into nitriles almost quantitatively upon long boiling in hexametapol (220°C–240°C) without the application of acidic reactants [380]. Dehydration of aldoximes in the presence of bases is underdeveloped in terms of preparative value. It has been noted [331] that the heating of oximes with metal oxides and hydroxides is accompanied by the ammonia release and the carbonyl compound is partially regenerated. Sometimes, nitriles are managed to be synthesized by the action of potassium amide in liquid ammonia on O-alkyl ethers of oximes [331]. Formation of nitrile is observed when organomagnesium compounds react with camphor oxime [381]. Other methods of aldoximes transformation into nitriles have also been developed [33,382–386].

In an attempt to extent the reaction of ketoximes with acetylene affording pyrroles over aldoximes, it has been revealed [4,7,387,388] that oximes of aliphatic, aromatic, and heteroaromatic aldehydes are easily transformed into the corresponding nitriles upon moderate heating (60°C–140°C) in the system KOH/DMSO (Scheme 1.179).

The yields of nitriles in optimum conditions can exceed 90% [387].

Benzaldoxime at 140°C under the same conditions is converted into benzamide (25% yield) (Scheme 1.180).

Since the primary dehydration of aldoximes in these conditions has been proved, the conclusion is made [7] that benzamide is generated not via the Beckmann rearrangement, but owing to hydration of the benzonitrile formed (Scheme 1.181).

Against the Beckmann rearrangement is also the fact that acetophenone oxime in the same conditions does not form amide and is recovered from the reaction intact.

This method can be used for the preparation of amides from nitriles as well as from aldoximes, especially when they contain acid-nonresistant fragments. In the absence of DMSO, the benzaldoximes are dehydrated by an alkali only at boiling temperature (200°C).

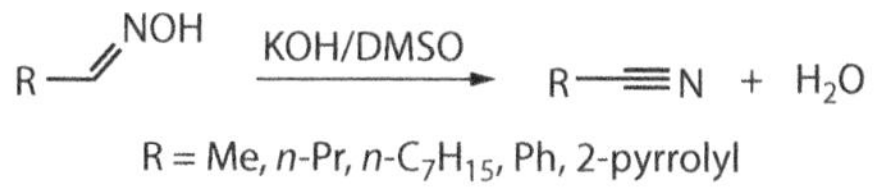

$$R \overset{NOH}{\diagup\!\!\!\diagup} \xrightarrow{\text{KOH/DMSO}} R \!-\!\!\equiv\!\! N \ + \ H_2O$$

R = Me, n-Pr, n-C$_7$H$_{15}$, Ph, 2-pyrrolyl

SCHEME 1.179 Synthesis of nitriles from aldoximes in the KOH/DMSO system.

$$PhCH\!=\!NOH \xrightarrow[\text{140°C, 3 h}]{\text{KOH/DMSO}} PhCONH_2$$

SCHEME 1.180 Transformation of benzaldoxime to benzamide in the KOH/DMSO system.

$$PhCN \ + \ H_2O \xrightarrow[\text{140°C, 3 h}]{\text{KOH/DMSO}} PhCONH_2$$

SCHEME 1.181 Hydration of the benzonitrile to benzamide in the KOH/DMSO system.

Dehydration of aldoximes into nitriles under the action of alkalis was described by Hantzsch [389,390] who discovered that the heating of *E*-thiophene aldoxime with soda [389] and *E*-mesityl aldoximes with hot alkali [390] affords nitriles. Later on, Reissert [391] observed the formation of nitrile from *o*-nitrobenzaldoxime in the presence of hydroxide ions.

However, at that time, preparative methods for the synthesis of nitriles have not been developed on the basis of these reactions. The main obstacle is, apparently, further transformation of the nitriles into acid salts.

In certain cases, when aldoximes are treated with alkalis, the nitriles are not detected at all due to their instant conversion into acids. Interestingly, *E*-isomers show here the increased reactivity [392].

Undoubtedly, the formation of carbonic acid salts from aldoximes via intermediate nitriles and amides also takes place in the KOH/DMSO system. Evidently this is a reason explaining why the corresponding nitrile is not generated in the aforementioned system from furfurol oxime [387]. Later, it has been found that the dehydration of aldoximes into nitriles can be carried out in much softer conditions, without using the autoclave (60°C–100°C, acetylene, atmospheric pressure) [7]. In no case, pyrroles and O-vinyl oximes are identified in the reaction products.

Thus, the reason hampering the formation of pyrroles from aldoximes and acetylene in the systems MOH/DMSO is obvious. This is their ready dehydration under the reaction conditions. Probably, the interaction of aldoximes with acetylene could lead to O-vinyl derivatives. However, due to base-catalytic elimination of vinyl alcohol (acetaldehyde) from them, this process should also result in nitriles (Scheme 1.182).

Dehydration of aldoximes in the system MOH/DMSO definitely starts with abstraction of the proton, closest to the oxime moiety (Scheme 1.183).

In the case of ketoximes in the system MOH/DMSO, one can expect that in the same conditions, the 1,3-dehydration should occur producing such unstable intermediates as C,O-dianion **A**, zwitterion **B**, azirine **C**, vinylnitrene **D**, or yielding nitrile, a stable product of one of these intermediate rearrangements (Scheme 1.184).

The nitrile in the reaction conditions can be quickly converted into a salt of the corresponding carbonic acid. This, for example, can rationalize nonrecoverable

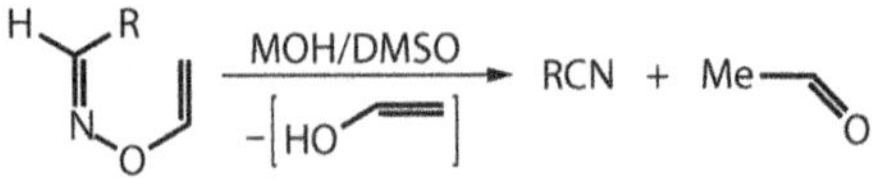

SCHEME 1.182   Possible formation of nitriles via O-vinyl aldoximes.

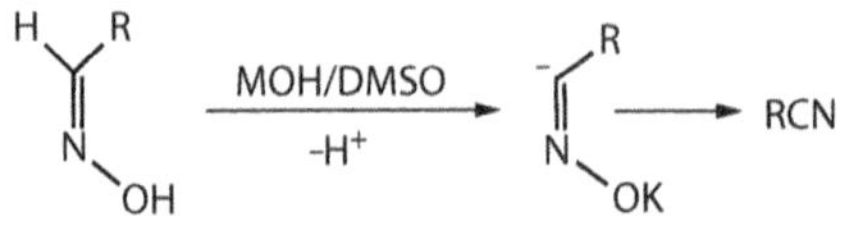

SCHEME 1.183   Tentative carbanionic intermediates in transformation of aldoximes to nitriles.

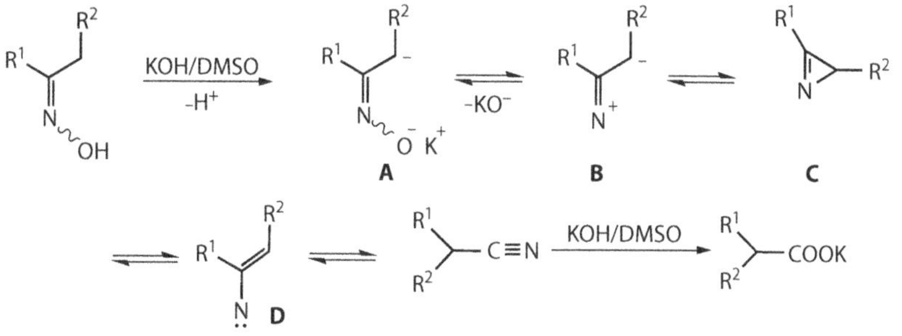

SCHEME 1.184 Expected side transformations of ketoximes in the KOH/DMSO system.

losses of ketoximes (without noticeable signs of resinification) sometimes observed during the synthesis of pyrroles from ketoximes and acetylene.

Thus, the investigations aimed to expand the scope of oxime–acetylene synthesis have resulted in a convenient method for the dehydration of aldoximes in nitriles [387,388] as well as a protocol for base-promoted transformations of aldoximes into amides [387].

1.9 MECHANISM OF PYRROLE SYNTHESIS FROM KETOXIMES AND ACETYLENE

1.9.1 OXIMES AS NUCLEOPHILES IN THE REACTION WITH ACETYLENES: LITERATURE ANALYSIS

The results of early works on the interaction of oximes with acetylenes are contradictory and often are inconsistent among themselves. Oximes, that is, tridentate O-, N-, and C-nucleophiles, theoretically can add to the triple C–C bond via any of these nucleophilic centers.

For instance, with ethoxyacetylene, they behave as O-centered nucleophiles to form ortho ester-like adducts (Scheme 1.185) [393].

For aldoximes in the same conditions, along with formation of the O-adduct like ortho ester, sometimes, dehydrations to the nitriles are observed, the second product of the reaction being ethyl acetate (Scheme 1.186) [393].

Nitriles are also formed by the interaction of aldoximes with 1-(N,N-diethylamino) propyne (Scheme 1.187) [394].

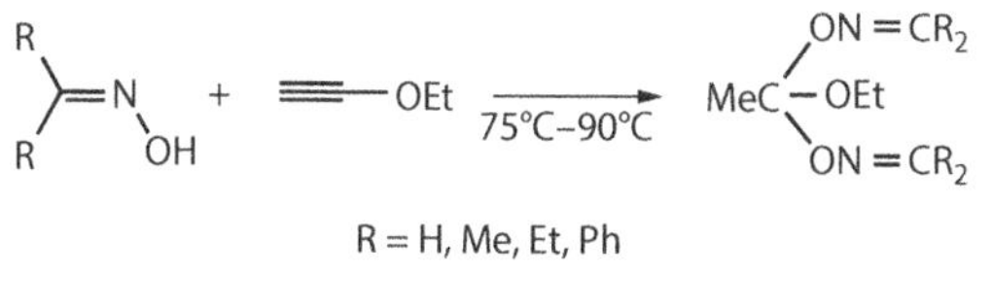

$$R = H, Me, Et, Ph$$

SCHEME 1.185 Ketoximes as O-nucleophiles in the reaction with ethoxyacetylene.

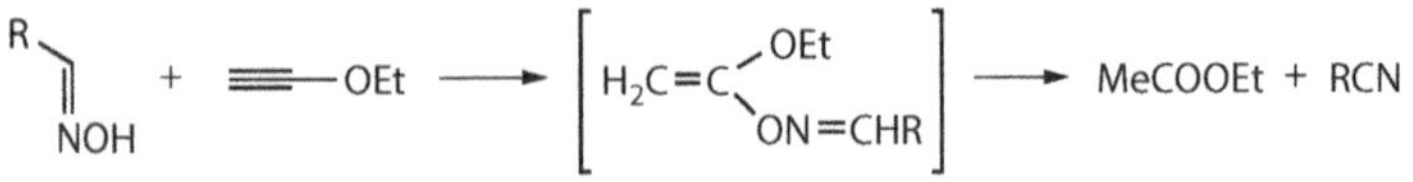

SCHEME 1.186 Reaction of aldoximes with ethoxyacetylene.

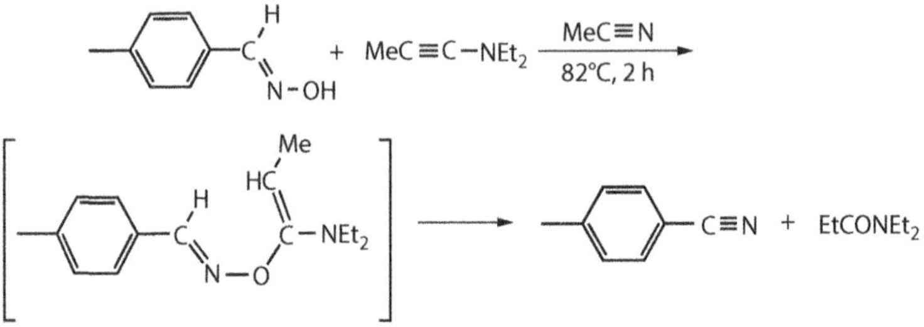

SCHEME 1.187 Formation of nitriles from aldoximes and 1-(N,N-diethylamino)propyne.

Acetoxime is added to methyl ether of acetylenedicarboxylic and propargylic acids in polar non hydroxylic solvents to deliver derivative of 1,2-oxazole (Scheme 1.188) [306–308].

It is presumed [306–308] that addition across the C≡C bond generates a zwitterion, in which intramolecular proton transfer occurs through the five-membered intermediate to give nitrone. The latter enters the 1,3-dipolar addition reaction with the second molecule of dimethyl acetylenedicarboxylate affording oxazole derivative.

The reaction of acetone, cyclopentanone, and cyclohexanone oximes with methyl propynoate (MeOH, 55°C–60°C) affords a mixture of products among which

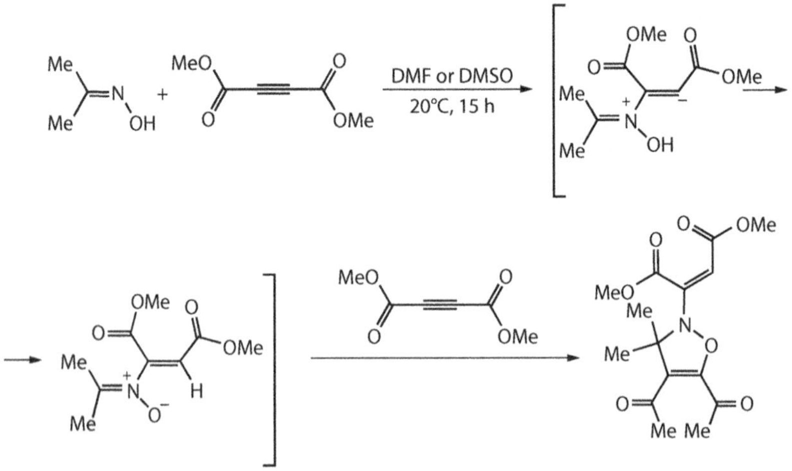

SCHEME   1.188   Formation   of   oxazoles   from   acetoxime   and   dimethyl acetylenedicarboxylate.

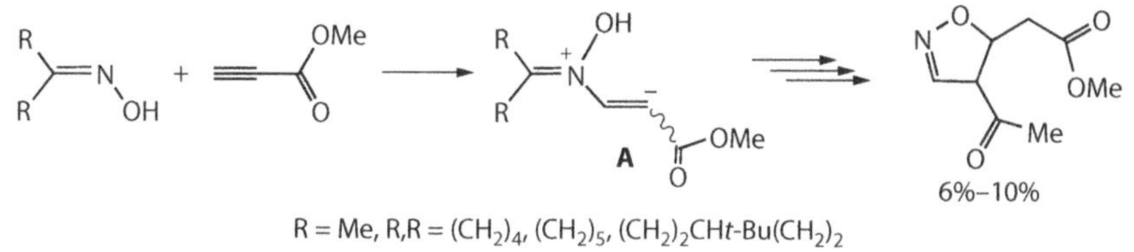

R = Me, R,R = (CH$_2$)$_4$, (CH$_2$)$_5$, (CH$_2$)$_2$CHt-Bu(CH$_2$)$_2$

SCHEME 1.189 Formation of isoxazolines from ketoximes and methyl propionate.

substituted isoxazolines together with O-vinyl oximes have been isolated (Scheme 1.189) [395]. The authors suppose [395] that this finding confirms the attack of acetylene by ketoxime nitrogen atom to generate zwitterion.

At the same time, according to data given in [304], the interaction of ketoximes with dimethyl acetylenedicarboxylate esters in the presence of sodium methylate leads to have formation of O-adduct (Scheme 1.190), which is isolated as a mixture of Z- and E-isomers (1:2).

Apart from chemical shift values for olefinic and methyl protons, no other characteristics of the adduct are given.

By analogy with pyrrole synthesis, the reaction of amidoximes with acetylene could lead to imidazoles (Scheme 1.191).

In attempting to prepare imidazole derivatives using this protocol, it has been found [396,397] that amidoximes quickly react with acetylene (5–7 min, 75°C) under pressure (12–14 atm) in the superbase system KOH/DMSO to deliver O-vinylamidoximes in up to 90% yield (Scheme 1.192). A peculiarity of this reaction is that it (like the reaction of ketoximes with acetylene) proceeds unusually fast, almost instantly (5–7 min) and at a temperature of only 75°C. From two conjugated competing nucleophilic centers, amino and hydroxyl groups, the latter selectively participates in the reaction.

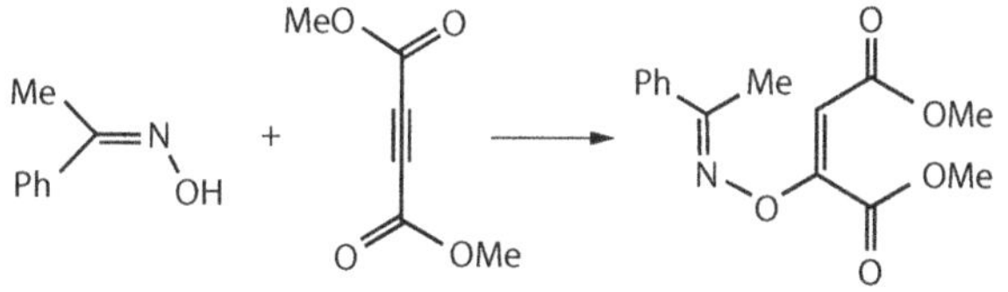

SCHEME 1.190 Addition of acetophenone oxime as O-nucleophile to dimethyl acetylenedicarboxylate.

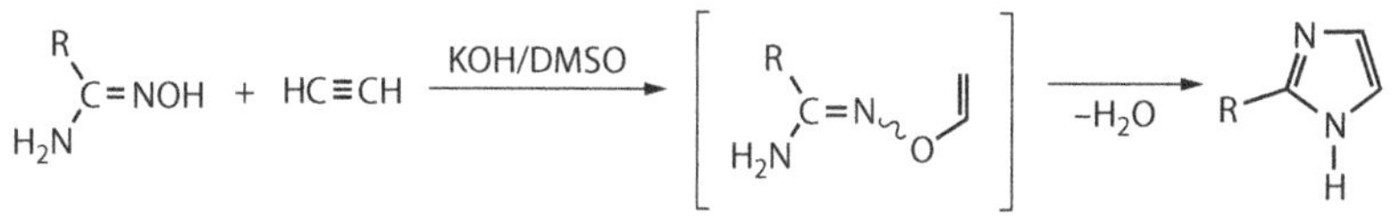

SCHEME 1.191 Possible formation of imidazoles from amidoximes and acetylene.

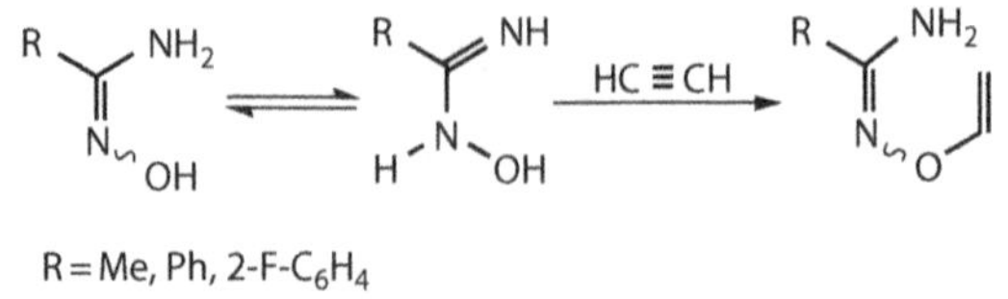

R = Me, Ph, 2-F-C₆H₄

SCHEME 1.192   Selective O-vinylation of amidoximes with acetylene in the KOH/DMSO system.

Thus, nitrogen-centered anion particles of both the initial amidoximes and their imine tautomers are not capable to compete with oxygen-centered anions during the nucleophilic addition to acetylene under the aforementioned conditions.

The O-vinylamidoximes formed are stable in the reaction conditions; imidazoles are not produced even at increased temperatures.

At the same time, the rearrangement and cyclization leading to imidazole derivatives has been previously observed in the reaction of amidoximes with alkyl propynoates (Scheme 1.193) [398].

It is assumed [398] that here unusual Claisen-type rearrangement of the intermediate O-adduct involving three heteroatoms takes place. However, the primary O-adduct is not identified unambiguously. It is only reported that this adduct is of the Z-configuration, that is, the reaction in this case, unlike that of described in the work [304], is stereospecific. The specified structure is attributed to the adduct in view of the fact that in acylation and alkylation reactions, the amidoxime oxygen function is the most nucleophilic center [399] and that O-methyl ethers of amidoximes do not react with methylpropiolate.

The interaction of amidoximes with diacetylene in the presence of KOH in aqueous DMSO furnishes O-adducts, but their rearrangement into the corresponding ethynylimidazoles does not occur (Scheme 1.194) [400].

As already mentioned (Section 1.3.), the interaction of ketoximes with diacetylene [293,294] and cyanoacetylene [296] results in O-adducts.

However, formaldoxime with ethylpropyolate gives 3,5-dicarboethoxypyridine (Scheme 1.195) [309], that is, the oxime hydroxyl group remains here inert and the nitrogen atom acts as the nucleophilic center.

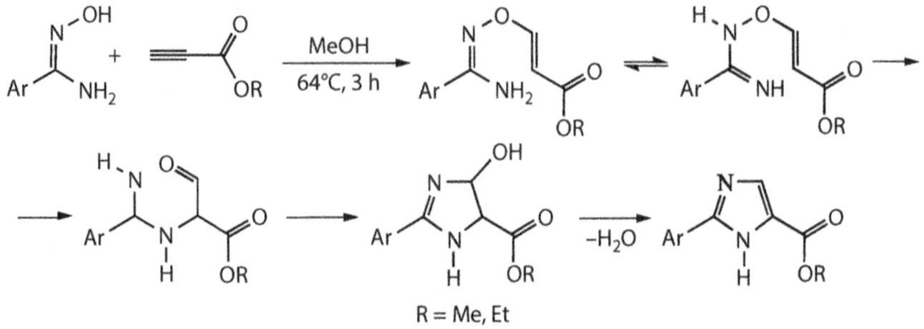

R = Me, Et

SCHEME 1.193   Formation of imidazoles from amidoximes and alkyl propynoates.

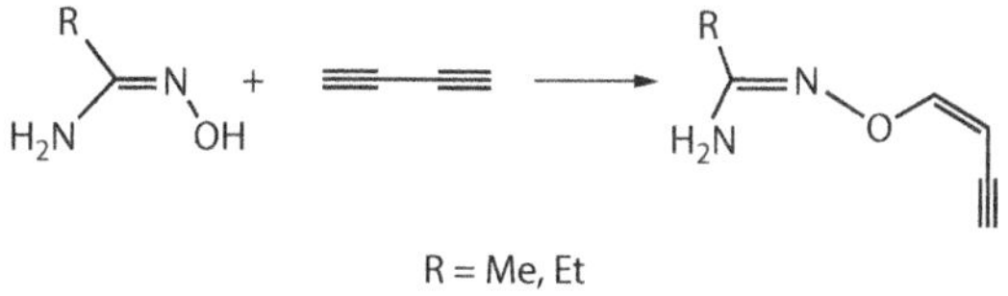

R = Me, Et

SCHEME 1.194 Addition of amidoximes as O-nucleophiles to diacetylene.

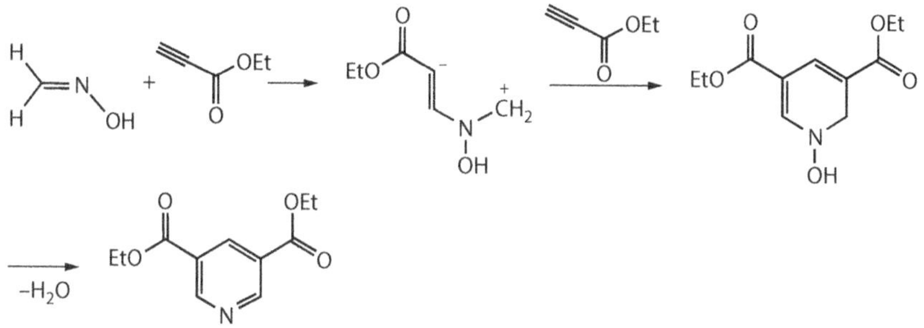

SCHEME 1.195 Formaldoxime as N-nucleophile in the reaction with ethyl propiolate.

Besides, it has been shown [309] that formaldoxime can behave as a typical 1,3-dipole relative to α,β-unsaturated nitriles.

1.9.2 POSSIBLE MECHANISMS OF PYRROLE SYNTHESIS FROM KETOXIMES AND ACETYLENE

Ketoximes are known to behave in some cases as C–H acids, for example, they are metalated across the α-CH$_3$ and CH$_2$ groups (Scheme 1.196) [278,281,401,402].

Regioselective deprotonation of ketoxime methyl ethers, namely, proton abstraction from the *cis*-position relative to the methoxy group, has been noted (Scheme 1.197) [278].

The authors admit [278] a paradoxical attractive interaction between the carbanionic center and the oxygen atom in the formed system of six π-electrons. Such stabilizing interaction has been previously predicted using quantum-chemical calculations and confirmed experimentally for Z-1,2-difluoroethylene [403]. The rough energy value of this stabilization for carbanion **A** is 1.5 kcal/mol [278].

It is not inconceivable that in the superbase system KOH/(CD$_3$)$_2$SO, CH acidity of ketoximes plays an important role (Scheme 1.198).

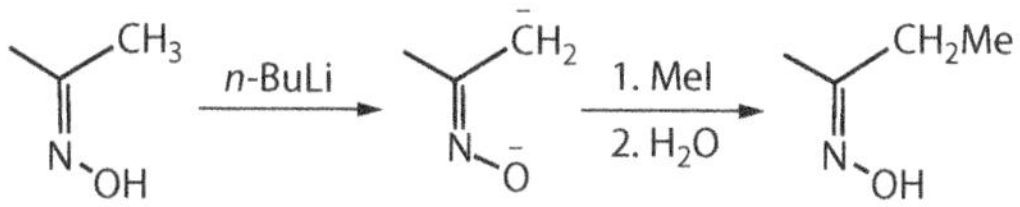

SCHEME 1.196 Ketoximes as C–H-acids.

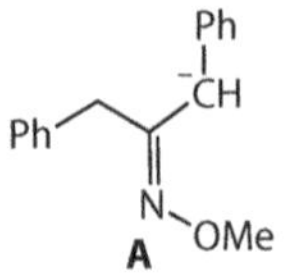

SCHEME 1.197   Regioselective deprotonation of O-methyl dibenzyl ketoxime.

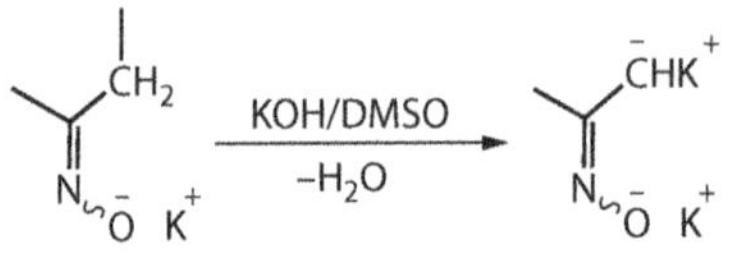

SCHEME 1.198   Possible deprotonation of potassium oximates by the KOH/DMSO system.

It stands for reason that the existence of carbanions in noticeable concentration in the presence of water, though in small amounts, seems to be improbable. Meanwhile, one should bear in mind that in this specific system (KOH/DMSO), the part of potassium hydroxide applied in high excess is usually insoluble and represents a suspension. In such form, alkali can play a role of drying medium. Besides, DMSO itself forms strong complexes with water such as $Me_2SO·2H_2O$ [176,177]. The hydrogen bonding between DMSO and water is stronger than that of water associates [404]. All this should lower the activity of water considerably. Finally, in such two-phase systems, there are conditions for the formation of high-basic complex aggregates of various natures [405].

The generation of carbanions from ketoximes and DMSO in the presence of high amounts of KOH is verified by deuterium exchange between DMSO and ketoxime α-hydrogen atoms found in the conditions of pyrrole synthesis (Scheme 1.199) [4,7].

Tridentate character of oximate anion, that is, its ability to act as O-, N-, and C-nucleophiles, complicates considerably the analysis of the mechanism of base-catalytic heterocyclization of ketoximes with acetylene (Scheme 1.200).

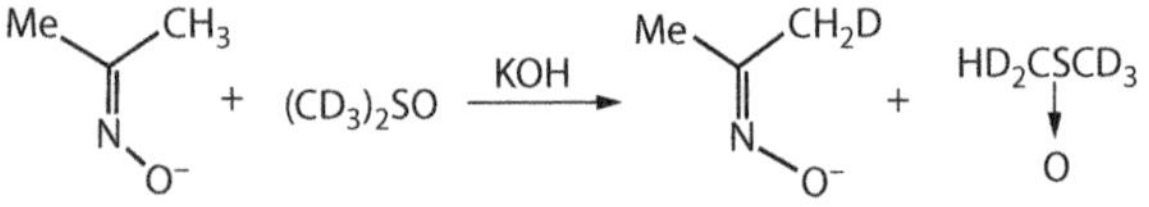

SCHEME 1.199   Deuterium exchange between $(CD_3)_2SO$ and acetoxime.

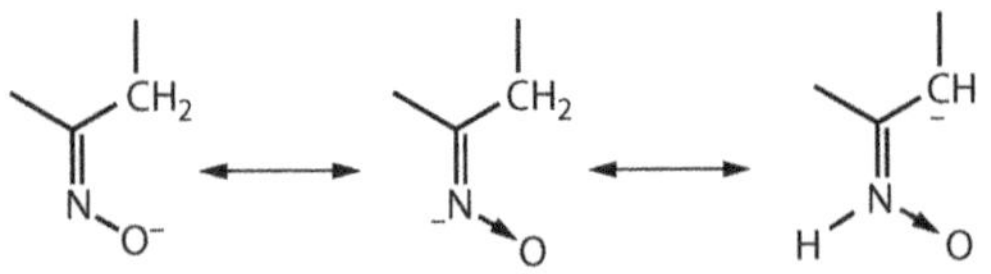

SCHEME 1.200   Oximate-anion as tridentate nucleophile.

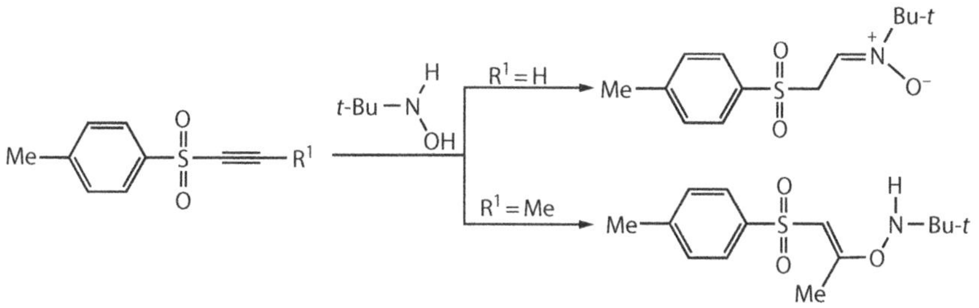

SCHEME 1.201 N-*t*-butylhydroxylamine as O- and N-nucleophile depending on acetylene structure.

The capacity of hydroxylamines derivatives to be either O- or N-nucleophiles in the reactions with acetylene compounds often depends even upon the structure of the latter. For example, N-*t*-butylhydroxylamine can add to acetylenic sulfones through nitrogen or oxygen atom depending on the character of the second radical at the triple bond (Scheme 1.201) [406].

In this case, differences in the reaction directions could be explained by steric hindrances relative to attack of the triple bond by nitrogen atom when R^1=Me, as it takes place, for example, in acylation of N-monosubstituted hydroxylamine [407]. In this process, nitrogen is usually a more active nucleophilic center, but if the access to it is sterically hindered, the acylation can occur at the oxygen atom also. However, actually, the situation is even more complicated: when R^1=Ph, that is, in the case of even higher steric encumbrances, the same reaction leads to a mixture of N- and O-adducts (Scheme 1.202) [406].

Therefore, the direction of hydroxylamine derivatives addition to acetylene is defined by not only steric but also electronic factors. This conclusion is supported also by other authors [408] that have established that the addition of monosubstituted hydroxylamines to various acetylenes affords nitrones (Scheme 1.203).

However, the reaction with cyanophenylacetylene yields the diadduct owing to the participation of N- and O-nucleophilic centers in the addition (Scheme 1.204) [408].

If R^1=CO$_2$Et, R^2=Ph, isoxazolone is formed (Scheme 1.205).

Aromatic hydroxylamines (HONHAr) give in this reaction either nitrones or side-products [408]. Noteworthy, that the structure of diadduct entirely corresponds to

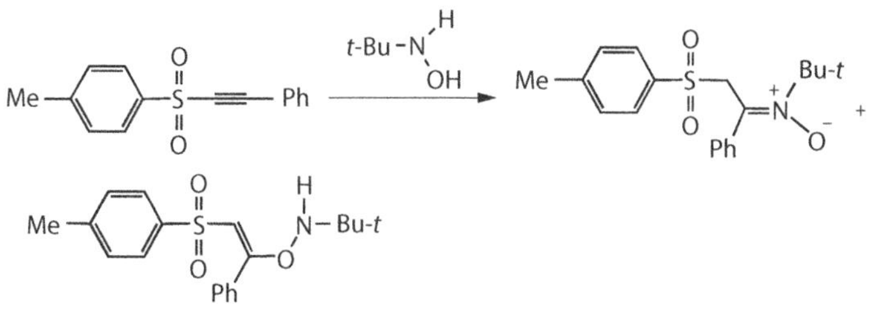

SCHEME 1.202 Simultaneous O- and N-nucleophilic behavior of N-*t*-butylhydroxylamine towards the triple bond.

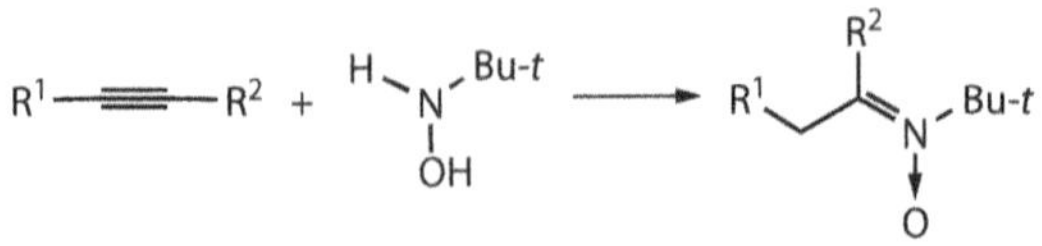

$$R^1 = PhSO_2, PhSO_2, CO_2Me; R^2 = H, Ph, CO_2Me$$

SCHEME 1.203 Formation of nitrones from N-*t*-butylhydroxylamine and acetylenes.

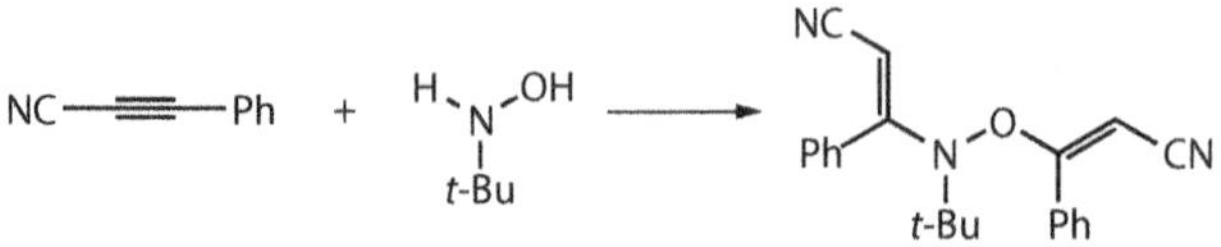

SCHEME 1.204 Simultaneous participation of N- and O-nucleophilic centers of N-*t*-butylhydroxylamine in the reaction with cyanophenylacetylene.

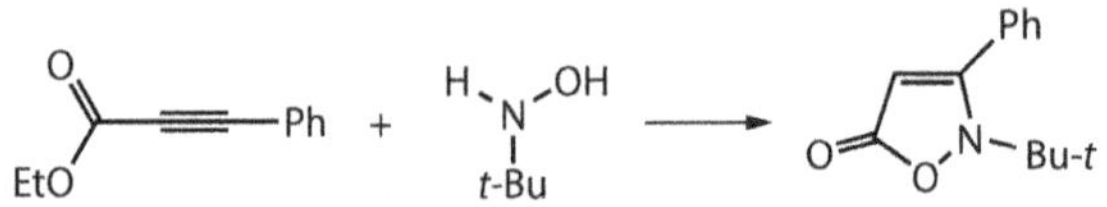

SCHEME 1.205 Formation of isoxazolone from N-*t*-butylhydroxylamine and ethyl phenyl carboxylate.

that of the assumed intermediate detected in the synthesis of pyrroles via O-vinyl oximes. However, its conversion into the corresponding pyrrole is not observed (Scheme 1.206) [408].

The discussion of the mechanism of pyrrole synthesis from ketoximes and acetylene should also account for a possibility of azirine intermediate (or its open-chain forms, zwitterion and vinyl nitrene) participation in the reaction, since ketoximes are known to undergo 1,3-dehydration under the influence of the strong bases to produce azirines (the Hoch–Campbell reaction) (Scheme 1.207) [344–348].

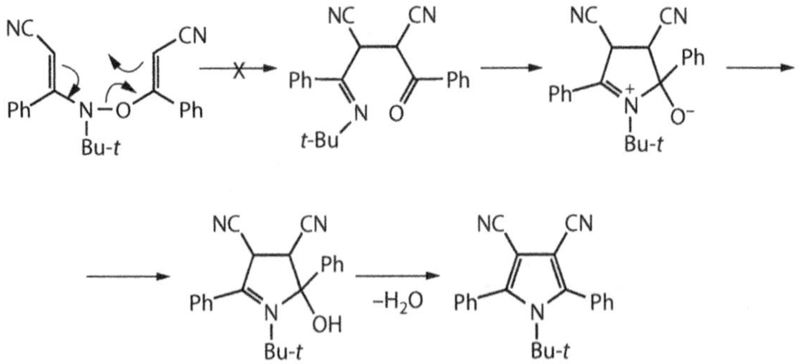

SCHEME 1.206 Unrealizable expected rearrangement of the functionalized N,O-divinyl hydroxylamine to pyrrole.

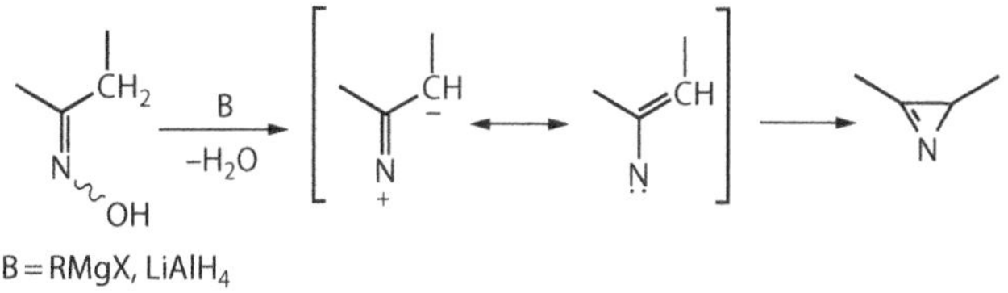

SCHEME 1.207 Formation of azirine from ketoximes under the action of bases.

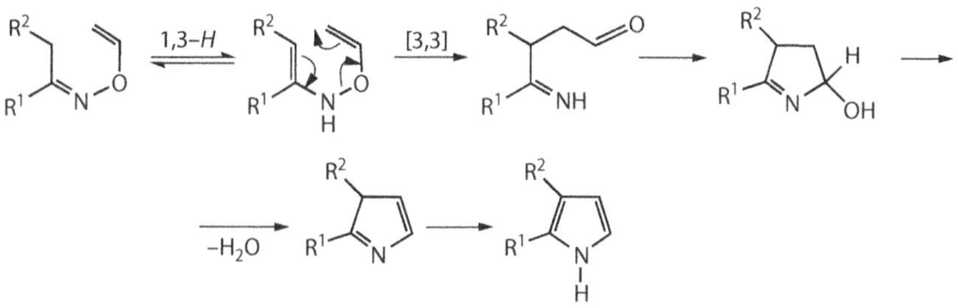

SCHEME 1.208 Formation of pyrroles from ketoximes and acetylene via O-vinyl oximes.

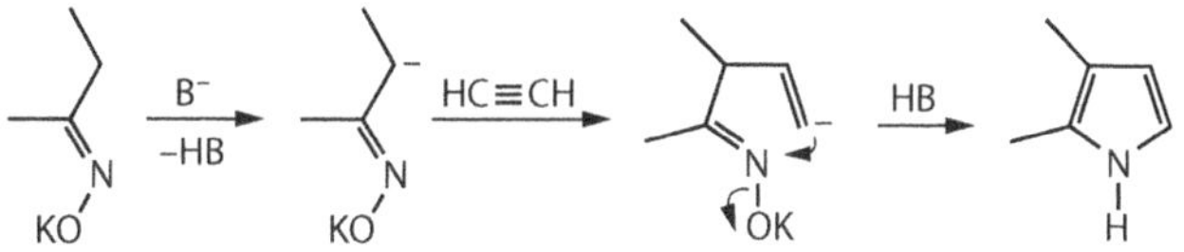

SCHEME 1.209 Alternative formation of pyrroles via nucleophilic attack of ketoxime carbanions to acetylene.

In the light of the previous analysis, one can discuss the following mechanisms of pyrroles formation from ketoximes and acetylene:

Mechanism 1. 1,3-Prototropic shift in O-vinyl oximes and the subsequent 3,3-sigmatropic rearrangement of O-vinylalkenylhydroxylamines (Scheme 1.208)

Mechanism 2. Nucleophilic attack of ketoxime carbanions to acetylene (only key stages are depicted on Scheme 1.209)

Mechanism 3. 1,3-Dehydration of ketoximes and addition of zwitterions (or the corresponding vinyl nitrene) across the triple bond (Scheme 1.210)

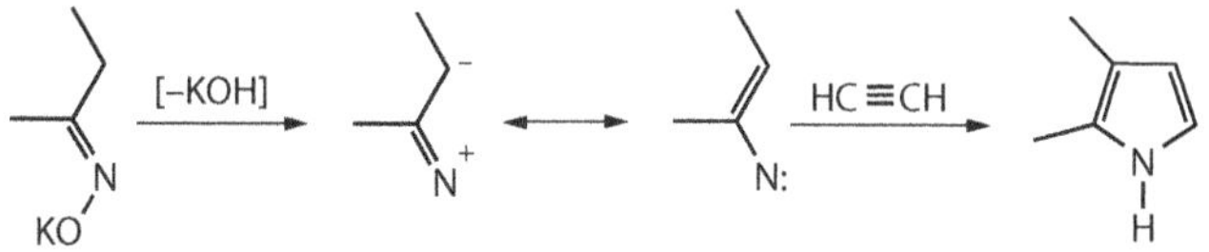

SCHEME 1.210 Alternative formation of pyrroles via addition of zwitterions, generated by 1,3-dehydration of ketoximes, to acetylene.

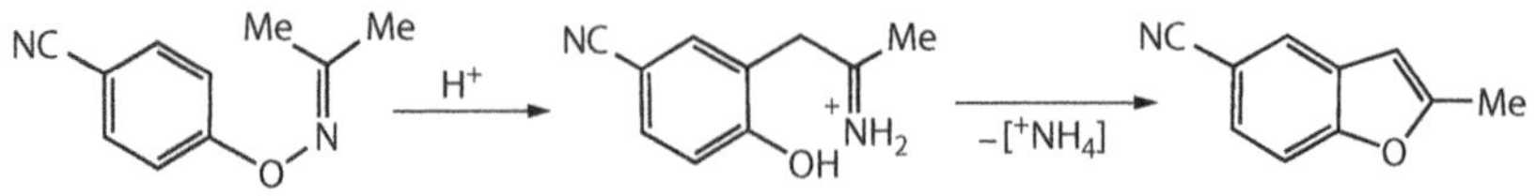

SCHEME 1.211 Rearrangement of O-aryl oximes to furans.

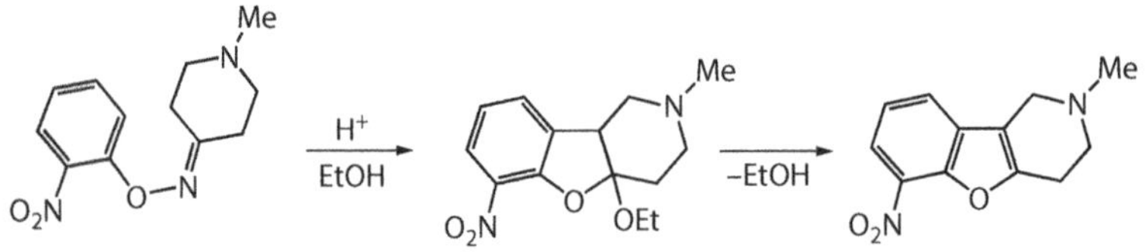

SCHEME 1.212 Formation of 2-methyl-6-nitro-1,2,3,4-tetrahydrobenzofuro[3,2-c]pyridine from the functionalized O-aryl oxime.

Now, the first mechanism has been experimentally confirmed (see Sections 1.5.1 through 1.5.3). It is indisputable that formation of pyrroles can proceed through O-vinyl derivatives of oximes. However, it still remains obscure whether this is the only route and whether mechanisms 2 and 3 are switched on in certain cases.

It is also unclear why one fails to prepare the expected ethynylpyrroles using diacetylene, though the corresponding Z-O-(β-ethynylvinyl) oximes are formed easily (see Section 1.3).

Interestingly, O-aryl oximes, close congeners of O-vinyl oximes, rearrange not into the pyrroles but into the corresponding furans (Scheme 1.211) [409–414]. The intermediate product of this rearrangement, 4-hydroxy-3-(2-iminopropyl)benzonitrile, has been isolated [413].

Similarly, 2-methyl-6-nitro-1,2,3,4-tetrahydrobenzofuro[3,2-c]pyridine has been synthesized from 1-methyl-4-nitrophenyloxyiminopiperidine under the action of HCl in ethanol, the intermediate acetal being also isolated (Scheme 1.212) [415,416].

Nevertheless, concluding this section, one can state that all now available experimental data correlate better with the O-vinyl oxime mechanism. At the same time, it should be remembered that all, even the most plausible, mechanistic schemes cannot be considered adequate if they do not explain why the pyrrole synthesis from ketoximes and acetylene successfully proceeds only in the presence of specific superbase systems KOH/DMSO.

2 Novel Aspects of NH- and N-Vinylpyrroles Reactivity

2.1 REACTIONS WITH PARTICIPATION OF THE PYRROLE RING

2.1.1 PROTONATION

The behavior of NH- and N-vinylpyrroles in the reactions with electrophiles is defined by the competition of several nucleophilic centers, namely, α- and β-positions of the pyrrole ring, nitrogen atom, and, in the case of N-vinylpyrrole, β-carbon atom of the vinyl group. In this line, the issues of peculiarities of pyrroles and N-vinylpyrroles behavior in acidic media, possibilities and routes of pyrrolium and N-pyrrolium ions generation, degrees of their stability, chemical properties, electronic and conformational structure, and synthetic potential represent urgent challenge.

2.1.1.1 Electron Structure of N-Vinylpyrrolium Ions

The nuclear magnetic resonance (NMR) studies [21,24,36,37] allow detecting the N-vinylpyrroles protonated at both α- and β-positions of the heterocycle and β-carbon atom of the double bond (separately and in pairs) except for both centers simultaneously. The direction of protonation and stability of the cations formed depends upon the strength and nature of the acid, temperature, and character of substitution in the pyrrole cycle.

At −80°C, irrespective of the nature of acid and substituents in the pyrrole ring, only α-position of the pyrrole cycle is protonated, the vinyl group remaining intact (Scheme 2.1) [21,24,417,418].

The increase in temperature leads to various transformations of these cations; however, N-vinylpyrrolium ions **1** remain as the key particles defining these transformations.

^{1}H and ^{13}C NMR spectra of a wide series of N-vinylpyrrolium **1** fluorosulfonates ($X = SO_3F$), generated at −50°C and quite stable up to 50°C, have been studied [419].

Comparison of the chemical shifts of the terminal carbon atoms C_β of the vinyl group in N-vinylpyrrolium cations **1** and the initial N-vinylpyrroles (110–115 ppm and 95–100 ppm, respectively) [420,421] shows that there is less electron density on this atom in the cations **1**. Nevertheless, the shielding of β-carbon atom in the cations is greater than that of the carbon nuclei in ethylene (123.3 ppm). This means that the protonated pyrrole ring, despite its total positive charge, is a π-donor for the

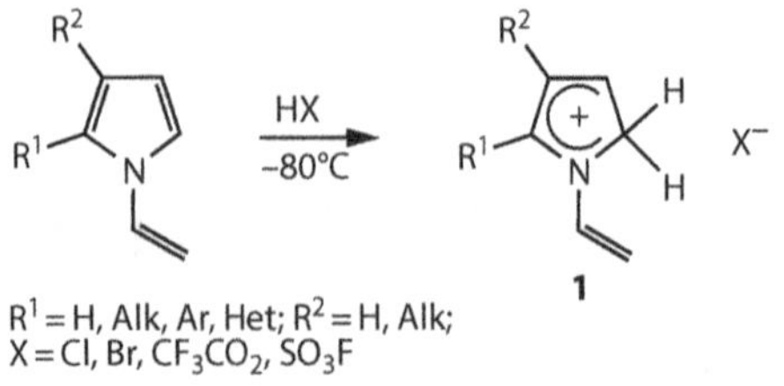

$R^1 = H$, Alk, Ar, Het; $R^2 = H$, Alk;
$X = Cl$, Br, CF_3CO_2, SO_3F

SCHEME 2.1   Protonation of N-vinylpyrroles.

N-vinyl group, though weaker compared to the nonprotonated ring. Thus, the extent of bonding between the electrons of unshared pair of the nitrogen atom and the vinyl group in the cations and initial pyrroles differs insignificantly (accounting for the essentially increased electronegativity of the pyrrole ring upon protonation).

There is a satisfactory correlation between the chemical shifts of C_β atom in N-vinylpyrrolium cations and their nonprotonated predecessors:

$$\delta\left(C_\beta^+\right) = 39.4 + (0.76 \pm 0.14)\delta\left(C_\beta^o\right)$$

$$r = 0.98, \ s = 0.05, \ n = 10$$

This correlation emphasizes the generality of the factors that determine the electronic structures of the protonated and nonprotonated N-vinylpyrroles (p-π-conjugation of nitrogen atom with the double bond and steric hindrances of coplanarity) [419,421].

The total deshielding of carbon nuclei in the ring during the cation formation is expressed as follows: $\sum \delta C_i^+ - \sum \delta C_i^o$, where $\sum \delta C_i^+$ is the sum of the chemical shifts of C_2, C_3, and C_4 atoms (i.e., atoms that do not change their hybridization upon protonation) and $\sum \delta C_i^o$ is the same sum for neutral molecules.

This formula provides information about the distribution of positive charge in the cation.

In alkyl derivatives, the total deshielding is 120 ± 5 ppm, 2-aryl substituents decrease this value to 109 ± 2 ppm, and for heteroaryl derivatives, deshielding is even lesser (97 ± 5 ppm). These results show an increased degree of positive charge delocalization of the cations in the series: 2-alkyl < 2-aryl < 2-hetaryl.

2.1.1.2   Dimerization of Protonated N-Vinylpyrroles

N-Vinylpyrrolium ions **1**, generated by the initial protonation (−80°C) of N-vinylpyrroles with trifluoroacetic acid, are transformed to dimeric cations **2** upon heating (Scheme 2.2) [417].

Apparently, in vinylpyrrolium cations, the proton is transferred from carbon atom in α-position of the protonated pyrrole ring to β-carbon atom of the double bond. The immonium cation **A** attacks α-position of the nonprotonated pyrrole to give a dimeric cation **B** that rearranges further in more stable cation **2** (Scheme 2.3).

The rate of N-vinylpyrroles dimerization strongly depends on a structure of the substituent in position 2 of the pyrrole cycle.

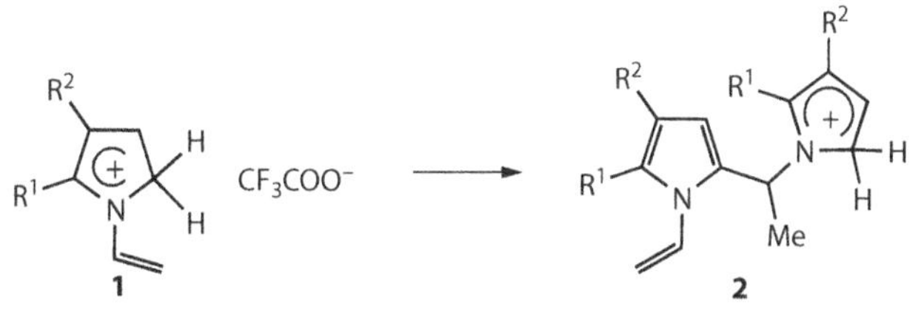

R^1 = Me, Ph, 2-furyl; R^2 = H, Et, *i*-Pr

SCHEME 2.2   Dimerization of protonated N-vinylpyrroles.

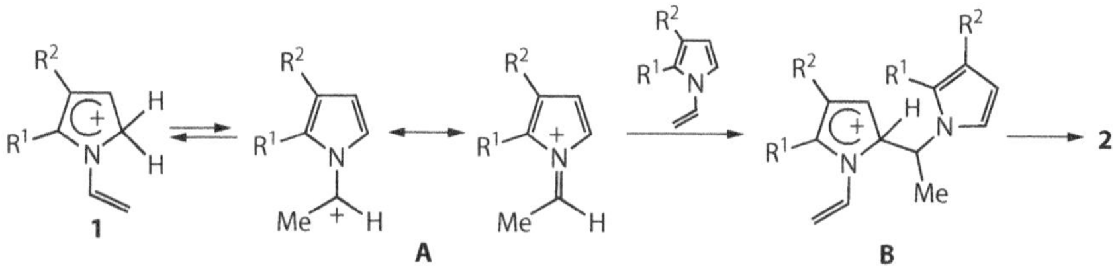

SCHEME 2.3   Plausible scheme of the dimerization of the N-vinylpyrrole cations.

2.1.1.3   Peculiarities of N-Vinylpyrroles Protonation with Hydrogen Halides

The direction of N-vinylpyrroles protonation with hydrogen halides depends on the temperature [21,24]. For example, at −80°C, the proton is added exclusively at the α-position of the pyrrole ring to afford N-vinylpyrrolium 1 halides 1 (X=Cl, Br) (Scheme 2.4) [418]. If the reaction is carried out at a higher temperature (−40°C), the second molecule of HX adds to the vinyl group to deliver N-(1-haloethyl)pyrrolium halides 3. At −40°C and equimolar amount of HX, only N-(1-haloethyl)pyrroles 4 are formed (Scheme 2.4) [418,422].

The formation of N-(1-haloethyl)pyrrolium halides 3 from vinylpyrrolium 1 proceeds slower if high excess of hydrogen halide (HX) is present in the solution [418]. All these features are obviously due to the interrelation of kinetic and thermodynamic factors. Primary (kinetic) protonation at low temperatures leads to the formation of α-protonated forms 1 that are converted (at higher temperatures) to thermodynamically more stable uncharged adducts, N-(1-haloethyl)pyrroles 4.

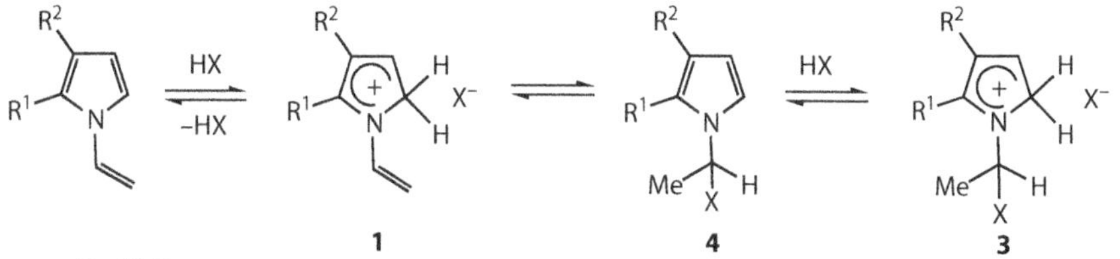

X = Cl, Br

SCHEME 2.4   Protonation of the N-vinylpyrroles with hydrogen halides.

The latter in the acid excess are again protonated at the α-position to generate cations **3**. Free hydrogen halide in solution hinders deprotonation (owing to binding of the anion X to less basic and nucleophilic complex anion HX_2^-) [423]. This explains the observed inverse dependence between HX concentration and rate of formation of N-(1-haloethyl)pyrrolium halides **3** [21,24,418].

2.1.1.4 Addition of Hydrogen Halides to the Pyrrole Ring

2-Substituted N-vinylpyrroles with excess hydrogen halides HX (X=Cl, Br) form (0°C, CD_2Cl_2) 3-halodihydropyrrolium ions **5** (Scheme 2.5). This reaction represents the first example of additive hydrohalogenation of the pyrrole ring [424].

Obviously, here there is a formation of covalent bonding between halogen anion and positively charged carbon atom C4 with simultaneous addition of the HX molecule across the vinyl group. The intermediate pyrroline is protonated by the third HX molecule generating dihydropyrrolium cation **5**.

It is intriguing that dihydropyrrolium cations are not detected in the case of the corresponding NH-pyrroles. It can be explained by the increase of electrophilicity of the pyrrole ring after the introduction of acceptor substituent CHXMe. Also, the addition of HX to the pyrrole ring is not observed for 3-alkyl-N-vinylpyrroles, likely due to electron-donating effect of the alkyl substituent lowering the electrophilicity of the pyrrole ring.

As far as regioselectivity and structure of the dihydropyrrolium cations **5** are concerned, the reactions of 2-substituted N-vinylpyrroles with HX (X=Cl, Br) give the same outcomes. This is confirmed by nearly complete concordance of their ^{1}H NMR parameters including spin-spin coupling constants. The main distinction is that the addition of HBr is almost exhaustive, whereas for HCl, owing to weaker nucleophilicity of chloride ion, the equilibrium is shifted toward pyrrolium cation **3**.

Thus, contrary to the common opinion about the inability of the pyrrole cycle to the addition reactions, the results discussed previously testify that in some special cases, such reactions are possible.

2.1.1.5 Protonation of N-Vinylpyrroles with Superacids

In superacidic system, $HSO_3F/SbF_5/SO_2FCl$ at −70°C 2-*tert*-butyl-N-vinylpyrrole gives β-protonated form **6** (primary kinetic intermediate), the double bond remaining intact (Scheme 2.6) [21,24]. Apparently, this is the first example of experimental observation (^{1}H NMR) of protonation of the pyrrole ring β-position.

At 0°C for several hours, cation **6** is transformed (via 1,2-hydride shift) into the expected thermodynamically more stable α-protonated form **7**. Before this work,

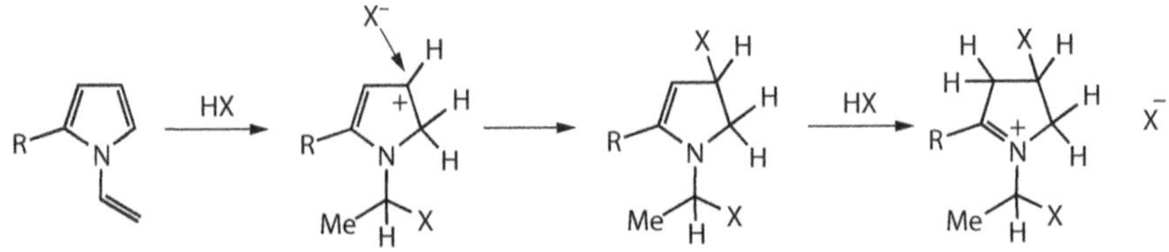

R = Me, *t*-Bu, Ph, 2-furyl, 2-thienyl; X = Cl, Br

SCHEME 2.5   Addition of hydrogen halides to the N-vinylpyrroles.

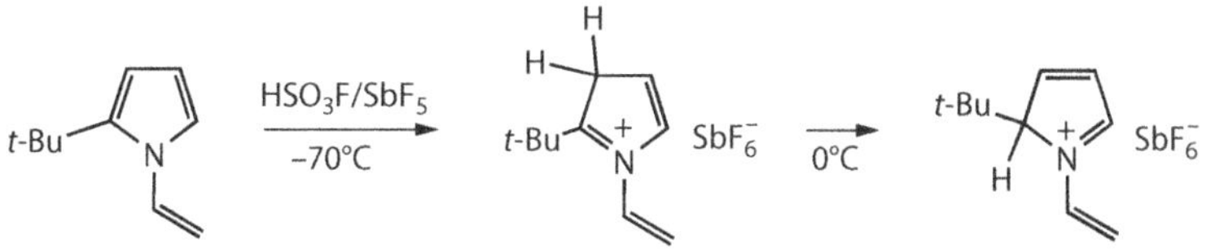

SCHEME 2.6 Protonation of N-vinylpyrroles with the superacidic system HSO_3F/SbF_5.

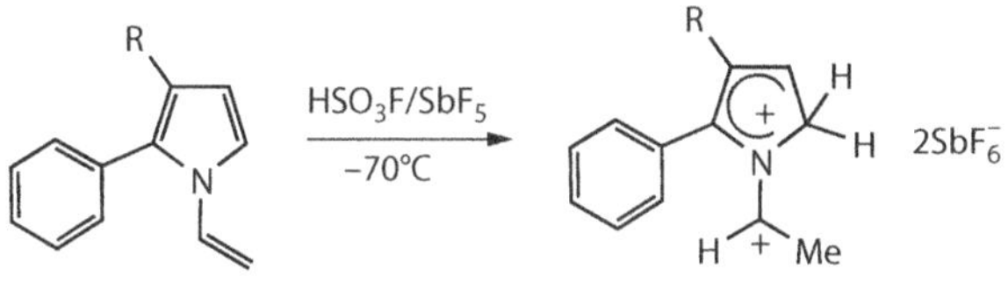

R = H, Me, Et, *i*-Pr

SCHEME 2.7 Formation of the double-charged cation from 2-phenyl-N-vinylpyrroles in the HSO_3F/SbF_5 system.

such intramolecular transformation had no direct experimental confirmation though it was theoretically predicted many times [2].

The interaction of 2-phenyl-N-vinylpyrroles with HSO_3F/SbF_5 ($-70°C$) leads to double-charged cations **8** (Scheme 2.7) [21].

2-*tert*-butyl-N-vinylpyrrole does not form dication, probably, owing to the absence of rather effective delocalization of the positive charge.

When temperature increases to $-30°C$, the gradual (to 1 h) transformation of dication **8** into cation **9** is observed (Scheme 2.8). This is the result of intramolecular attack of the carbenium center at the *ortho*-position of the phenyl ring [21,24].

After neutralization of the mixture with saturated aqueous K_2CO_3, a product of intramolecular cyclization, 5-methylindano[2,3-*a*]pyrrole (7% yield), has been isolated.

Apparently, partial transfer of the positive charge of the protonated heterocycle to the phenyl ring is a main factor ensuring relative stability of the dication. It is proved by the fact that the protonation of 1,4-*bis*(N-vinylpyrrol-2-yl)benzene with the system HSO_3F/SbF_5 leads to the formation of dication **10** (Scheme 2.9) with preservation of the vinyl groups and protonation of different positions of the pyrrole rings [21,24,425].

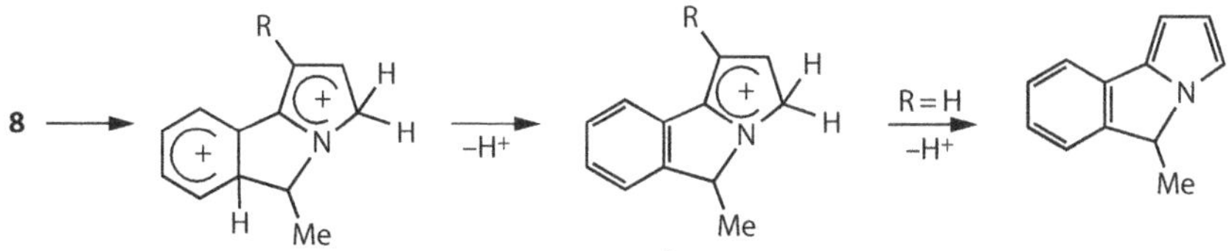

SCHEME 2.8 Annulation of the double-charged cation at $-30°C$.

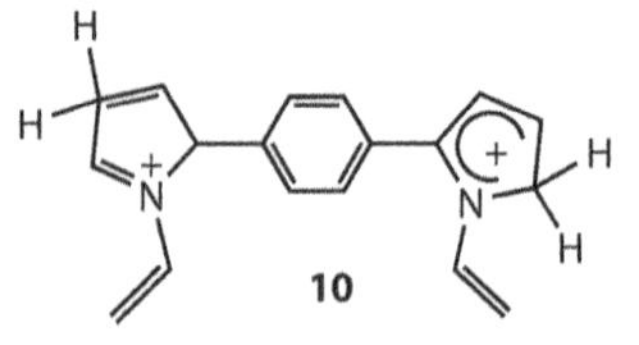

SCHEME 2.9   Dication of 1,4-*bis*(N-vinylpyrrol-2-yl)benzene.

SCHEME 2.10   Reversible charge distribution in the monoprotonated 1,4-*bis*(N-vinylpyrrol-2-yl)benzene.

Obviously, in intermediate with protonated α-position of one pyrrole ring, there is a deactivation of α-position of the second pyrrole ring (due to partial transfer of the positive charge through the benzene ring). Therefore, β-position of the second pyrrole ring becomes an object of the attack of the next proton (Scheme 2.10) [425].

Dication **10** is reported to be stable [425]. Its ^{1}H NMR spectrum does not change at the temperature increase to 0°C.

Coplanarity of the molecule is a prerequisite for effective mutual influence of the protonated pyrrole rings separated by 1,4-phenylene. This is confirmed by different results of this dipyrrole reaction with hydrogen halides. The reaction leads to the formation of symmetric dication **11** with α-protonated pyrrole cycles, both N-vinyl groups adding the hydrogen halide molecule (Scheme 2.11) [21,425].

Such direction of protonation testifies to relative independence of heterocycles upon each other in dication **11**, which is obviously explained by the steric hindrances of coplanarity created by α-haloethyl substituents at the nitrogen atoms.

Thus, in the superacidic medium, the direction and character of N-vinylpyrroles protonation change, that is, it results in the formation of stable dications, products of simultaneous protonation of the α-position of the pyrrole ring and β-position of the vinyl group, which are prone (in the case of phenyl substituent) to intramolecular cyclization, and "kinetic" cations having β-protonated pyrrole undergoing rearrangement via 1,2-hydride shift.

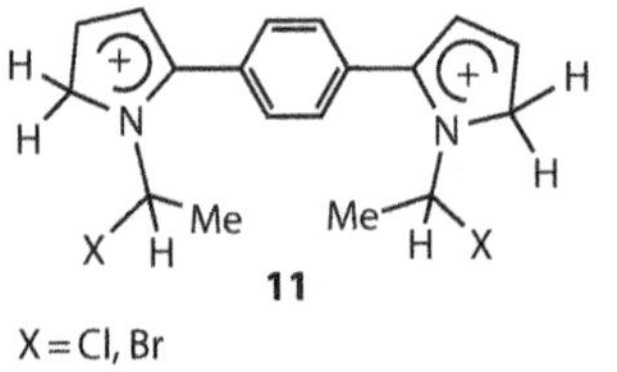

X = Cl, Br

SCHEME 2.11   Dication **11** formed by protonation of 1,4-*bis*(N-vinylpyrrol-2-yl)benzene with hydrogen halides.

2.1.1.6 Protonation of Hetarylpyrroles

2-Hetarylpyrroles at −80°C, regardless of the nature of second heterocycle and acid (except for superacidic system), are protonated at the C5 atom of the pyrrole ring (Scheme 2.12). When the pyrroles contain the vinyl group, the latter remains intact ([^{1}H NMR) [426–430].

The interaction of 2-furylpyrroles with hydrogen halides at −30°C affords a mixture of pyrrolium (**12**) and furanium (**13**) cations (Scheme 2.13). Thus, a formal transfer of the proton from protonated pyrrole ring to the furan cycle is realized (for N-vinylpyrroles, a simultaneous addition of hydrogen halides to the double bond takes place) [426,427,429].

In the same conditions, the thiophene cycle is inert relative to hydrogen halides [428,430].

The reaction of furylpyrroles with HBr does not stop at this step. In the [^{1}H NMR spectra, after further heating of samples to 0°C, signals of cations **12** and **13** gradually disappear (notably, their intensity drops synchronously that confirms equilibrium **12**⇌**13**) and signals of cations **14**, formed by the addition of HBr molecule to the furan ring, emerge.

At 20°C, furanium cations **14** are completely transformed into 4,5-dihydropyrrolium cations **15** with retroaromatization of the furan counterpart of the molecule. Obviously, it occurs through the dehydrobromination of cation **14** to cation **13** and its further deprotonation to neutral molecule, that is, ultimately, via cation **12** (Scheme 2.14).

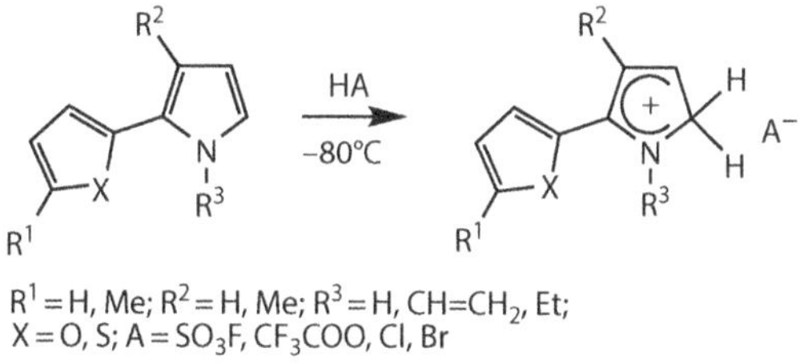

R^1 = H, Me; R^2 = H, Me; R^3 = H, CH=CH$_2$, Et;
X = O, S; A = SO$_3$F, CF$_3$COO, Cl, Br

SCHEME 2.12 Direction of protonation of 2-(2-furyl)-N-vinylpyrrole and 2-(2-thienyl)-N-vinylpyrrole at −80°C.

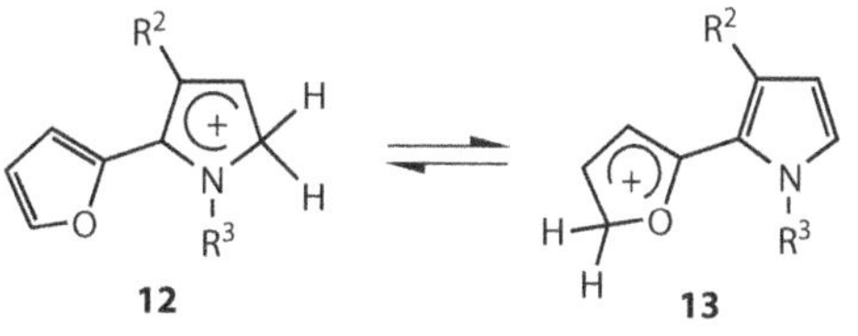

R^2 = H, Me; R^3 = H, CHHalMe, Et

SCHEME 2.13 Directions of protonation of 2-(2-furyl)pyrroles at −30°C.

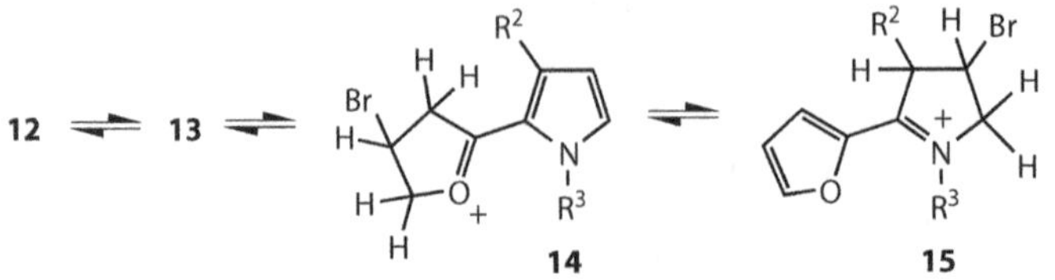

$R^2 = H$, Me; $R^3 = H$, CHBrMe

SCHEME 2.14 Evolution of cations **12**, **13** upon heating up to 0°C.

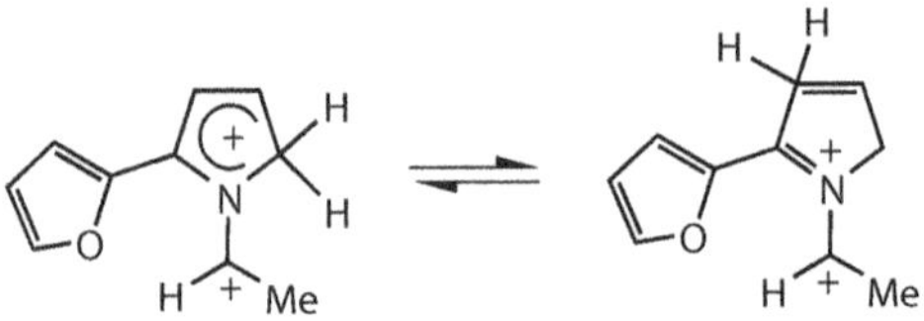

SCHEME 2.15 Equilibrium between dications of 2-(2-furyl)-N-vinylpyrrole generated in the HSO_3F/SbF_5 system.

In superacidic system, HSO_3F/SbF_5 2-(2-furyl)-N-vinylpyrrole forms an equilibrium mixture of dications with α- and β-protonated pyrrole rings in 3:1 ratio (Scheme 2.15).

In the case of 2-(2-thienyl)-N-vinylpyrrole in the same conditions, two dications **16** and **17** (in the ratio ~2:1) are detected (Scheme 2.16).

The reason for the different behaviors of 2-(2-furyl)pyrrolium cations in the superacidic medium and in the presence of hydrogen halides at temperature increase is likely the higher stability of the cations in HSO_3F and CF_3COOH as compared to systems containing hydrogen halides.

Selective protonation of the pyrrole ring at a low temperature can be considered as the kinetic result leading to thermodynamic nonequilibrium state with predominance of pyrrolium cations **12**, the energy value of which is comparable with that of their furanium isomers **13**. When temperature rises, the system reaches equilibrium and concentration of isomeric cations **12** and **13** levels off according to their energies.

2-(2-Thienyl)pyrrolium ions in the excess HBr do not undergo essential alterations [429,430]. This is apparently caused by the existence of higher barrier for protonation of the thiophene ring.

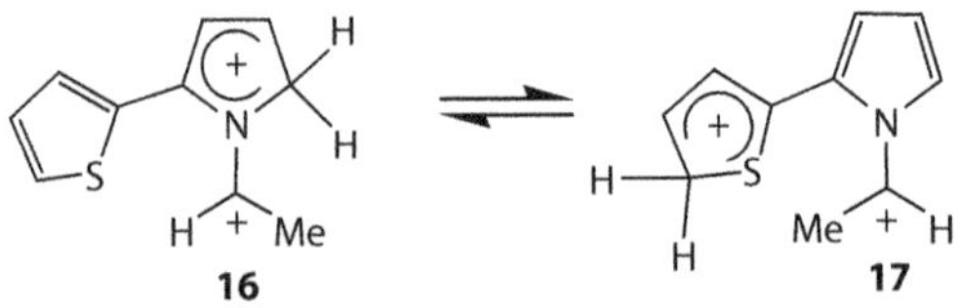

SCHEME 2.16 Equilibrium between dications of 2-(2-thienyl)-N-vinylpyrrole generated in the HSO_3F/SbF_5 system.

The theoretical analysis (MNDO) of 2-(2-hetaryl)pyrroles protonation shows [428–431] that the regioselectivity of the first stage of reaction (protonation at the C_5 of the pyrrole ring) is determined not by the charge distribution but HOMO partial electronic density. The contributions of AO to HOMO are highest for C_2 and C_5 atoms of the pyrrole ring, and they significantly exceed those for similar positions of the furan and thiophene cycles. Since the C_2 atom is sterically screened, the proton should attack mainly the C_5 atom, and that is confirmed experimentally. The subsequent protonation of the furan ring with partial deprotonation of the pyrrole moiety (formation of equimolar mixture of cations 12 and 13) is possible only if their energy values are equal. The balance of the calculated ΔH values (~176 kcal/mol for *cis*-conformations and ~178 kcal/mol for *trans*-conformations) [428] entirely proves this assumption and explains the unprecedented case of approximately equal reactivity of the pyrrole and furan cycles in thermodynamically controlled electrophilic process.

The protonation of the thiophene ring by superacidic system HSO_3F/SbF_5 evidences the contribution of not only orbital but also charge control. The highest charges in 2-(2-thienyl)pyrrole molecule belong to C_5 thiophene and C_4 pyrrole atoms. These atoms are the most probable centers of the proton attack that is proved by protonation of the thiophene ring in this case. Dication 17 with the protonated thiophene moiety is preserved upon temperature increase; moreover, its contents even increase. This is in full agreement with the calculated values of heat formation [430], which show that thermodynamic stability of thienylpyrrolium and pyrrolothiophenium ions is close (ΔH ~ 216 kcal/mol for *cis*-conformations and ~217 kcal/mol for *trans*-conformations, respectively).

Thus, the protonation of 2-(2-furyl)- and 2-(2-thienyl)pyrroles leads to, depending on the reaction conditions, the equilibrium mixtures of pyrrolium and furanium as well as pyrrolium and thiophenium ions that represent an example of successful competition of the pyrrole, furan, and thiophene cycles in thermodynamically controlled reaction with electrophiles.

2.1.2 HYDROGENATION AND DEHYDROGENATION

2.1.2.1 Hydrogenation

4,5,6,7-Tetrahydroindole and N-vinyl-4,5,6,7-tetrahydroindole, which are now readily prepared from cyclohexanone oxime and acetylene, may become a source of difficult-to-obtain octahydroindole and N-ethyloctahydroindole.

The catalytic hydrogenation of N-vinyl-4,5,6,7-tetrahydroindole over Raney Ni (ethanol, 50°C –90°C, hydrogen pressure 40–60 atm) proceeds selectively to deliver N-ethyl-4,5,6,7-tetrahydroindole in 90% yield (Scheme 2.17) [432]. When the reaction temperature increases up to 140°C, only N-ethyloctahydroindole is formed (96% yield) [433].

Hydrogenation at 120°C affords a mixture of N-ethyloctahydro- and N-ethyl-4,5,6,7-tetrahydroindoles. Raising the temperature to 200°C at a hydrogen pressure of 50–60 atm does not give rise to any significant hydrogenolysis: only traces of the cleavage product are present in the reaction mixture. N-Ethyloctahydroindole (89% yield) has been obtained instead of the expected unsubstituted NH-octahydroindole

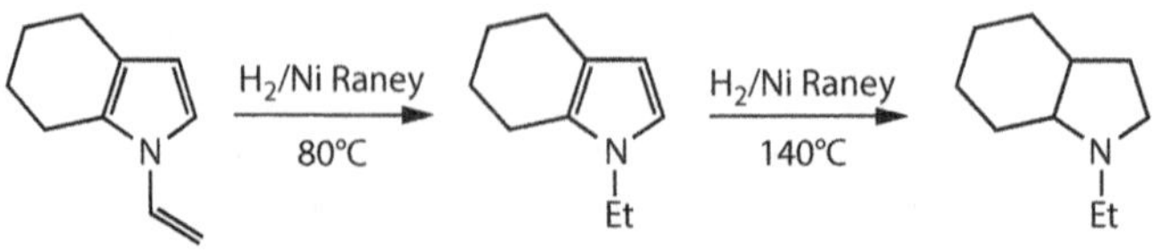

SCHEME 2.17   Catalytic hydrogenation of N-vinyl-4,5,6,7-tetrahydroindole.

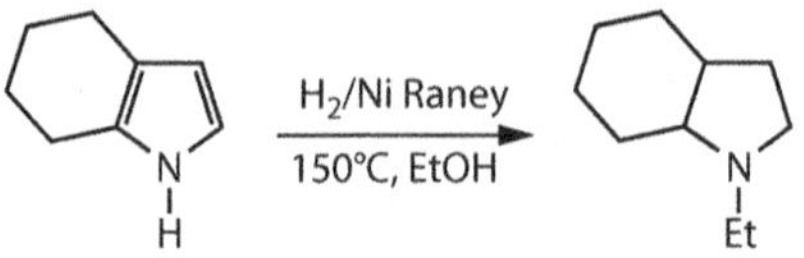

SCHEME 2.18   Ethylation of 4,5,6,7-tetrahydroindole during its catalytic hydrogenation in ethanol.

in the catalytic hydrogenation of 4,5,6,7-tetrahydroindole in ethanol, that is, alkylation with ethanol proceeds along with reduction (Scheme 2.18).

It has been reported [434,435] on the alkylation of ammonia, aliphatic, and some aromatic amines by primary and secondary alcohols upon heating under hydrogenation on Raney Ni. However, the data on alkylation of indole or pyrrole in similar conditions were absent until the work [433]. In this work, it has been shown that reductive alkylation may also be affected by other primary alcohols (e.g., butanol), although in the latter case, reduction proceeds with greater difficulty than in ethanol: N-butyloctahydroindole is detected in the reaction mixture only by chromatography.

In an effort to synthesize hardly available octahydroindole [436,437], free of its ethyl derivative, the hydrogenation of 4,5,6,7-tetrahydroindole has been carried out in tetrahydrofuran. However, the resulting reaction mixture (GLC) contains only 70% of octahydroindole along with N-butyloctahydroindole (21%) and 2-ethylcyclohexylamine (10%) (Scheme 2.19). N-Butyloctahydroindole is likely formed via butylation of octahydroindole with butanol (a product of tetrahydrofuran hydrogenolysis).

Octahydroindole is prepared by the hydrogenation of 4,5,6,7-tetrahydroindole over the Adams catalyst [438], the reaction in this case being accompanied by polymerization.

The hydrogenation of 2-phenyl- and 3-alkyl-2-phenylpyrroles and their N-vinyl derivatives under the aforementioned conditions gives a complex mixture of products [433].

Selective hydrogenation of the aforementioned N-vinylpyrroles at the vinyl group has been accomplished over Raney Ni under atmospheric pressure and at room

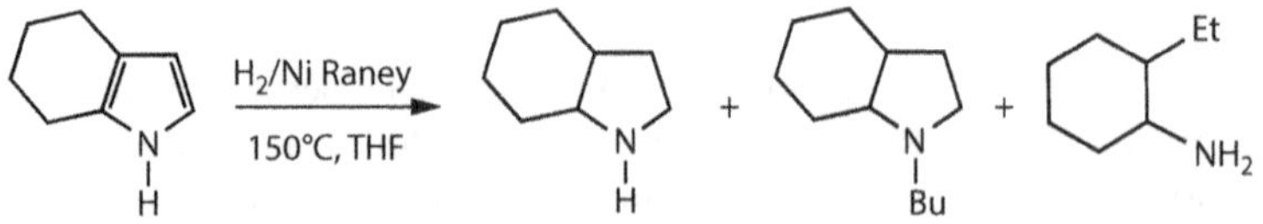

SCHEME 2.19   Products of catalytic hydrogenation of 4,5,6,7-tetrahydroindole in THF.

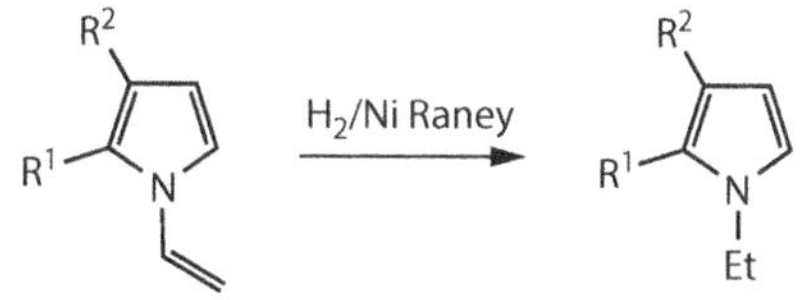

$R^1 = Ph; R^2 = H, Me, Et, n\text{-}Pr, i\text{-}Pr; R^1\text{-}R^2 = (CH_2)_4$

SCHEME 2.20   Synthesis of N-ethylpyrroles by selective hydrogenation of N-vinylpyrroles.

temperature as well as in autoclave at 50°C–90°C and initial hydrogen pressure 40–60 atm (Scheme 2.20) [432].

Yields and constants of products of N-vinylpyrrole hydrogenation are given in Table 2.1.

2.1.2.2   Selective Dehydrogenation of 4,5,6,7-Tetrahydroindole

The application of nickel sulfide deposited on γ-Al$_2$O$_3$ provides the dehydrogenation of 4,5,6,7-tetrahydroindole into indole in 96% yield and a selectivity close to 100% (Scheme 2.21) [136,137,439,440].

The novelty of this approach is in that the catalytic system, containing nano-sized nickel sulfide as the only active component, has been applied for the first time for dehydrogenation of 4,5,6,7-tetrahydroindole. Earlier, such catalytic system was not employed for the aromatization of nitrogen heterocycles bearing saturated fragments. Usually, palladium- or rare-earth element–based catalysts were used in such processes [441].

Further, this catalyst has been modified to improve its efficacy and facilitate the technology of its preparation. To reach these goals, hydrogen sulfide is replaced by aqueous solution of sodium sulfide, and nickel chloride is used instead of nickel acetate. An industrial surfactant, sodium dodecylsulfonate, is applied for better impregnation of aluminum oxide.

As a consequence, operation time of the catalyst increased by 4–5 times and its regeneration was significantly facilitated (heating at 370°C on air for 10 h). The catalyst allows carrying out the dehydrogenation of 4,5,6,7-tetrahydroindole into indole without the use of inert gas carrier only in a solvent vapor flow (toluene) that is important for technology.

This development should be considered as the first technologically real synthesis of indole on the basis of industrial cyclohexanone oxime.

2.1.2.3   Dehydrogenation of 4,5-dihydrobenz[g]indole

The catalytic system NiS/γ-Al$_2$O$_3$ is successfully employed for the preparation of benz[g]indole via the dehydrogenation of easily accessible (see Section 1.1.5) 4,5-dihydrobenz[g]indole [442]. Unexpectedly, on the "fresh" catalyst, benz[e] indole has been detected as the major product (70%). As the reaction time increases, the concentration of e-isomer in the catalysate drops and that of the target g-isomer increases, which after 2.5 h becomes the main reaction product (70%) (Scheme 2.22).

TABLE 2.1

Hydrogenated and Dehydrogenated Pyrrole Derivatives

No.	Structure	Yield (%)	Bp, °C/torr (Mp, °C)	d_4^{20}	n_D^{20}	References
1		40	41–42/2		1.4881	[433]
2		94	68/5	0.9008	1.4739	[433]
3		90	66–67/1	0.9783	1.5200	[433]
4		93	114/4–5	1.0175	1.5795	[432]
5		85	94/1	1.0169	1.5730	[432]
6		81	135–136/2	0.9931	1.5595	[432]
7		80	126/1	0.9846	1.5535	[432]
8		85	136–137/4	0.9760	1.5480	[432]
9		94	85–89/3	1.0416	1.5670	[628]

(Continued)

TABLE 2.1 (*Continued*)

Hydrogenated and Dehydrogenated Pyrrole Derivatives

No.	Structure	Yield (%)	Bp, °C/torr (Mp, °C)	d_4^{20}	n_D^{20}	References
10	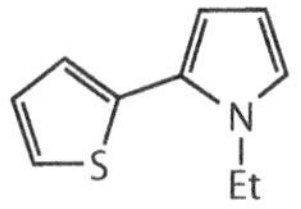	46	98–99/3	1.1118	1.6056	[628]
11	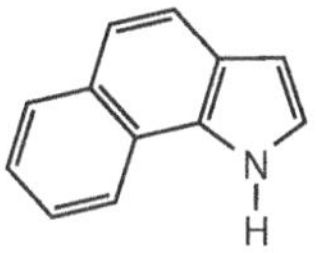	70	(181)			[442]
12	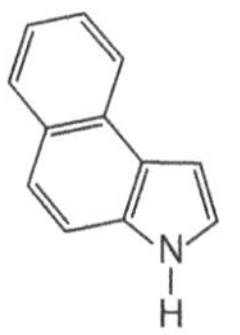	18	Oil			[442]

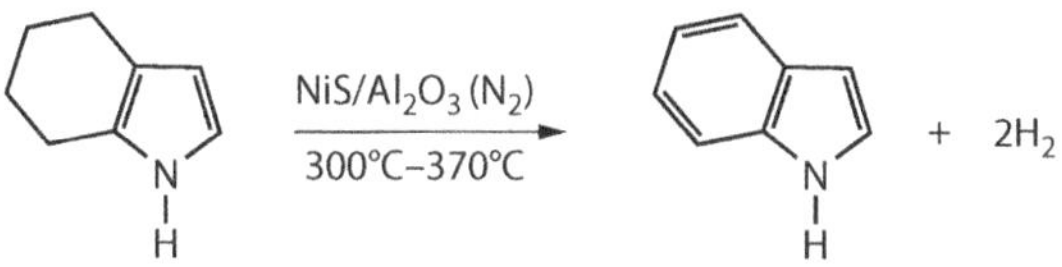

SCHEME 2.21   Selective synthesis of indole from 4,5,6,7-tetrahydroindole.

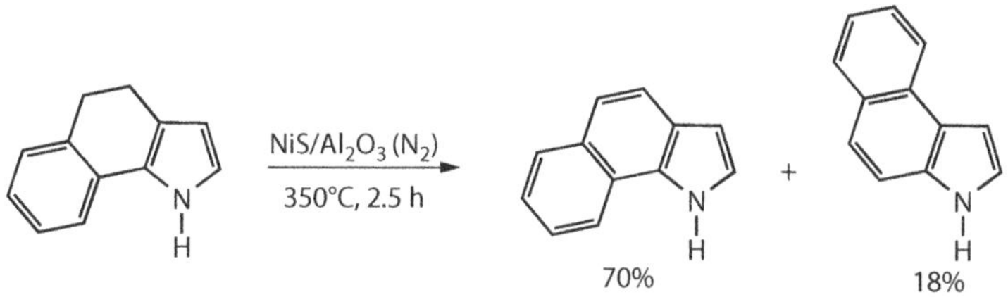

SCHEME 2.22   Rearrangement during dehydrogenation of 4,5-dihydrobenz[g]indole on the NiS/γ-Al$_2$O$_3$ catalyst.

The rearrangement is likely triggered by protonation of the pyrrole ring (Scheme 2.23). Positive charge is concentrated in its α′-position where it becomes most stable as "benzyl," "allyl," and "iminium" cation (due to the stabilization with pyrrole nitrogen). Further, synchronic transfer of electron pairs with cleavage of the C$_{\beta'}$–C4 and C$_{\alpha'}$–C9a bonds to form two novel C–C-bonds (C$_{\beta'}$–C9a and C$_{\alpha'}$–C4) follows. Deprotonation of the rearranged cation **A** and dehydrogenation of 4,5-dihydrobenz[e] indole afford benz[e]indole.

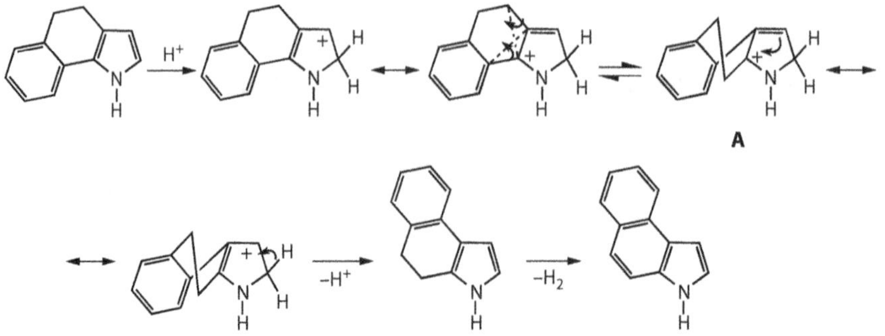

SCHEME 2.23 Plausible scheme of the rearrangement during the dehydrogenation of 4,5-dihydrobenz[g]indole on the NiS/γ-Al$_2$O$_3$ catalyst.

Carbocationic nature of the rearrangement can be caused by the acidic centers of the carrier [γ-Al$_2$O$_3$·(H$_2$O)$_n$], which are gradually quenched by the basic products of pyrrole ring side oligomerization.

A comparison of the experimental results with the quantum-chemical calculations data [B3LYP/6-31G(d) and MP2/6-311G(d,p)] confirms [442] that the reaction mixture does not reach an equilibrium and the ratio between benz[e]- and benz[g] indoles is controlled by kinetic factors such as the concentration of protogenic centers in the catalyst and their evolution during the reaction.

A synthetic advantage of this method is that it allows rare isomers of benzindole (Table 2.1) to be synthesized from readily available 1-tetralone (see Section 1.1.5) in two simple preparative steps.

The skeleton rearrangement occurring in the course of 4,5-dihydrobenz[g]indole dehydrogenation, apart from its practical utility, unveils novel facets of benzindoles behavior under heterogeneous catalysis conditions.

2.1.3 REACTIONS WITH ELECTROPHILIC ALKENES

2.1.3.1 Nucleophilic Addition to Vinyl Sulfones

Pyrroles easily add to divinyl sulfone in the presence of catalytic amounts of metal hydroxide(MOH) to form selectively di[2-(pyrrol-1-yl)ethyl]sulfones in up to 95% yields (Scheme 2.24, Table 2.2) [443].

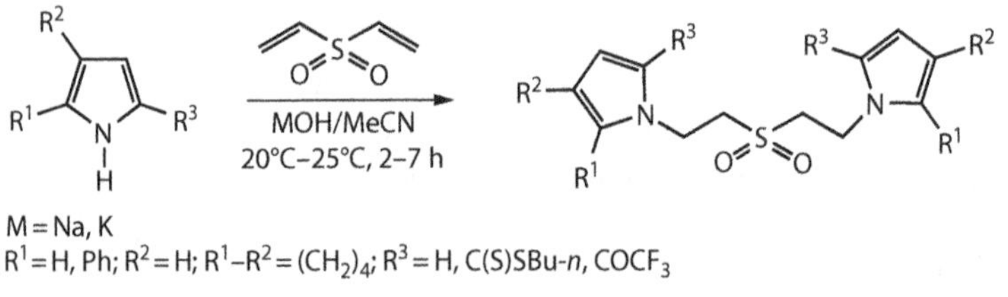

M = Na, K
R^1 = H, Ph; R^2 = H; R^1–R^2 = (CH$_2$)$_4$; R^3 = H, C(S)SBu-n, COCF$_3$

SCHEME 2.24 Addition of pyrroles to divinyl sulfone.

TABLE 2.2
Adducts of Pyrroles with Vinyl Sulfones

No.	Structure	Yield (%)	Mp (°C)	References
1		95	121	[443]
2		91	84–85	[443]
3		94	256	[443]
4		90	200 (decomp.)	[443]
5		78	142	[443]
6		76	135–136	[443]
7		16		[443]
8		68	83–85	[443]
9		95	95–96	[443]
10		45	82–83	[443]

(Continued)

TABLE 2.2 (*Continued*)
Adducts of Pyrroles with Vinyl Sulfones

No.	Structure	Yield (%)	Mp (°C)	References
11	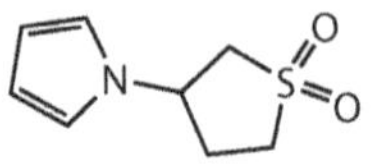	93	174–175	[443]
12	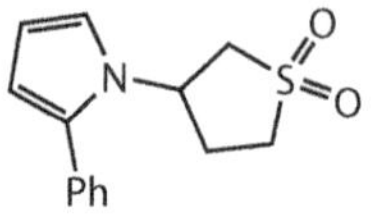	18	160–161	[443]

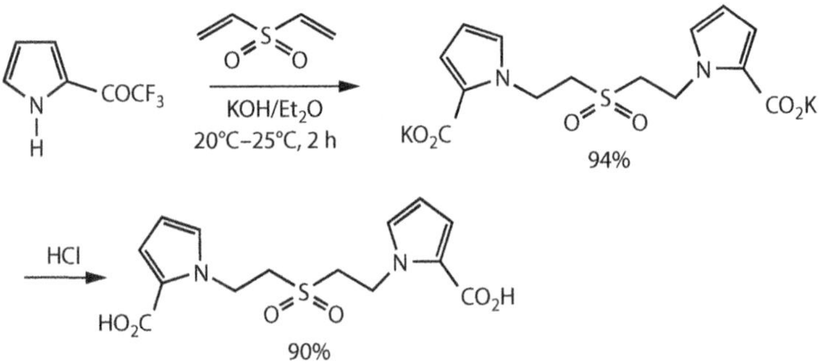

SCHEME 2.25 Haloform cleavage during nucleophilic addition of 2-trifluoroacetylpyrrole to divinyl sulfone.

The reaction of 2-trifluoroacetylpyrrole with divinyl sulfone in ether is accompanied by haloform splitting of the trifluoroacetyl moiety to furnish the corresponding salt and (after acidification) pyrrolecarboxylic acid (Scheme 2.25, Table 2.2).

Nucleophilic addition of pyrroles to phenylvinyl sulfone successfully proceeds at higher concentration of KOH (equimolar amount) and the increased reaction time (Scheme 2.26, Table 2.2) [443].

The reaction of pyrrole and 2-phenylpyrrole with 2,5-dihydrothiophene-1,1-dioxide, isomerized under the action of the base into 4,5-dihydrothiophene-1,1-dioxide

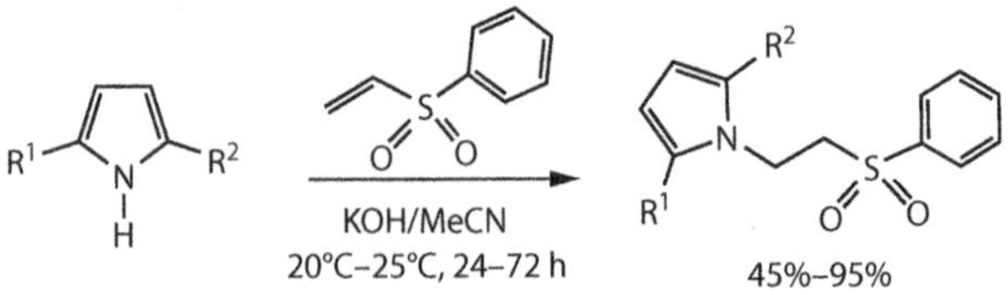

$R^1 = H, Ph; R^2 = COCF_3$

SCHEME 2.26 Addition of pyrroles to phenylvinyl sulfone.

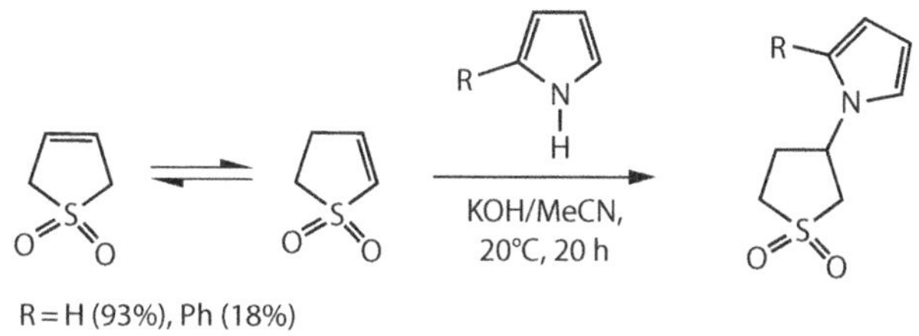

R = H (93%), Ph (18%)

SCHEME 2.27 Addition of pyrroles to 2,5-dihydrothiophene-1,1-dioxide.

with electrophilic double bond, leads to the corresponding 3-(pyrrol-1-yl)thiolane-1,1-dioxides (Scheme 2.27, Table 2.2) [443].

2.1.3.2 Reactions with Tetracyanoethylene

The interaction of pyrroles with tetracyanoethylene involves substitution of one of the nitrile groups by the pyrrole fragment that results in either 2-(R^3=H) [444] or 3-tricyanovinylpyrroles (R^3=Me) [445] in quantitative yields (Scheme 2.28, Table 2.3). The reaction likely proceeds via the addition–elimination mechanism.

2-Methyl-5-(2-thienyl)- and 5-(2-furyl)-2-methypyrrole can react with tetracyanoethylene both across the pyrrole and furan or thiophene ring.

It is known [446] that pyrrole is considerably more reactive toward electrophilic substitution than furan and even more reactive than thiophene. For instance, trifluoroacetylation of 2-(2-furyl)- and 2-(2-thienyl)pyrroles with trifluoroacetic acid anhydride proceeds selectively to form the corresponding 5-trifluoroacetyl-substituted pyrroles; although under the same conditions, 2-(2-furyl)-N-vinylpyrrole is selectively acetylated at the furan ring [447].

It is interesting to note in this connection that the reaction of 2-methyl-5-(2-thienyl)pyrrole with tetracyanoethylene gives rise to both 3- and 4-tricyanovinyl-5-(2-thienyl)pyrroles (90% total yield, 5:1 isomer ratio [Scheme 2.29, Table 2.3]) [445].

5-(2-Furyl)-2-methylpyrrole behaves differently in this reaction: the main direction becomes the attack at the furan ring α-position [445]. The ratio of the tricyanovinylation products at the furan ring α-position and the pyrrole ring β-positions is 2:1 (Scheme 2.30, Table 2.3).

Regioselectivity of the reaction can be rationalized taking into account the following factors: the considerably greater nucleophilicity of the furan ring as compared with that of the thiophene cycle and the steric hindrance of the 5-methyl group

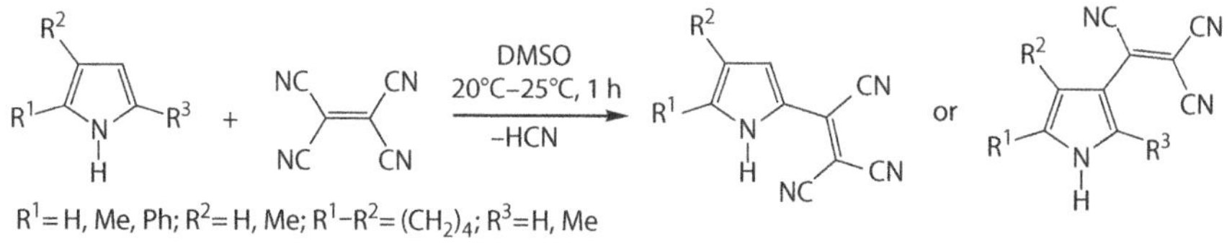

R^1=H, Me, Ph; R^2=H, Me; R^1–R^2=(CH$_2$)$_4$; R^3=H, Me

SCHEME 2.28 Tricyanovinylation of pyrroles with tetracyanoethylene.

TABLE 2.3
Products of Reaction of Pyrroles with Tetracyanoethylene

No.	Structure	Yield (%)	Mp (°C)	References
1	2	3	4	5
1		92	211–212	[444]
2		91	202–203	[444]
3		88	218	[444]
4		91	148	[448,449]
5		51	109	[449]
6		89	148	[448]
7		88	182	[448]
8		91	209	[448]

(Continued)

TABLE 2.3 (*Continued*)

Products of Reaction of Pyrroles with Tetracyanoethylene

No.	Structure	Yield (%)	Mp (°C)	References
1	2	3	4	5
9		89	170–172	[448]
10		7	101	[453]
11		7	155	[453]
12		6	163	[453]
13		85	152	[445]
14		86	210	[445]
15		94	241–242	[445]
16		18	230	[445]

(*Continued*)

TABLE 2.3 (*Continued*)

Products of Reaction of Pyrroles with Tetracyanoethylene

No.	Structure	Yield (%)	Mp (°C)	References
1	**2**	**3**	**4**	**5**
17		75	240	[445]
18		9	130	[445]
19		15	180	[445]
20		57	240	[445]
21		66	186–188	[448]
22		36	148	[448]
23		32	167	[448]

(Continued)

TABLE 2.3 (*Continued*)
Products of Reaction of Pyrroles with Tetracyanoethylene

No.	Structure	Yield (%)	Mp (°C)	References
1	2	3	4	5
24		25	190	[448]
25		28	191	[448]
26		92	132	[445]
27		91	149–150	[445]
28		9	132–133	[453]
29		8	160	[453]
30		7	175	[453]

(Continued)

TABLE 2.3 (*Continued*)
Products of Reaction of Pyrroles with Tetracyanoethylene

No.	Structure	Yield (%)	Mp (°C)	References
1	2	3	4	5
31		95	236	[453]
32		89	238–239	[453]
33		92	252	[453]
34		78	137	[453]
35		71	137	[453]
36		89	159–160	[453]

(Continued)

TABLE 2.3 (*Continued*)
Products of Reaction of Pyrroles with Tetracyanoethylene

No.	Structure	Yield (%)	Mp (°C)	References
1	2	3	4	5
37	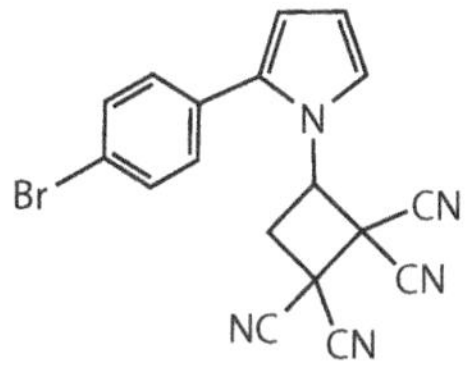	92	179–180	[453]
38	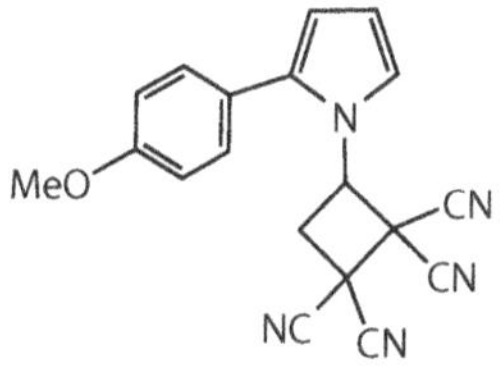	94	144–145	[453]
39	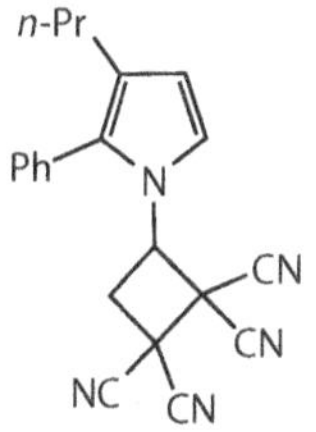	88	130–131	[453]
40	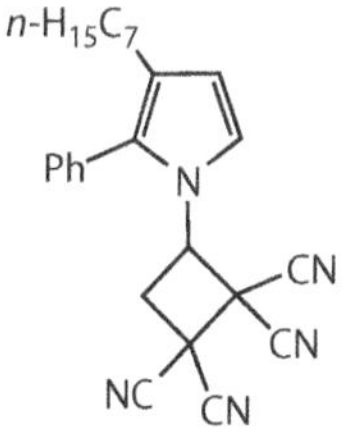	88	119–120	[453]

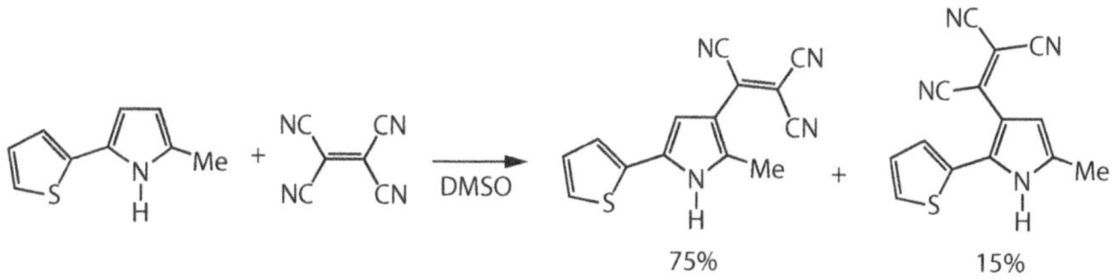

SCHEME 2.29 Tricyanovinylation of 2-methyl-5-(2-thienyl)pyrrole.

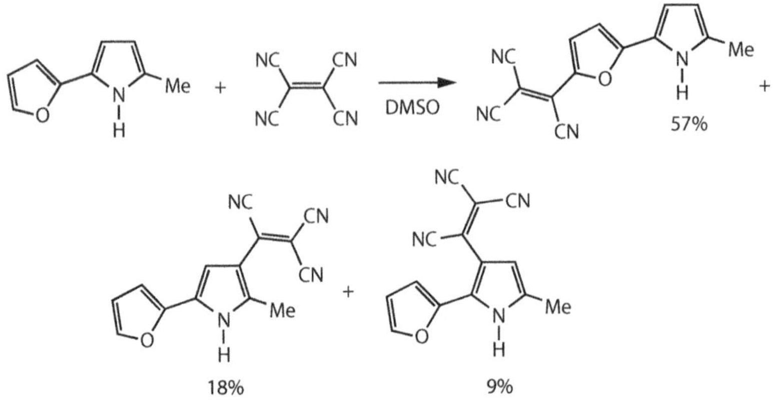
Me
NC
CN
DMSO
57%
18%
9%

SCHEME 2.30 Tricyanovinylation of 2-methyl-5-(2-furyl)pyrrole.

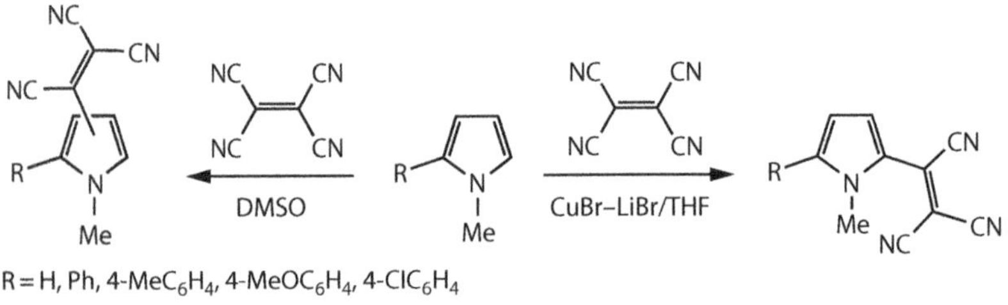
DMSO
CuBr–LiBr/THF
R = H, Ph, 4-MeC6H4, 4-MeOC6H4, 4-ClC6H4

SCHEME 2.31 Selective tricyanovinylation of N-methylpyrroles in the presence of complex CuBr–LiBr.

impeding the attack at neighbor β-positions of the pyrrole ring. Double tricyanovinylation does not occur, probably, due to strong deactivating effect of the tricyanovinyl substituent that is transmitted from the pyrrole or furan ring to the furan and pyrrole cycle, respectively.

N-Methylpyrroles react with tetracyanoethylene in various solvents (acetone, tetrahydrofuran (THF), dimethylsulfoxide(DMSO)) nonselectively to form mixtures of 3-, 4-, and 5-tricyanovinylpyrroles (Scheme 2.31, Table 2.3) [448,449]. However, in the presence CuBr–LiBr complex (soluble in THF), the reaction proceeds regioselectively to give exclusively the products of tricyanovinylation at α-position of the pyrrole ring [448].

When vinyl or isopropenyl groups are introduced in the pyrrole molecule, apart from the previous reaction, the [2+2]-cycloaddition becomes also possible. This process, typical for N-vinylheterocycles, yields cyclobutane derivatives of type **18** (Scheme 2.32) [450–452].

However, N-vinyl- and N-isopropenyl-3-tricyanovinyl derivatives (more than 90% yields) are the only products of the reaction between N-vinyl- and N-isopropenyl-2-methyl-4,5,6,7-tetrahydroindoles with tetracyanoethylene in DMSO (Scheme 2.32, Table 2.3) [445].

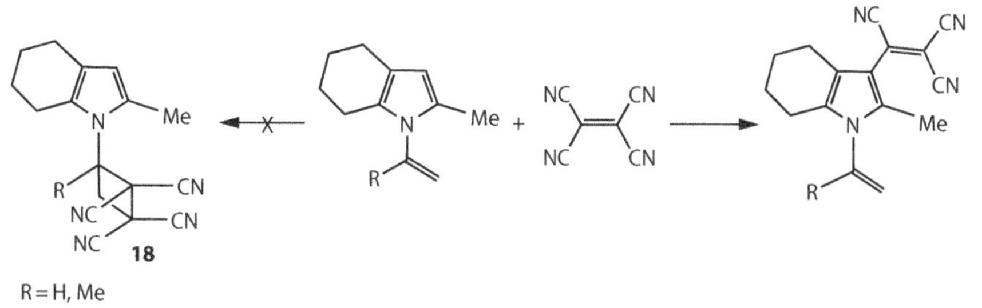

SCHEME 2.32 Tricyanovinylation of N-vinyl- and N-isopropenyl-2-methyl-4,5,6,7-tetrahydroindoles.

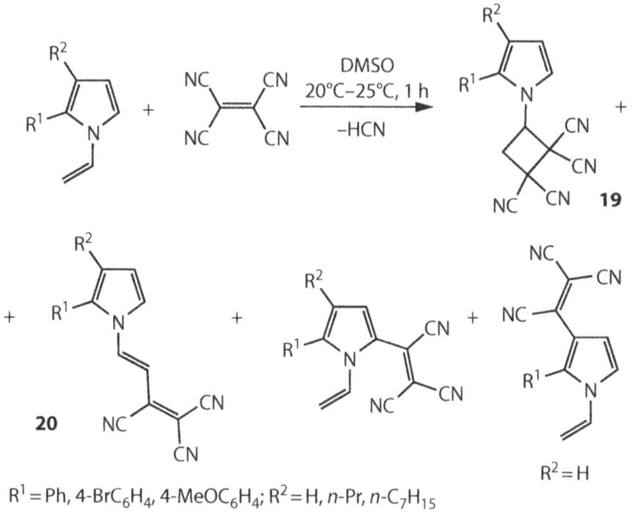

$R^1 = Ph, 4\text{-}BrC_6H_4, 4\text{-}MeOC_6H_4; R^2 = H, n\text{-}Pr, n\text{-}C_7H_{15}$

SCHEME 2.33 Tricyanovinylation of N-vinylpyrroles.

2-Aryl-N-vinylpyrroles in acetone, THF, and benzene react with tetracyano-ethylene chemo- and regiospecifically across the vinyl group to afford 3-(2-aryl-pyrrol-1-yl)-1,1,2,2-cyclobutanetetracarbonitriles **19** (Scheme 2.33, Table 2.3). The latter, upon recrystallization from EtOH, eliminate hydrogen cyanide(HCN) and stereospecifically rearrange to *trans*-(3E)-4-(2-arylpyrrol-1-yl)-1,3-butadiene-1,1,2-tricarbonitriles **20** (Table 2.3) [453]. In DMSO, along with the previous [2 + 2]-cyclo-addition delivering pyrroles **19** and **20**, tricyanovinylation of the pyrrole ring occurs to form the corresponding 3- and 5-tricyanovinylpyrroles, the products ratio being dependent on the substituents in the pyrrole ring and the reaction conditions.

2- and 3-Tricyanovinylpyrroles possess electroconductivity and photosensitizing properties with respect to polymeric conductors. They are also paramagnetic in the solid state. In their electron-spin resonance (ESR) spectra, a singlet with concentration of paramagnetic centers of 10^{18} spin/g is observed. These properties are due to a strong charge transfer and layered packaging of crystals of these compounds [445].

2.1.4 Reactions with Acetylene

2.1.4.1 N-Vinylation

N-Vinyl derivatives of nitrogen-containing heterocycles are valuable monomers and semiproducts. They find application in the production of plastics and synthetic fibers in radio engineering and medicine (see, e.g., [96,109,454,455]). Besides, they are extensively used in the design of electrophotographic materials [456,457]. In particular, N-vinylpyrroles are frequently employed as synthetic intermediates and monomers for various polymeric materials [36,39].

The conventional methods for the synthesis of N-vinyl nitrogen-containing heterocycles have been covered in monographs [96,107–109] and reviews [93,95] and involve either dehydration of β-oxyethyl compounds [92,94,458–461], dehydrohalogenation of haloethyl derivatives [462,463], or the vinylation of NH heterocycles with acetylene [93,95,96,109], vinyl chloride [95,464,465], vinyl bromide [466], dihaloalkanes [467–469], and vinyl ethers [470]. All these processes have been used mainly for carbazole. Sirotkina et al. have successfully elaborated efficient methods for the preparation of N-vinylcarbazoles via isomerization of N-allyl derivatives [471] and catalytic decomposition of available N-(1-alkoxyalkyl)carbazoles [472]. The credit for the systematic development of direct methods for vinylation of various NH heterocycles, including indole, with acetylene is given to Skvortsova et al. [93,473,474].

N-Vinylpyrroles with functional substituents in the ring are synthesized by elimination of methanol or vinyl alcohol (Scheme 2.34) from N-(2-methoxyethyl)- [475] and N-(2-vinyloxyethyl)pyrroles, respectively [476,477].

Synthesis of N-vinylpyrroles, bearing some functional substituents in the vinyl group, from ethoxyalkenes [478] or vinyl trifluoromethanesulfonates [479] has been disclosed (Scheme 2.35).

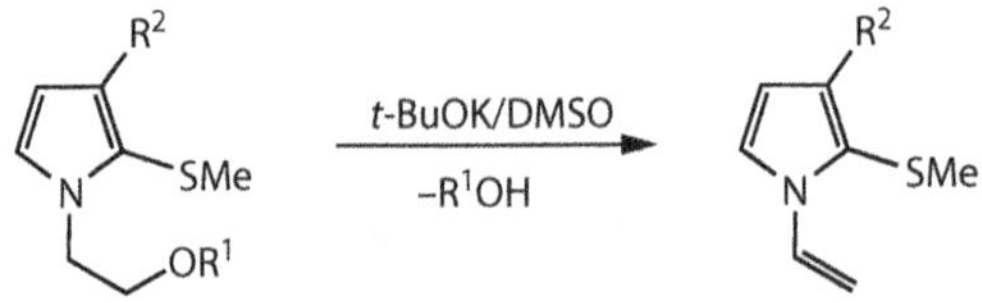

$R^1 = Me, CH=CH_2; R^2 = MeO, MeCH(OEt)O$

SCHEME 2.34 Formation of N-vinylpyrrole via superbase-induced elimination of alcohols from N-(2-methoxyethyl)- and N-(2-vinyloxyethyl)pyrroles.

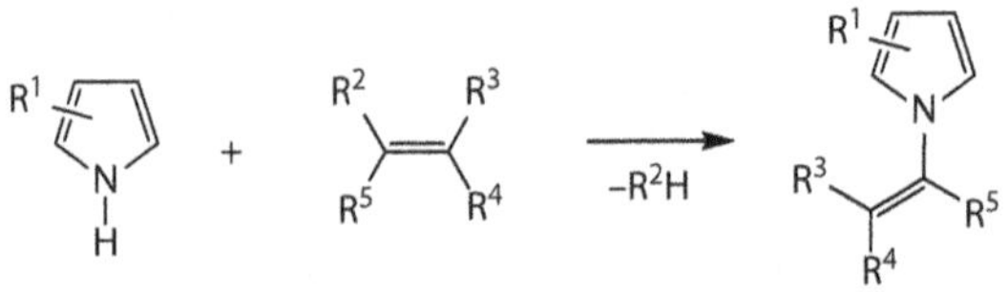

$R^1=H, Et, EtCO, CN; R^2=EtO, TfO; R^3=H, CO_2Et, CN;$
$R^4 = H, CN, CO_2Et; R^5 = H, Me, Ph; R^4-R^5 = (CH_2)_3, (CH_2)_4$

SCHEME 2.35 N-Vinylation of pyrroles with functionalized alkenes.

However, direct vinylation with acetylene remains the most widely accepted procedure for introduction of the vinyl group to heterocycle nitrogen atom [93,105,480,481].

In this line, one should recall that although N-vinylcarbazole was first prepared in 1924 without acetylene [463], its application in the polymer production became possible only after the development of acetylenic vinylation procedure. Commercial production of monomeric N-vinylcarbazole and its polymers dates back to the first years of World War II, when, for military purposes, Germany speeded up large-scale work with acetylene under pressure [95,96,109].

However, the use of pressure still restricts the introduction of such technology into practice since there are hard constraints for application of compressed acetylene in industry.

Pyrrole and its benzo derivatives are also vinylated in the gas phase under normal pressure [482,483], but this procedure requires temperature of 250°C–320°C, and the product yields are considerably lower than those obtained by liquid-phase vinylation.

Data from a systematic study of the vinylation of a number of pyrroles obtained by the reaction of acetylene with ketoximes are presented in the paper [484]. N-Vinylpyrroles are prepared in up to 97% yield (Scheme 2.36, Table 1.3).

The reaction proceeds effectively in the presence of 30% KOH in superbase systems (KOH in aprotic polar solvents such as DMSO, sulfolane, and hexametapol). KOH/DMSO turned out to be the most effective system. The use of the latter allows decreasing the reaction temperature to 80°C–100°C, which is almost 100°C lower than the temperature of classical vinylation of NH heterocycles. The application of this system makes it also possible to carry out the vinylation of pyrroles at the close-to-atmospheric acetylene pressure (1.1–1.5 atm).

As mentioned earlier, in the system KOH/DMSO, there is extra activation of pyrrole anions with a simultaneous increase in their concentration, as well as activation of acetylene (see Section 1.1.1). These effects lead to a significant increase in the reaction rate and allow reducing excessive acetylene pressure.

The method used for vinylation of pyrroles in the system alkali metal hydroxide/ DMSO has been extended to phenylacetylene, the latter being successfully employed for vinylation of indole and carbazole (Scheme 2.37) [489].

The reaction proceeds regio- and stereospecifically under mild conditions (100°C–110°C, 20% KOH from the vinylated compound weight) to afford Z-isomers of N-(2-phenylvinyl)indole and N-(2-phenylvinyl)carbazole in 80% and 97% yields, respectively.

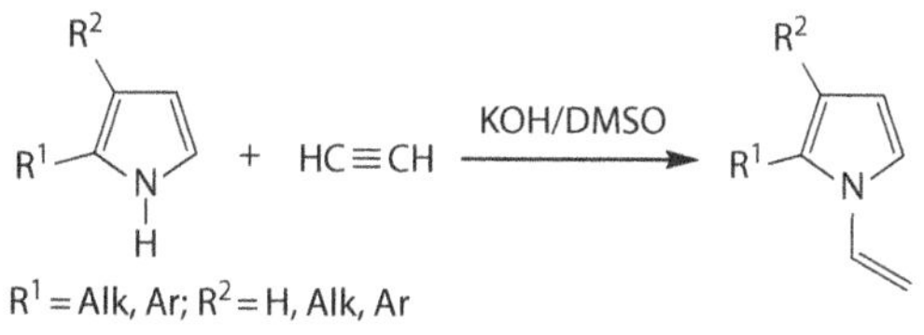

$R^1 = $ Alk, Ar; $R^2 = $ H, Alk, Ar

SCHEME 2.36 N-Vinylation of pyrroles with acetylene in the KOH/DMSO system.

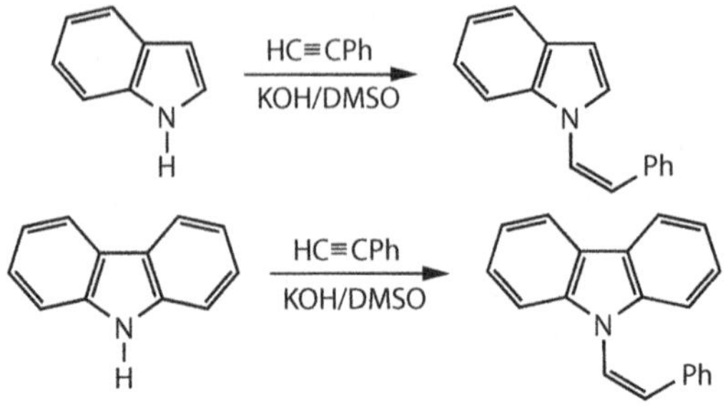

SCHEME 2.37 N-Vinylation of indole and carbazole with phenylacetylene.

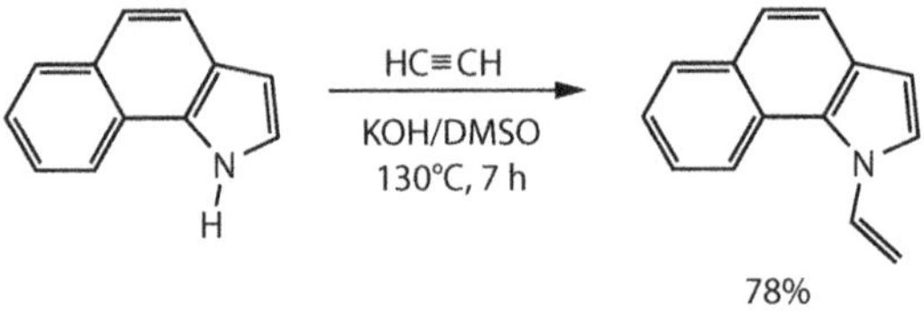

SCHEME 2.38 N-Vinylation of benz[g]indole with acetylene.

The discovery of highly reactive superbase catalytic system KOH/DMSO has provided new wide and fundamental opportunities for direct vinylation of NH heterocycles with acetylene. Using this method, large pilot batches of N-vinyl-4,5,6,7-tetrahydroindole have been produced in Angarsk plant of chemicals [179,490]. Direct vinylation of carbazole with acetylene under atmospheric pressure (100°C, 5–7 h) in the system KOH/DMSO furnishes especially pure N-vinylcarbazole in a high yield [491–493]. Its structural isomer, N-vinylbenz[g]indole, is synthesized in 78% yield by vinylation of benz[g]indole (85% conversion) under atmospheric pressure in the same system (Scheme 2.38, Table 2.3) [442].

Systematic studies of acetylenic vinylation of pyrroles in the system KOH/DMSO have shown that the reaction slows down in the presence of nitroxyl radicals and other inhibitors of radical processes [205,494]. This indicates a possibility of single electron transfer in nucleophilic addition of pyrroles to acetylene in superbase media.

On the example of the reaction of 4,5,6,7-tetrahydroindole with acetylene and phenylacetylene in the presence of radical trap, 2-methyl-2-nitrosopropane (Scheme 2.39), it has been established that vinylation includes a stage of the

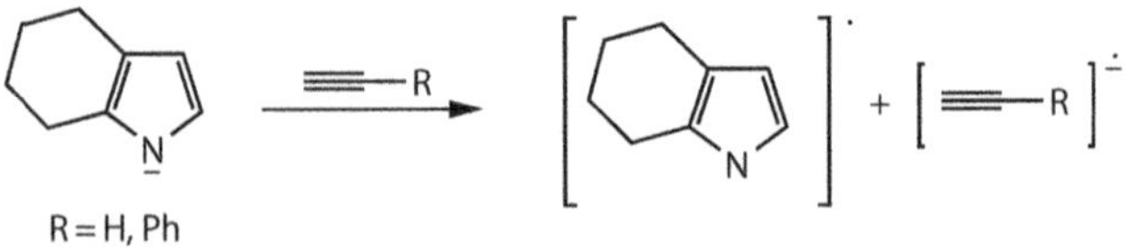

R = H, Ph

SCHEME 2.39 Formation of 4,5,6,7-tetrahydroindolyl radical in the vinylation of 4,5,6,7-tetrahydroindole with acetylenes.

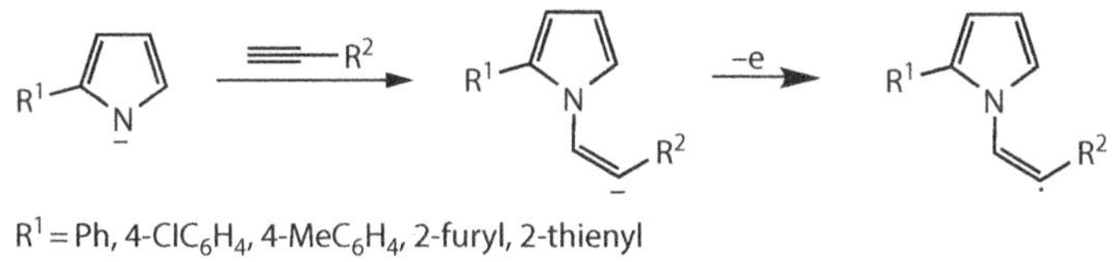

$R^1 = Ph, 4-ClC_6H_4, 4-MeC_6H_4, 2-furyl, 2-thienyl$

SCHEME 2.40 Formation of free radical adducts of 2-arylpyrroles to acetylenes.

pyrrole radical formation (probably due to the electron transfer from the pyrrole anion to antibonding π-orbitals of acetylene [495]).

During the vinylation of 2-arylpyrroles in the presence of the same radical trap, the ESR signals of vinyl-*tert*-butylnitroxyl and spin adducts of *t*-BuNO with the adducts of 2-substituted pyrrole radicals to acetylenes have been detected. N-Centered radicals of 2-arylpyrrole are quite stable and are observed directly in the ESR spectra (Scheme 2.40) [496].

Also, the mechanism of pyrrole vinylation with acetylenes in the system KOH/DMSO was studied using ESR technique on the example of the addition of unsubstituted pyrrole to cyanophenylacetylene [497,498].

2.1.4.2 Reactions with Electrophilic Acetylenes

Pyrrole and 2-phenylpyrrole react with acyl- and cyanophenylacetylenes (KOH/DMSO, 20°C–25°C, 3 h) exclusively as N-nucleophiles to afford, depending on the starting pyrrole structure, either predominantly *E*- or only *Z*-adducts (in 32%–69% yields, Scheme 2.41, Table 2.4) [28,499].

The reaction between unsubstituted pyrrole and benzoylphenylacetylene is complicated by alkaline hydrolysis to give 1,3-diphenylpropan-1,3-dione in enol form, phenylacetylene, and benzoic acid (Scheme 2.42).

Under similar conditions, 4,5,6,7-tetrahydroindole selectively adds to benzoylacetylene to deliver C-adduct (*E*/*Z* = 5:1) (Scheme 2.43) [500]. This is the first example of C-vinylation of pyrroles in superbase medium.

The same pyrrole reacts with benzoylphenylacetylene under similar conditions to afford both C- (*Z*-isomer) and N-adducts (*E*- and *Z*-isomers) in 56% overall yield (Scheme 2.44) [499,500]. Besides, a condensed tricyclic system is isolated (24% yield), which is formed due to cyclization of diadduct **A** [499,501]. The latter is not detected among the reaction products.

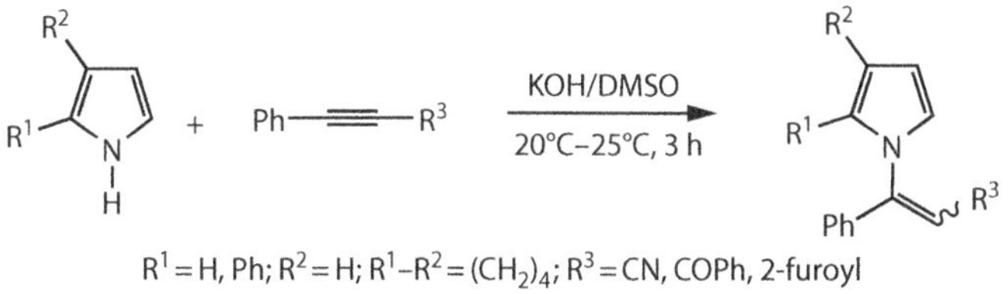

$R^1 = H, Ph; R^2 = H; R^1-R^2 = (CH_2)_4; R^3 = CN, COPh, 2-furoyl$

SCHEME 2.41 Nucleophilic addition of pyrroles to acyl- and cyanophenylacetylenes.

TABLE 2.4

Adducts of Pyrroles with Electrophilic Acetylenes

No.	Structure	Yield (%)	Mp (°C)	References
1	2	3	4	5
1		90	Oil	[499]
2		89 48	Oil (E/Z 1:1.2) 95–96 (Z)	[499]
3		88 78	106–107 (Z) Oil (E/Z 1:4)	[499]
4		32	Oil (E/Z 4:1) 82–83 (E/Z 9:1)	[499]
5		54	92–93 (E/Z 10:1)	[499]
6		26	Oil (E/Z 2:1)	[499,500]
7		68	138–139	[499]
8		57	158–159	[499]

(Continued)

TABLE 2.4 (*Continued*)
Adducts of Pyrroles with Electrophilic Acetylenes

No.	Structure	Yield (%)	Mp (°C)	References
1	2	3	4	5
9		72	131–134 (*E*) 98–99 (*Z*)	[503,504]
10		61	134–140 (*E/Z* 2:1)	[504]
11		6	72–76	[511]
12		59	72–74 (*E/Z* 5:1)	[504]
13		54	136–138 (*E/Z* 5:1)	[500,504]
14		87	152 (*E/Z* 5:1)	[504]
15		80	190–191 (*E*) 68–69 (*Z*)	[503,504]
16		90	199–201 (*E*)	[504]

(Continued)

TABLE 2.4 (*Continued*)
Adducts of Pyrroles with Electrophilic Acetylenes

No.	Structure	Yield (%)	Mp (°C)	References
1	**2**	**3**	**4**	**5**
17		6	166–168	[511]
18		90	128–129 (*E*) Oil (*Z*)	[504]
19		20	121–123 (*E/Z* 4:1) 127 (*E*)	[510]
20		19	91–92 (*E*) 89 (*Z*)	[510]
21		43	100–101	[510]
22		74	161–162 (*E*) 112–113 (*Z*)	[499,500,510]
23		34	Oil	[510]
24		13	113–116	[510]

(Continued)

TABLE 2.4 (*Continued*)

Adducts of Pyrroles with Electrophilic Acetylenes

No.	Structure	Yield (%)	Mp (°C)	References
1	2	3	4	5
25		90	170–171	[510]
26		89	160–161	[510]
27		24	154–155	[499,501]
28		77	148–149	[505,510]
29		54	Oil	[508]
30		87	187	[505,510]

(Continued)

TABLE 2.4 (*Continued*)
Adducts of Pyrroles with Electrophilic Acetylenes

No.	Structure	Yield (%)	Mp (°C)	References
1	2	3	4	5
31		89	178	[505,510]
32		33	125–126	[510]
33		37	153–154	[506]
34		50	84	[511]
35		72	80–81	[511]
36		54	Oil	[511]
37		81	Oil	[511]
38		84	80–81	[511]

(Continued)

TABLE 2.4 (*Continued*)

Adducts of Pyrroles with Electrophilic Acetylenes

No.	Structure	Yield (%)	Mp (°C)	References
1	2	3	4	5
39	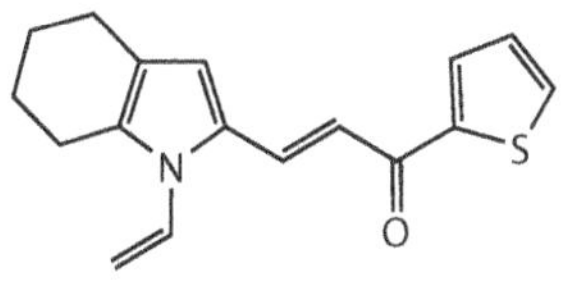	78	111–112	[512]
40	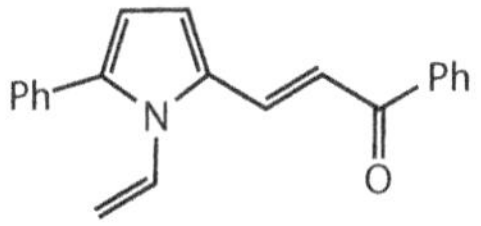	68	Oil	[511]
41	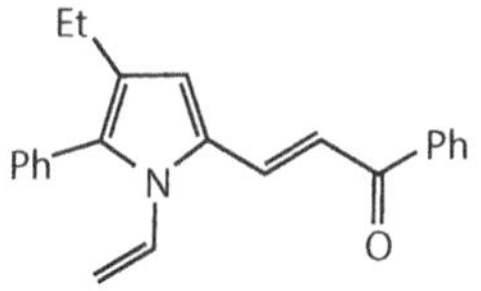	60 (*E*)	Oil	[511]
42	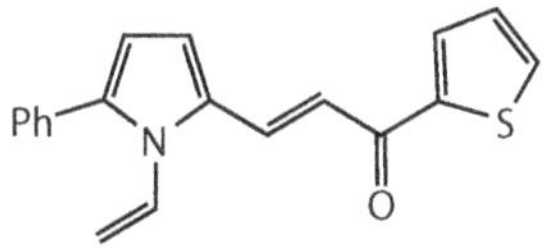	57	117–118	[512]
43	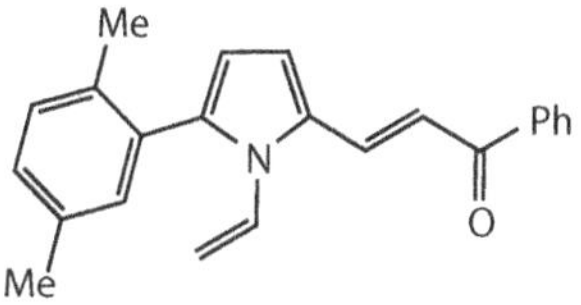	56	Oil	[511]

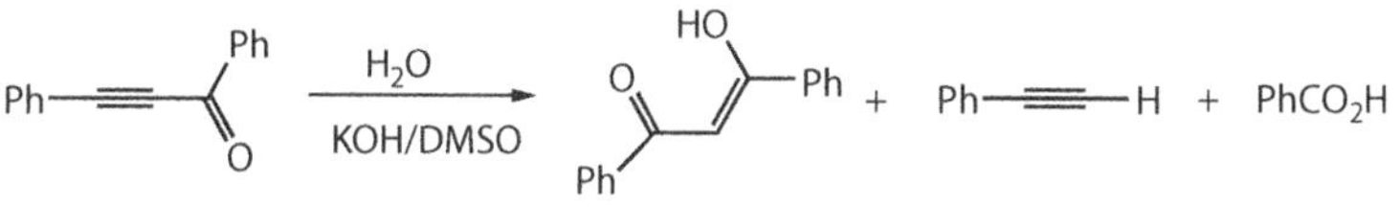

SCHEME 2.42 Side products formed from benzoylphenylacetylene during its reaction with pyrroles in the KOH/DMSO system.

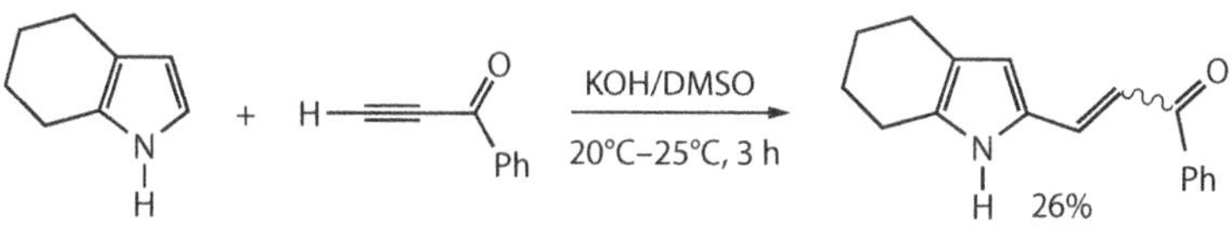

SCHEME 2.43 C-Vinylation of 4,5,6,7-tetrahydroindole with benzoylacetylene in the KOH/DMSO system.

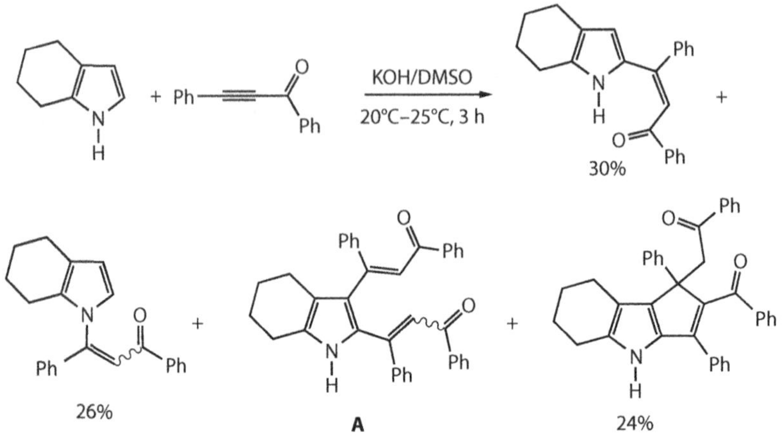

SCHEME 2.44 Reaction of 4,5,6,7-tetrahydroindole with benzoylphenylacetylene in the KOH/DMSO system.

C-Adducts are paramagnetic in solid state. Their ESR spectra show single asymmetric narrow singlets (concentration of paramagnetic centers 10^{17} spin/g, g-factor 2.004–2.005). Paramagnetism in this case is rationalized [499] by intermolecular donor–acceptor interaction involving the charge transfer. This rationale is supported by disappearance of the paramagnetism during the C-adducts dissolution as well as by its absence in N-adducts.

ESR study of the mechanism of pyrrole addition to acyl- and cyanophenylacetylenes in the superbase system KOH/DMSO evidences the formation of radicals [502]. One of the signals has been assigned to N-adducts of pyrrolyl radicals with acetylenes.

Noncatalytic addition of pyrroles as C-nucleophiles to electron-deficient triple bond ensures the most direct route to functionally substituted C-vinylpyrroles (Scheme 2.45, Table 2.4) [28,503,504].

The reaction proceeds under mild conditions (room temperature) in the absence of a solvent as well as in protic (methanol, ethanol) and aprotic (ether, benzene, hexane, acetonitrile) solvents to furnish predominantly Z-isomers of 2-(2-acylvinyl) pyrroles in 52%–87% yields. However, the Z-isomer is readily converted into the E-isomer in the course of isolation and purification or even upon storage in a solvent.

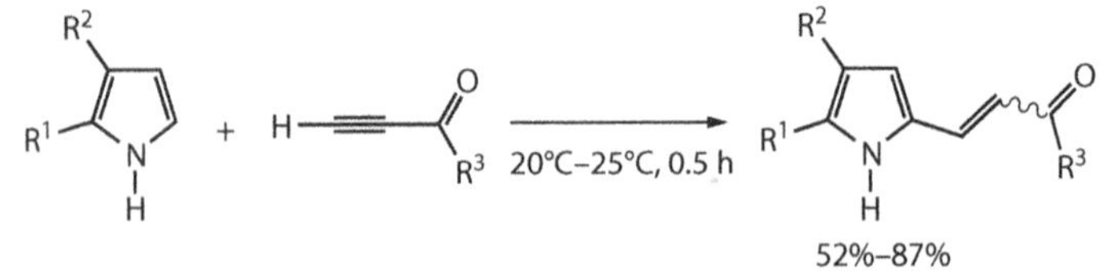

$R^1 = H,\ i\text{-Pr, Ph};\ R^2 = H,\ Me,\ n\text{-}C_7H_{15};\ R^1\text{–}R^2 = (CH_2)_4;\ R^3 = Ph,\ 2\text{-thienyl}$

SCHEME 2.45 Noncatalytic C-vinylation of pyrroles with acylacetylenes.

Monitoring of the addition of unsubstituted pyrrole to benzoylacetylene in methanol (an equimolar ratio of the reagents, room temperature) by ultraviolet (UV) spectroscopy reveals that a band (λ_{max} 396 nm) characteristic of the Z-isomer is observed
in the initial stage of the reaction [504]. Gradually, this band is replaced by another
one with λ_{max} 385 nm assigned to E-adducts. Therefore, the Z-isomer is a kinetically
controlled product that is in agreement with the rule of *trans*-nucleophilic addition
to acetylenes.

The mechanism of pyrrole interaction with acylacetylenes has been investigated
using the spin-capture technique [504]. Based on the analysis of the ESR spectra, it
can be suggested that ion-radical pair **B**, formed in the reaction under study, is not
a stable intermediate species in nonpolar medium. Proton transfer from the cation
radical of pyrrole to the anion radical of benzoylacetylene occurs, and free radicals
thus generated are captured by the spin trap (ONBu-*t*) (Scheme 2.46).

The analysis of the hyperfine structure constants indicates the formation of spin
adduct **C** (consecutive addition of two molecules of ONBu-*t* to benzoylacetylene
radical) and spin adduct **D** (capture of the pyrrole radical by the spin trap).

Thus, the experimental data confirm the presence of a one-electron transfer route
in this reaction.

Unlike benzoylacetylene, acylphenylacetylenes are less reactive and do not react
with pyrroles at room temperature. This reaction is carried out upon heating the
equimolar amounts of reagents in the presence of 5–10-fold excess of silica gel to
yield Z-2-(2-acyl-1-phenylvinyl)pyrroles (Scheme 2.47, Table 2.4) [28,505–510].

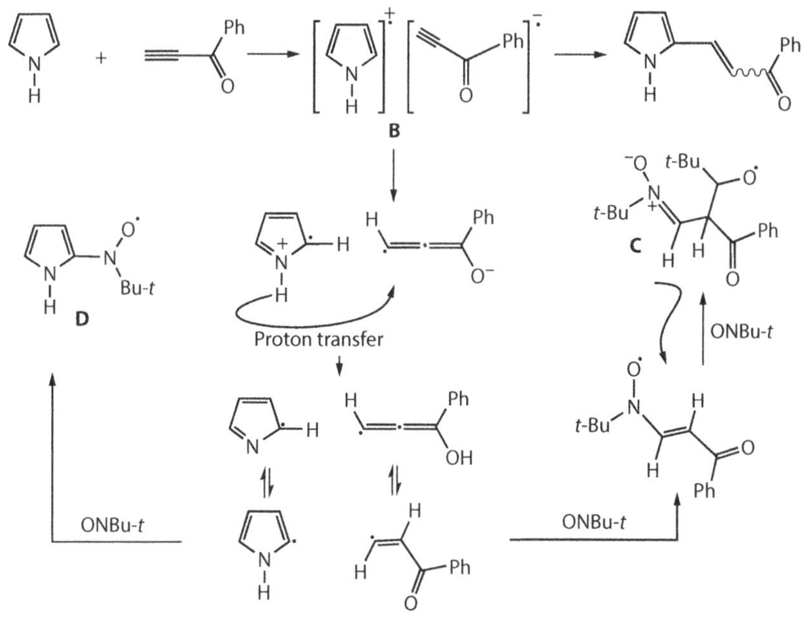

SCHEME 2.46 Evidences for the single electron transfer in the noncatalytic vinylation of
pyrrole with benzoylacetylene.

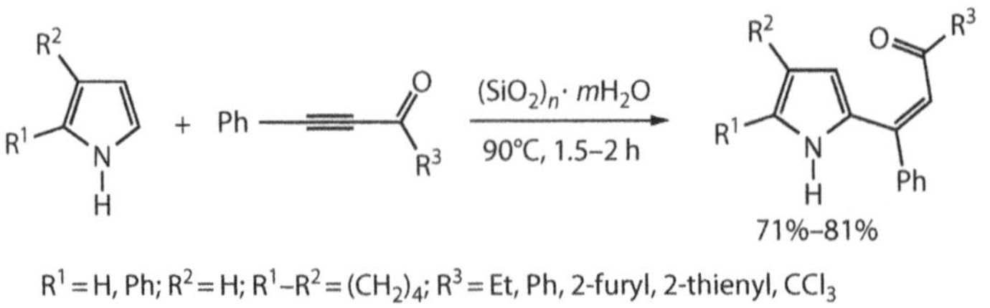
(SiO2)n· mH2O
90°C, 1.5–2 h
71%–81%
R1 = H, Ph; R2 = H; R1–R2 = (CH2)4; R3 = Et, Ph, 2-furyl, 2-thienyl, CCl3

SCHEME 2.47 C-Vinylation of pyrroles with benzoylphenylacetylene on SiO_2.

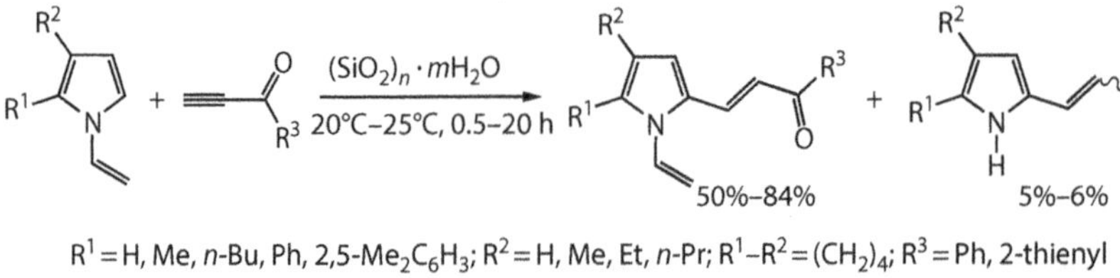
(SiO2)n·mH2O
20°C–25°C, 0.5–20 h
50%–84%
5%–6%
R1 = H, Me, n-Bu, Ph, 2,5-Me2C6H3; R2 = H, Me, Et, n-Pr; R1–R2 = (CH2)4; R3 = Ph, 2-thienyl

SCHEME 2.48 C-Vinylation of N-vinylpyrroles with benzoylacetylene on SiO_2.

Regio- and stereospecific C-vinylation of N-vinylpyrroles with benzoylacetylene proceeds at room temperature on a silica gel surface to form *E*-adducts (Scheme 2.48, Table 2.4) [28,511,512].

The reaction of pyrroles (R^1=Me, R^2=*n*-Pr or R^1=Ph, R^2=Et) with benzoylacetylene gives, along with the target adducts, C-vinylpyrroles with free NH group. They are likely formed due to the hydrolysis of the N-vinyl group of both the starting N-vinylpyrroles and the target products on $(SiO_2)_n \cdot mH_2O$ [511].

The structure of 2-(2-acylvinyl)- and 2-(2-acyl-1-phenylvinyl)pyrroles is specific: their *Z*-isomers have a strong intramolecular hydrogen bond between NH-proton and oxygen atom of the carbonyl group [506–510].

The most characteristic feature of the ^{1}H NMR spectra of 2-(2-acylvinyl)- and 2-(2-acyl-1-phenylvinyl)pyrroles is a strong downfield shift of the NH-proton signal in the Z-isomers relative to the *E*-isomers. In the case of 2-(2-acylvinyl)pyrroles, the δ NH values range from 12.8 to 13.7 ppm, while in analogous 2-(2-acyl-1-phenylvinyl) pyrroles, the signal of the NH-proton is displaced even more downfield (up to 14.8 ppm). In the spectra of the *E*-isomers, the NH-proton signal appears at 8.3–8.9 ppm. Such a strong downfield shift of the NH-proton signals results from its participation in hydrogen bond formation, which can be realized only in the Z-isomer [509,510].

Two dimensional ^{1}H NMR study has shown that both isomers of 2-(2-acylvinyl) pyrroles are in *cis*-conformation (Scheme 2.49). For example, high-field doublet of the *E*-isomer, assigned to H_α, has cross peaks with a proton of the NH group and with ortho-proton of the benzene ring, and downfield H_β doublet displays only cross peak with H_3. In the Z-isomer, H_α proton signal has cross peaks with the signal of H_β and H-*ortho*-benzene ring. Thus, *E*- and Z-isomers exist in different rotamer forms: the carbonyl moiety of the *E*-isomer is in *anti* position relative to nitrogen atom (*anti-s-cis*-rotamer), whereas in the Z-isomer, these fragments are very close to each other

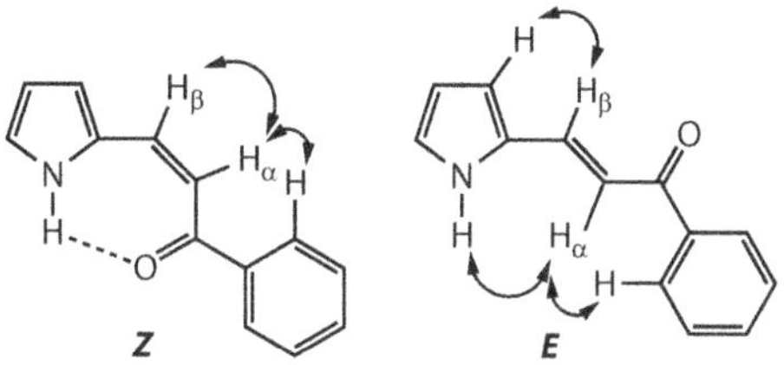

SCHEME 2.49   2D ^{1}H NMR assignment of E-and Z-isomers of 2-(2-acylvinyl)pyrroles.

(sin-s-cis-rotamer) that also confirms the existence of strong intramolecular hydrogen bond in this isomer [509].

The infrared (IR) spectra of the E-isomers of 2-(2-acylvinyl)pyrroles (CCl$_4$ solution) display only one narrow absorption band in the region 3470–3480 cm^{-1}, which is indicative of the presence of a free NH group [504].

A different situation is observed in the IR spectra of the Z-isomers. At concentrations, which completely exclude intermolecular hydrogen bonding ($C = 1 \times 10^{-4}$ mol/L), the spectra show a broad band at 3150–3180 cm^{-1}, which is attributable to the NH group with very strong intramolecular hydrogen bonding [504].

In the IR spectra of solid samples and solutions (CCl$_4$) of 2-(2-acyl-1-phenylvinyl)pyrroles, a stretching vibration band of the associated NH group at 3150–3400 cm^{-1} is absent. Frequency shift (300–400 cm^{-1}) of the stretching vibration of the N–H bond in the IR spectra of these compounds is detected only in CS$_2$ solution at 165°K–298°K [506].

The results of x-ray diffraction study of Z-2-(2-acyl-1-phenylvinyl)pyrroles [507,510] show the increased length of the C=O bond (1.244–1.249 Å) as compared to that characteristic of a free carbonyl bond (1.208 Å) in the –C=C–C=O fragment. The C=C bond in molecule also gets longer (from 1.325 Å to 1.372–1.378 Å) and the =C–C= bond is shortened (from 1.486 Å to 1.443–1.448 Å). These changes in the lengths of bonds may be ascribed to the delocalization of electron density along the 7-membered cycle formed due to intramolecular interaction N–H⋯O.

2.1.4.3   Reactions with 1-Alkylthio-2-chloroacetylenes

The main direction of the reaction of pyrroles with 1-alkylthio-2-chloroacetylenes (30°C–35°C, 1.5 h) in the system KOH/DMSO is nucleophilic substitution of chlorine atom with pyrrolate anions to afford N-(alkylthioethynyl)pyrroles in 14%–42% yield (Scheme 2.50) [513]. Apart from these pyrroles, the reaction mixture contains 2-alkylthio-1,1-bis- (45%–51%) and 1-alkylthio-1,2-bis-(pyrrol-1-yl)ethenes (3%–5%). When pyrrole–acetylene molar ratio is 2:1, 2-alkylthio-1,1-bis (pyrrol-1-yl) ethenes become major reaction products (up to 51% yield) (Table 2.5).

The reaction of 4,5,6,7-tetrahydroindole with alkylthiochloroacetylenes stops at the stage of acetylene formation (41%–42% yield, Table 2.5). It proceeds with much more difficulties than the reaction with unsubstituted pyrrole: in 5 h, the conversion of 4,5,6,7-tetrahydroindole is 56%–58%. Such distinction in behavior of these pyrroles is probably due to the steric factors. 2-Phenylpyrrole interacts with 1-propylthio-2-chloroacetylene (1:1 ratio, the same conditions) forming a mixture of

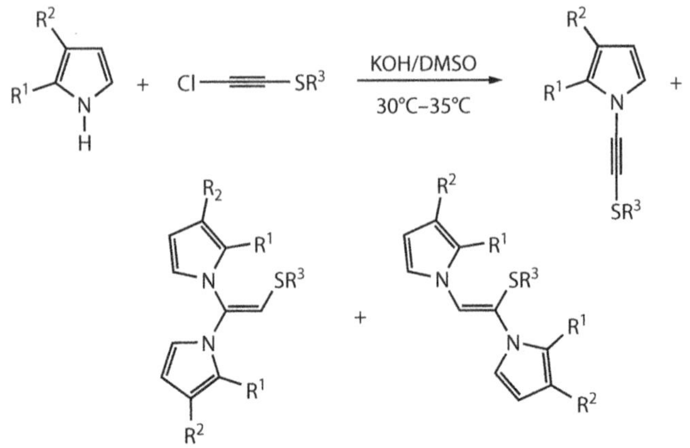

$R^1 = H, Ph; R^2 = H; R^1-R^2 = (CH_2)_4; R^3 = Et, n$-Pr

SCHEME 2.50   Reaction of pyrroles with alkylthiochloroacetylenes.

two products, 1-(2-propylthioethynyl)-2-phenylpyrrole (24%) and 2-propylthio-1,1-*bis*(2-phenylpyrrol-1-yl)ethene (yield 45%, Scheme 2.50, Table 2.5).

2.1.5   CROSS-COUPLING OF PYRROLES WITH HALOACETYLENES

2.1.5.1   Ethynylation of Pyrroles

Pyrroles undergo cross-coupling with acylbromoacetylenes on the surface of Al_2O_3 to regioselectively afford 2-(2-acylethynyl)pyrroles in good yields (Scheme 2.51, Table 2.6). This reaction represents the first example of a palladium-, copper-, base-, and solvent-free ethynylation of pyrroles [38,514–516]. In the absence of Al_2O_3 (or with Al_2O_3 in a solvent), the cross-coupling does not take place.

The mechanism of the reaction involves the addition–elimination sequence [514,516].

The intermediates, 2-(2-acyl-1-bromovinyl)pyrroles (Table 2.6), isolated from the reaction mixture, are converted into the corresponding 2-(2-acylethynyl)pyrroles over Al_2O_3 surface [516].

Side products of the cross-coupling, 2-acyl-1,1-di(pyrrol-2-yl)ethenes (Table 2.6), are likely formed via the substitution of bromine atom by the second pyrrole molecule in 2-(2-acyl-1-bromovinyl)pyrroles [516]. Their yields are normally up to 20% but expectedly increase (to 40%), when excess pyrrole, higher temperatures, or longer reaction times are employed [514].

Ethyl bromopropynoate undergoes similar cross-coupling with 4,5,6,7-tetrahydroindole on the surface of Al_2O_3 to give 3-(4,5,6,7-tetrahydroindol-2-yl)-2-propynoate (R = H) [38,517] and ethyl-3,3-di(4,5,6,7-tetrahydroindol-2-yl)acrylate (Scheme 2.52). In some cases, the yield of 3-(4,5,6,7-tetrahydroindol-2-yl)-2-propynoate reaches 46% (Table 2.6). Unlike bromopropynoate, ethyl iodopropynoate reacts with 4,5,6,7-tetrahydroindole in a selective manner to afford ethyl-3,3-di(4,5,6,7-tetrahydroindol-2-yl) acrylate in high yield (79%).

TABLE 2.5
Products of Reaction of Pyrroles with
1-Alkylthio-2-Chloroacetylenes

No.	Structure	Yield (%)	Mp (°C)	References
1		14	Oil	[513]
2		21	Oil	[513]
3		41	Oil	[513]
4		42	Oil	[513]
5		24	Oil	[513]
6		49	Oil	[513]
7		51	Oil	[513]
8		45	Oil	[513]
9		4	48	[513]

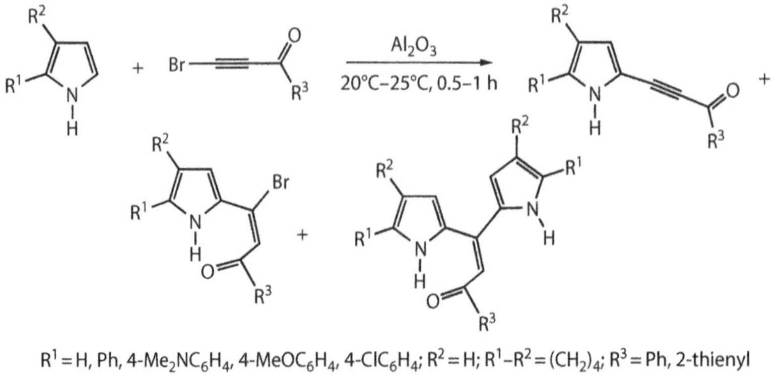

$R^1 = H$, Ph, 4-$Me_2NC_6H_4$, 4-$MeOC_6H_4$, 4-ClC_6H_4; $R^2 = H$; R^1–$R^2 = (CH_2)_4$; $R^3 = $ Ph, 2-thienyl

SCHEME 2.51 Solvent-free reaction of pyrroles with acylhaloacetylenes on the surface of Al_2O_3.

N-Vinyl-4,5,6,7-tetrahydroindole reacts with both ethyl bromopropynoate and ethyl iodopropynoate to furnish a cross-coupling product, ethyl 3-(1-vinyl-4,5,6,7-tetrahydroindol-2-yl)-2-propynoate (R=CH=CH$_2$, up to 71% yield, Table 2.6) [38,517].

N-Vinyl- and N-isopropenylpyrroles react with acylbromoacetylenes on Al_2O_3 surface to regioselectively give the corresponding 2-(acylethynyl)pyrroles (Scheme 2.53, Table 2.6) [38,518]. Ethynylation of N-isopropenylpyrroles is accompanied by depropenylation delivering minor amounts of NH-2-acylethynylpyrroles. This is likely caused by HBr-catalyzed hydrolysis (due to hydrate water in Al_2O_3) of N-isopropenyl group both in starting compounds and target products.

As mentioned earlier, this new cross coupling has been first realized on alumina surface. Naturally, the questions arise of whether Al_2O_3 is the unique metal oxide to affect this reaction and whether other active surfaces of metal oxides or salts can also stimulate the ethynylation. To answer these questions, the solvent-free interaction of 2-phenylpyrrole with benzoylbromoacetylene on active surfaces of metal oxides (MgO, CaO, ZnO, BaO, Al_2O_3, TiO_2, ZrO_2) and salts ($CaCO_3$, $ZrSiO_4$) has been investigated [519]. The reaction leads, depending upon the active surface employed, to either the cross-coupling product, 2-(2-benzoylethynyl)-5-phenylpyrrole, or the product of 2-phenylpyrrole C-vinylation, 2-(2-benzoyl-1-bromovinyl)-5-phenylpyrrole (Scheme 2.54). The activity of the metal oxides in the ethynylation reaction falls in the following order (in the brackets, the content of 2-benzoylethynyl-5-phenylpyrrole in the reaction mixture is given): ZnO (81%), BaO (73%), Al_2O_3 (71%), MgO (69%), and CaO (50%). The oxides SiO_2 [520], TiO_2, and ZrO_2 and salts $CaCO_3$ and $ZrSiO_4$ are inactive in the ethynylation reaction affording exclusively 2-(2-benzoyl-1-bromovinyl)-5-phenylpyrrole, with ZrO_2 the isolated yield of this pyrrole reaching 60% [519].

2-(2-Benzoylethynyl)-5-phenylpyrrole, synthesized from 2-phenylpyrrole and benzoylbromoacetylene on alumina surface, crystallizes (after chromatography on Al_2O_3) in two visually distinctive forms (prisms and needles), which represent the *cis*- and *trans*-rotamers relative to mutual disposition of the nitrogen atom and the carbonyl group (x-ray analysis data) [515].

TABLE 2.6
Products of Reaction of Pyrroles with Haloacetylenes

No.	Structure	Yield (%)	Mp (°C)	References
1	2	3	4	5
1		55	141–142	[514,520]
2		70	143–144	[514,520]
3		66	167–168	[514,520]
4		60	147–148	[514,520]
5		68	116–117	[521]
6		71	106–107	[521]
7		69	182–183	[514,515,519,520]
8		63	183–184	[514,520]

(Continued)

TABLE 2.6 (*Continued*)

Products of Reaction of Pyrroles with Haloacetylenes

No.	Structure	Yield (%)	Mp (°C)	References
1	**2**	**3**	**4**	**5**
9		62	169–170	[516]
10		45	192	[516]
11		94	188	[516]
12		46	Oil	[518]
13		70	82–83	[518]
14		50	96–98	[518]
15		65	100–101	[518]
16		43	85–86	[518]
17		58	Oil	[518]

(*Continued*)

TABLE 2.6 (*Continued*)
Products of Reaction of Pyrroles with Haloacetylenes

No.	Structure	Yield (%)	Mp (°C)	References
1	2	3	4	5
18		42	77–78	[518]
19		39	118–120	[518]
20		68	82–83	[518]
21		90	114–115	[517,522]
22		80	68–69	[522]
23		82	26–27	[517,522]
24		90	69–70	[522]
25		64	Oil	[522]

(*Continued*)

TABLE 2.6 (*Continued*)

Products of Reaction of Pyrroles with Haloacetylenes

No.	Structure	Yield (%)	Mp (°C)	References
1	**2**	**3**	**4**	**5**
26		62	Oil	[522]
27		71	Oil	[522]
28		81	Oil	[522]
29		68	178	[523]
30		22	215–216	[523]
31		72	178–179	[523]
32		76	167–168	[523]
33		23	226–227	[523]

(Continued)

TABLE 2.6 (*Continued*)

Products of Reaction of Pyrroles with Haloacetylenes

No.	Structure	Yield (%)	Mp (°C)	References
1	2	3	4	5
34		17	100–102	[524]
35		12	136–138	[524]
36		3	127–128	[520]
37		60	124–125	[519,520]
38		19	138–139	[516]
39		3	119–120	[516]
40		11	160–161	[520]

(Continued)

TABLE 2.6 (*Continued*)

Products of Reaction of Pyrroles with Haloacetylenes

No.	Structure	Yield (%)	Mp (°C)	References
1	2	3	4	5
41		17	Oil	[514,520]
42		19	186–187	[514,520]
43		57	210–211	[514,520]
44		39	204–205	[514,515,519,520]
45		60	195–198	[520]
46		5	208–209	[516]
47		5	257–258	[516]

(*Continued*)

TABLE 2.6 (*Continued*)

Products of Reaction of Pyrroles with Haloacetylenes

No.	Structure	Yield (%)	Mp (°C)	References
1	2	3	4	5
48		9	223–224	[523]
49		79	140	[517]
50		31	Oil	[517]
51		6	146–148	[523]
52		41	210–212	[524]
53		57	240–242	[524]

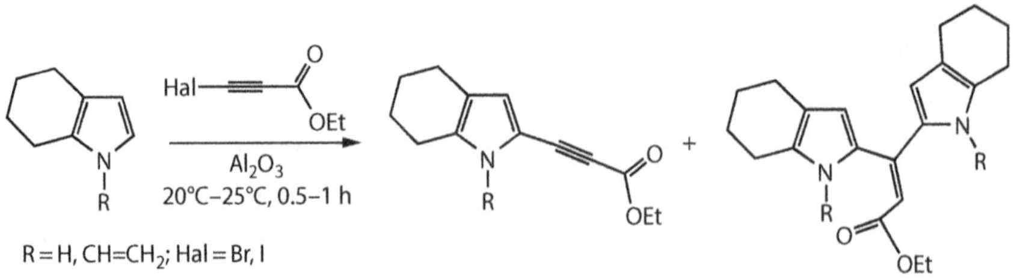

SCHEME 2.52 Solvent-free reaction of 4,5,6,7-tetrahydroindoles with halopropynoates on the surface of Al_2O_3.

$R^1 = H$, Me; $R^2 = H$, Ph; $R^3 = H$; $R^2-R^3 = (CH_2)_4$; $R^4 = Ph$, 2-thienyl

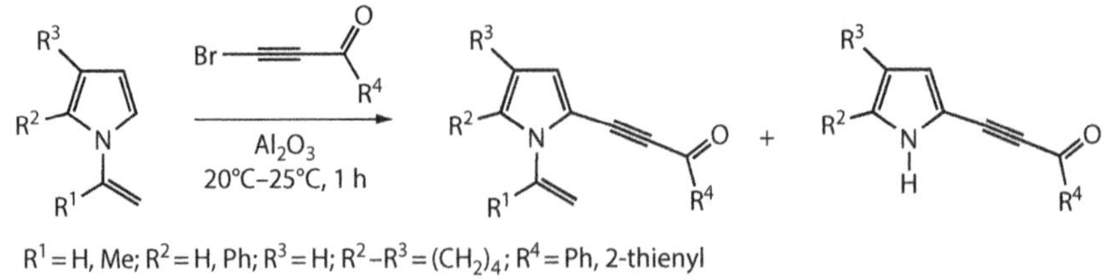

SCHEME 2.53 Solvent-free reaction of N-vinyl- and N-isopropenylpyrroles with acylbromoacetylenes on the surface of Al_2O_3.

AS—active surface

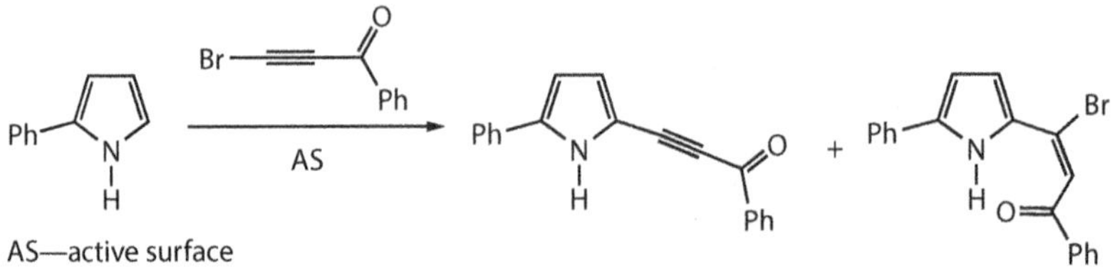

SCHEME 2.54 Solvent-free reaction of 2-phenylpyrrole with benzoylbromoacetylene on surfaces of metal oxides and salts.

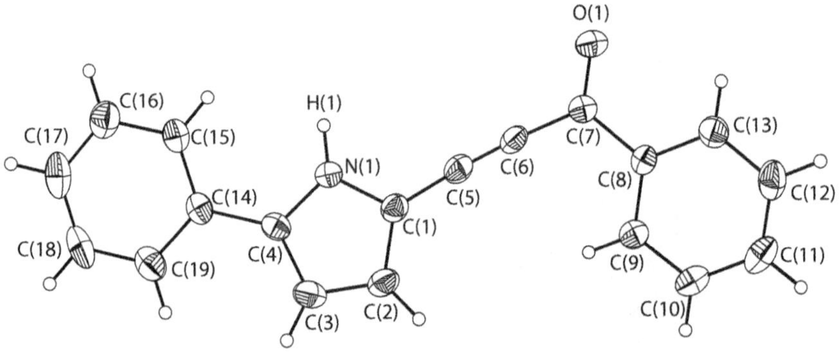

The general view of *cis*-rotamer of 2-(2-benzoylethynyl)-5-phenylpyrrole

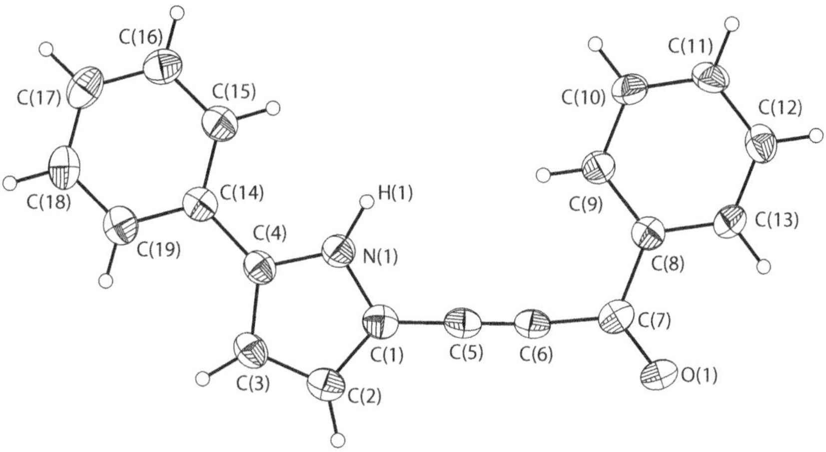

The general view of *trans*-rotamer of 2-(2-benzoylethynyl)-5-phenylpyrrole

The prisms, formed in small amounts, are less stable and, upon recrystallization, are transformed to the needles [515]. This is probably the first experimental observation of the hindered rotation around the C_{sp}–C_{sp2} bond.

2-(2-Benzoylethynyl)-5-phenylpyrrole, prepared from 2-(2-benzoyl-1-bromovinyl)-5-phenylpyrrole on the ZrO_2, crystallizes mostly as prisms (*cis*-rotamer, x-ray analysis) upon passing through Al_2O_3 [519].

4,5,6,7-Tetrahydroindole reacts with ethyl bromopropynoate on all the mentioned active surfaces to deliver the corresponding ethynyl derivative (Scheme 2.55) [521]. The reaction mixtures obtained on some of these surfaces, namely, CaO, BaO, MgO, and ZnO contain several times more 2-ethynyl-4,5,6,7-tetrahydroindole than a mixture of products formed on Al_2O_3. The conversion of 4,5,6,7-tetrahydroindole on all the studied surfaces is close to 100%.

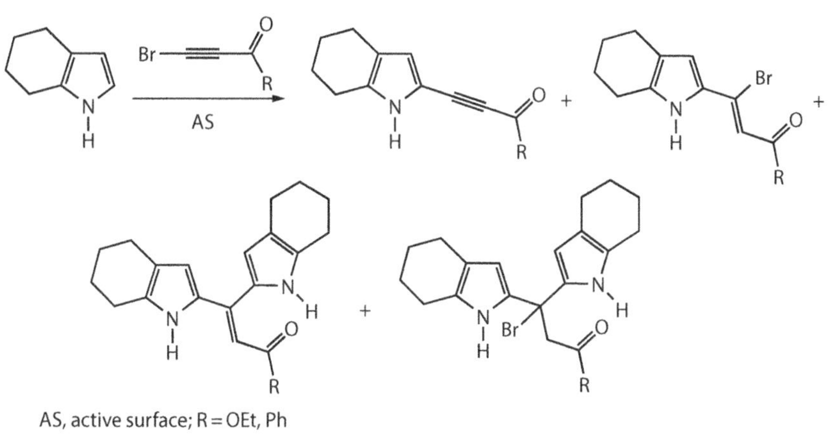

AS, active surface; R = OEt, Ph

SCHEME 2.55 Solvent-free reaction of 4,5,6,7-tetrahydroindole with benzoylbromoacetylene and ethyl bromopropynoate on surfaces of metal oxides and salts.

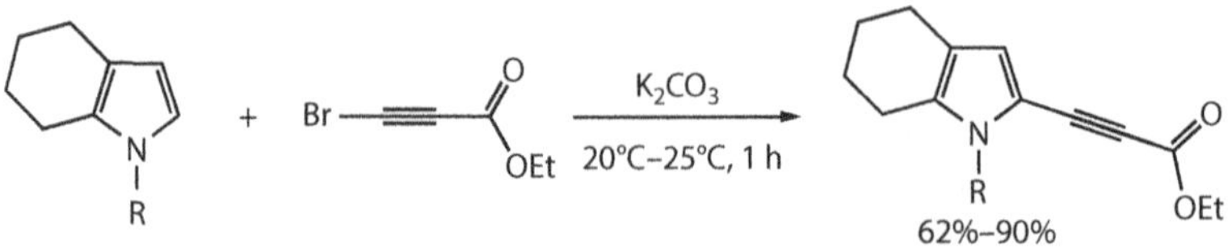

R = H, Me, Bn, CH = CH$_2$, Me(CH)OPr-i, Me(CH)OBu-n, (CH$_2$)$_2$SEt, (CH$_2$)$_2$SPr-n

SCHEME 2.56 Selective ethynylation of 4,5,6,7-tetrahydroindole with ethyl bromopropynoate on the surface of K$_2$CO$_3$.

Ethynylation of 4,5,6,7-tetrahydroindole also proceeds on surfaces of oxides (SiO$_2$, TiO$_2$, ZrO$_2$) and salts (CaCO$_3$, ZrSiO$_4$), though the yields of the products are low (13%–24%).

In these cases, mainly unstable di(4,5,6,7-tetrahydroindol-2-yl)bromoethane is formed due to the nucleophilic addition of 4,5,6,7-tetrahydroindole to the intermediate 2-(1-bromovinyl)-4,5,6,7-tetrahydroindole. This adduct is generated also without participation of the active surface [517].

4,5,6,7-Tetrahydroindoles undergo a rapid and selective ethynylation with ethyl bromopropynoates upon grinding with solid K$_2$CO$_3$ to afford ethyl 3-(4,5,6,7-tetrahydroindol-2-yl)-2-propynoates in high yields (Scheme 2.56, Table 2.6). Interestingly, this cross-coupling does not occur in solution (diethyl ether, CHCl$_3$) both with and without K$_2$CO$_3$ [522].

Formally, the ethynylation can be rationalized both as electrophilic substitution of hydrogen in the pyrrole ring and nucleophilic addition of the electron-rich pyrrole moiety to the electron-deficient triple bond. In both cases, the one-electron transfer from pyrrole to acetylenic fragment is plausible. This is confirmed by the fact that in the ESR spectrum of the reaction mixture, a singlet ($g = 2.0023$, $\Delta H = 1.8$ mT) is observed [522].

The refluxing of ethyl 3-(N-methyl-4,5,6,7-tetrahydroindol-2-yl)-2-propynoate in o-xylene with Pd/C (10% Pd, 12 h) leads to a mixture of (indol-2-yl)propanoate and (indol-2-yl)acrylate (4:1) resulting from the hydrogen redistribution between the cyclohexane ring and the triple bond (Scheme 2.57) [522]. Thus, 2-functionalized indoles become readily accessible.

Indoles smoothly react with benzoylbromoacetylene on Al$_2$O$_3$ to give 3-(benzoylethynyl)indoles chemo- and regioselectively in high yields (Scheme 2.58, Table 2.6), thus representing the first example of direct ethynylation of the indole ring [523].

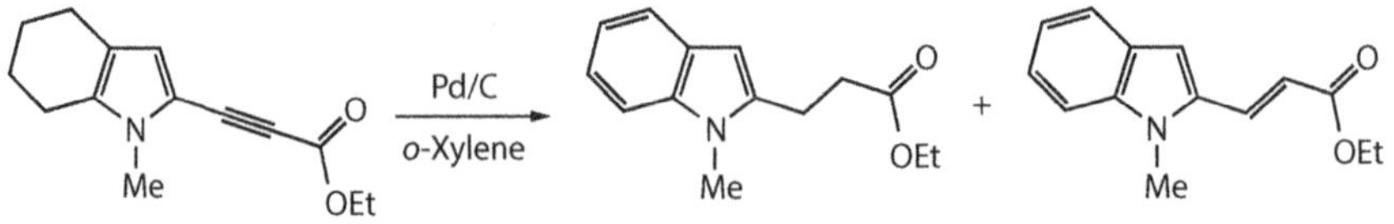

SCHEME 2.57 Hydrogenation of ethyl 3-(N-methyl-4,5,6,7-tetrahydroindol-2-yl)-2-propynoate on Pd/C.

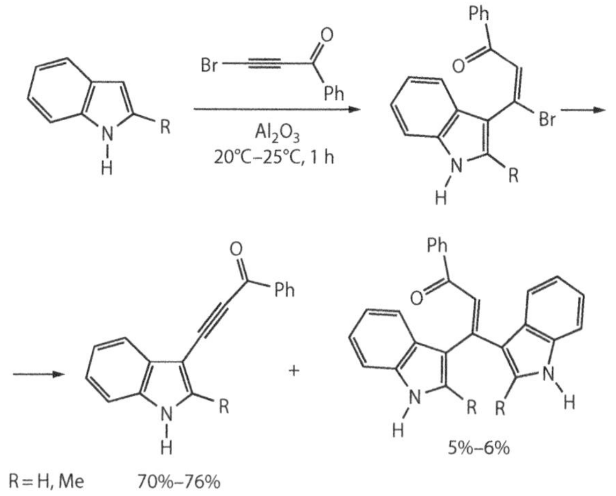

SCHEME 2.58   Solvent-free ethynylation of indoles with benzoylbromoacetylenes on the surface of Al$_2$O$_3$.

The only side products of the reaction, 2-benzoyl-1,1-di(indol-3-yl)ethenes (5%–6% yields), are probably formed via the nucleophilic substitution of bromine atom with indole in the intermediates, 3-(2-benzoyl-1-bromovinyl)indoles. 3-Benzoylethynylindoles do not add indoles under the reaction conditions.

The condensed indole, benz[g]indole, under the same conditions, is coupled with benzoylbromoacetylene to furnish a mixture of 3- and 2-benzoylethynylindoles in a 45% overall yield (Scheme 2.59, Table 2.6). In this case, 2-(2-benzoyl-1-bromovinyl) benz[g]indole, stabilized by strong intramolecular hydrogen bonding between the NH and C=O groups (δ NH 14.59 ppm), is also detected (^{1}H NMR) [523].

4,5-Dihydrobenz[g]indole, being actually a pyrrole, upon reacting with bromobenzoylacetylene on alumina, is readily converted to the corresponding

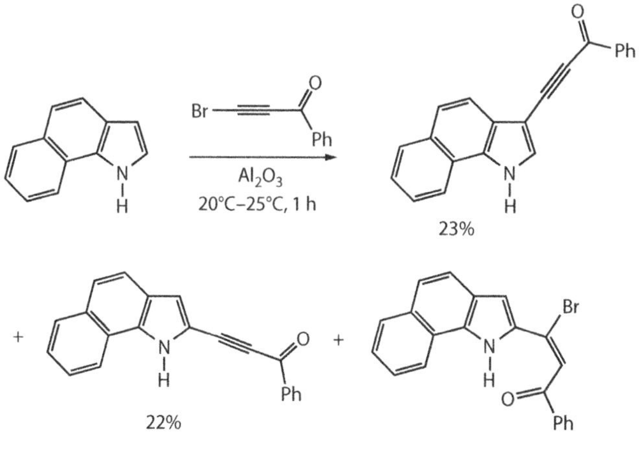

SCHEME 2.59   Ethynylation of benz[g]indole with benzoylbromoacetylenes on Al$_2$O$_3$.

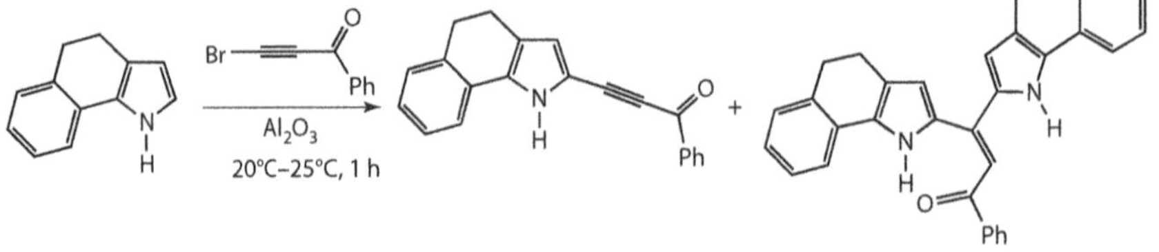

SCHEME 2.60 Ethynylation of 4,5-dihydrobenz[g]indole with benzoylbromoacetylenes on Al_2O_3.

2-ethynyl derivative in 68% yield. A side product of the reaction, 2-benzoyl-1,1-di(4,5-dihydrobenz[g]indol-2-yl)ethene, is isolated in 9% yield (Scheme 2.60, Table 2.6) [523].

Indoles are coupled with ethylbromopropynoate on alumina to give mainly ethyl 3,3-di(indol-3-yl)acrylates (in 41% and 57% yields, Table 2.6). 3-(Indol-3-yl)propynoates are formed with insignificant yields (17% and 12%) [524].

Thus, this reaction paves a facile one-pot route to earlier hardly accessible pyrroles and indoles with functional acetylene substituents, highly reactive building blocks for drug design and materials for molecular electronics.

The discussed variants of novel ethynylation of the pyrrole nucleus look even more attractive taking into account that the standard transition metal-catalyzed reactions, including the Sonogashira cross-coupling, do not allow the ethynylpyrroles with electron-withdrawing functions in acetylene substituent to be synthesized [525].

2.1.5.2 Reaction of 2-Ethynylpyrroles with 2,3-Dichloro-5,6-dicyano-1,4-benzoquinone

2,3-Dichloro-5,6-dicyano-1,4-benzoquinone (DDQ), an effective reagent for aromatization, is often employed for dehydrogenation of hydroaromatic compounds [526] including those bearing the pyrrole fragment. For example, the treatment of 4,7-dihydroindol-2-yl-fumarate with DDQ (benzene, 20°C, 1 h) gives the expected indole in 94% yield [527]. However, in the case of 2-ethynyl-4,5,6,7-tetrahydroindoles $[R^2–R^3=(CH_2)_4]$, the anticipated aromatization does not occur under analogous conditions. Instead, [2+2]-cycloaddition to the triple bond takes place to afford bicyclo[4.2.0]octadienediones **21** (81%–93% yields, Scheme 2.61, Table 2.7) [528].

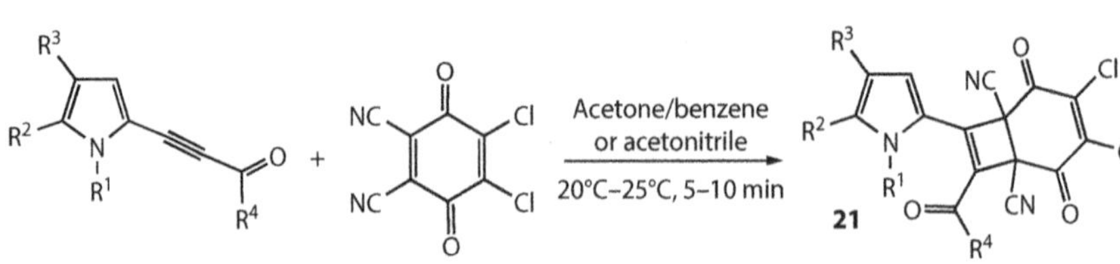

$R^1 = H, Me, Bn; R^2 = Ph, 4-ClC_6H_4, 4-MeOC_6H_4; R^3 = H; R^2–R^3 = (CH_2)_4; R^4 = OEt, Ph, 4-NO_2C_6H_4$

SCHEME 2.61 Cycloaddition of DDQ to 2-acylethynylpyrroles.

TABLE 2.7
Cycloadducts of C-Ethynylpyrroles and Indoles with DDQ and Products of Their Alcoholysis

No.	Structure	Yield (%)	Mp (°C)	References
1	2	3	4	5
1		87	207–208	[528]
2		93	163–164	[528]
3		81	181–182	[528]
4		88	>210 (decomp.)	[528]
5		82	215–216	[528]
6		76	>300	[528]

(Continued)

TABLE 2.7 (*Continued*)

Cycloadducts of C-Ethynylpyrroles and Indoles with DDQ and Products of Their Alcoholysis

No.	Structure	Yield (%)	Mp (°C)	References
1	2	3	4	5
7		92	>300	[528]
8		87	>300	[528]
9		84	>300	[528]
10		7	239–240	[531]
11		39	230–231	[529]

(*Continued*)

TABLE 2.7 (*Continued*)

Cycloadducts of C-Ethynylpyrroles and Indoles with DDQ and Products of Their Alcoholysis

No.	Structure	Yield (%)	Mp (°C)	References
1	2	3	4	5
12	(structure: tetrahydroindole, N–H; O=C–OPr-*n*, NC, Cl, NC, O, O=C–OEt)	2	220–221	[531]
13	(structure: tetrahydroindole, N–H; O=C–OPr-*i*, NC, Cl, NC, O, O=C–OEt)	33	223–224	[531]
14	(structure: tetrahydroindole, N–Me; O=C–OEt, NC, Cl, NC, O, O=C–OEt)	43	158–160	[531]
15	(structure: tetrahydroindole, N–CH2Ph; O=C–OEt, NC, Cl, NC, O, O=C–OEt)	26	135–136	[531]
16	(structure: tetrahydroindole, N–H; O=C–OEt, NC, Cl, NC, O, O=C–Ph)	71	233–235	[529]
17	(structure: Ph–pyrrole, N–H; O=C–OEt, NC, Cl, NC, O, O=C–Ph)	55	234–236	[529]

(*Continued*)

TABLE 2.7 (*Continued*)

Cycloadducts of C-Ethynylpyrroles and Indoles with DDQ and Products of Their Alcoholysis

No.	Structure	Yield (%)	Mp (°C)	References
1	2	3	4	5
18		57	233–234	[529]
19		78	279–280	[529]
20		24	133–134	[531]
21		58	209–210	[531]
22		44	211–212	[529]
23		59	169–170	[531]

(Continued)

TABLE 2.7 (*Continued*)
Cycloadducts of C-Ethynylpyrroles and Indoles with DDQ and Products of Their Alcoholysis

No.	Structure	Yield (%)	Mp (°C)	References
1	2	3	4	5
24	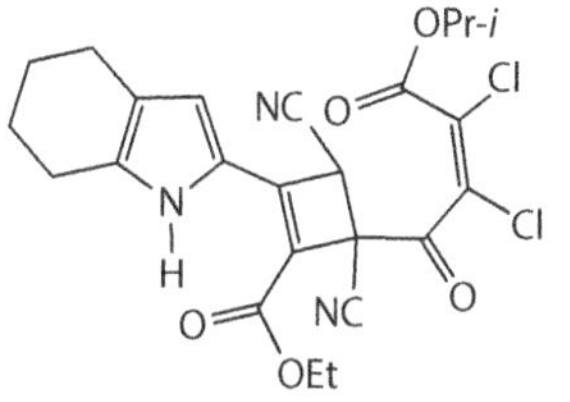	33	172–173	[531]

The reaction is general in character. Apart from 2-ethynyl-4,5,6,7-tetrahydroindoles, 5-aryl-2-ethynylpyrroles and 3-ethynylindoles also undergo [2+2]-cycloaddition. This is the first example of nonphotochemical [2+2]-cycloaddition of DDQ to the triple bond. The mechanism of the reaction involves one-electron transfer to generate ion-radical pair containing ethynylpyrrole radical cation (detected in the UV spectra of the reaction mixture at −30°C) and DDQ radical anion (identified by ESR).

The adducts synthesized are charge-transfer complexes that are paramagnetic in the solid state [528].

Cycloadducts **21**, upon ethanolysis (reflux, 15 min or room temperature, 24 h), rearrange to bicyclo[3.2.0]heptadienones **22** and cyclobutenyl-dihydrofuranone ensembles in 55%–83% total yields (Scheme 2.62, Table 2.7), the former rearrangement being the major direction of the reaction (selectivity 85%–90%) [529]. The only exception is the ethanolysis of the cycloadduct containing ethoxycarbonyl substituent and the tetrahydroindole counterpart [R^1–R^2=$(CH_2)_4$, R^3=OEt]. In this case, the rearrangement products are formed in 1:1.2 ratio with 83% total yield.

It is interesting to note that unlike this rearrangement, similar bicyclo[4.2.0]octadienediones, obtained by photoaddition of diphenylacetylene to benzo- and naphthoquinones, upon alcoholysis, rearrange with splitting of the cyclobutene ring [530].

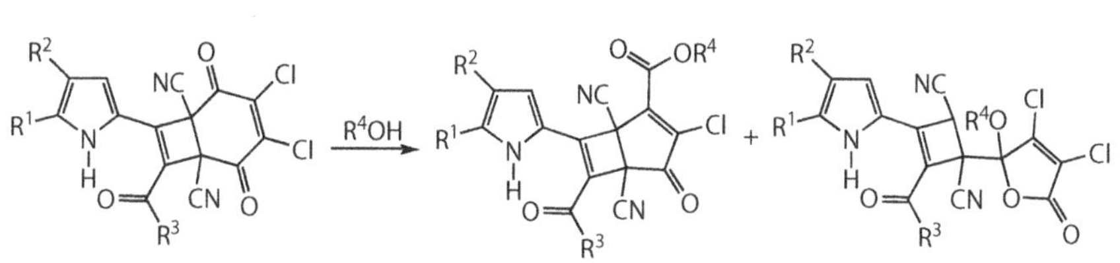

R^1 = Ph, 4-ClC_6H_4, 4-$MeOC_6H_4$; R^2 = H; R^1–R^2 = $(CH_2)_4$; R^3 = OEt, Ph, 4-$NO_2C_6H_4$; R^4 = Et

SCHEME 2.62 Rearrangement of the cycloadducts of 2-acylethynylpyrroles with DDQ under the action of EtOH.

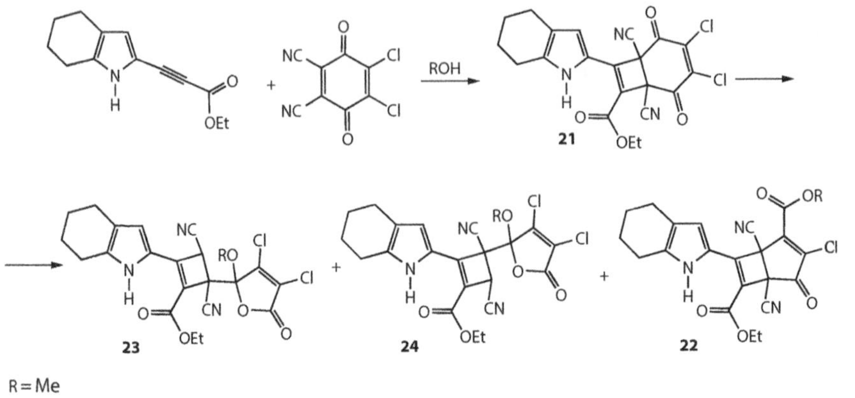

R = Me

SCHEME 2.63 The three-component reaction between ethyl 3-(4,5,6,7-tetrahydroindol-2-yl)propynoate, DDQ, and methanol.

The three-component reaction between ethyl 3-(4,5,6,7-tetrahydroindol-2-yl)pro-pynoate, DDQ, and methanol affords furan-2-on-4,5,6,7-tetrahydroindolecyclobu-tene ensemble **23** in 53% yield (Scheme 2.63) [531]. As minor products, furan-2-one isomer **24** along with bicycloheptadienone **22** has been either isolated or identified. Like in the case of benzene or acetone, the corresponding [2+2]-cycloadduct **21** is the intermediate of this reaction.

The ratio of 3- (**24**) and 4- (**23**) dihydrofuranone isomers changes in the course of the reaction. In the beginning of the rearrangement, isomer **24** prevails but becomes with time a minor product (ratio **24:23**—1.7 [15 min], 1.4 [30 min], 1.2 [60 min], 0.8 [90 min], 0.7 [120 min]). Thus, dihydrofuranone **24** appears to be a kinetic prod-uct and its isomer, dihydrofuranone **23**, is a thermodynamic one. This is supported by isolation of almost pure dihydrofuranone **23** when a mixture of isomers was chroma-tographed on Si$_2$O, the mass of the sample remaining close to the initial one.

Obviously, the formation of furanones **23** and **24** represents (Scheme 2.64) a typical domino process, triggered by the addition of methanol to the carbonyl group adjacent

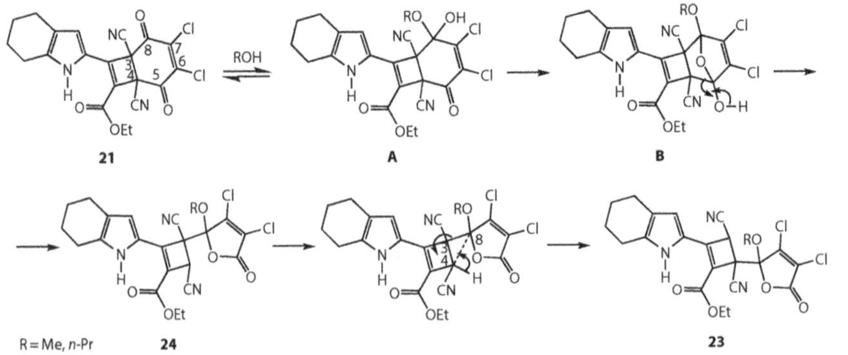

R = Me, n-Pr

SCHEME 2.64 Tentative mechanism of the pyrrole-dihydrofuranone formation from DDQ-ethyl 3-(4,5,6,7-tetrahydroindol-2-yl)propynoate adduct with under the action of alcohol.

to position 3 of the cyclobutene ring. Hemiacetal **A** attacks by its hydroxyl group the opposite carbonyl function (at carbon atom C5), and in hemiacetal **B**, thus formed, the cleavage of the C4–C5 bond occurs. This is accompanied by the simultaneous proton transfer from hydroxyl to anionic carbon atom C4 to finally give the labile (kinetic) isomer **24**. The latter rearranges to the more stable isomer **23** via the cleavage of the C3–C8 and C4–H bonds and the concerted formation of the C4–C8 and C3–H bonds.

The cycloadduct **21** undergoes the similar rearrangements upon heating (55°C –56°C, 2 h) in *n*-propanol [531].

However, the reaction with *i*-propanol (20°C–25°C, 7 days or 55°C–56°C, 2 h) leads to bicycloheptadienone **22** and its precursor, the open-chained ester **25** (Scheme 2.65) [531]. When their 1:1 mixture is passed through the SiO₂ column, the intermediate **25** is completely transformed to bicycloheptadienone **22**. Evidently, in this case, the intermediate hemiacetal **C** rearranges via the cleavage of the C3–C8 bond with simultaneous proton transfer from the hydroxyl group to C3 anionic center to furnish the open-chained ester **25**. The proton abstraction from position 3 of ester **25** (under the action of *i*-propanol as a base) followed by the nucleophilic attack at the C–Cl bond and release of the chlorine anion (as a good fugitive group) leads to bicycloheptadienone **22**.

t-Butanol does not react with [2 + 2]-cycloadduct **21** under even harsher conditions (reflux, 2 h). In this case, instead of the expected products, unstable furan-2-one, bearing the hydroxy group in place of the *t*-butoxy substituent, has been isolated in 24% yield (Scheme 2.66) [531]. Apparently, this may result from the reactions with trace water that is supported by the formation of the same hemiacetal upon refluxing the cycloadduct **21** in aqueous dioxane (24 h).

The only isolable products of the ethanolysis of N-substituted (R=Me, Bn) cycloadducts are bicycloheptadienones **22** (Scheme 2.67). Apart from the latter, the reaction mixtures contain also their precursors **25**, which are transformed on SiO₂ to bicycloheptadienones **22**.

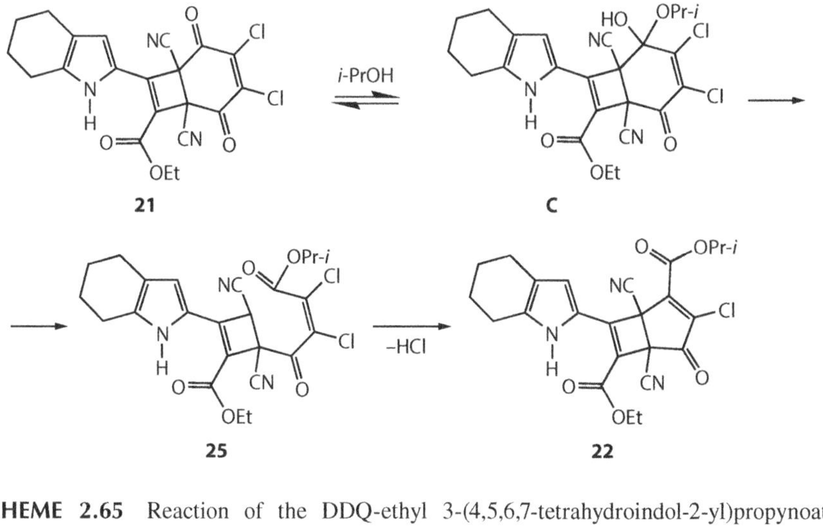

SCHEME 2.65 Reaction of the DDQ-ethyl 3-(4,5,6,7-tetrahydroindol-2-yl)propynoate adduct with *i*-propanol.

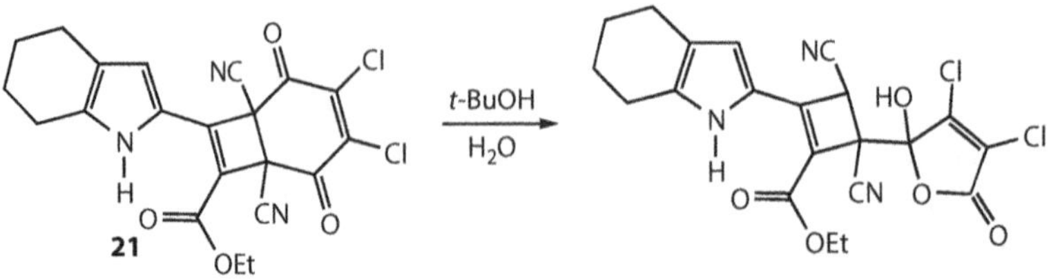

SCHEME 2.66 Reaction of the DDQ-ethyl 3-(4,5,6,7-tetrahydroindol-2-yl)propynoate adduct with *t*-butanol.

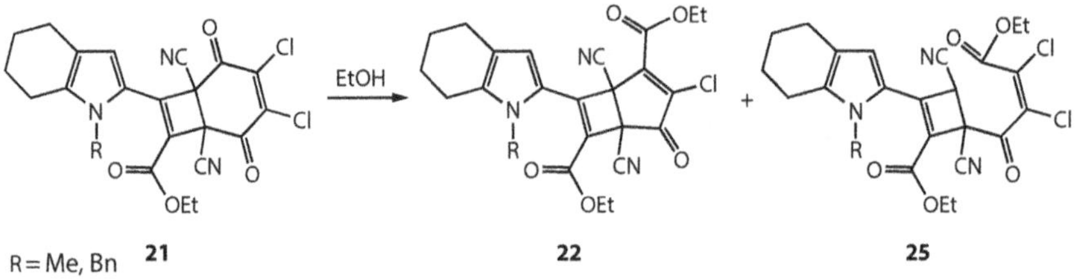

R = Me, Bn **21**   **22**   **25**

SCHEME 2.67 Reaction of the DDQ-ethyl 3-(N-methyl-and N-benzyl-4,5,6,7-tetrahydroindol-2-yl)propynoate adducts with ethanol.

The aforementioned rearrangements are diastereospecific: in all cases, only one diastereomer is formed, thus indicating the concerted character of the cleavage and formation of all C–C- and C–H bonds involved into the domino reactions.

The yields and constants of cycloadducts and products of their alcoholysis are given in Table 2.7.

2.1.5.3 Hydroamination of 2-Ethynylpyrroles

Hydroamination of 2-benzoylethynyl-4,5,6,7-tetrahydroindoles with dialkylamines proceeds under mild conditions (room temperature, aqueous ethanol, 1 h) to afford the corresponding *E*-3-dialkylamino-(4,5,6,7-tetrahydroindol-2-yl)-2-propenones in 64%–88% yields and 72%–92% stereoselectivity (Scheme 2.68, Table 2.8) [532].

Under the same conditions, dimethyl- and diethylamines react with ethyl 3-(4,5,6,7-tetrahydroindol-2-yl)-2-propynoates in a different manner. The former converts the ester function into an amide, giving the corresponding N,N-dimethyl-3-(4,5,6,7-tetrahydroindol-2-yl)-2-propynamides in 70%–86% yields (Scheme 2.69, Table 2.8),

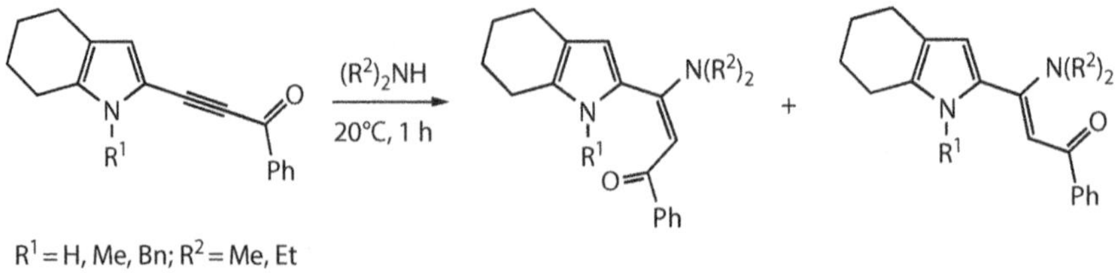

$R^1 = H, Me, Bn; R^2 = Me, Et$

SCHEME 2.68 Addition of secondary amines to 2-benzoylethynyl-4,5,6,7-tetrahydroindoles.

TABLE 2.8
Products of Reaction of 2-Ethynylpyrroles with Amines

No.	Structure	Yield (%)	Mp (°C)	References
1	2	3	4	5
1		88	144–145	[532]
2		87	Oil	[532]
3		89	Oil	[532]
4		64	128–129	[532]
5		83	94–96	[532]
6		81	106–108	[532]
7		70	220–221	[532]

(*Continued*)

TABLE 2.8 (*Continued*)

Products of Reaction of 2-Ethynylpyrroles with Amines

No.	Structure	Yield (%)	Mp (°C)	References
1	2	3	4	5
8		77	107–108	[532]
9		86	94–95	[532]
10		85	Oil	[532]
11		84	Oil	[532]
12		34	100–101	[532]
13		64	148–149	[532]
14		6	200–202	[532]

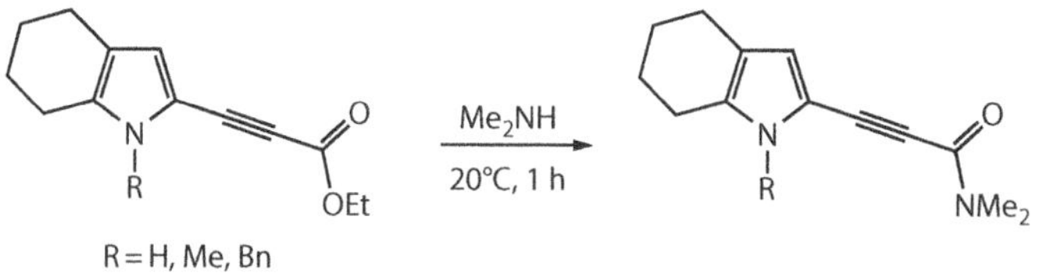

SCHEME 2.69 Addition of dimethylamine to ethyl-3-(4,5,6,7-tetrahydroindol-2-yl)-2-pro-pynoates.

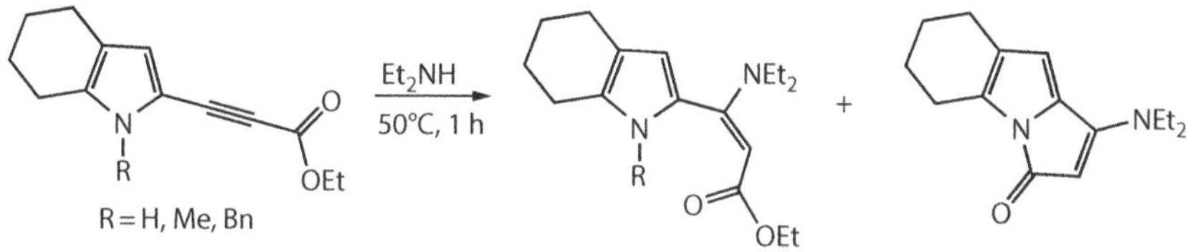

SCHEME 2.70 Addition of diethylamine to ethyl-3-(4,5,6,7-tetrahydroindol-2-yl)-2-pro-pynoates.

whereas diethylamine adds to the triple bond to deliver 3-(diethylamino)-3-(4,5,6,7-tetrahydroindol-2-yl)-2-propenoates with 100% stereoselectivity and up to 85% yield (Scheme 2.70, Table 2.8). These differences have been rationalized in terms of steric factors [532].

The reaction of ethyl 3-(NH-4,5,6,7-tetrahydroindole-2-yl)-2-propynoate with diethylamine is accompanied by intramolecular cyclization of the primary adduct to pyrroloindole, their ratio being ~1:1. Upon separation of this mixture (Al$_2$O$_3$), the adduct cyclizes completely to pyrroloindole.

2.1.6 REACTIONS WITH CARBON DISULFIDE

2.1.6.1 Synthesis of Pyrrolecarbodithioates

Previously, it was assumed that the addition of pyrrole anions to carbon disulfide leads mainly to pyrrole-1-carbodithioates [533–536].

Systematic investigations of the reaction of pyrroles with carbon disulfide in the superbase system KOH/DMSO [537–541] have shown that pyrrole anions, generated in this system, attack CS$_2$ (20°C–25°C, 2 h) exclusively or mainly by the position 2 to afford pyrrole-2-carbodithioates. The latter, after alkylation with alkylhalides (20°C–25°C, 2 h), give the corresponding pyrrole-2-carbodithionic acid esters in 46%–75% yields (Scheme 2.71, Table 2.9) [537–540]. The only exception is unsubstituted pyrrole, which gives only pyrrole-1-carbodithioate [540].

The ratio between substitution. When only one methyl group is introduced into the α-position of the pyrrole ring, pyrrole-2-carbodithioate becomes the single reaction product (46% yield), while N-isomer is not detected in the reaction mixture [542]. Any other combinations of alkyl substituents in the pyrrole moiety lead to selective formation of pyrrole-2-carbodithioates in up to 71% yield (Table 2.9) [537–542].

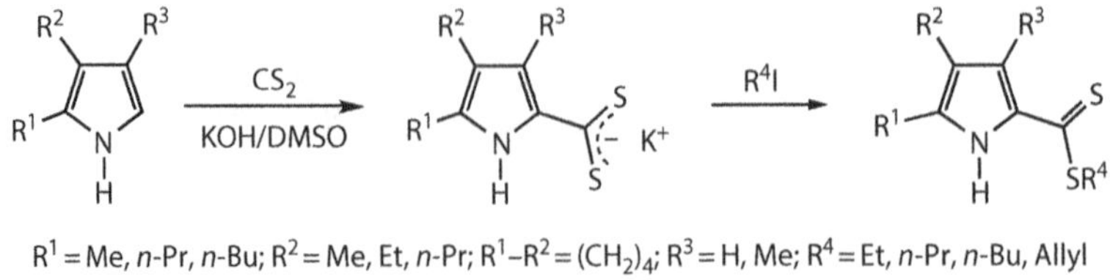

R^1 = Me, n-Pr, n-Bu; R^2 = Me, Et, n-Pr; R^1–R^2 = (CH$_2$)$_4$; R^3 = H, Me; R^4 = Et, n-Pr, n-Bu, Allyl

SCHEME 2.71 Synthesis of pyrrole-2-carbodithioates from pyrroles and CS$_2$ in the KOH/DMSO system.

The regioselectivity of the reaction is breached when aryl substituent is introduced into the pyrrole α-position: along with the major products, pyrrole-2-carbodithioate (44%–59% yields), N-isomers are also formed (24%–33% yields, Scheme 2.72, Table 2.9) [538–540].

Contrary to data [533,534], it has been established that if the pyrrole contains substituents at both α-positions, the reaction exclusively delivers pyrrole-3-carbodithioates in 36%–61% yield (Scheme 2.73, Table 2.9) [542]. Pyrrole-1-carbodithioates have not been found among the reaction products.

In the case of 2-aryl(hetaryl)-5-methylpyrroles, from two possible isomers, pyrrole-3- and pyrrole-4-carbodithioates, only the latter are formed (44%–61% yields, Table 2.9), that is, the isomers having the dithioate function adjacent to the methyl group (Scheme 2.74) [542]. Possibly, such regiospecificity is caused by steric shielding of the pyrrole position 3 by *ortho*-hydrogen atom of the aromatic or heteroaromatic substituent.

Thus, the introduction of a single methyl substituent into the pyrrole ring position 2 is enough to change the direction of multidentate pyrrole anion attack at carbon disulfide. These data are in agreement with quantum-chemical calculations of the pyrrole anion and its 2-methyl- and 2,3-dimethyl derivatives [543–547].

According to the calculations [543–545], the highest negative charge (~0.4) is located on nitrogen atom of the pyrrole anions, and it practically does not depend upon substitution with the methyl groups in the ring (Scheme 2.75). Consequently, if the reaction is controlled by the charges, the pyrrole anion in all the cases will attack CS$_2$ by its nitrogen atom to give mainly 1(N)-isomers. At the same time, charges on carbon atoms in the pyrrole ring significantly change upon introduction of the substituents. The methyl group in the pyrrole α-position considerably increases a negative charge in the position 4, and its value becomes comparable with that of the nitrogen atom. Therefore, if the reaction is controlled by charges, 2-methylpyrrole anion would probably attack carbon disulfide by its β-position.

In 2,3-dimethylpyrrole anion, the charges on all carbon atoms significantly alter, and position 4 is more negatively charged than nitrogen atom. However, actually, a considerable difference in the behavior of these two pyrroles is not observed (yields of the corresponding pyrrole-2-carbodithioates are 46% and 51% [543], i.e., within experimental error, 2%–3%). Despite the high negative charges in β-positions of 2-methyl- and 2,3-dimethylpyrrole anions (which are much higher than charges in α-positions), instead of the expected 3-isomers, pyrrole-2-carbodithioates are exclusively formed.

TABLE 2.9
Pyrrolecarbodithioates

No.	Structure	Yield (%)	Mp (°C) (d_4^{20})	References
1	2	3	4	5
1		63	(1.1625)	[540]
2		95	110–111	[548,549]
3		62	44–45	[549,554]
4		65	88–89	[549,554]
5		24	(1.1902)	[538–540]
6		29	(1.0677)	[538,539]
7		37	29–30	[540]
8		44	Oil	[540]
9		36	50–51	[540]

(*Continued*)

TABLE 2.9 (*Continued*)
Pyrrolecarbodithioates

No.	Structure	Yield (%)	Mp (°C) (d_4^{20})	References
1	2	3	4	5
10		46	36	[542]
11		51	80	[542]
12		48	96	[542]
13		62	105	[542]
14		69	(1.1009)	[540]
15		75	(1.1132)	[540]
16		71	(1.1006)	[540]
17		69	(1.0374)	[540]
18		71	69	[537–539]

(Continued)

TABLE 2.9 (*Continued*)
Pyrrolecarbodithioates

No.	Structure	Yield (%)	Mp (°C) (d_4^{20})	References
1	2	3	4	5
19		44	37–38	[538,539]
20	SBu-*n*	54	46–47	[538,539]
21	CN	62	123–124	[548,549]
22	CONH$_2$	52	147–148	[548,549]
23	CO$_2$Me	36	66–67	[548,549]
24	CO$_2$Et	51	118	[549,554]
25	COPh	3	Oil	[549,554]
26	Ph, SEt	59	67–68	[538,539]
27	Ph, SBu-*n*	52	48–49	[538,539]

(*Continued*)

TABLE 2.9 (*Continued*)
Pyrrolecarbodithioates

No.	Structure	Yield (%)	Mp (°C) (d_4^{20})	References
1	**2**	**3**	**4**	**5**
28		15	83–84	[548]
29		38	60–61	[540]
30		30	32–33	[540]
31		30	62–63	[540]
32		47	193–194	[549,554]
33		46	158–159	[549,554]
34		52	60	[542]
35		43	38	[542]

(Continued)

TABLE 2.9 (*Continued*)
Pyrrolecarbodithioates

No.	Structure	Yield (%)	Mp (°C) (d_4^{20})	References
1	2	3	4	5
36		36	60	[542]
37		18	88	[549]
38		21	130	[549]
39		61	108	[542]
40		44	52	[542]
41		46	64	[542]

Therefore, either orbital control of the reaction or rearrangement of initially formed 1- or 3-isomers into 2-isomers takes place. As follows from quantum-chemical calculations [543,544,546], energies of highest occupied molecular orbital (HOMO) and shapes of the orbitals for the studied pyrrole anions differ slightly. Thus, the rearrangement remains the most probable explanation of the observed regiochemistry.

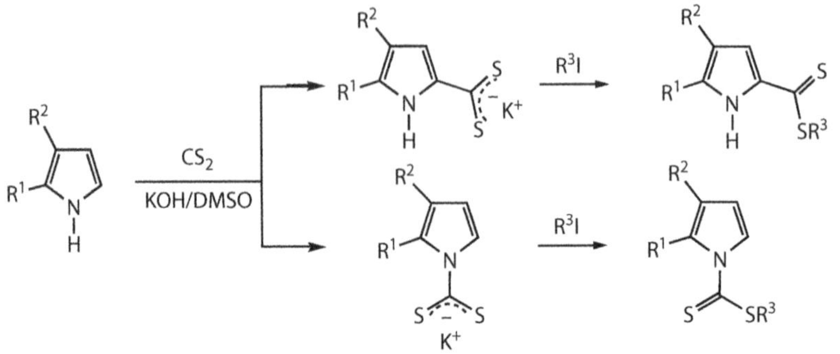

R^1 = Ph, 4-ClC$_6$H$_4$, 2-naphthyl; R^2 = H; n-Pr; R^3 = Et, n-Bu

SCHEME 2.72 Formation of pyrrole-1- and pyrrole-2-carbodithioates from 2-arylpyrroles and CS$_2$ in the KOH/DMSO system.

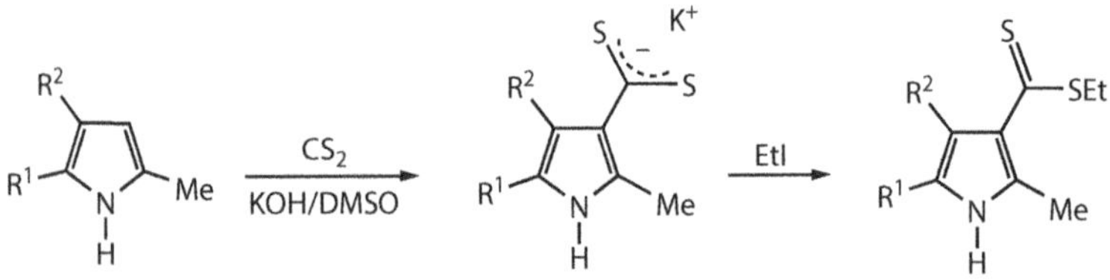

R^1 = Me; R^2 = H, Me; R^1–R^2 = (CH$_2$)$_4$

SCHEME 2.73 Synthesis of pyrrole-3-carbodithioates from pyrroles and CS$_2$ in the KOH/DMSO system.

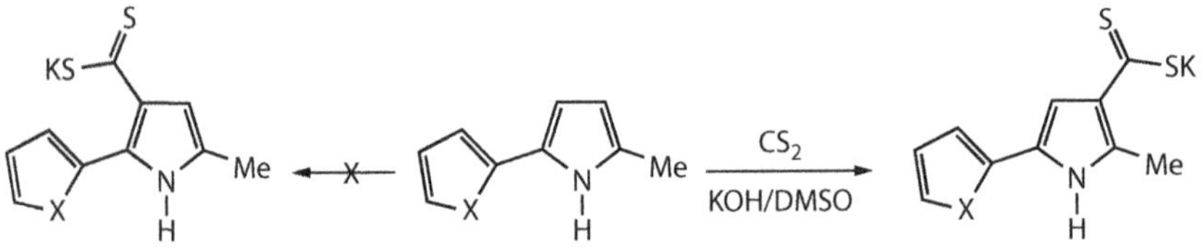

X = –HC=CH–, S, O

SCHEME 2.74 Regioselective carbodithioation of 2-aryl(hetaryl)-5-methylpyrroles with CS$_2$ in the KOH/DMSO system.

In 2,5-dimethylpyrrole anion, the negative charge in β-position also insignificantly diverges from the charge on nitrogen atom. It can be assumed that the β-position will participate in the reaction with CS$_2$, especially bearing in mind that nitrogen atom is sterically deshielded by two methyl groups. In such a case, theoretical speculations do correlate with the experimental data.

The calculated values of energy difference (the same basis) show that the unsubstituted pyrrole-2-carbodithioate anion is by 5.47 kcal/mol more stable than the

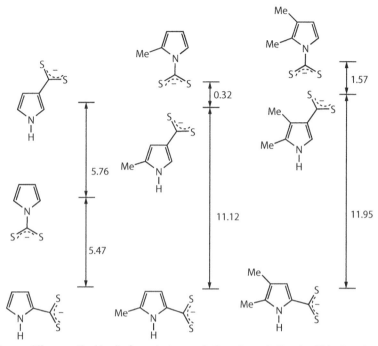

Energy differences (kcal/mol) of pyrrole-1-, pyrrole-2- and pyrrole-3-carbodithioate anions
(MP 2/6-31+G**): effect of the methyl substituent [543, 544, 546]

SCHEME 2.75   Energy differences (kcal/mol) of pyrrole-1-, pyrrole-2-, and pyrrole-3-car-bodithioate-anions (MP 2/6-31 + G**): effect of the methyl substituent.

corresponding 1-isomer [543,544,546]. This does not correlate with the experiment (only pyrrole-1-carbodithioate is formed); therefore, it can be considered as a kinetic product.

The methyl group in α-position increases the stability of pyrrole-2-carbodithio-ate anion by 11.12 kcal/mol as compared to the corresponding 1-isomer. The second methyl substituent in β-position augments this value only negligibly (to 11.95 kcal/mol). High energy preference of the methyl-substituted pyrrole-2-carbodithioate anions rationalizes their observed regioselective formation, which is most likely a thermodynamic result, while 1- or 3-isomers can be kinetic products.

It is obvious that in the case of unsubstituted pyrrolecarbodithioate anions, the energy difference (5.47 kcal/mol) is not high enough for an easy rearrangement to occur.

The energy of unsubstituted pyrrole-3-carbodithioate anion is by 5.76 kcal/mol higher than that of its 1-isomer, while 3-carbodithioate anions of 2-methyl- and 2,3-dimethylpyrroles slightly differ in energy from the 1-carbodithioate isomers: 2-methylpyrrole-3-carbodithioate   and   2,3-dimethylpyrrole-1-carbodithioate   are more preferable by 0.32 kcal/mol and 1.57 kcal/mol, respectively [543,544,546]. Relying on the calculation results, one can expect that 2,5-dimethylpyrrole anion will attack CS_2 by 3(4) position (at thermodynamic stage of the reaction) to afford the corresponding 3(4)-carbodithioate, which is by 6.93 kcal/mol more stable than 1-isomer [543,544,546]. Indeed, this is in agreement with the experimental data.

2.1.6.2 Addition of Pyrrolecarbodithioate Anions to the Multiple Bond

Pyrrole-1-carbodithioate, generated in situ from pyrrole and carbon disulfide in the system KOH/DMSO, does not almost add to electrophilic alkenes, such as acrylonitrile, acrylamide, or methyl acrylate [548,549]. However, its treatment with excess HCl in the specified system leads to the acrylamide adduct in quantitative yield (Scheme 2.76, Table 2.9).

Under the same conditions, pyrrole-1-carbodithioate does not react with acrylonitrile and methyl acrylate; only pyrrolyldisulfide is isolated in both cases (Scheme 2.77).

In contrast to pyrrole-1-carbodithioates, pyrrole-2-carbodithioate anions easily add to acrylic acid derivatives to afford the corresponding pyrrole-2-carbodithioate in up to 62% yield (Scheme 2.78, Table 2.9) [548,549].

Under the same conditions, pyrrole-3-carbodithioate reacts with acrylonitrile delivering the corresponding adduct in a lower yield (18%, Scheme 2.79, Table 2.9) [549].

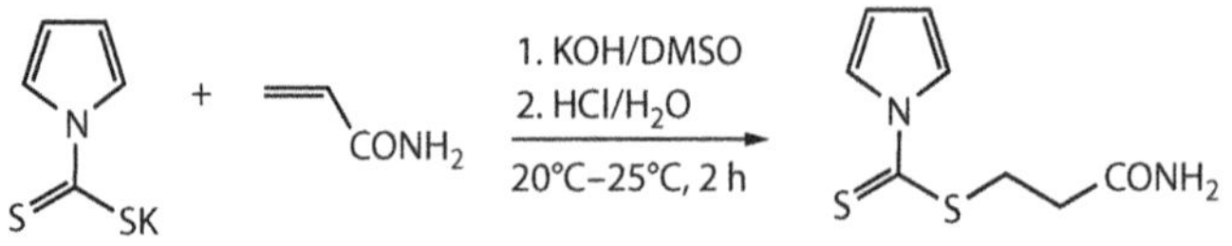

SCHEME 2.76 Addition of pyrrole-1-carbodithioate to acrylamide.

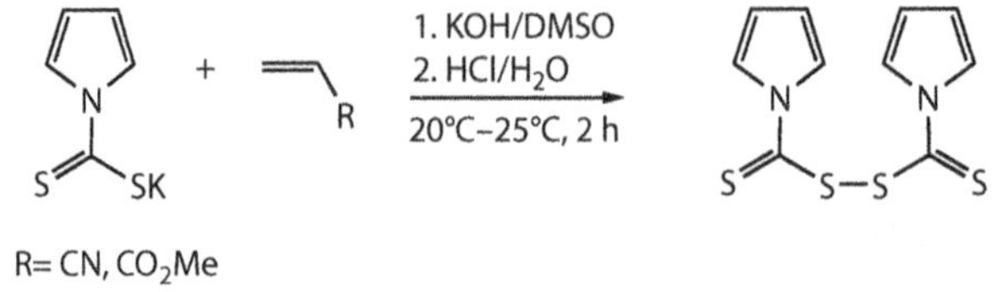

R= CN, CO$_2$Me

SCHEME 2.77 Formation of pyrrolyldisulfide in the attempted reaction of pyrrole-1-carbodithioate with acrylonitrile and methyl acrylate.

R^1 = Ph, R^2 = H; R^1–R^2 = (CH$_2$)$_4$; R^3 = CN, CONH$_2$, CO$_2$Me

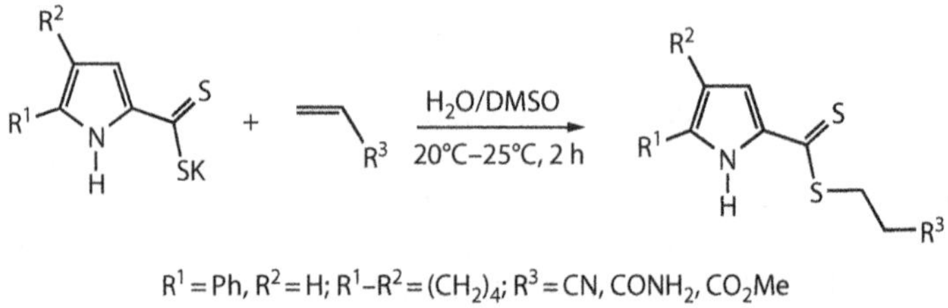

SCHEME 2.78 Addition of pyrrole-2-carbodithioates to acrylic acid derivatives.

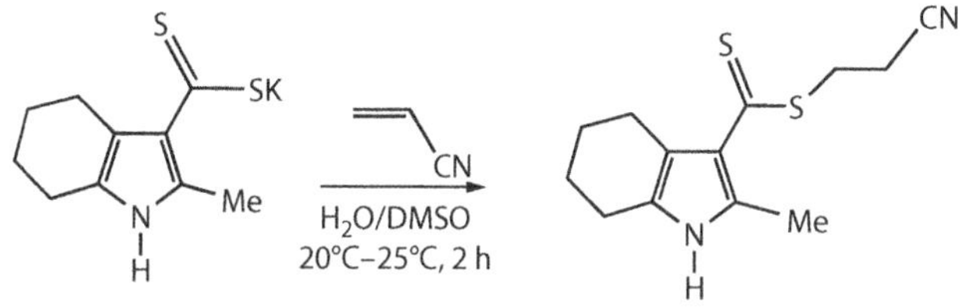

SCHEME 2.79 Addition of 2-methyl-(4,5,6,7-tetrahydroindolyl-2-yl)-3-carbodithioate to acrylonitrile.

With electron-deficient acetylenes, the dithioate anions can be involved in the reactions of both 1,3-anionic cyclo- and nucleophilic additions [550]. The reactions of 1,3-anionic cycloaddition are typical provided that the central atom of [S–C–S]⁻ has aromatic substituents ensuring the anion stabilization [551,552]. For instance, potassium pyrrole-1-carbodithioate selectively reacts (acetonitrile, −30°C, CH₃COOH) with dimethyl ester of acetylenedicarboxylic acid to form rapidly polymerizing 2-(pyrrol-1-yl)-4,5-dimethoxycarbonyl-1,3-dithiol (Scheme 2.80) [552,553].

The reaction of pyrrole-1-carbodithioate with ethyl propynoate and benzoylacetylene (aqueous DMSO, KOH, room temperature, 2 h) leads to normal adducts, S-vinylpyrrole-1-carbodithioates in a low yield (10%–18%, Scheme 2.81, Table 2.9) [549,554]. Among the major reaction products are substituted divinyl sulfides as a mixture of E,E-, Z,E-, and Z,Z-isomers.

In the presence of acetic acid preventing both the cleavage of the C–S bond under the action of hydroxide anions and formation of divinyl sulfides, the yields of the adducts reach 65%.

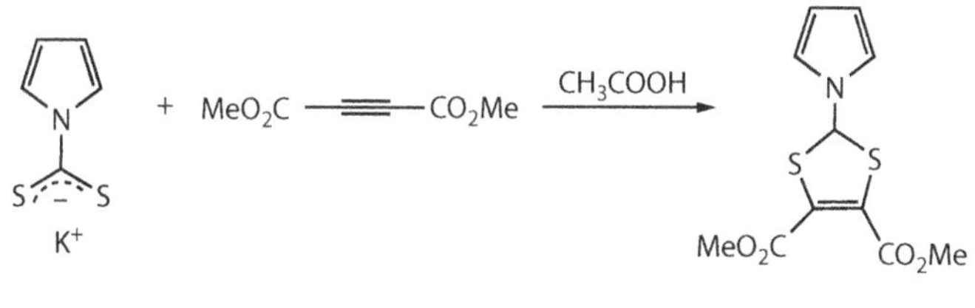

SCHEME 2.80 Reaction of pyrrole-1-carbodithioates with dimethyl ester of acetylenedicarboxylic acid.

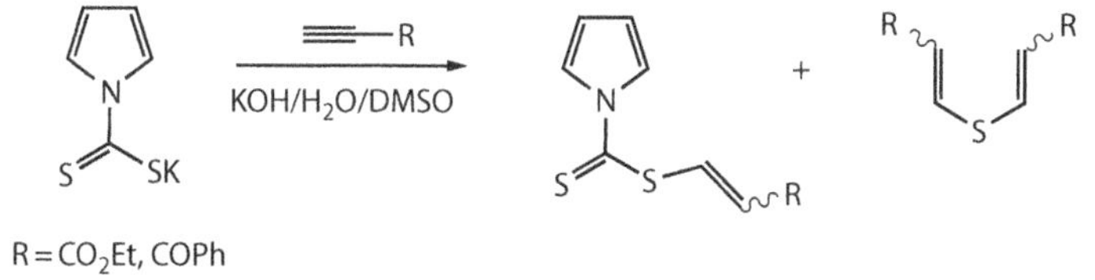

SCHEME 2.81 Reaction of pyrrole-1-carbodithioate with terminal acetylenes.

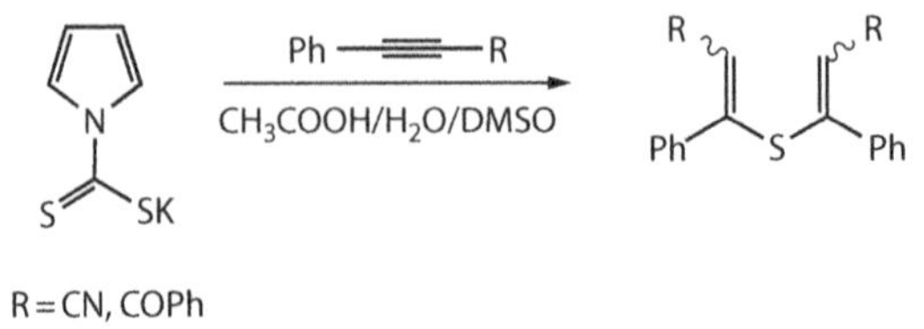

R = CN, COPh

SCHEME 2.82   Reaction of pyrrole-1-carbodithioate with disubstituted acetylenes.

The reaction of pyrrole-1-carbodithioate with disubstituted electron-deficient acetylenes exclusively gives the corresponding divinyl sulfides, which are also formed in the presence of acetic acid (Scheme 2.82).

Obviously, due to the steric hindrances in the disubstituted acetylenes, the competing solvolysis of pyrrole-1-carbodithioate anions results in sulfide ions that react with acetylenes to furnish divinyl sulfides.

Unlike pyrrole-1-carbodithioate, 4,5,6,7-tetrahydroindole-2-carbodithioate reacts with acylacetylenes giving rise to pyrrolothiazolidines (46%–47%, Scheme 2.83, Table 2.9), the products of the intramolecular cyclization of the intermediate 2-acylvinylpyrrole-2-carbodithioates. The last are detected in the reaction mixture only in trace amounts [549,554]. The reaction proceeds readily in a two-phase system (aqueous DMSO/diethyl ether) at 20°C–25°C.

With ethyl propynoate, pyrrole-2-carbodithioate reacts selectively under the aforementioned conditions to produce divinyl sulfide (Scheme 2.84, 42% yield).

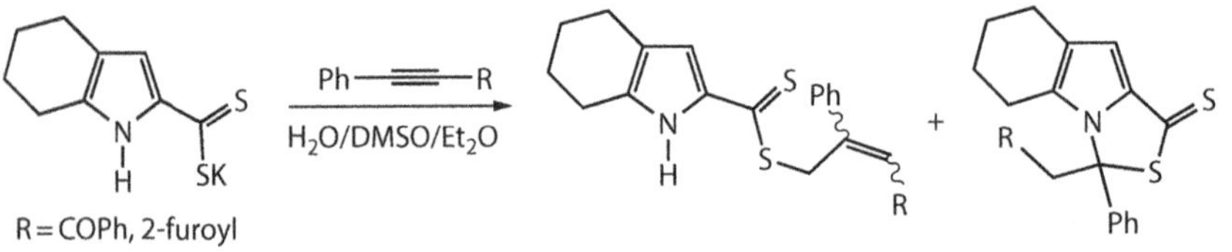

R = COPh, 2-furoyl

SCHEME 2.83   Cyclization of 4,5,6,7-tetrahydroindole-2-carbodithioate with acylphenylacetylenes.

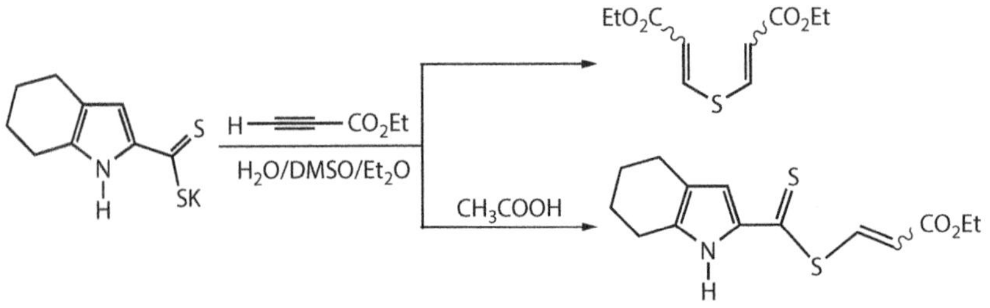

SCHEME 2.84   Reactions of 4,5,6,7-tetrahydroindole-2-carbodithioate with ethyl propynoate.

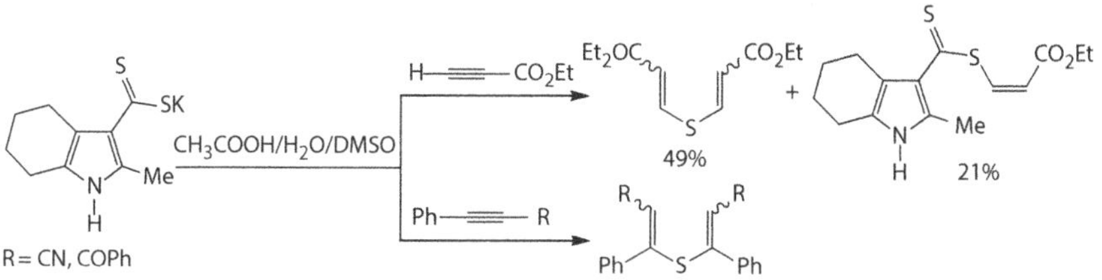

SCHEME 2.85 Reactions of 4,5,6,7-tetrahydroindole-3-carbodithioate with acetylenes.

The primary adduct, ethyl-3-[(4,5,6,7-tetrahydro-1*H*-indole-2-carbothioyl)sulfanyl]prop-2-enoate, has been isolated only in the presence of acetic acid (51% yield, Scheme 2.84, Table 2.9).

Pyrrole-3-carbodithioate reacts with ethyl propynoate in the presence of acetic acid to give a mixture of divinyl sulfide and the adduct of *Z*-configurations (Scheme 2.85, Table 2.9), while with disubstituted electron-deficient acetylenes, only divinyl sulfides are formed [549].

2.1.6.3 Reactions of S-Alkylpyrrolecarbodithioates

S-Alkylpyrrolecarbodithioates are usually more stable than acids. Therefore, they are the most employed derivatives of pyrrolecarbodithioic acids. Owing to their thioamide nature, like the corresponding acids and their salts, they are prone to solvolysis across the C–N bond. For example, ethyl 2-phenylpyrrole-1-carbodithioate in the system KOH/DMSO undergoes a cleavage of the C–N bond already at room temperature (2 h) to afford 2-phenylpyrrole (95% yield, Scheme 2.86). Under similar conditions, pyrrole-2-carbodithioate is stable (shown on the example of 4,5,6,7-tetrahydroindole-2-carbodithioate) [539].

The ability of S-alkylpyrrole-2-carbodithioates to be condensed with nitriles representing CH-acids is widely used in organic synthesis for preparation of functionally substituted S-vinylpyrroles [555–567].

Thus, carbanions, generated from various methylene-active nitriles such as malononitrile, and cyanoacetamide. In the system KOH/DMSO add to thiocarbonyl group of pyrrole-2-carbodithioates at 100°C–110°C to furnish vinyl thiolates. The latter are alkylated giving functionalized 2-vinylpyrroles in 50%–90% yields (depending on substituents in the pyrrole ring, functions of the starting CH acids, and structure of the alkylating agent) (Scheme 2.87, Table 2.10) [28,555,556,558,559].

The reaction of pyrrole-2-carbodithioates with malononitrile and cyanoacetamide leads to, along with 2-vinylpyrroles, a small amount (5%) of their intramolecular

SCHEME 2.86 Solvolysis of 2-phenylpyrrole-1-carbodithioate in the KOH/DMSO system.

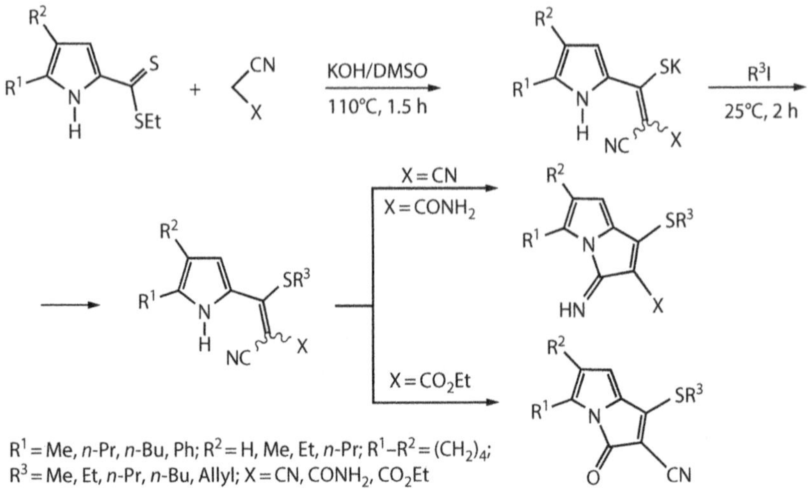

R^1 = Me, n-Pr, n-Bu, Ph; R^2 = H, Me, Et, n-Pr; R^1–R^2 = $(CH_2)_4$;
R^3 = Me, Et, n-Pr, n-Bu, Allyl; X = CN, $CONH_2$, CO_2Et

SCHEME 2.87   Reactions of pyrrole-2-carbodithioates with methylene active nitriles in the KOH/DMSO system.

cyclization products, 3-iminopyrrolizines. With cyanoacetate, the condensation is accompanied by annulation of 2-vinylpyrrole into the corresponding Pyrrolizin-3-one (the major product, 61% yield) [556].

Pyrrole-3-cardodithioates selectively react under similar conditions with malononitrile, cyanoacetamide, and cyanoacetate to form the corresponding 3-vinylpyrroles in 28%–58% yields (Scheme 2.88, Table 2.10) [560].

Condensation of pyrrole-3-carbodithioates with CH-acids is chemoselective (in this case, intramolecular cyclization involving the NH function is impossible) and stereoselective: 3-vinylpyrroles are always formed as single isomer, which probably have less sterically strained E-configuration with syn-orientation of the NH pyrrole and CN group.

2.1.6.4   Synthesis of Pyrrolothiazolidines

Pyrrole-2-vinylthiolates, formed from pyrrole-2-carbodithioates and CH-acid anions, are highly reactive mild nucleophiles that readily (room temperature) substitute halogen even at the sp-carbon atom (in bromobenzoyl- or chloroethylthioacetylenes) to afford ethynyl sulfides, which further cyclize to functionally substituted pyrrolothiazolidines (37%–85% yield, Scheme 2.89, Table 2.11) [561,562].

2.1.6.5   Synthesis of Pyrrolizin-3-One

The reaction of ethylcyanoacetate, ethylacetoacetate, or diethylmalonate as CH acids with pyrrole-2-carbodithioates leads to vinylpyrroles bearing the ester function (Scheme 2.90) [563,564].

The latter cyclize much easier than 2-vinylpyrroles containing only nitrile or amide functions, this often occurring already at a stage of their synthesis [563,564].

TABLE 2.10

Functional C-Vinylpyrroles as Pyrrolecarbodithioate Derivatives

No.	Structure	Yield (%)	Mp (°C)	References
1	2	3	4	5
1		60	142	[558,559]
2		58	153	[559]
3		69	112	[559]
4		64	140	[559]
5		48	54–55	[556]
6		90	109–110	[556]
7		51	Oil	[556]
8		91	79–80	[556]

(Continued)

TABLE 2.10 (*Continued*)
Functional C-Vinylpyrroles as Pyrrolecarbodithioate Derivatives

No.	Structure	Yield (%)	Mp (°C)	References
1	2	3	4	5
9		88	138–139	[555,556]
10		90	134–135	[555,556]
11		29	95–96	[556,563,564]
12		70	93–94	[556]
13		64	114	[556]
14		63	109–110	[556]
15		52	129	[556]
16		80	90–91	[555,556]

(*Continued*)

TABLE 2.10 (*Continued*)

Functional C-Vinylpyrroles as Pyrrolecarbodithioate Derivatives

No.	Structure	Yield (%)	Mp (°C)	References
1	2	3	4	5
17		82	148	[555,556]
18		85	68	[568,576]
19		75	192 (*E/Z* 1:5)	[568,576]
20		78	104–105	[568]
21		84	183–184 (*E/Z* 1:5)	[568]
22		58	100	[560]
23		37	190	[560]
24		41	153	[560]

(Continued)

TABLE 2.10 (*Continued*)
Functional C-Vinylpyrroles as Pyrrolecarbodithioate Derivatives

No.	Structure	Yield (%)	Mp (°C)	References
1	2	3	4	5
25		41	153	[560]
26		38	100	[560]
27		30	132	[560]
28		78	150–151	[568]
29		92	148–149	[568]
30		90	148–149	[568]
31		92	154–155	[568]

(*Continued*)

TABLE 2.10 (*Continued*)

Functional C-Vinylpyrroles as Pyrrolecarbodithioate Derivatives

No.	Structure	Yield (%)	Mp (°C)	References
1	2	3	4	5
32		94	203–204	[568]
33		53	183	[449]
34		86	193–194	[568]
35		50	114–115	[568,569]
36		85	161–162	[568]
37		86	204	[449]
38		85	172–173	[568,569]
39		84	191–192	[568]

(Continued)

TABLE 2.10 (*Continued*)
Functional C-Vinylpyrroles as Pyrrolecarbodithioate Derivatives

No.	Structure	Yield (%)	Mp (°C)	References
1	2	3	4	5
40		79	170–171	[449]
41		79	204–206	[449]
42		96	166–167	[449]
43		90	147	[449]
44		90	179–180	[569]
45		74	176–177	[576]
46		68	135–136	[576]

(Continued)

TABLE 2.10 (*Continued*)

Functional C-Vinylpyrroles as Pyrrolecarbodithioate Derivatives

No.	Structure	Yield (%)	Mp (°C)	References
1	2	3	4	5
47	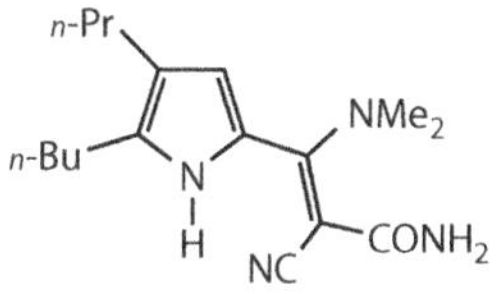	87	164–165	[576]
48	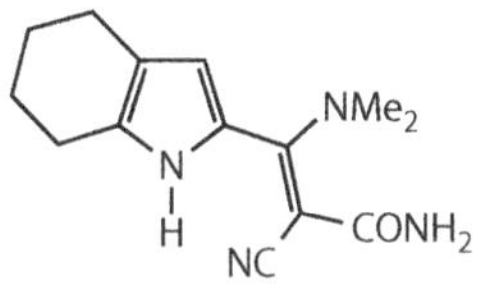	78	219–220	[571,576]
49	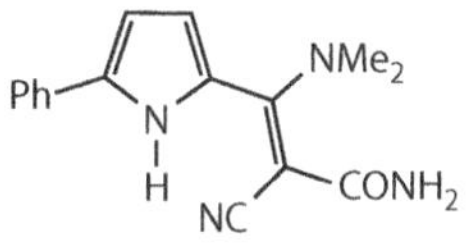	80	228–229	[571,576]
50	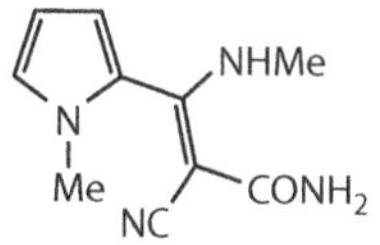	73	212–213	[576]
51	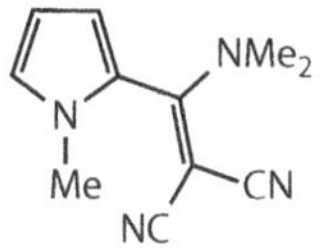	65	116–117	[576]
52	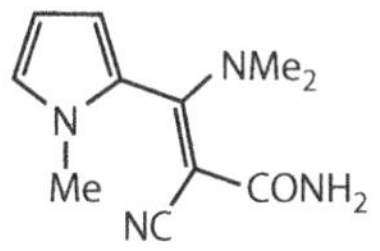	69	174–175	[576]

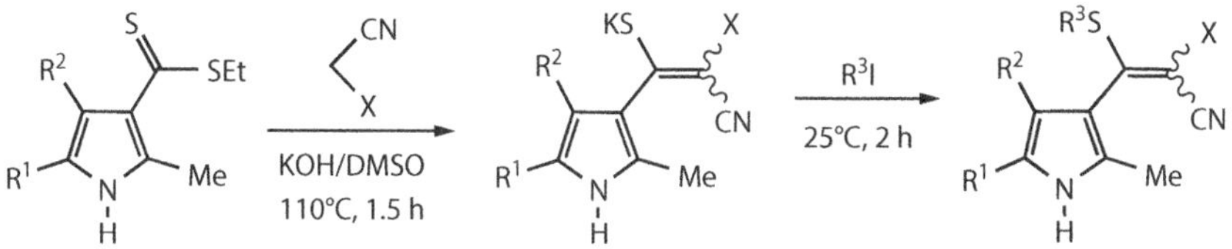

$R^1 = Me, Ph; R^2 = H, Me; R^3 = Me, Et; R^1–R^2 = (CH_2)_4; X = CN, CONH_2, CO_2Et$

SCHEME 2.88 Synthesis of functionalized 3-vinylpyrroles from pyrrole-3-cardodithioates with methylene active nitriles.

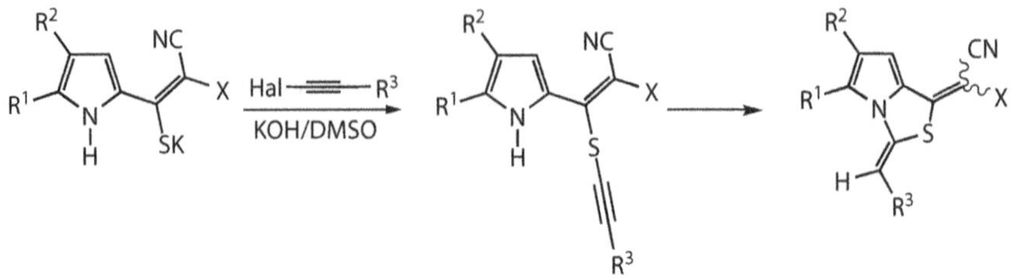

R¹ = Me, n-Pr, n-Bu; R² = Me, Et, n-Pr; R¹–R² = (CH₂)₄; R³ = SEt, COPh; X = CN, CONH₂; Hal = Cl, Br

SCHEME 2.89 Synthesis of pyrrolothiazolidines from pyrrole-2-carbodithioate and halo-acetylenes via the intermediate pyrrole-2-vinylthiolates.

The yields of pyrrolizin-3-ones range from 61% to 75% (Table 2.11). The decrease in yields is due to instability of the esters in KOH/DMSO system as well as configurational hindrances.

This reaction is chemoselective. In cases when other functions (such as nitrile or acetyl) capable of interacting with the NH-pyrrole moiety are adjacent to the double bond, only ethoxycarbonyl group takes part in the cyclization.

Pyrrolizin-3-ones can also be synthesized by the acid hydrolysis of 3-iminopyrrolizines, the cyclization product of the corresponding 2-(cyanovinyl) pyrroles (Scheme 2.91) [564].

2.1.6.6 Reactions of Functionally Substituted C-Vinylpyrroles with Hydroxide Anion: Synthesis of Stable Enols

The ethylthio group of 2-(1-ethylthio-2-carbamoyl-2-cyanovinyl)pyrroles is readily (NaOH, H₂O/MeOH, 1:2, 40°C–45°C, 1–4 h) substituted by the hydroxyl group (Scheme 2.92) [568]. The enols obtained are stable and can be isolated in almost quantitative yields (86%–94%, Table 2.10).

The reaction is chemoselective and stereospecific: only one enol isomer with *syn*-disposition of the carbamoyl and hydroxyl groups, linked by hydrogen bond, is formed.

The first stage of the reaction involves the cyclization to 1-ethylthio-3-iminopyrrolizines, which disappear by the end of the reaction. In the case of 2-(1-ethylthio-2-carbamoyl-2-cyanovinyl)-4,5,6,7-tetrahydroindole (R¹–R² = (CH₂)₄), 3-iminopyrrolizine is precipitated after the addition of alkali and transformed into the enol by the subsequent treatment with NaOH (H₂O/MeOH, 40°C–45°C, 1 h). Apparently, the formation of enols occurs through nucleophilic substitution of the ethylthio group by hydroxyl moiety accompanied by the pyrrolizine ring opening.

Such reaction pathway is supported by the synthesis of enols from 1-ethylthio-3-iminopyrrolizine-2-carbonitriles under the same conditions (Scheme 2.93) [568].

Meanwhile, 1-ethylthio-3-iminopyrrolizines are prone to the reversible ring opening. Therefore, the parallel enol formation via the direct nucleophilic substitution of the ethylthio group in C-vinylpyrroles cannot be ruled out.

Indeed, in N-methyl-2-vinylpyrroles having no free NH function and hence not capable of cyclizing to pyrrolizines, the exchange of ethylthio groups by the hydroxyl moiety takes place (Scheme 2.94) [568].

TABLE 2.11
Condensed Pyrrole Systems as Functional C-Vinylpyrrole Derivatives

No.	Structure	Yield (%)	Mp (°C)	References
1	2	3	4	5
1		42	160 (E/Z 4.7:1)	[562]
2		37	260 (E/Z 8:1)	[562]
3		68	160 (E/Z 1.2:1)	[562]
4		67	222 (E/Z 2.4:1) 243–244 (E)	[562]
5		59	202	[561,562]
6		63	260	[561,562]

(Continued)

TABLE 2.11 (*Continued*)
Condensed Pyrrole Systems as Functional C-Vinylpyrrole Derivatives

No.	Structure	Yield (%)	Mp (°C)	References
1	2	3	4	5
7		65	196	[561,562]
8		85	244	[561,562]
9		67	265	[561,562]
10		84	292	[561,562]
11		61	163–164	[556,563,564]
12		62	142–143	[563,564]
13		75	105–106	[563,564]

(Continued)

TABLE 2.11 (*Continued*)
Condensed Pyrrole Systems as Functional C-Vinylpyrrole Derivatives

No.	Structure	Yield (%)	Mp (°C)	References
1	2	3	4	5
14		51	139–140	[564]
15		59	160–161	[564]
16		10	148	[559]
17		62	132–133	[564]
18		48	178	[564]
19		68	128	[564]
20		10	174–175	[576]

(*Continued*)

TABLE 2.11 (*Continued*)

Condensed Pyrrole Systems as Functional C-Vinylpyrrole Derivatives

No.	Structure	Yield (%)	Mp (°C)	References
1	2	3	4	5
21		85	185–186	[576]
22		90	171–172	[564]
23		91	111–112	[564]
24		92	186–187	[564]
25		94	153–154	[564]
26		94	152	[564]
27		20	140–142	[572]

(*Continued*)

TABLE 2.11 (*Continued*)

Condensed Pyrrole Systems as Functional C-Vinylpyrrole Derivatives

No.	Structure	Yield (%)	Mp (°C)	References
1	2	3	4	5
28		30	124–126	[572]
29		70	152–153	[564]
30		46	190–192	[571]
31		93	191–192	[576]
32		90	154–155	[576]
33		88	215–216	[576]
34		91	225–226	[571,576]

(Continued)

TABLE 2.11 (*Continued*)
Condensed Pyrrole Systems as Functional C-Vinylpyrrole Derivatives

No.	Structure	Yield (%)	Mp (°C)	References
1	2	3	4	5
35		80	148	[576]
36		65	184	[576]
37		77	202–203	[576]
38		90	219–220	[577]
39		91	232–233	[577]
40		90	197–198	[576]
41		56	147–148	[574]

(*Continued*)

TABLE 2.11 (*Continued*)
Condensed Pyrrole Systems as Functional C-Vinylpyrrole Derivatives

No.	Structure	Yield (%)	Mp (°C)	References
1	2	3	4	5
42	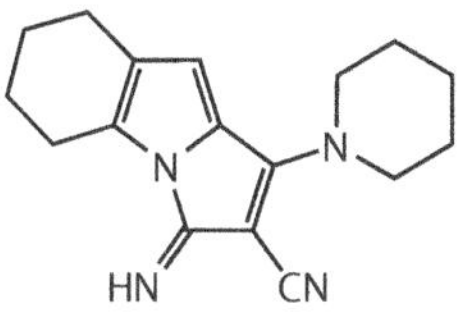	86	141–142	[573,574]
43	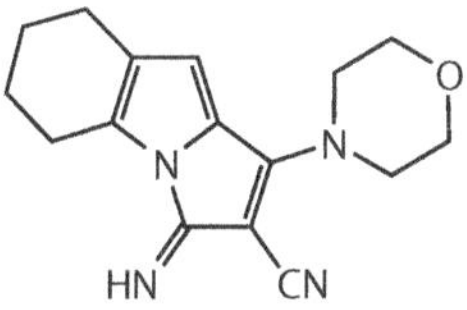	94	160–161	[573,574]
44	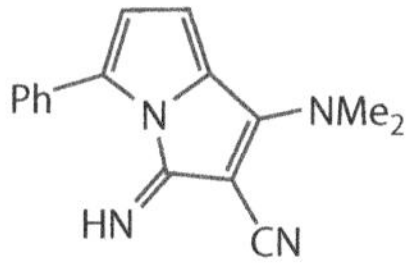	85	167–168	[576]

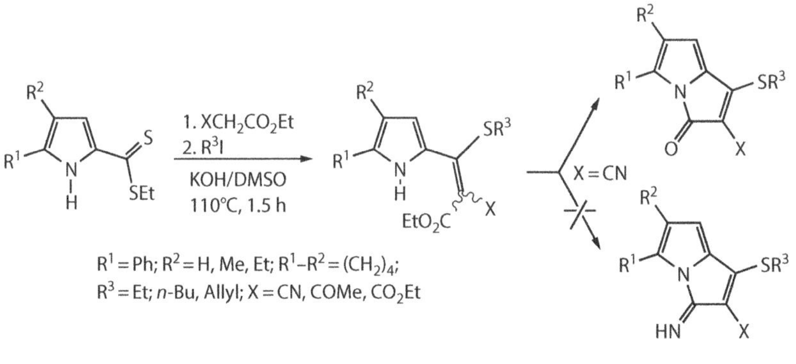

SCHEME 2.90 Synthesis of pyrrolizin-3-ones from pyrrole-2-carbodithioates and methylene active esters.

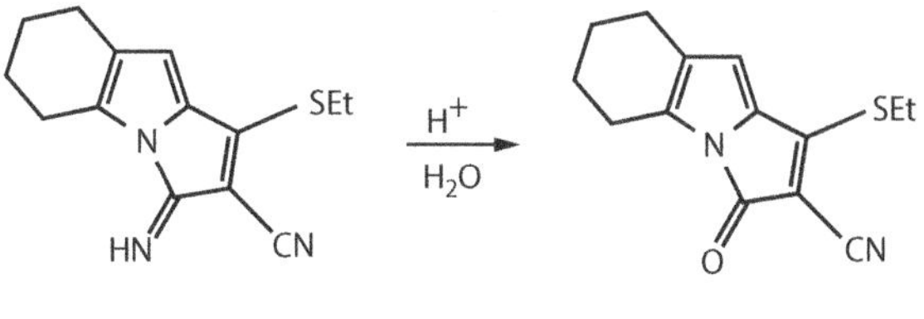

SCHEME 2.91 Synthesis of pyrrolizin-3-one by the acid-catalyzed hydrolysis of 3-iminopyrrolizine.

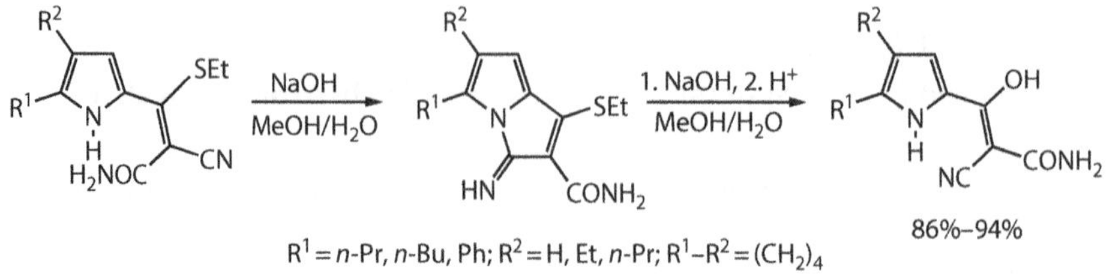

$R^1 = n\text{-Pr}, n\text{-Bu, Ph}; R^2 = H, Et, n\text{-Pr}; R^1–R^2 = (CH_2)_4$

SCHEME 2.92 Synthesis of stable enols from 2-(1-ethylthio-2-carbamoyl-2-cyanovinyl) pyrroles.

$R^1 = n\text{-Pr}, R^2 = Et (78\%); R^1 = n\text{-Bu}, R^2 = n\text{-Pr} (90\%)$

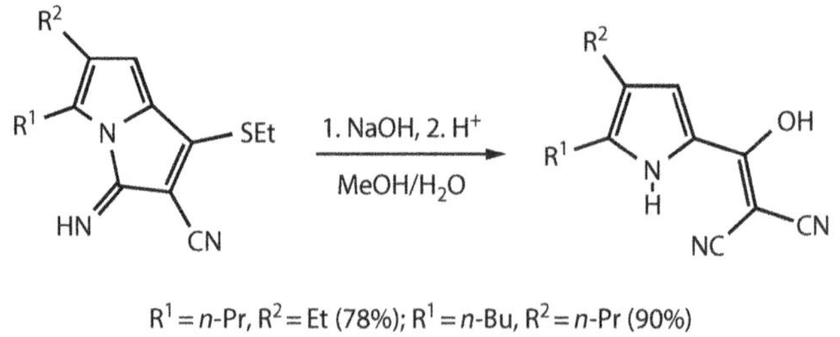

SCHEME 2.93 Synthesis of enols from 1-ethylthio-3-iminopyrrolizine-2-carbonitriles.

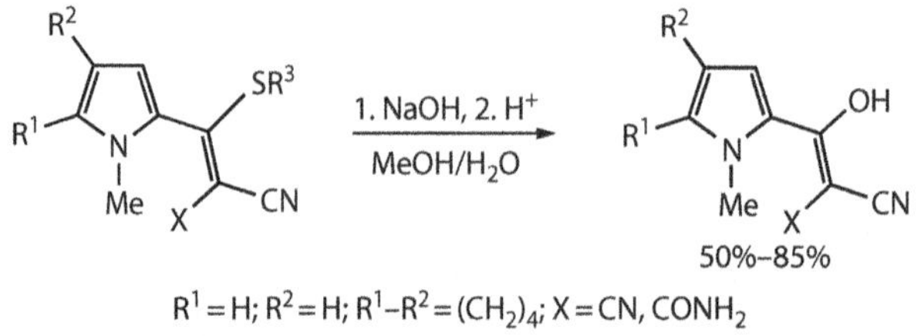

$R^1 = H; R^2 = H; R^1–R^2 = (CH_2)_4; X = CN, CONH_2$

SCHEME 2.94 Synthesis of enols from N-methyl-2-(1-ethylthio-2-cyanovinyl)pyrroles.

A peculiarity of 2-(1-hydroxy-2-carbamoyl-2-cyanovinyl)pyrroles is a strong intramolecular hydrogen bond between the hydroxyl and carbonyl groups showing up in the ^{1}H NMR spectra a strong downfield shifted signal of OH hydrogen (CDCl$_3$, 16.4–17.07 ppm). X-ray analysis data evidence that in the solid state, the compounds synthesized exist solely in the enol form stabilized by intramolecular hydrogen bonding [568].

Upon heating (toluene, xylene, 75°C–135°C), enols **26** unexpectedly readily rearrange to their 3-isomers in high yields (Scheme 2.95, Table 2.10) [449,569].

The observed rearrangement is unique for pyrrole chemistry since the thermal migration of the substituents occurs at significantly higher temperature and, as a rule, leads to the prevailing formation of α-isomers [2]. The migration of substituents from the α- to the β-position of the pyrrole ring proceeds relatively easily only in the presence of acids and is mainly typical for electron-withdrawing groups [570].

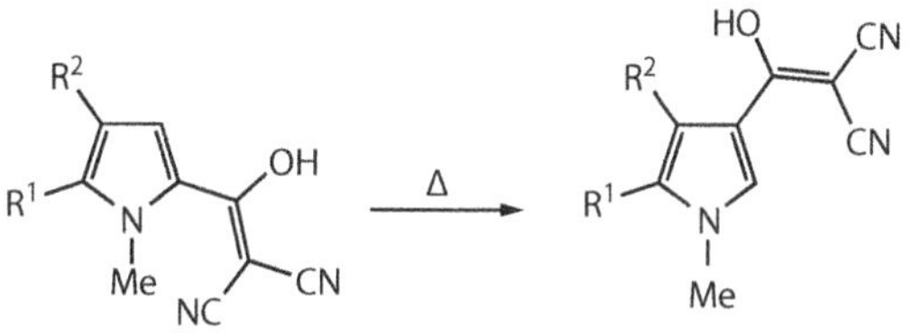

26

$R^1 = H, n\text{-}Pr; R_2 = H, Et; R^1\text{-}R^2 = (CH_2)_4$

SCHEME 2.95 Easy migration of the enol fragment from α- to β-position of pyrrole ring.

Apparently, the easy migration of the enol fragment from the α- to the β-position may be rationalized by sensitivity of the pyrrole ring to protonation combined with the enhanced acidity of the hydroxyl function of the functionalized enol substituent (owing to the effect of two strong electron-withdrawing groups). Actually, the enols synthesized appear to be strong acids (pK_a ~ 5–6) comparable in the acidity with the picric acid. The role of the hydroxyl proton in the rearrangement is confirmed by the stability of ethylthio or nitrile analogs under similar conditions.

The rearrangement is likely triggered by intramolecular protonation of the α-position of the pyrrole ring to generate zwitterion **A** that through the isomer form **B** is transformed into zwitterion **C** (Scheme 2.96). The process is completed by a proton transfer in zwitterion **C** from position 3 of the pyrrole ring to the oxygen atom. An intermolecular protonation of the pyrrole ring is also not excluded.

The energy difference (MP2/6-311G**) of 2- and 3-isomers of (1-hydroxy-2,2-dicyanovinyl)-1-methylpyrroles is 2.2 kcal/mol in favor of the 3-isomer [449]. One of the reasons of 2-isomer destabilization is a probable steric repulsion of the N-methyl group and functionalized vinyl substituent. High stability of 3-isomers is in agreement with decrease of their hydroxyl group acidity (by ~0.5 un. pK_a) as compared to

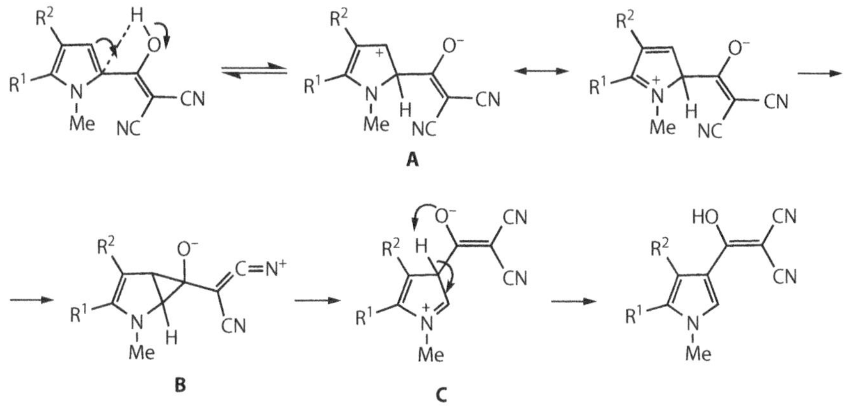

SCHEME 2.96 Mechanism of migration of the enol moiety from α- to β-position of the pyrrole ring.

2-isomers. This also means that the 3-pyrrolyl fragment is more donating relative to the enol substituent than the corresponding 2-pyrrolyl counterpart.

Under similar conditions, migration of the vinyl fragment does not occur in 2-(1-hydroxy-2-carbamoyl-2-cyanovinyl)-1-methylpyrroles. The most probable cause is a strong intramolecular hydrogen bond between the hydroxyl proton and carbonyl group (x-ray diffraction (XRD) data), considerably lowering the acidity of the hydroxyl (~3.3 un. pK_a) [568].

The N-unsubstituted pyrrole is stable under the rearrangement conditions [449]. The absence of the enol substituent migration from position 2 to position 3 is due to the lack of steric strain in this case (the interaction of N-methyl and functionalized vinyl groups), as well as decrease in the π-donor effect of the N-unsubstituted fragment (in comparison with the N-substituted one) relative to the π-acceptor enol system. This weakens intramolecular donor–acceptor interaction (intramolecular charge transfer) in the molecule and results in higher thermodynamic stability of 2-isomer.

The enol substituents migration in the pyrrole ring fundamentally supplements the existing concepts of the rearrangement processes involving pyrrolium cations and significantly extends the synthetic prospects of functionalized C-vinylpyrroles.

2.1.6.7 Reactions of Functionalized 2-Vinylpyrroles with Amines

1-Alkylthio-2-cyanovinylpyrroles **27a,b** are stable in the system KOH/DMSO and do not cyclize into 3-iminopyrrolizines **28a,b** even at 100°C–110°C. However, in methanol, in the presence of trace amounts of KOH (0.2%) at 50°C, the cyclization occurs almost instantly (1–2 min) (Scheme 2.97). It means that assistance of the alcohol proton is necessary for successful intramolecular nucleophilic addition of the NH-pyrrole function to the nitrile group.

1-Alkylthio-3-iminopyrrolizines **28a,b** contain in this case small amounts of methoxy- and hydroxypyrrolizines, products of nucleophilic substitution of alkylthio group by methoxide and hydroxide ions.

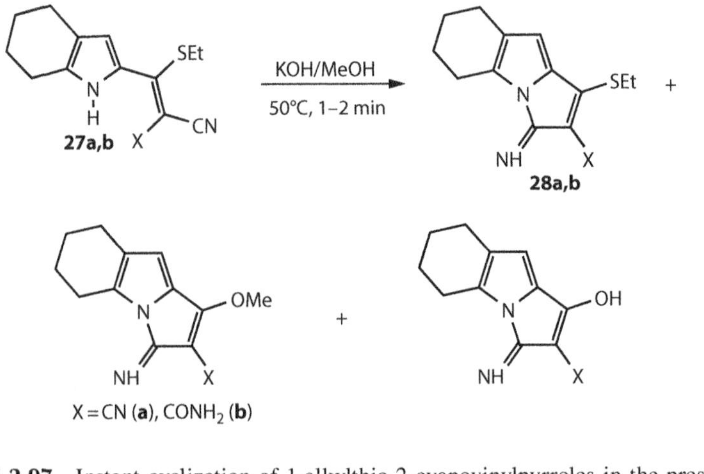

X = CN (**a**), CONH$_2$ (**b**)

SCHEME 2.97 Instant cyclization of 1-alkylthio-2-cyanovinylpyrroles in the presence of KOH in MeOH.

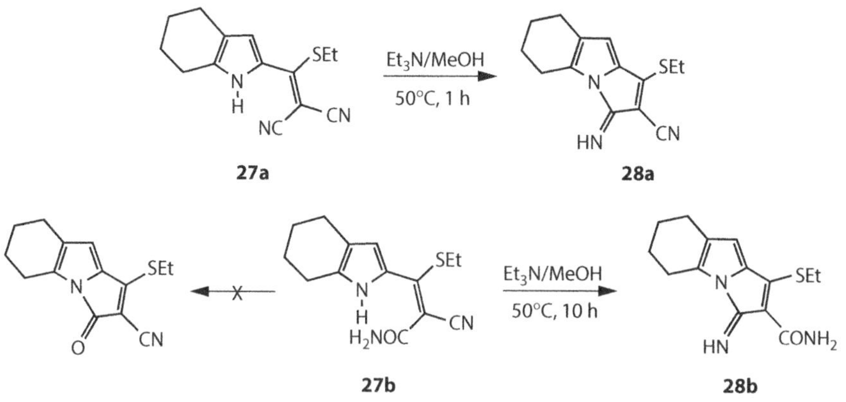

SCHEME 2.98   Cyclization of functionalized 2-(1-ethylthio-2-cyanovinyl)-4,5,6,7-tetrahy-droindoles in the presence of Et_3N.

2-(1-Alkylthio-2,2-dicyanovinyl)pyrroles selectively cyclize into 1-alkylthio-3-iminopyrrolizines (Scheme 2.98, Table 2.11) upon refluxing in methanol in the presence of catalytic amounts of triethylamine [564]. 2-(1-Ethylthio-2,2-dicyanovinyl)pyrrole **27a** cyclizes much more readily than 2-(1-ethylthio-2-carbamoyl-2-cyanovi-nyl)pyrroles **27b** [571,572]. The reason is, first of all, reversibility of the reaction and configuration of the vinyl substituents in pyrrole **27b** (according to ^{1}H NMR data, the carbamoyl group is *syn*-oriented with respect to the pyrrole NH group).

Despite the poor yield of pyrrolizine with carbamoyl moiety **28b** and the cor-responding configuration of vinylpyrrole **27b**, the reaction remains chemoselective: among two functional groups capable of participating in the cyclization (amide and nitrile), only the latter is active.

Vinylpyrrole **27a** is treated with alcohol solutions of secondary amines (reflux-ing, 4 h) to give the corresponding aminopyrrolizines (56%–89% yields, Scheme 2.99, Table 2.11) [573,574]. It is shown [574] that the first stage of reaction involves cyclization of pyrrole **27a** into iminopyrrolizine **28a** followed by the substitution of the ethylthio group in cyclic derivative. Such reaction route is supported by direct synthesis of aminopyrrolizines from pyrrolizine **28a** and secondary amines.

Similar aminopyrrolizines have been obtained by substitution of the methylthio group in 2-(2,2-dicyano-1-methylthiovinyl)pyrrole with secondary amine and the subsequent cyclization of 2-(1-amino-2,2-dicyanovinyl)pyrrole **29** in the presence of tertiary amine [575].

Analogously, vinylpyrroles **27a,b** and 3-iminopyrrolizines **28a,b** are coupled with aqueous methyl- [571] and *n*-butylamines [576] to furnish 1-alkylamino-3-imi-nopyrrolizines (73%–93% yields, Scheme 2.100, Table 2.11).

However, upon refluxing with aqueous dimethylamine (0.5 h), 2-(1-ethylthio-2-carbamoyl-2-cyanovinyl)pyrroles **27b** simply exchange the ethylthio group for the amine to deliver dimethylaminovinylpyrroles **30b** (68%–87% yields, Scheme 2.101, Table 2.10) [571,576].

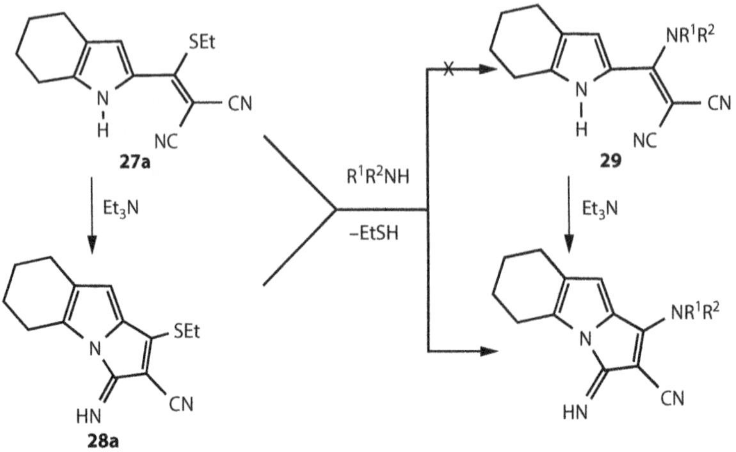

$R^1 = R^2 =$ Et; R^1–$R^2 = (CH_2)_5$, $(CH_2)_2O(CH_2)_2$

SCHEME 2.99 Synthesis of 1-amino-3-iminopyrrolizines from 2-(1-ethylthio-2,2-dicyanovinyl)pyrroles or 1-ethylthio-3-iminopyrrolizine-2-carbonitriles under the action of the secondary amines.

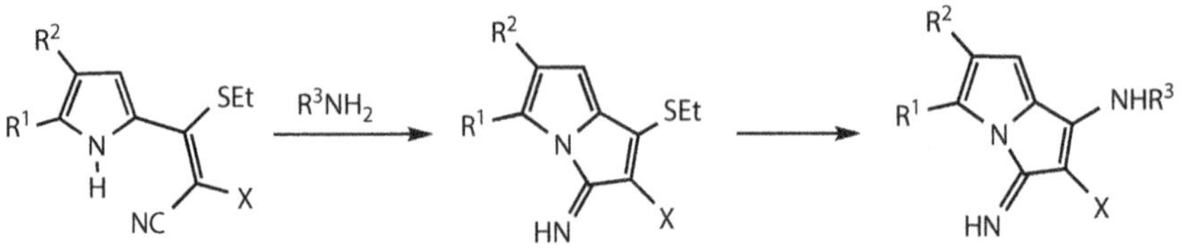

27a,b **28a,b**

$R^1 =$ *n*-Pr, *n*-Bu, Ph; $R^2 =$ H, Et, *n*-Pr; R^1–$R^2 = (CH_2)_4$; $R^3 =$ Me, *n*-Bu; X = CN, $CONH_2$

SCHEME 2.100 Formation of 1-amino-3-iminopyrrolizines from 2-(1-ethylthio-2-cyanovinyl)pyrroles or 1-ethylthio-3-iminopyrrolizines and primary amines in water.

Under the same conditions, pyrrolizines **28b** undergo recyclization into vinylpyrroles **27b** followed by the substitution of the alkylthio group with dimethylamine to form dimethylaminovinylpyrroles **30b** in up to 78% yields (Table 2.10).

Thus, pyrroles **27b** react with aqueous dimethyl- and methylamine differently: with the former they only exchange the SEt-group for the NMe_2 moiety without cyclization, whereas with the latter, the cyclization of pyrroles **27b** to pyrrolizine **28b** occurs and then substitution of the SEt-group by the NHMe fragment takes place. Obviously, pyrroles **30b** preferably exist in configuration with *cis*-disposition of the carbamoyl and pyrrole NH groups stabilized by the intramolecular H-bond, which is unfavorable for cyclization. Intermediate **32b** is capable of isomerizing to intermediate **33** (Scheme 2.102). The latter, owing to free rotation around the C_1–C_2-bond, can cyclize to pyrrolizine **31b**. In the case of dimethylamine, such transformations are impossible.

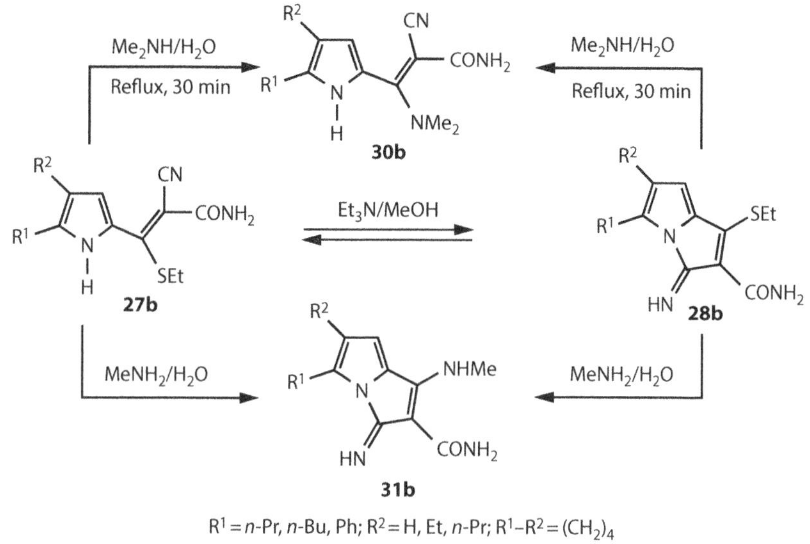

SCHEME 2.101   Reaction of 2-(1-ethylthio-2-carbamoyl-2-cyanovinyl)pyrroles with aqueous Me$_2$NH.

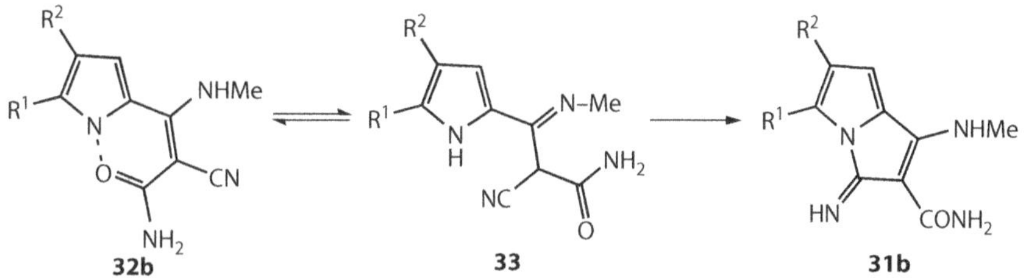

SCHEME 2.102   Rationale of the reaction diversity of 2-(1-ethylthio-2-carbamoyl-2-cyano-vinyl)pyrroles with methyl-and dimethylamines.

As mentioned earlier, 3-iminopyrrolizine exchanges alkylthio group for the secondary amine upon refluxing in methanol for 4 h. In pyrrolizine-3-ones, such exchange proceeds much easier (room temperature, 15 min) (Scheme 2.103) [564]. The yields and constants of 1-aminopyrrolizin-3-ones are given in Table 2.11.

2.1.6.8   Synthesis of 5-Amino-3-(pyrrol-2-yl)pyrazoles

2-(1-Ethylthio-2-cyanovinyl)pyrroles smoothly react with hydrazine hydrate (refluxing in methanol, 0.5 h) to afford 3-amino-5-(pyrrol-2-yl)pyrazoles (according to the specified data) in 86%–90% yields (Scheme 2.104, Table 2.12) [577,578]. The reaction is chemo- and regioselective: other products including 3-iminopyrrolizines are not detected in the reaction mixture. The reaction is likely triggered by the addition of hydrazine to the double bond. Further, after ethanethiol elimination, the

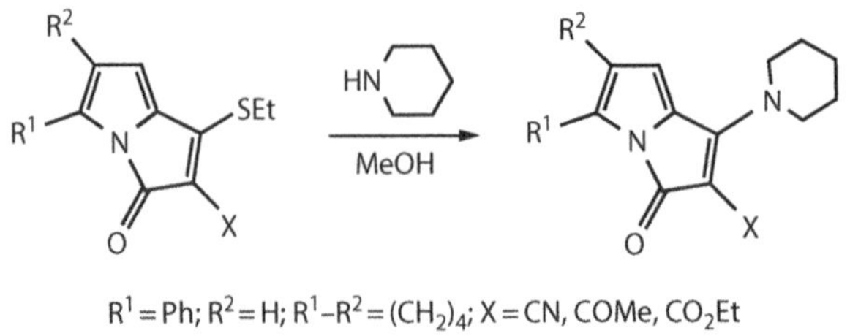

R2
R1
N
O
SEt
X
HN
MeOH
R2
R1
N
O
N
X
R1 = Ph; R2 = H; R1–R2 = (CH2)4; X = CN, COMe, CO2Et

SCHEME 2.103 Instant replacement of ethylthio substituent in pyrrolizin-3-ones by amino group.

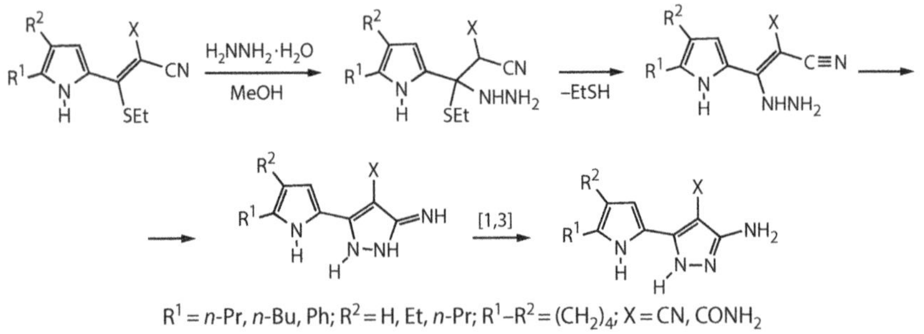

H2NNH2·H2O
MeOH
-EtSH
[1,3]
R1 = n-Pr, n-Bu, Ph; R2 = H, Et, n-Pr; R1–R2 = (CH2)4; X = CN, CONH2

SCHEME 2.104 Synthesis of 5-amino-3-(pyrrol-2-yl)pyrazoles from 2-(1-ethylthio-2-cyanovinyl)pyrroles and hydrazine hydrate.

intramolecular nucleophilic addition of the hydrazine fragment to the nitrile group follows. The prototropic rearrangement of iminodihydropyrazole intermediate leads to 3-aminopyrazoles (x-ray analysis data).

In the case of 2-(1-ethylthio-2-cyanovinyl)pyrrole with CO_2Et group, the formation of the pyrazole cycle is accompanied by hydrazinolysis of the ester function, and the only product is pyrazole with X = $CONHNH_2$ (54%, Table 2.12).

Under similar conditions the treatment of 1-alkylthio-3-iminopyrrolizines with hydrazine hydrate involves ready and almost quantitative replacement of the alkylthio substituent by the hydrazine function to produce 1-hydrazinyl-3-iminopyrrolizines (90–91%, Scheme 2.105, Table 2.11).

The reaction of 2-(1-ethylthio-2-cyanovinyl)pyrroles with ethylhydrazine proceeds much slower (80% conversion of reagents is reached by boiling in methanol for 10 h) (Scheme 2.106). The predominant direction of the reaction of 2-(1-ethylthio-2,2-dicyanovinyl)pyrrole **27a** is intramolecular cyclization into 1-ethylthio-3-iminopyrrolizine **28a** including the exchange of 1-ethylthio group for ethylhydrazine moiety. Ultimately, a mixture of pyrrolizine and pyrazole in 3:1 ratio is formed (48% overall yield). In the case of 2-(1-ethylthio-2-carbamoyl-2-cyanovinyl)pyrrole **27b** on the contrary, the main reaction route is formation of pyrazole (pyrrolizine–pyrazole ratio is 1:2.5).

TABLE 2.12
Pyrrolylpyrazoles and Their Derivatives

No.	Structure	Yield (%)	Mp (°C)	References
1	2	3	4	5
1		86	192–193	[577,578]
2		91	147–148	[577,578]
3		90	228–229	[577,578]
4		92	299–300	[577,578]
5		54	>300	[577,578]
6		87	234	[578]
7		95	>300	[577,578]

(*Continued*)

TABLE 2.12 (*Continued*)

Pyrrolylpyrazoles and Their Derivatives

No.	Structure	Yield (%)	Mp (°C)	References
1	2	3	4	5
8		85	>300	[578]
9		63	>300	[578]
10		79	>300	[578]
11		86	>300	[578]
12		70	>300	[578]
13		70	>300	[578]

(Continued)

TABLE 2.12 (*Continued*)
Pyrrolylpyrazoles and Their Derivatives

No.	Structure	Yield (%)	Mp (°C)	References
1	2	3	4	5
14		89	252	[580]
15		62	230	[580]
16		74	>300	[580]
17		88	290	[580]
18		28	>300	[580]
19		90	>300	[580]

(*Continued*)

TABLE 2.12 (*Continued*)
Pyrrolylpyrazoles and Their Derivatives

No.	Structure	Yield (%)	Mp (°C)	References
1	2	3	4	5
20		65	286	[580]
21		71	288	[580]
22		65	>300	[580]
23		73	>300	[580]
24		53	249–252	[581]
25		65	250–253	[581]

(*Continued*)

TABLE 2.12 (*Continued*)
Pyrrolylpyrazoles and Their Derivatives

No.	Structure	Yield (%)	Mp (°C)	References
1	2	3	4	5
26		95	271–274	[581]
27		89	267–269	[581]
28		65	>350	[581]
29		79	>350	[581]
30		47	274–275	[581]
31		90	267–270	[581]

Obviously, the reason is configuration of 2-(1-ethylthio-2-carbamoyl-2-cyanovinyl)pyrrole stabilized by intramolecular H-bond, which is favorable for the pyrazole formation.

In the presence of fluorine anions, low-saturated (5 µM) solutions of one of the pyrazoles, namely, 3-amino-5-(5-phenylpyrrol-2-yl)pyrazole-4-carboxamide, intensively fluoresce in blue region (λ_{max} 424 nm, quantum yield 0.74), while other anions (Cl⁻, Br⁻, I⁻, HSO_4^-, $H_2PO_4^-$, AcO⁻) do not affect the emission parameters [579]. The results indicate this compound is suitable for fluoride sensing by a naked eye detection.

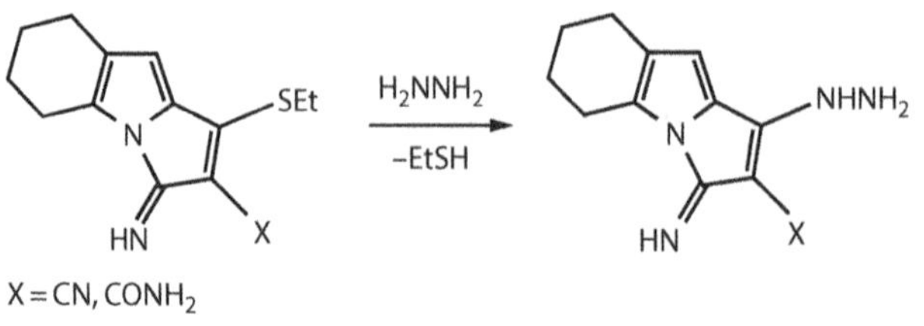

SCHEME 2.105 Replacement of ethylthio substituent in 1-ethylthio-3-iminopyrrolizines by hydrazine moiety.

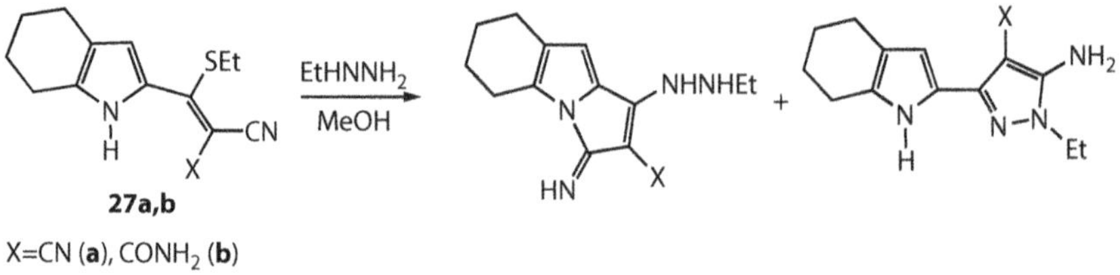

X=CN (**a**), CONH$_2$ (**b**)

SCHEME 2.106 Reaction of 2-(1-ethylthio-2-cyanovinyl)pyrroles with ethylhydrazine.

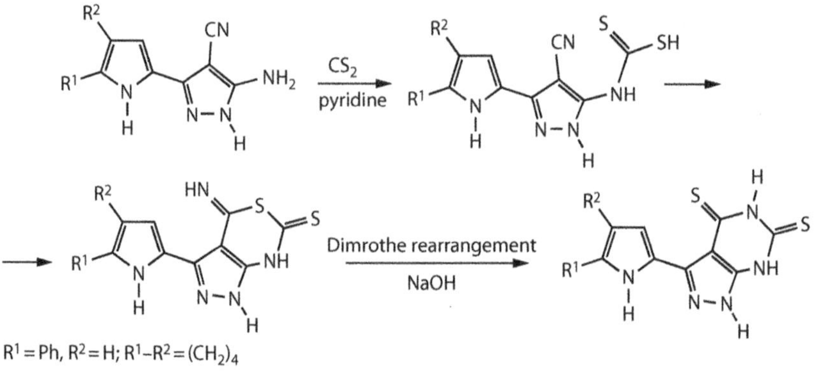

R^1 = Ph, R^2 = H; R^1–R^2 = (CH$_2$)$_4$

SCHEME 2.107 Synthesis of pyrimidinedithiones from 5-amino-3-(pyrrol-2-yl)pyrazole-2-carbonitriles by their reactions with CS$_2$.

According to the NMR data, the fluorescence can be explained by deprotonation of the pyrazole counterpart of the molecule under the action of fluoride anion.

Thus, pyrrolylpyrazoles are promising compounds that can be applied in medicine and for solution of environmental problems, since fluoride anions possess neurotoxicity and cause osteosarcoma and DNA mutation.

The reaction of pyrazolecarbonitriles with carbon disulfide in boiling pyridine gives pyrimidinedithiones in up to 80% yield (Scheme 2.107, Table 2.12) [578].

Intermediate iminothiazines are not detected in the reaction mixture. Obviously, it is pyridine that acts as the base catalyzing the Dimroth rearrangement (opening the thiazine ring and subsequent cyclization into pyrimidinedithione cycle).

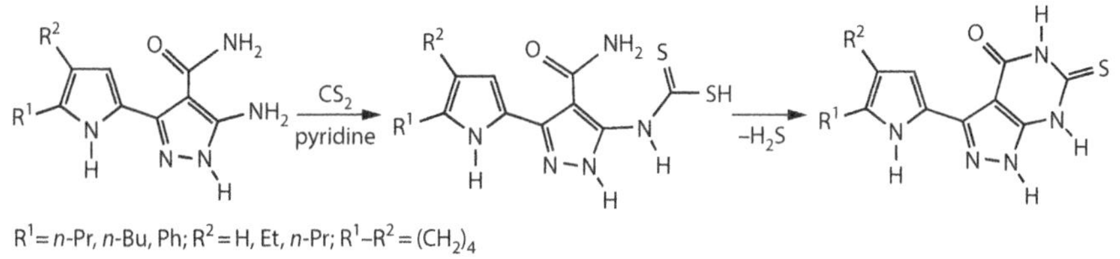

R^1 = n-Pr, n-Bu, Ph; R^2 = H, Et, n-Pr; R^1–R^2 = (CH$_2$)$_4$

SCHEME 2.108   Synthesis of 6-thioxo-4,5,6,7-tetrahydro-4H-pyrazolo[3,4-d]pyrimidine-4-ones from 5-amino-3-(pyrrol-2-yl)pyrazole-2-carboxamides by their reactions with CS$_2$.

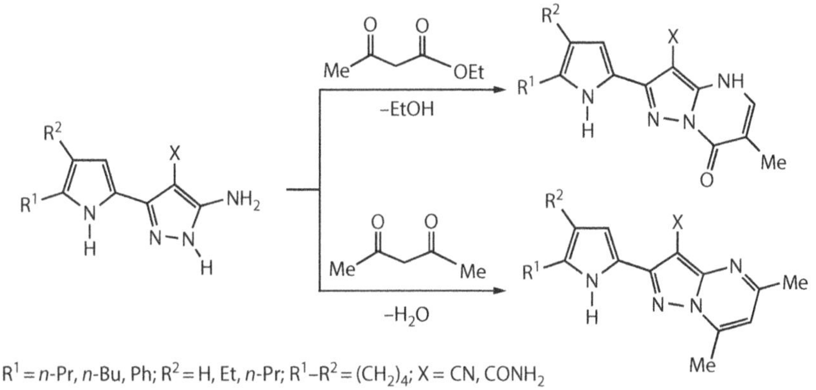

R^1 = n-Pr, n-Bu, Ph; R^2 = H, Et, n-Pr; R^1–R^2 = (CH$_2$)$_4$; X = CN, CONH$_2$

SCHEME 2.109   Synthesis of pyrazolo[1,5-a]pyrimidines from 5-amino-3-(pyrrol-2-yl)pyrazoles by their reaction with dicarbonyl compounds.

Pyrazolecarboxamides react with carbon disulfide in boiling pyridine with a loss of hydrogen sulfide molecule to give 6-thioxo-4,5,6,7-tetrahydro-4H-pyrazolo[3,4-d] pyrimidine-4-ones (70%–86% yields, Scheme 2.108, Table 2.12) [578].

The reaction of pyrazolecarboxamides with carbon disulfide proceeds much slower than an analogous reaction of pyrazolecarbonitriles (the complete conversion is achieved for 10–40 h).

6-Thioxo-4,5,6,7-tetrahydro-4H-pyrazolo[3,4-d]pyrimidine-4-ones inhibit DNA topoisomerase of plant origin (data of the Siberian Institute of Plant Physiology and Biochemistry, SB RAS).

Aminopyrazoles react with dicarbonyl compounds (acetylacetone and ethyl acetoacetate) upon heating (130°C–155°C, 2–7 h) to afford pyrazolo[1,5-a]pyrimidines (Scheme 2.109, Table 2.12) [580].

Currently, such planar polyfused heterocycles with pyrrole or indole fragments capable of charge transfer, hydrogen bond formation, and interplane self-assembling via π–π interaction (π-stacking) attract great interest as potential DNA-interactive reagents.

Noncatalytic reaction of aminopyrazoles with N-vinyl- and N-ethylimidazole (benzimidazole)carbaldehydes leads to the corresponding azomethines (Scheme 2.110, Table 2.12) [581].

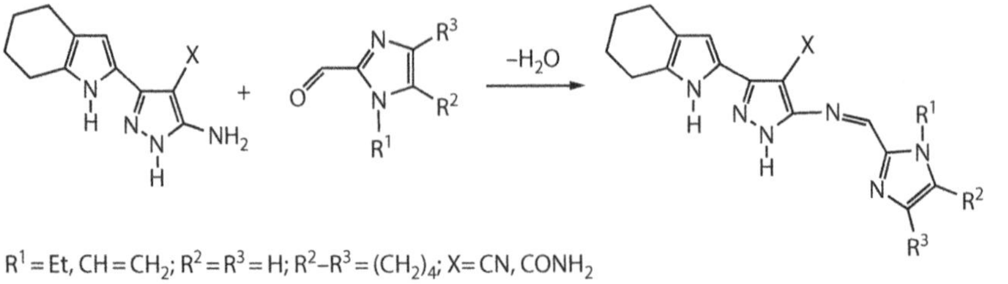

$R^1 = Et, CH=CH_2; R^2=R^3=H; R^2-R^3=(CH_2)_4; X=CN, CONH_2$

SCHEME 2.110   Reaction of 5-amino-3-(pyrrol-2-yl)pyrazoles with imidazolecarbalde-hydes.

Such systems represent promising multidentate ligands for chelating the heavy metals (Ni, Co, Fe, Cu, Pd, etc.) and preparation of biologically and catalytically active complexes like vitamin B_{12}.

2.1.6.9   Synthesis of 5(3)-Amino-3(5)-(pyrrol-2-yl)isoxazoles

2-(1-Ethylthio-2,2-dicyanovinyl)pyrroles upon heating (40°C–45°C, 30 min) in methanol with aqueous hydroxylamine readily exchange their ethylthio group for a hydroxylamine moiety to form, after the subsequent cyclization of interme-diates **A**, 5-amino-3-(pyrrol-2-yl)isoxazoles in 56%–81% yields (Scheme 2.111, Table 2.13) [582].

Under similar conditions, with 2-(1-ethylthio-2-carbamoyl-2-cyanovinyl)pyr-roles, which exist exclusively as isomers with the *cis*-disposition of carbamoyl and NH-pyrrole groups, the reaction chemoselectivity breaks [582]: along with major 5-amino-3-(pyrrol-2-yl)isoxazoles (the content in the reaction mixture is 95%–97%), their structural isomers, 3-amino-5-(pyrrol-2-yl)isoxazoles, are also formed (Scheme 2.112, Table 2.13).

2-(1-Ethylthio-2-carbamoyl-2-cyanovinyl)pyrroles selectively react with hydrox-ylamine upon refluxing in tetrahydrofuran to furnish 5-aminoisoxazoles as sole products. However, in this case, despite the longer reaction time (1 h), the yields of 5-aminoisoxazoles are lower (45%–67%) than those in methanol.

Formation of two isomeric aminoisoxazoles in these reactions may be rationalized by the existence of intermediates **B** as two isomers with *cis-* and *trans*-disposition of hydroxylamine and nitrile groups (Scheme 2.113). While the *cis*-isomers readily

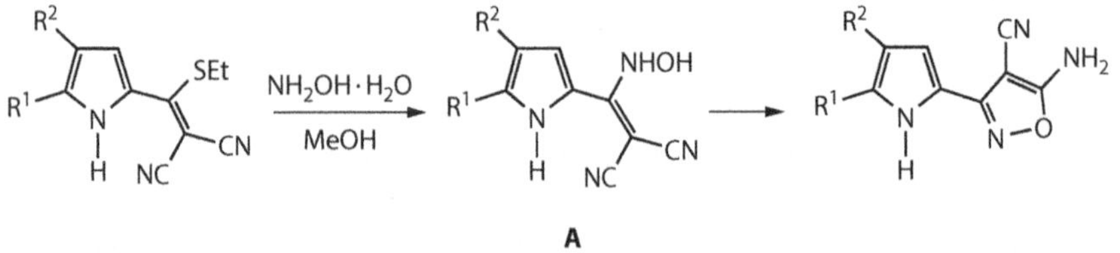

A

$R^1 = n\text{-}Pr, n\text{-}Bu, Ph; R^2=H, Et, n\text{-}Pr; R^1-R^2=(CH_2)_4$

SCHEME 2.111   Synthesis of 5-amino-3-(pyrrol-2-yl)isoxazoles from 2-(1-ethylthio-2,2-dicyanovinyl)pyrroles and hydroxylamine.

TABLE 2.13
Products of Reaction of Functional C-Vinylpyrroles with Hydroxylamine

No.	Structure	Yield (%)	Mp (°C)	References
1	2	3	4	5
1		68	128–129	[582]
2		69	147–148	[582]
3		78	163–164	[582]
4		56	121–122	[582]
5		76	103	[582]
6		81	>350	[582]
7		89	>350	[582]

(*Continued*)

TABLE 2.13 (*Continued*)
Products of Reaction of Functional C-Vinylpyrroles with Hydroxylamine

No.	Structure	Yield (%)	Mp (°C)	References
1	2	3	4	5
8		71	>350	[582]
9		73	182–183	[582]
10		49	207–208	[582,584]
11		33	213–214	[582,584]
12		40	231–232	[582,584]
13		48	252	[582]
14		35	223–224	[582,584]

(*Continued*)

TABLE 2.13 (*Continued*)
Products of Reaction of Functional C-Vinylpyrroles with Hydroxylamine

No.	Structure	Yield (%)	Mp (°C)	References
1	2	3	4	5
15	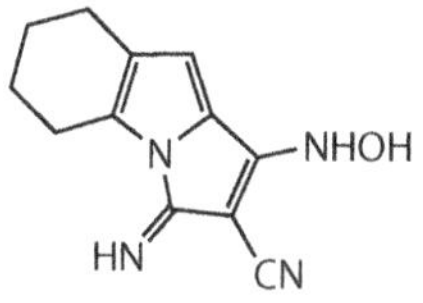	41	243–244	[582,584]
16	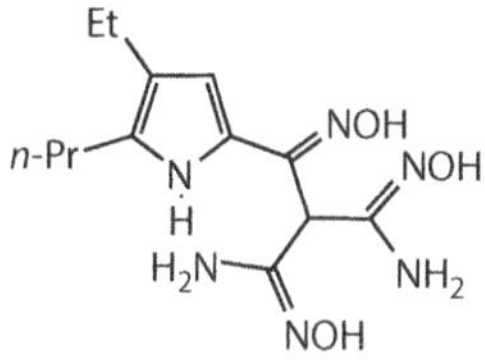	61	>350	[582]

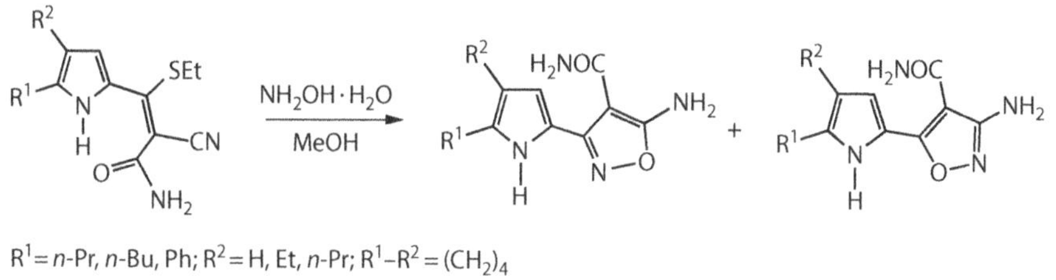

$R^1 = n\text{-}Pr, n\text{-}Bu, Ph; R^2 = H, Et, n\text{-}Pr; R^1\text{-}R^2 = (CH_2)_4$

SCHEME 2.112 Synthesis of 5-amino-3-(pyrrol-2-yl)isoxazoles from 2-(1-ethylthio-2-carbamoyl-2-cyanovinyl)pyrroles and hydroxylamine.

cyclize to 5-aminoisoxazoles, the *trans*-isomers either partially transform to the *cis*-form or add a second hydroxylamine molecule to the nitrile carbon to generate the *bis*-adducts **C**, which further cyclize with elimination of a hydroxylamine molecule giving the minor reaction products, 3-aminoisoxazoles.

pK_a values for the first and the second steps of dissociation of hydroxylamine (6.00 and 13.7, respectively) [583] indicate that in highly basic media, the latter exists mostly as the aminohydroxy anion, whereas under neutral or weakly basic conditions, it presents as free hydroxylamine. Therefore, in the presence of NaOH, hydroxylamine reacts with 2-(1-ethylthio-2-carbamoyl-2-cyanovinyl) pyrroles as an O-nucleophile giving exclusively 3-aminoisoxazoles (12%–41% yields) (Scheme 2.114).

3-Aminoisoxazoles are also synthesized by the reaction of 3-imino-2-pyrrolizinecarboxamides with hydroxylamine (Scheme 2.115) [582,584]. A side pathway of the reaction is the exchange of ethylthio group in the initial pyrrolizines

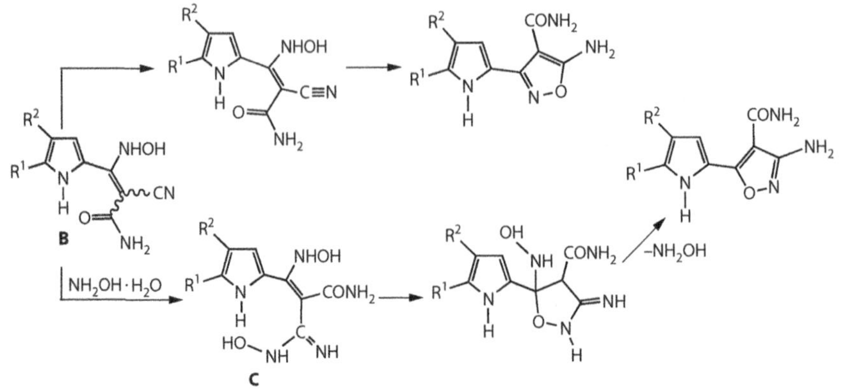

SCHEME 2.113 Rationale of the formation of two isomeric pyrrolylisoxazoles from 2-(1-ethylthio-2-carbamoyl-2-cyanovinyl)pyrroles and hydroxylamine.

$R^1 = n\text{-Pr}, n\text{-Bu}, \text{Ph}; R^2 = H, \text{Et}, n\text{-Pr}; R^1\text{–}R^2 = (CH_2)_4$

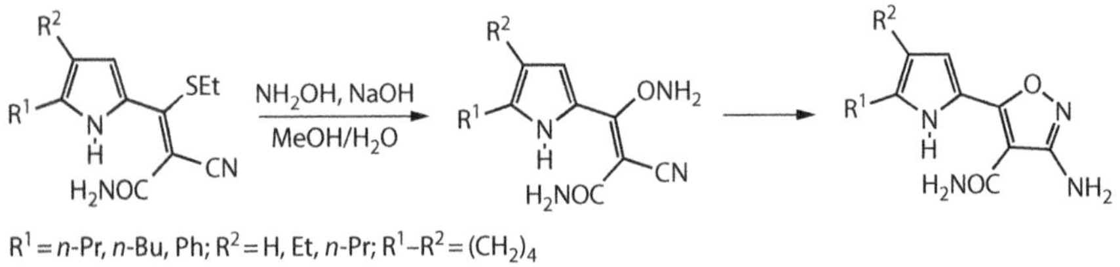

SCHEME 2.114 Selective synthesis of 3-amino-5-(pyrrol-2-yl)isoxazoles from 2-(1-ethylthio-2-carbamoyl-2-cyanovinyl)pyrroles and hydroxylamine in the presence of NaOH in MeOH.

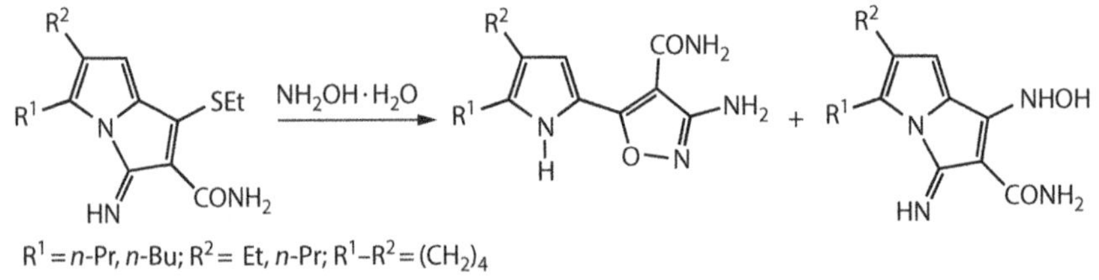

$R^1 = n\text{-Pr}, n\text{-Bu}; R^2 = \text{Et}, n\text{-Pr}; R^1\text{–}R^2 = (CH_2)_4$

SCHEME 2.115 Synthesis of 3-amino-5-(pyrrol-2-yl)isoxazoles from 1-ethylthio-3-imino-pyrrolizine-2-carboxamides and hydroxylamine.

for hydroxylamine to afford 1-hydroxylamino-3-imino-2-pyrrolizinecarboxamides, the ratio of products being 2.5:1.

In this case, 3-aminoisoxazoles are likely formed due to the opening of the pyrrolizine ring that leads to 2-(1-ethylthio-2-carbamoyl-2-cyanovinyl)pyrroles with the *cis*-configuration of nitrile and NH groups and (after reaction with hydroxylamine) intermediates **B**, which further cyclize to the isoxazoles (Scheme 2.116).

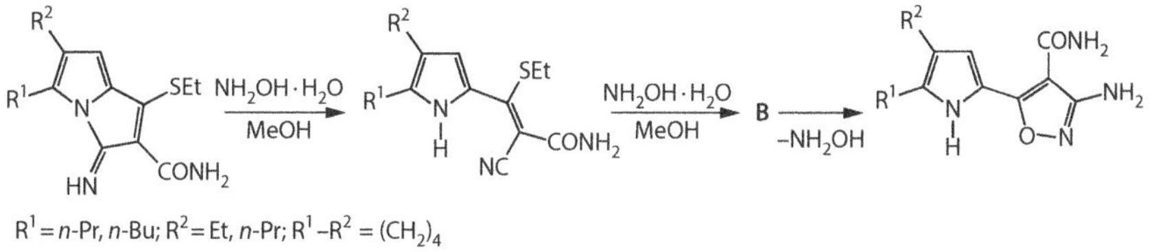

$R^1 = n$-Pr, n-Bu; $R^2 = $ Et, n-Pr; R^1–$R^2 = $ (CH$_2$)$_4$

SCHEME 2.116 Possible steps of the formation of 3-amino-5-(pyrrol-2-yl)isoxazoles from 1-ethylthio-3-iminopyrrolizine-2-carboxamides and hydroxylamine.

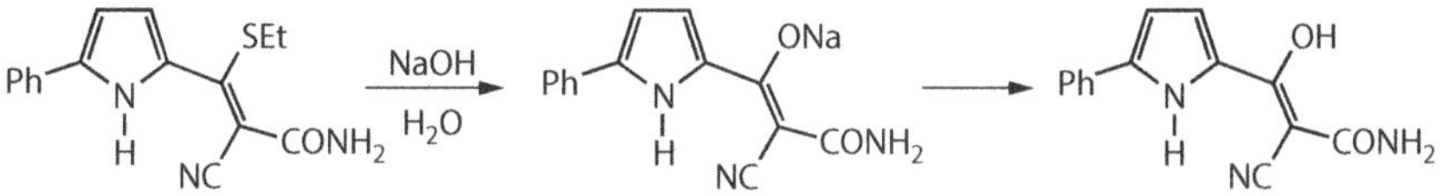

SCHEME 2.117 Synthesis of stable enol from 2-(1-ethylthio-2-carbamoyl-2-cyanovinyl)-5-phenylpyrrole in the presence of NaOH in aqueous MeOH.

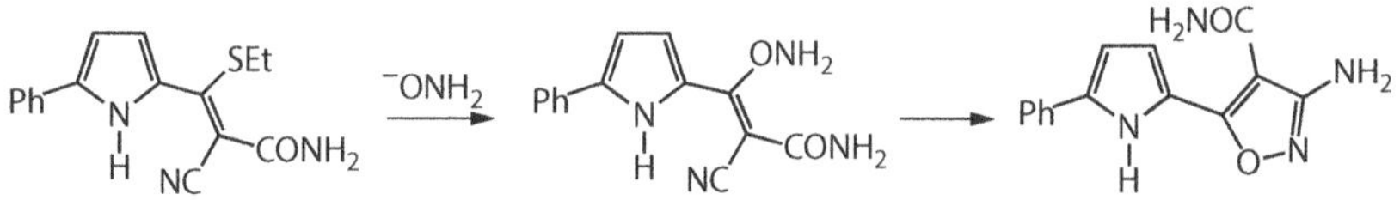

SCHEME 2.118 Rationale of the formation of 3-amino-5-(5-phenylpyrrol-2-yl)isoxazole from 2-(1-ethylthio-2-carbamoyl-2-cyanovinyl)-5-phenylpyrrole and hydroxylamine.

2-(1-Ethylthio-2-carbamoyl-2-cyanovinyl)-5-phenylpyrrole, unlike analogously functionalized pyrroles bearing alkyl or cycloalkyl substituents, does not cyclize to 3-iminopyrrolizine upon refluxing in methanol in the presence of triethylamine. Upon heating, in methanol (4 h) with aqueous NaOH, this pyrrole transforms, after acidification of the corresponding salt, into 2-(1-hydroxy-2-carbamoyl-2-cyanovinyl)-5-phenylpyrrole (Scheme 2.117, Table 2.10) [582].

Therefore, the probable route of the formation of 3-aminoisoxazole with phenyl substituent seems to be the exchange of the ethylthio group for aminohydroxy anion (Scheme 2.118).

The only pathway of the reaction of pyrrolizines bearing a nitrile group, 3-imino-2-pyrrolizinecarbonitriles, with hydroxylamine in methanol is the substitution of an ethylthio moiety for hydroxylamine to deliver 1-hydroxylamino-3-imino-2-pyrrolizinecarbonitriles (Scheme 2.119, Table 2.13) [582]. The latter are unstable in DMSO solutions and gradually transform to 2-(1-hydroxylamino-2,2-dicyanovinyl)pyrroles.

The treatment of 1-ethylthio-6-ethyl-5-propyl-3-iminopyrrolizine with hydroxylamine in THF (50°C–55°C, 1 h) results in trioxime **34** in 61% yield (Scheme 2.120, Table 2.13).

The latter is likely formed via the pyrrolizine ring opening, exchange of the ethylthio group for hydroxylamine, and addition of another two molecules of hydroxylamine to both nitrile functions.

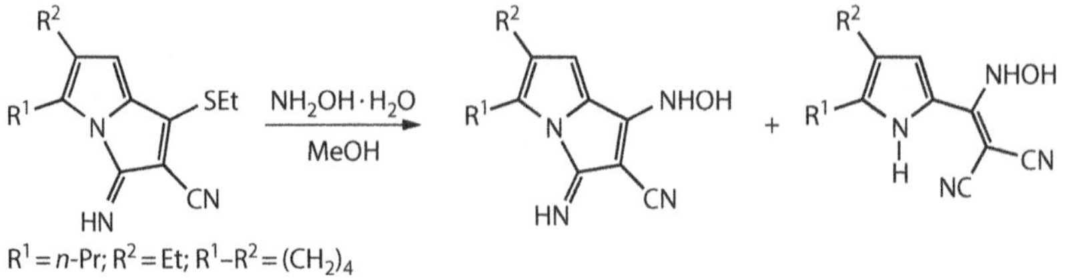

SCHEME 2.119 Synthesis of 1-hydroxylamino-3-iminopyrrolizine-2-carbonitriles from 1-ethylthio-3-iminopyrrolizine-2-carbonitriles.

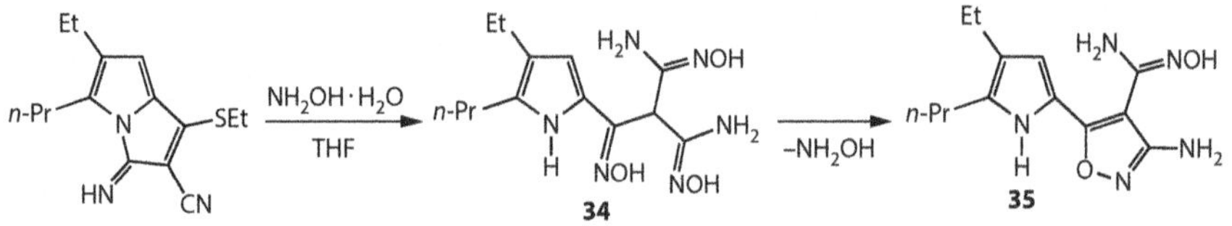

SCHEME 2.120 The reaction diversity of 1-ethylthio-6-ethyl-5-propyl-3-iminopyrrolizine with hydroxylamine in THF.

Upon longer heating (50°C–55°C, 4 h and 20°C–25°C, 12 h), this reaction affords 3-aminoisoxazole with the amidoxime moiety **35** (Scheme 2.120, Table 2.13), a product of intramolecular cyclization of trioxime **34** with elimination of a hydroxylamine molecule.

The pyrrolylisoxazoles obtained can be applied as highly effective and selective sensors for fluorine anions. For example, in the UV spectrum of 3-amino-5-(4,5,6,7-tetrahydroindol-2-yl)isoxazole-4-carboxamide (MeCN) upon addition of fluorine anions, the absorption band is shifted to red region by 35 nm [585]. Fluorescence intensity of the same isoxazole solution (excited at 340 nm) in the presence of F⁻ sharply drops and emission peak is shifted to the red region (by 32 nm). Other anions (Cl⁻, Br⁻, I⁻, HSO_4^-, $H_2PO_4^-$, AcO⁻) do not cause any changes in the spectra.

2.1.7 ALKYLATION OF PYRROLES WITH FUNCTIONAL ORGANIC HALIDES

2.1.7.1 Alkylation of Pyrroles with Allyl Halides

N-(2-Propenyl)pyrrole and isomeric N-(E,Z-1-propenyl)pyrroles, promising monomers and building blocks, have been synthesized by allylation of pyrrole with allyl chloride in the system KOH/DMSO. For instance, the pyrrole reacts with allyl chloride (room temperature, 1.5 h, molar ratio of pyrrole–allyl chloride–KOH–DMSO = 1:2:2.5:14) to give N-(2-propenyl)pyrrole in 88% yield (Scheme 2.121, Table 2.14) [586]. Apart from the major product, N,2- and N,3-di(2-propenyl)pyrroles in 8% overall yield are identified in the reaction mixture.

In the system KOH/DMSO (15°C, 10 min), N-(2-propenyl)pyrrole quantitatively isomerizes into N-(1-propenyl)pyrrole (E/Z 1:4) and at 60°C stereoselectively (96%) transforms to N-E-(1-propenyl)pyrrole (Table 2.14).

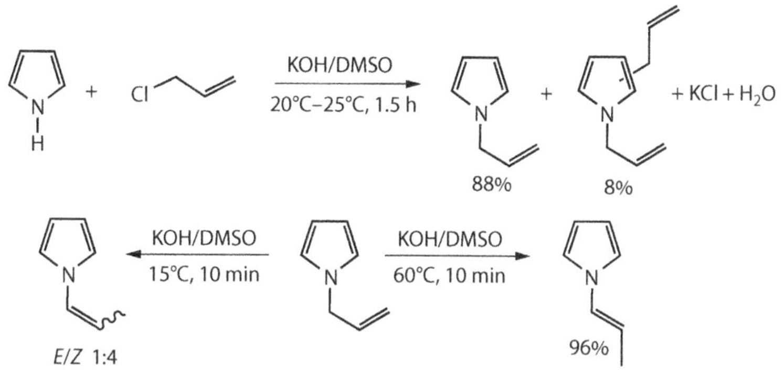

SCHEME 2.121   Synthesis of N-(2-propenyl)- and N-(E,Z-1-propenyl)pyrroles from pyrrole and allyl chloride in the KOH/DMSO system.

2.1.7.2   Alkylation of Pyrroles with Propargyl Halides

The interest in N-propargylpyrrole is motivated, in particular, by its application in the synthesis of diacetylene derivatives, which are promising for the preparation of highly ordered polymeric structures possessing specific optical, electronic, and magnetic properties [587,588].

N-Propargylpyrrole has been synthesized in 70% by the reaction of pyrrole with propargyl bromide in the system KNH_2/NH_3 (liq.) (Scheme 2.122, Table 2.14) [589].

Also, 2-propargylpyrrole and N,2- and N,3-dipropargylpyrroles are isolated from the reaction mixture (preparative GLC), total yield of the side-products being about 12%. The same reaction in $LiNH_2/NH_3$ (liq.) delivers N-propargylpyrrole in only 48%. In $NaNH_2/NH_3$ (liq.) (1 h), a mixture of N-propargylpyrrole and its allenic isomer (10:1) in 43% yield is formed.

The reaction of pyrroles with propargyl halides in KOH/DMSO suspension leads to N-allenylpyrroles in up to 68% yield (Scheme 2.123, Table 2.14) [590–592].

Allenylation of pyrrole with propargyl bromide proceeds easier than with propargyl chloride: in the former case, calcinated KOH or additives (in the case of $KOH \cdot 0.5H_2O$) of t-BuOK are not required. N-Allenylpyrrole has been synthesized from pyrrole and propargyl bromide in 68% yield and 98% purity (2% of propargyl isomer) [592].

Before the works [590,591], N-allenylpyrrole was prepared by laborious two-stage protocol: first, N-propargylpyrrole was synthesized from pyrrole and propargyl bromide in the system $NaNH_2/NH_3$ (liq.), and then acetylene–allene isomerization in the presence of superbases KOH (t-BuOK)/DMSO [593] and $NaNH_2/Al_2O_3$ [594] was carried out.

2.1.7.3   Allenylation of Pyrroles with 2,3-Dichloro-1-propene

N-Allenylpyrroles (Table 2.14) have been synthesized by the reaction of pyrroles with 2,3-dichloro-1-propene (readily obtainable from 1,2,3-trichloropropane, side product in the epichlorohydrin manufacture) in the KOH/DMSO suspension (KOH is used in fourfold molar excess relative to 2,3-dichloro-1-propene) (Scheme 2.124) [591,595].

TABLE 2.14

Products of Pyrroles Alkylation by Functional Organic Halides

No.	Structure	Yield (%)	Bp (°C/torr)	n_D^{20}	References
1	2	3	4	6	7
1		88	147/760	1.4944	[586]
2		87	70–72/50	1.5230	[586]
3		90	45–46/50	1.5190	[586]
4		80	107–109/760	1.5130	[597,598]
5		70	60/15	1.5122	[589]
6		80	53/15	1.5590	[589,592,686,687]
7		68	—	1.5520	[590,591]
8		49	—	1.5568	[591]
9		88	—	1.5790	[590,591,595]

(Continued)

TABLE 2.14 (*Continued*)
Products of Pyrroles Alkylation by Functional Organic Halides

No.	Structure	Yield (%)	Bp (°C/torr)	n_D^{20}	References
1	2	3	4	6	7
10	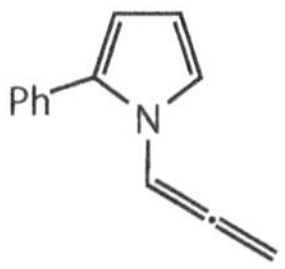	67	—	1.6370	[590,591,595,596]
11	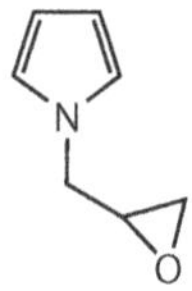	39	—	—	[600]
12	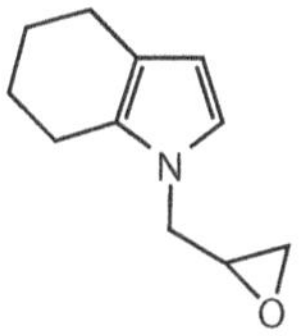	68	105–108/1	1.5308	[599]

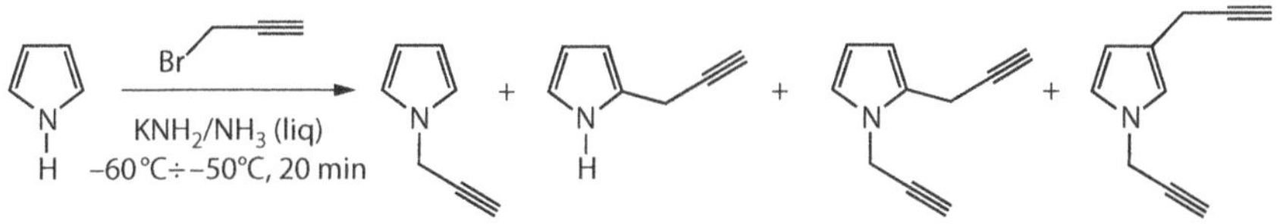

SCHEME 2.122 Reactions of pyrrole with propargyl bromide in KNH_2/NH_3 (liquid) system.

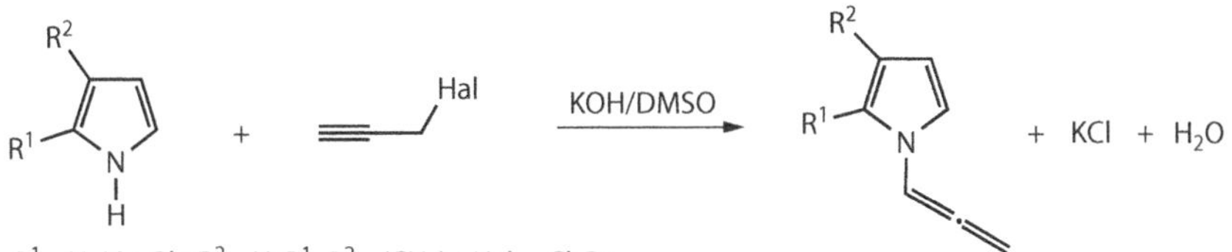

$R^1 = H, Me, Ph; R^2 = H; R^1–R^2 = (CH_2)_4; Hal = Cl; Br$

SCHEME 2.123 Synthesis of N-allenylpyrroles from pyrroles and propargyl halides in KOH/DMSO system.

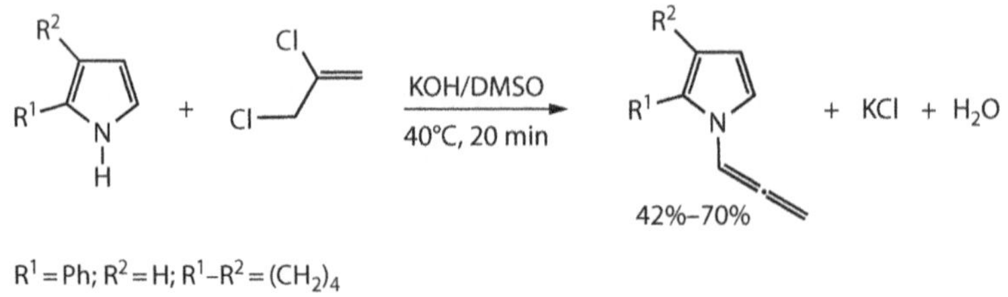

$R^1 = Ph; R^2 = H; R^1 - R^2 = (CH_2)_4$

SCHEME 2.124   Synthesis of N-allenylpyrroles from pyrroles and 2,3-dichloro-1-propene in KOH/DMSO system.

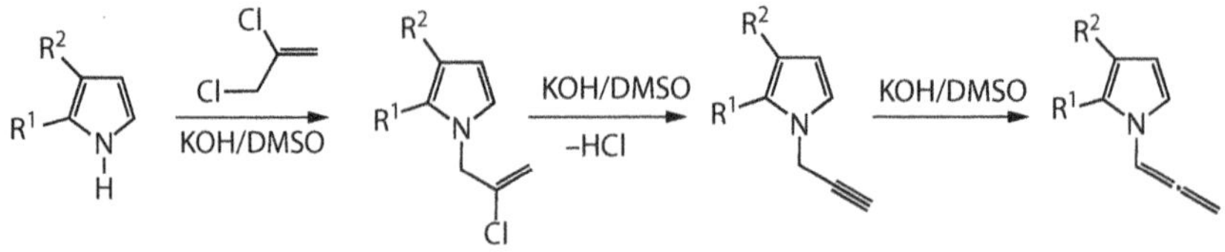

SCHEME 2.125   Expected steps of the pyrroles allenylation by 2,3-dichloro-1-propene.

The reaction, apparently, proceeds through formation of the intermediate N-(2-chloro-2-propenyl)pyrroles, which are dehydrochlorinated to N-propargylpyrroles further isomerizing to N-allenylpyrroles (Scheme 2.125).

Dehydrochlorination of 2,3-dichloro-1-propene to propargyl chloride, which reacts with pyrroles as described previously, is not also excluded. Obviously, these two directions can occur in parallel.

2.1.7.4   Allenylation of Pyrroles with 1,2,3-Trichloropropane

Pyrroles are efficiently allenylated with 1,2,3-trichloropropane in the KOH/DMSO suspension, when KOH is used in amounts significantly exceeding those needed for elimination of all chlorine atoms (5 mol per 1 mol of trichloropropane) [591,596]. The double dehydrochlorination, nucleophilic substitution of the chlorine atom by the pyrrolate anion, and complete prototropic isomerization of the intermediate N-propargylpyrrole (one of possible products of N-(2-chloro-2-propenyl)pyrrole dehydrochlorination) to N-allenylpyrrole are accomplished in a single preparative step with almost 100% selectivity (Scheme 2.126).

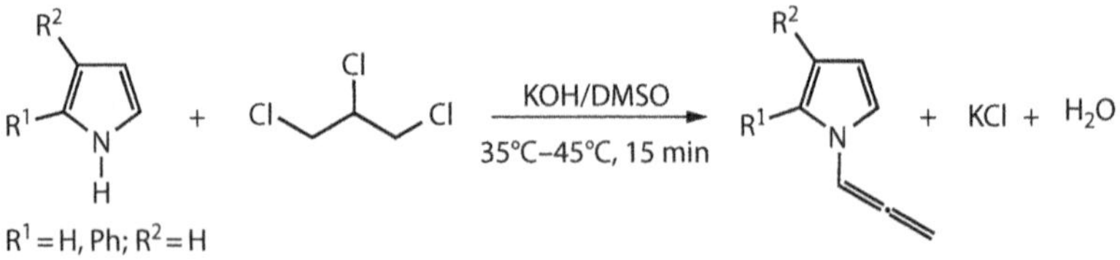

$R^1 = H, Ph; R^2 = H$

SCHEME 2.126   Synthesis of N-allenylpyrroles from pyrroles and trichloropropane in KOH/DMSO system.

In light of the availability of 1,2,3-trichloropropane, a large-scale waste of epi-chlorohydrin production, the simple and efficient method for pyrroles allenylation under mild conditions holds promise for further applications in the synthesis of various heterocyclic compounds with a labile hydrogen atom.

2.1.7.5 Ethynylation of Pyrroles with 1,2-Dichloroethene

N-Ethynylpyrrole (Table 2.14) is a unique acetylene. In particular, it is a starting product for the synthesis of pyrrolyldiacetylenes, which form polymeric films with promising optical properties. N-Ethynylpyrrole is synthesized from a pyrrole and 1,2-dichloroethene in the system KOH/DMSO (Scheme 2.127) [597].

The reaction proceeds with release of heat, when 1,2-dichloroethene (*E/Z* 16:84) adds to a stirred suspension pyrrole/KOH/DMSO. The temperature of the reaction is maintained within 45°C–50°C by cooling.

2.1.7.6 Ethynylation of Pyrroles with Trichloroethene

N-Ethynylpyrrole has been prepared [598] by reacting pyrrole with trichloroethene and potassium *tert*-butoxide in tetrahydrofuran and subsequent dechlorinating the adduct with methyllithium or *n*-butyllithium in diethyl ether (Scheme 2.128). It is assumed that N-(1,2-dichlorovinyl)pyrrole is formed via nucleophilic addition of pyrrole to dichloroacetylene, a product of trichloroethene dehydrochlorination.

2.1.7.7 Epoxymethylation of Pyrroles with Epichlorohydrin

A method for epoxymethylation of pyrroles with epichlorohydrin has been elaborated. N-Glycidyl-4,5,6,7-tetrahydroindole (Table 2.14), a perspective modifying agent of epoxy resins, is synthesized by the reaction of sodium or potassium derivatives (obtained from 4,5,6,7-tetrahydroindole and NaOH or KOH with azeotropic removal of water) and epichlorohydrin (toluene, benzene, xylene) (Scheme 2.129) [599].

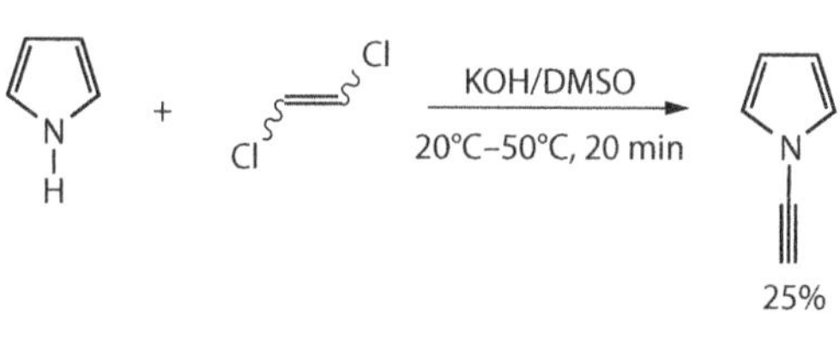

SCHEME 2.127 Synthesis of N-ethynylpyrrole via the reaction of pyrrole with 1,2-dichloroethene in the KOH/DMSO system.

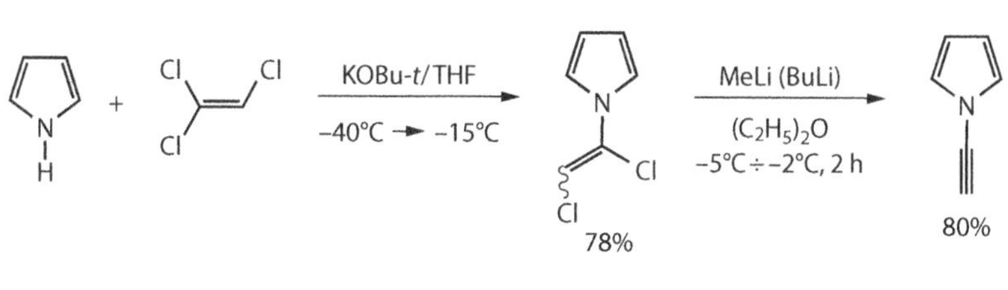

SCHEME 2.128 Synthesis of N-ethynylpyrrole from pyrrole and trichloroethene.

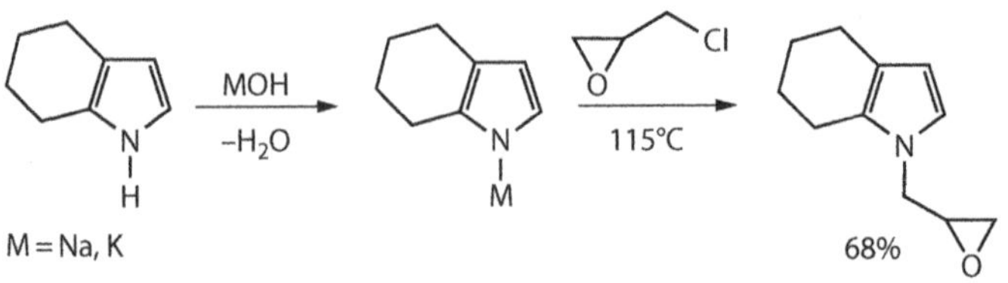

SCHEME 2.129 Synthesis of N-glycidyl-4,5,6,7-tetrahydroindole by epoxymethylation of 4,5,6,7-terahydroindole with epichlorohydrin.

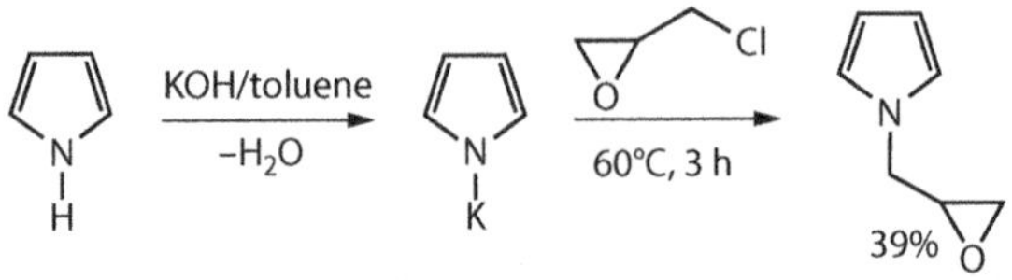

SCHEME 2.130 Synthesis of N-glycidylpyrrole by epoxymethylation of pyrrole with epichlorohydrin.

Alkylation of 4,5,6,7-tetrahydroindole with epichlorohydrin in the presence of alkaline metal hydroxides MOH (M = Na, K) proceeds with low efficiency (the yield of the products does not exceed 25%).

Similarly, N-glycidylpyrrole is prepared in 39% yield by the reaction of pyrrole and epichlorohydrin (60°C, 3 h) (Scheme 2.130, Table 2.14). N-Glycidylpyrrole is used as intermediate in the synthesis of activating additives to electrolytes of Li/S batteries [600].

2.1.8 FORMYLATION OF PYRROLES AND REACTIONS OF N-VINYLPYRROLE-2-CARBALDEHYDES

2.1.8.1 Synthesis of N-Vinylpyrrole-2-carbaldehydes

Formylation of N-vinylpyrroles proceeds under conditions of the Vilsmeier–Haack classical reaction (dimethylformamide (DMF)/POCl$_3$, 1,2-dichloroethane, reflux) to afford mixtures of N-vinyl- and NH-pyrrole-2-carbaldehydes in low yields. For example, from N-vinyl-4,5,6,7-tetrahydroindole, a mixture of the expected N-vinyl-4,5,6,7-tetrahydroindole-2-carbaldehyde and 4,5,6,7-tetrahydroindole-2-carbalde-hyde is formed (~1:1) (Scheme 2.131) [87].

Thus, N-vinylpyrroles are unstable under the conventional conditions of the Vilsmeier–Haack reaction due to the anticipated high sensitivity of electron-saturated

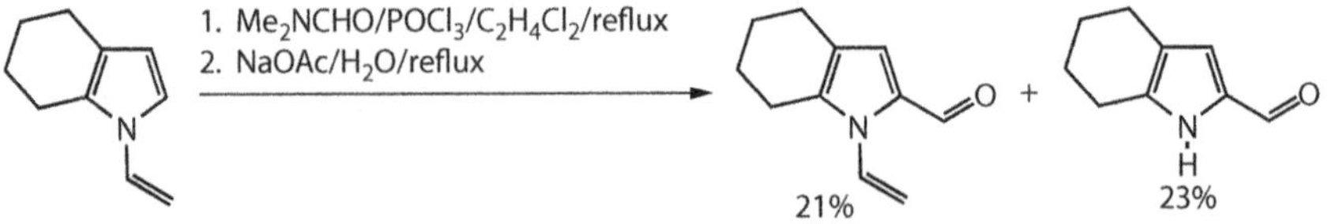

SCHEME 2.131 Reaction of N-vinyl-4,5,6,7-tetrahydroindole with DMF/POCl$_3$ complex.

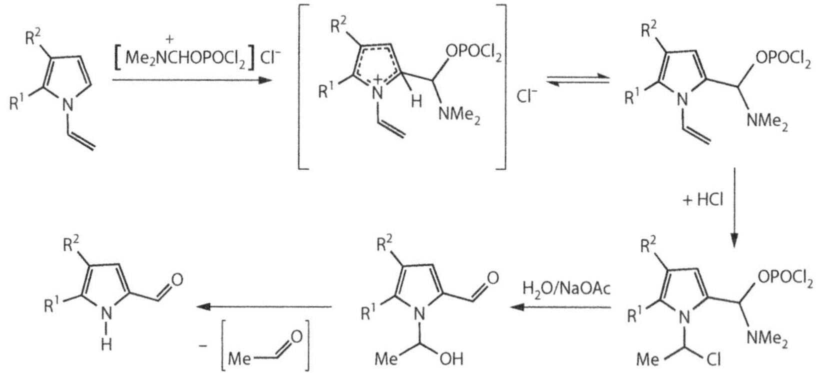

SCHEME 2.132 Rationale of the devinylation of N-vinylpyrroles during their formylation with DMF/POCl$_3$ complex.

N-vinyl group toward electrophilic reagents. Apparently, electrophilic attack on the pyrrole ring by the cationic complex DMF/POCl$_3$ is accompanied by addition of the released HCl to the vinyl group (Scheme 2.132).

Moreover, upon refluxing in 1,2-dichloroethane, electrophilic attack of the N-vinyl moiety by the cationic complex DMF/POCl$_3$ may lead to diadducts **A**, which on boiling in H$_2$O/NaOAc afford pyrrole-2-carbaldehydes and malonic aldehyde through unstable dialdehyde intermediate **B** (Scheme 2.133).

Decreasing the temperature of N-vinylpyrroles formylation to −78°C and performing the hydrolysis of the intermediate complex at room temperature (instead of refluxing) allow removal of the N-vinyl group to be avoided. As a result, N-vinylpyrrole-2-carbaldehydes bearing aromatic and heteroaromatic substituents have been synthesized in up to 88% yield (Scheme 2.134, Table 2.15) [87].

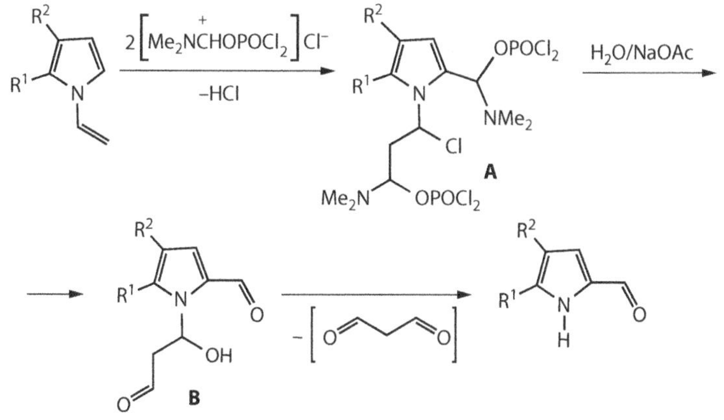

SCHEME 2.133 Alternative route to NH-pyrrrole-2-carbaldehydes in the formylation of N-vinylpyrroles with the DMF/POCl$_3$ complex.

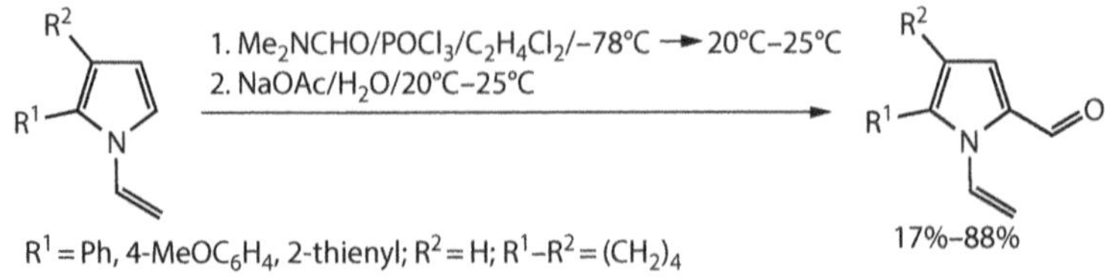

$R^1 = Ph$, 4-MeOC$_6$H$_4$, 2-thienyl; $R^2 = H$; $R^1-R^2 = (CH_2)_4$ 17%–88%

SCHEME 2.134 Selective synthesis of N-vinylpyrrole-2-carbaldehydes using with the DMF/POCl$_3$ complex.

The best yield (91%) is achieved in the case of N-vinyl-4,5-dihydrobenz[g]indole (Scheme 2.135, Table 2.15).

In the case of N-vinylpyrroles containing donor (alkyl) substituents, the yields of formylpyrroles are significantly reduced (17%–56%). This could be explained by the fact that aromatic substituents decrease the sensitivity of the N-vinyl group toward electrophilic reagents.

The replacement of phosphoryl chloride by oxalyl chloride allows the target N-vinylpyrrole-2-carbaldehydes to be selectively synthesized in high yields with complete conversion of the starting N-vinylpyrroles [601]. Besides, it becomes possible to perform all stages of the reaction at room temperature (instead of −78°C) and for shorter time (40 min instead of 3 h).

This method is applicable to various N-vinylpyrroles containing aliphatic, aromatic, condensed aromatic, and heteroaromatic substituents, as well as to N-vinylpyrroles condensed with cycloaliphatic and dihydronaphthalene systems (Schemes 2.136 and 2.137, Table 2.15).

The yields of N-vinylpyrrole-2-carbaldehydes bearing alkyl substituents increase approximately by 20%. Moreover, the yields of N-vinylpyrrole-2-carbaldehydes with aryl and heteroaryl substituents also augment (by 9%–27%).

2.1.8.2 Reactions of Pyrrole-2-carbaldehydes with Aromatic Di- and Tetraamines

The reaction of pyrrole-2-carbaldehydes with benzene-1,4-diamine or 4,4′-diphenylenediamine (ethanol, trifluoroacetic acid catalysis [TFAA]) has been implemented to synthesize a series of novel dipyrrole monomers separated by aromatic spacers. Consequently, the corresponding Schiff dibases have been obtained in 88% and 83%, respectively (Scheme 2.138, Table 2.15) [602].

The reaction of N-vinylpyrrole-2-carbaldehydes with benzene-1,2-diamine yields 2-(N-vinylpyrrol-2-yl)benzimidazoles (Schemes 2.139 and 2.140, Table 2.15). When needed, the reaction can be stopped at the stage of formation of the intermediate Schiff monobases **36**. Interestingly, the latter exist as E-isomers, which are predisposed to further cyclization into the intermediate imidazoline **C** [603].

One-pot protocol for the synthesis of 2-(N-vinylpyrrol-2-yl)benzimidazoles requires air bubbling through a reaction mixture (DMSO, 70°C–80°C, 1 h).

All of the 2-(N-vinylpyrrol-2-yl)benzimidazoles fluoresce intensely, including in practically important blue region of the spectrum (λ_{max} 343–417 nm, Stokes shift 990–6880 cm^{-1}).

TABLE 2.15

Pyrrole-2-carbaldehydes and Their Derivatives

No.	Structure	Yield (%)	Mp, °C (n_D^{20})	References
1	2	3	4	5
1		23	27–29	[87]
2		48	Oil	[601,604]
3		72	Oil	[87,601]
4		93	Oil	[87,601]
5		82	74–76	[87]
6		83	Oil	[601]
7		97	35–37	[601]
8		91	136–138	[87,601]
9		81	Oil	[601,606]

(Continued)

TABLE 2.15 (*Continued*)

Pyrrole-2-carbaldehydes and Their Derivatives

No.	Structure	Yield (%)	Mp, °C (n_D^{20})	References
1	2	3	4	5
10		87	(1.3840)	[612]
11		68	(1.3712)	[612]
12		77	(1.4113)	[612]
13		89	23–25	[612]
14		80	56–58	[612]
15		90	(1.4640)	[612]
16		91	(1.4857)	[612]

(Continued)

TABLE 2.15 (*Continued*)

Pyrrole-2-carbaldehydes and Their Derivatives

No.	Structure	Yield (%)	Mp, °C (n_D^{20})	References
1	2	3	4	5
17		99	(1.4998)	[612]
18		88	(1.4586)	[612]
19		94	(1.5101)	[612]
20		93	Oil	[89]
21		83	Oil	[89]
22		85	47–48	[89,388]
23		84	34–36	[89]
24		92	56–58	[89]
25		89	144–146	[89]

(*Continued*)

TABLE 2.15 (*Continued*)

Pyrrole-2-carbaldehydes and Their Derivatives

No.	Structure	Yield (%)	Mp, °C (n_D^{20})	References
1	2	3	4	5
26		95	65–90 (50% Z-isomer)	[604]
27		96	94–149 (58% Z-isomer)	[604]
28		99	85–128 (66% Z-isomer) 135–137 (decomp., Z)	[604]
29		97	148–180 (84% Z-isomer)	[604]
30		99	131–174 (decomp., 42% Z-isomer)	[604]
31		94		[605]
32		77		[605]
33		94		[605]

(*Continued*)

TABLE 2.15 (*Continued*)
Pyrrole-2-carbaldehydes and Their Derivatives

No.	Structure	Yield (%)	Mp, °C (n_D^{20})	References
1	2	3	4	5
34		94		[605]
35		88		[605]
36		72	197–199	[606]
37		68	221–223	[606]
38		76	234–236	[606]
39		71	203–205	[606]
40		72	206–208	[606]
41		91	238–240	[606]
42		69	186–188 (decomp.)	[606]

(*Continued*)

TABLE 2.15 (*Continued*)
Pyrrole-2-carbaldehydes and Their Derivatives

No.	Structure	Yield (%)	Mp, °C (n_D^{20})	References
1	2	3	4	5
43		81	192–194	[606]
44		86	210–212	[606]
45		69	150–153 (decomp.)	[607]
46		78	192–195 (decomp.)	[607]
47		89	179–183 (decomp.)	[607]
48		90	221–225 (decomp.)	[607]
49		77	208–211 (decomp.)	[607]
50		85	206–210 (decomp.)	[607]

(*Continued*)

TABLE 2.15 (*Continued*)
Pyrrole-2-carbaldehydes and Their Derivatives

No.	Structure	Yield (%)	Mp, °C (n_D^{20})	References
1	2	3	4	5
51		42	Oil	[608]
52		14	Oil	[608]
53		33	Oil	[608]
54		45	Oil	[608]
55		43	Oil	[608]

(Continued)

TABLE 2.15 (*Continued*)

Pyrrole-2-carbaldehydes and Their Derivatives

No.	Structure	Yield (%)	Mp, °C (n_D^{20})	References
1	2	3	4	5
56		86	216–218	[602]
57		88	224–226	[602]
58		83	>260	[602]
59		84	>300 (decomp.)	[602]
60		98	116–118	[603]
61		91	117–119	[603]
62		92	159–162	[603]
63		96	93–96	[603]

(*Continued*)

TABLE 2.15 (*Continued*)
Pyrrole-2-carbaldehydes and Their Derivatives

No.	Structure	Yield (%)	Mp, °C (n_D^{20})	References
1	2	3	4	5
64		98	Oil	[603]
65		97	102–104	[603]
66		96	123–125	[603]
67		95	177–180	[603]
68		97	125–128	[603]
69		83	261–262	[603]
70		77	120–123	[603]
71		61	75–78	[603]

(*Continued*)

TABLE 2.15 (*Continued*)

Pyrrole-2-carbaldehydes and Their Derivatives

No.	Structure	Yield (%)	Mp, °C (n_D^{20})	References
1	2	3	4	5
72		42	145–147	[603]
73		89	Oil	[603]
74		86	146–148	[603]
75		65	237–240	[603]
76		74	149–151	[603]
77		83	Oxid.	a
78		86	284–286	a
79		65	292–294	a

a Unpublished.

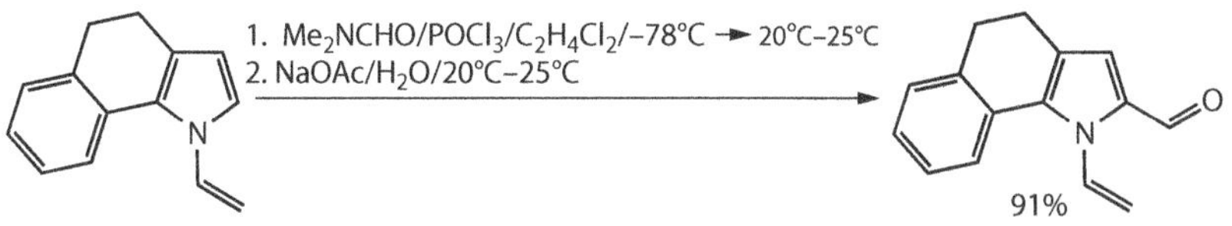

SCHEME 2.135 Selective formylation of N-vinyl-4,5-dihydrobenz[g]indole with the DMF/POCl$_3$ complex.

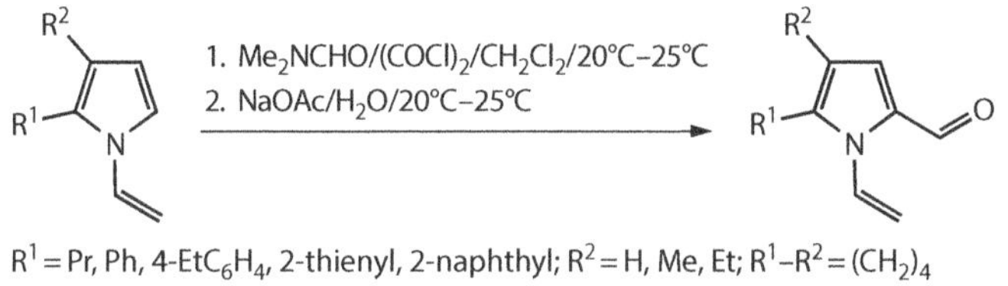

R^1 = Pr, Ph, 4-EtC$_6$H$_4$, 2-thienyl, 2-naphthyl; R^2 = H, Me, Et; R^1–R^2 = (CH$_2$)$_4$

SCHEME 2.136 Synthesis of N-vinylpyrrole-2-carbaldehydes using the DMF/(COCl)$_2$ complex.

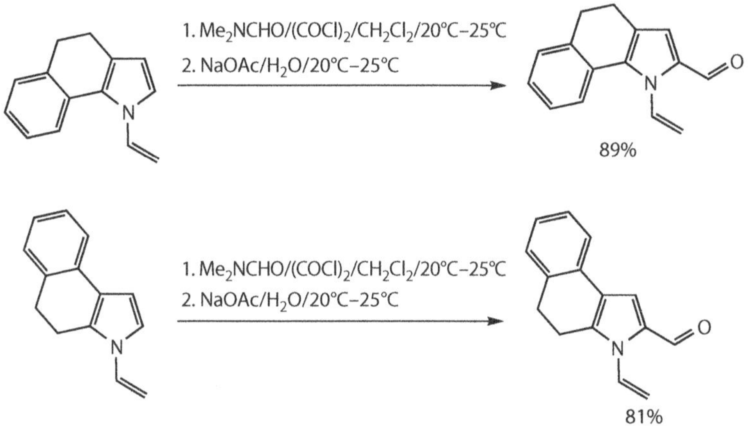

SCHEME 2.137 Formylation of N-vinyl-4,5-dihydrobenz[e]- and N-vinyl-4,5-dihydrobenz[g]indoles with the DMF/(COCl)$_2$ complex.

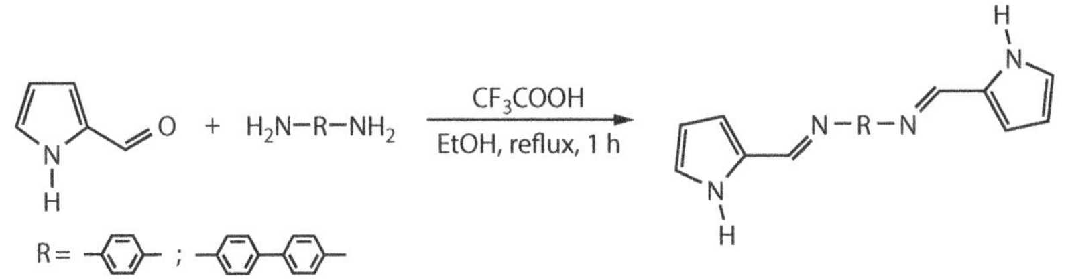

SCHEME 2.138 Reactions of pyrrole-2-carbaldehyde with benzene-1,4-diamine or 4,4′-diphenylenediamine.

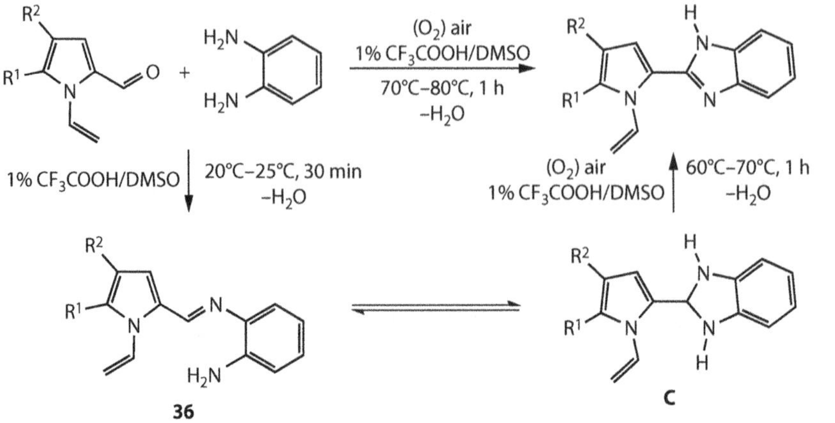

$R^1 = H$, Ph, 4-$MeOC_6H_4$, 2-naphthyl; $R^2 = H$; $R^1–R^2 = (CH_2)_4$

SCHEME 2.139 Reactions of N-vinylpyrrole-2-carbaldehydes with benzene-1,2-diamine.

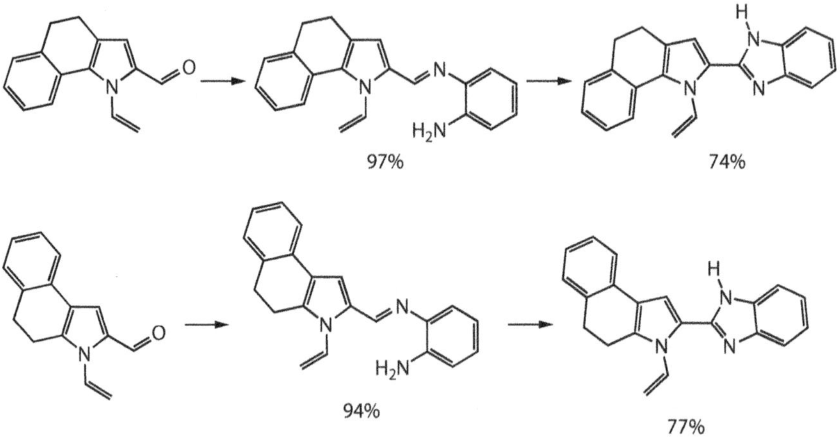

SCHEME 2.140 Reaction of N-vinyl-4,5-dihydrobenz[*e*]- and N-vinyl-4,5-dihydrobenz[*g*] indoles with benzene-1,2-diamine.

Condensation of NH- and N-vinylpyrrole-2-carbaldehydes with benzene-1,2,4,5-tetraamine, proceeding through formation of Schiff dibases, has been studied (Schemes 2.141 and 2.142).

In the case of N-vinylpyrrole-2-carbaldehydes, the reaction occurs in the presence of catalytic amounts of TFAA. If the molecule of the starting pyrrole-2-carbaldehyde contains no N-vinyl group, the catalyst is not required. Obviously, N-vinyl substituent sterically shields the carbonyl group toward the bulky nucleophile, thus significantly inhibiting the reaction. The acidic catalyst (TFAA) protonates the C=O bond, enhances its electrophilicity, and thus facilitates addition of nucleophile. The formyl moiety of the initial carbaldehydes regioselectively reacts with the amino

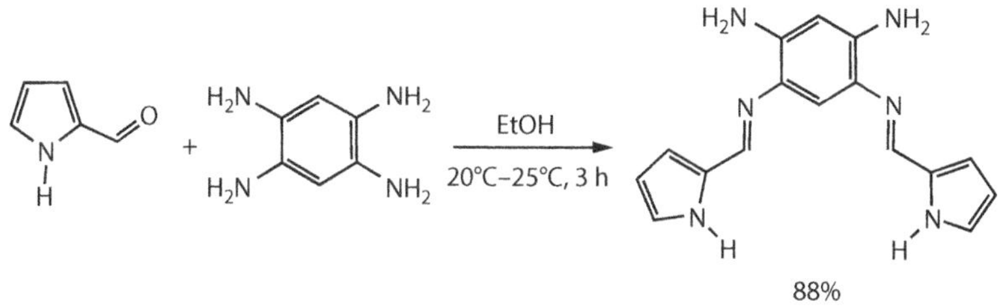

SCHEME 2.141   Condensation of pyrrole-2-carbaldehyde with benzene-1,2,4,5-tetramine.

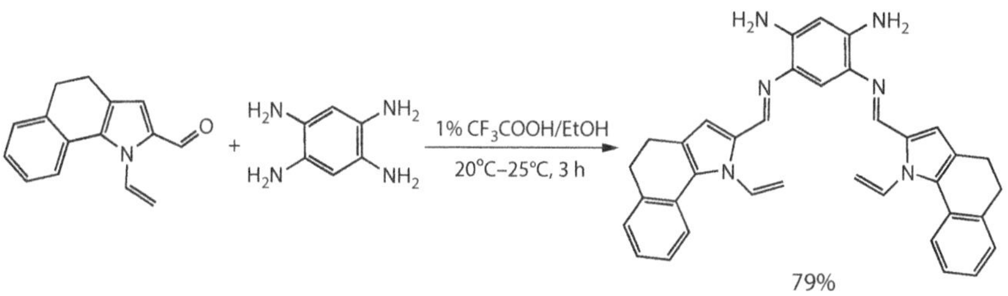

SCHEME 2.142   Condensation of N-vinyl-4,5-dihydrobenz[*g*]indole-2-carbaldehyde with benzene-1,2,4,5-tetramine.

groups located in positions 1 and 5 of the starting 1,2,4,5-benzenetetramine to furnish exclusively non-centrosymmetric products.

In DMSO solution, Schiff dibases selectively and almost quantitatively cyclize (room temperature) with the subsequent aromatization to 2,6-di(pyrrol-2-yl)-1,7-dihydroimidazo[4,5-*f*] [1,3]-benzimidazoles (up to 94% yield, Scheme 2.143).

Such dipyrroles separated by polycondensed benzimidazole spacers represent promising monomers for the design of electroconductive electrochromic polymers

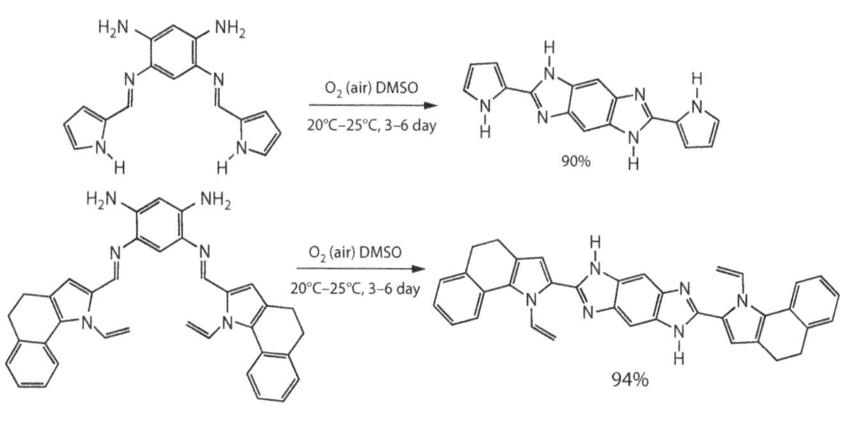

SCHEME 2.143   Synthesis of di(pyrrolyl)-dihydroimidazobenzimidazoles by oxidative aromatization of the Schiff bases, derivatives of pyrrole-2-carbaldehydes.

of new generation with controlled (through a variation of pyrrole counterpart structure) energy gap. Optoelectronic characteristics of the materials based on these precursors can also be controlled by vinyl polymerization and addition reactions of various reagents to the N-vinyl groups.

2.1.8.3 Reactions of Pyrrole-2-carbaldehydes with Hydroxylamine, Semicarbazide, Thiosemicarbazide, and Aminoguanidine

The reaction of N-vinylpyrrole-2-carbaldehydes with hydroxylamine in pyridine or ethanol affords N-vinylpyrrole-2-carbaldehyde oximes in up to 99% yields (Schemes 2.144 and 2.145, Table 2.15) [604].

The *E*- and *Z*-isomers of pyrrole-2-carbaldehyde oxime exist in preferable conformation with *cis*-disposition of the oxime function relative to the pyrrole ring (NMR, MP2). The conformation is stabilized by N–H···N (*E*) and N–H···O (*Z*) intramolecular hydrogen bonds (Scheme 2.146).

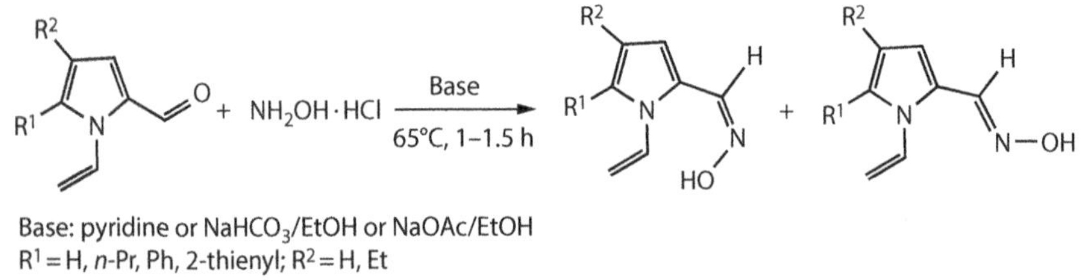
R2
R1
N
O
+ NH2OH·HCl
Base
65°C, 1–1.5 h
H
HO
+
N–OH
Base: pyridine or NaHCO3/EtOH or NaOAc/EtOH
R1 = H, n-Pr, Ph, 2-thienyl; R2 = H, Et

SCHEME 2.144 The reaction of N-vinylpyrrole-2-carbaldehydes with hydroxylamine.

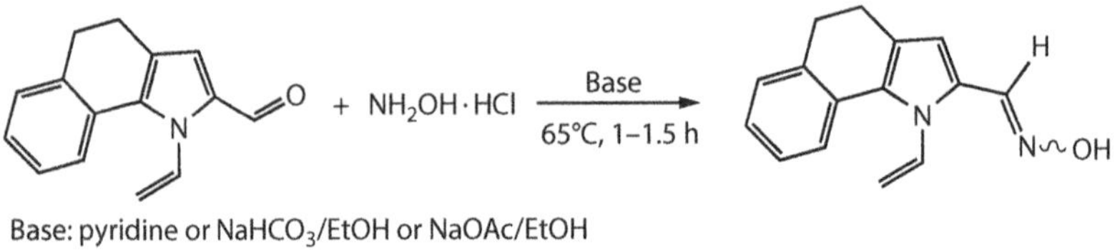
O
+ NH2OH·HCl
Base
65°C, 1–1.5 h
N
H
OH
Base: pyridine or NaHCO3/EtOH or NaOAc/EtOH

SCHEME 2.145 The reaction of N-vinyl-4,5-dihydrobenz[g]indole-2-carbaldehyde with hydroxylamine.

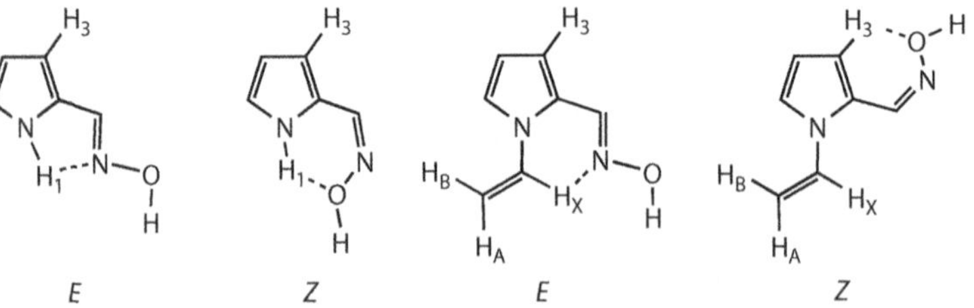
H3
N
H1
O
H
HB
HX
HA
E
Z
E
Z

SCHEME 2.146 *E*-and *Z*-isomerism of pyrrole-2-carbaldehyde oximes.

The *E*-isomer of N-vinylpyrrole-2-carbaldehyde oxime has *trans*-orientation of the vinyl group relative to the oxime function and *cis*-orientation of the oxime moiety with respect to the pyrrole nitrogen atom. In the Z-isomer, the oxime group is *trans*-oriented toward the pyrrole ring nitrogen.

Pyrrole-2-carbaldehyde oximes are readily protonated with acids (CF_3SO_3H, TFAA, HCl), the *E*-isomer being transformed to the Z-isomer to give the salts of Z-configuration only in up to 94% yields (Scheme 2.147) [605].

Stereospecific protonation of pyrrole-2-carbaldehyde oximes may be rationalized by through-space stabilization of positive charge in the pyrrole ring with hydroxyl group (^{1}H, ^{13}C, ^{15}N NMR, quantum chemical calculations using B3LYP/6-311G (d, p) basis set). The essence of the phenomenon is a deep charge transfer from the protonated oxime function to the pyrrole nucleus (Scheme 2.148).

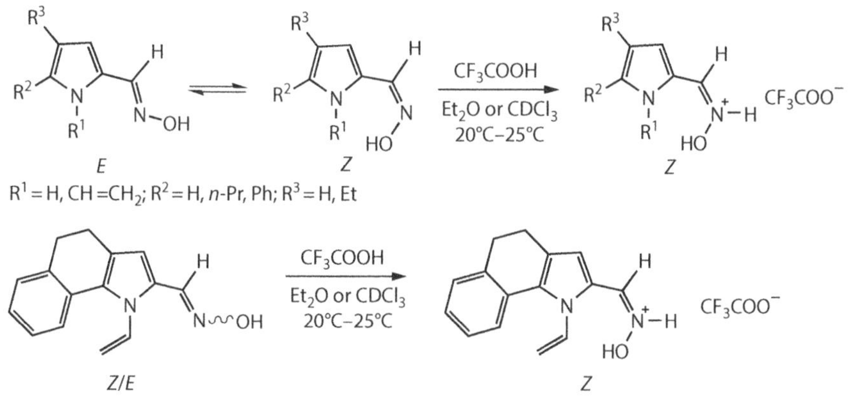

SCHEME 2.147   Stereoselective protonation of pyrrole-2-carbaldehyde oximes.

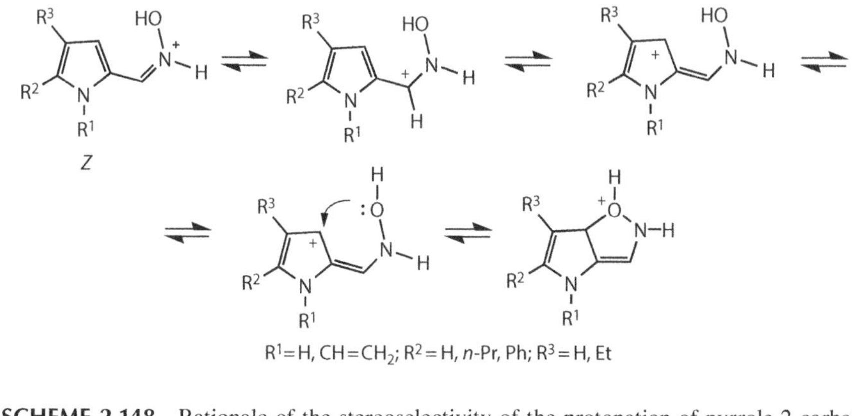

SCHEME 2.148   Rationale of the stereoselectivity of the protonation of pyrrole-2-carbaldehyde oximes.

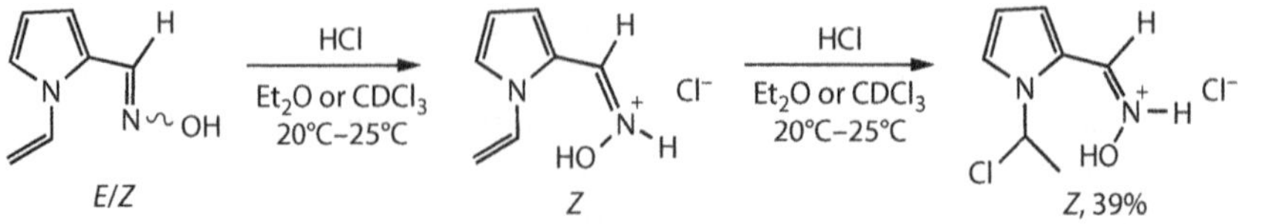

SCHEME 2.149 Addition of HCl to N-vinyl group upon protonation of N-vinylpyrrole-2-carbaldehyde oximes.

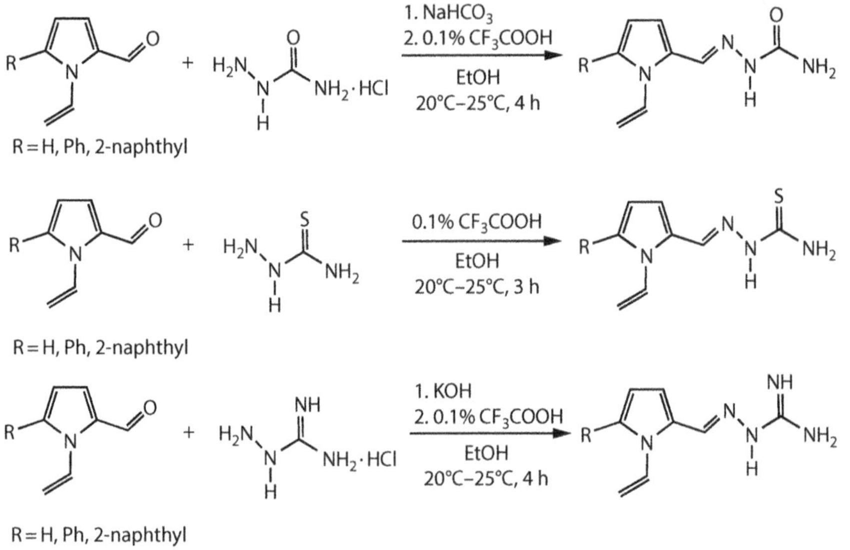

SCHEME 2.150 Reactions of N-vinylpyrrole-2-carbaldehydes with semicarbazide, thiosemicarbazide, and aminoguanidine.

Such stabilization is characteristic of pyrrole-2-carbaldehyde oximes and does not occur in other aromatic and aliphatic oximes. The stereospecific facile transformation of *E-/Z*-isomer mixtures to pure salts of *Z*-isomer is the first known method for the synthesis of individual *Z*-isomers of aldoxime salts.

When N-vinylpyrrole-2-carbaldehyde oxime is protonated with gaseous HCl, the latter adds to the vinyl group (Scheme 2.149).

N-Vinylpyrrole-2-carbaldehydes react with semicarbazide, thiosemicarbazide, and aminoguanidine to furnish the corresponding semicarbazones, thiosemicarbazones, and guanylhydrazones in high yields (up to 90%, Scheme 2.150, Table 2.15) [606].

2.1.8.4 Reactions of N-Vinylpyrrole-2-carbaldehydes with L-Lysine

Unnatural, optically active amino acids containing the N-vinylpyrrole moiety have been synthesized by condensation of N-vinylpyrrole-2-carbaldehydes with L-lysine (up to 90% yields, Schemes 2.151 and 2.152, Table 2.15) [607].

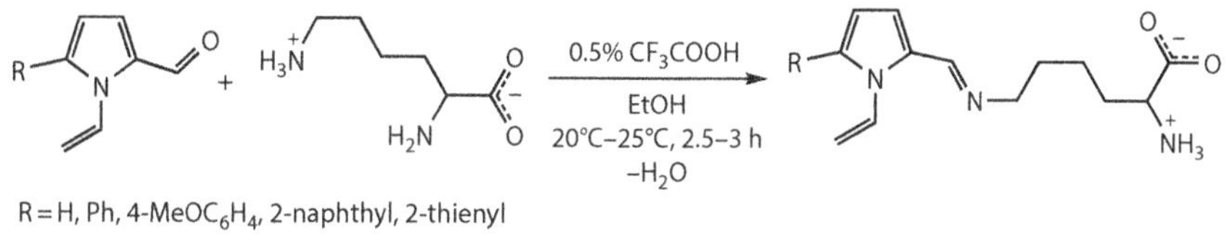

R = H, Ph, 4-MeOC$_6$H$_4$, 2-naphthyl, 2-thienyl

SCHEME 2.151 Selective reactions of N-vinylpyrrole-2-carbaldehydes with L-lysine.

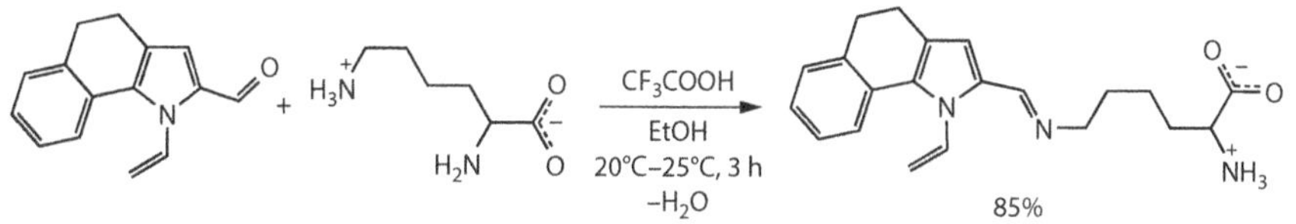

SCHEME 2.152 Selective reaction of N-vinyl-4,5-dihydrobenz[g]indole-2-carbaldehyde with L-lysine.

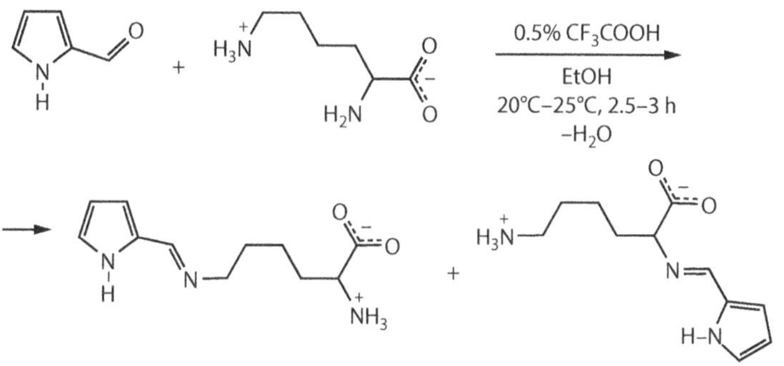

SCHEME 2.153 Nonselective reaction of pyrrole-2-carbaldehyde with L-lysine.

Noteworthy, under the same conditions, pyrrole-2-carbaldehyde reacts with lysine in non selective mode, that is, not only at the ε- but also at the α-amino group to afford both expected isomers in 3:1 ratio (Scheme 2.153).

The synthesized amino acids containing aromatic, condensed aromatic, and hydronaphthalene substituents in the pyrrole ring fluoresce in UV and visible spectral regions (350–382 nm, Stokes shift 6150–7800 cm^{-1}) and can be employed in the design of novel biological labels, prospective for monitoring of living processes. They also represent a possible alternative to dangerous radioactive labels. The fluorescence parameters of such amino acid can be controlled by introducing diverse aromatic substituents in the molecules of initial N-vinylpyrroles-2-carbaldehydes.

A rare combination of pharmacophoric N-vinylpyrrole and indispensable amino acid moieties makes this novel class of amino acids important building blocks for drug design. Moreover, the presence of highly reactive N-vinyl group in molecules of the amino acids synthesized fundamentally extends their prospects in the aforementioned applications.

2.1.8.5 Three-Component Reaction of N-Vinylpyrrole-2-carbaldehydes with N-Methylimidazole and Cyanophenylacetylene

It is found that N-vinylpyrrole-2-carbaldehydes react with N-methylimidazole and cyanophenylacetylene (20°C–25°C, MeCN) to stereoselectively give N-vinylpyrrole-imidazole ensembles **37** with a (*Z*,*Z*)-di-(2-cyano-1-phenylvinyl)oxy function in up to 45% yield (Schemes 2.154 and 2.155, Table 2.15) [608], analogs of pyrrole-imidazole alkaloids [609–611].

The initial zwitterion **A** generated from imidazole and cyanophenylacetylene is thought to be transformed into carbene **B**, which further inserts into the C=O bond of N-vinylpyrrolecarbaldehyde (shown on the example of N-vinylpyrrole-2-carbaldehyde). Two subsequent rearrangements of adduct **C** furnish intermediate **D**, which in its enol form adds to the second molecule of cyanophenylacetylene (Scheme 2.156).

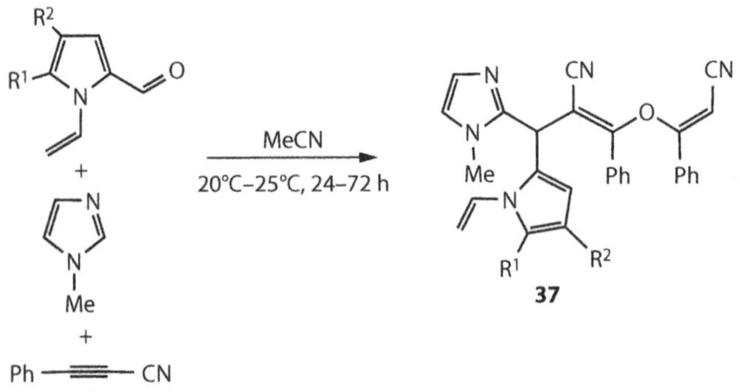

$R^1 = H, Ph, 4\text{-}MeOC_6H_4; R^2 = H; R^1\text{–}R^2 = (CH_2)_4$

SCHEME 2.154 Reaction of N-vinylpyrrole-2-carbaldehydes with N-methylimidazole and cyanophenylacetylene.

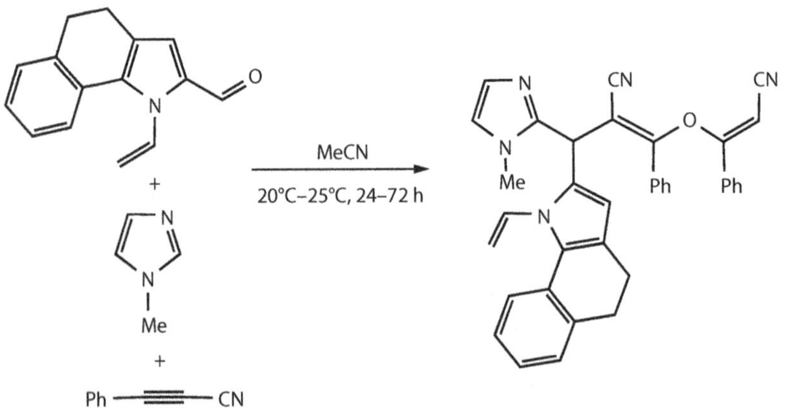

SCHEME 2.155 Reaction of N-vinyl-4,5-dihydrobenz[*g*]indole-2-carbaldehyde with N-methylimidazole and cyanophenylacetylene.

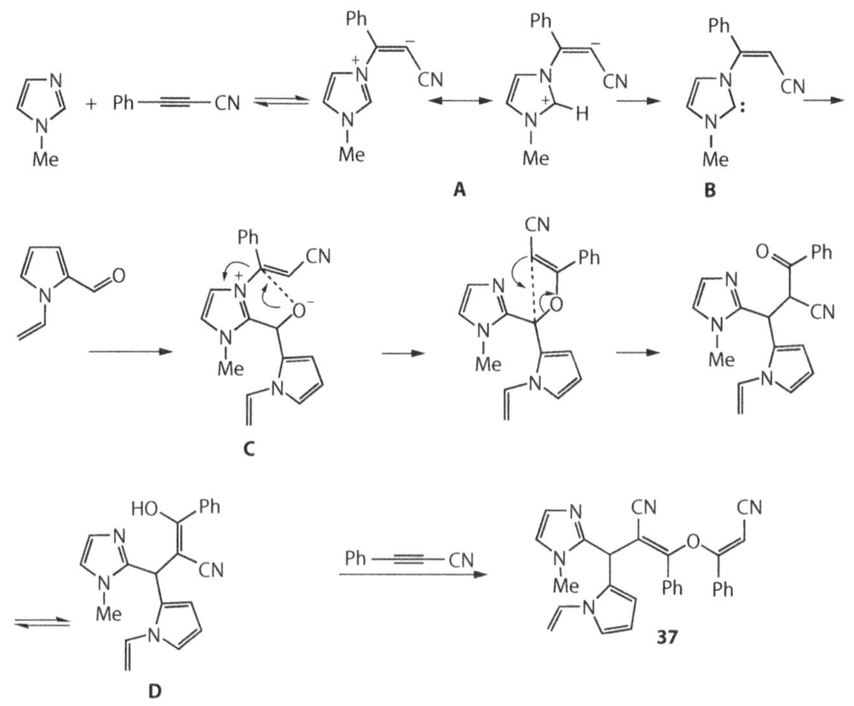

SCHEME 2.156 Zwitterionic intermediates in the reaction between N-vinylpyrrole-2-carbaldehydes, N-methylimidazole, and cyanophenylacetylene.

2.1.8.6 Thiylation of N-Vinylpyrrole-2-carbaldehydes

N-Vinylpyrrole-2-carbaldehydes react with excess ethanethiol in the presence of TFAA in benzene to afford thioacetals in up to 99% yield (Scheme 2.157, Table 2.15) [612].

Neither the expected addition of thiol across the vinyl group nor the cationic oligomerization of the initial and target products is observed. Such processes take

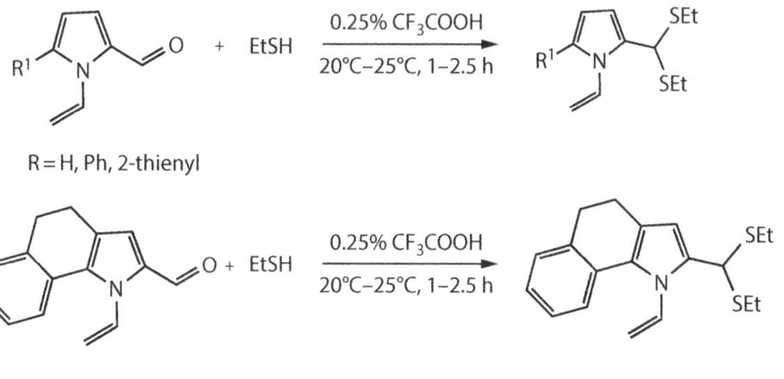

SCHEME 2.157 Electrophilic thiylation of N-vinylpyrrole-2-carbaldehydes.

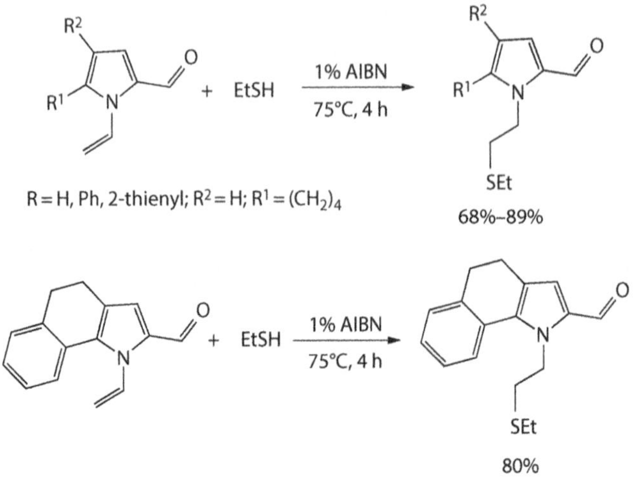

R = H, Ph, 2-thienyl; R^2 = H; R^1 = $(CH_2)_4$

68%–89%

80%

SCHEME 2.158   Free-radical thiylation of N-vinylpyrrole-2-carbaldehydes.

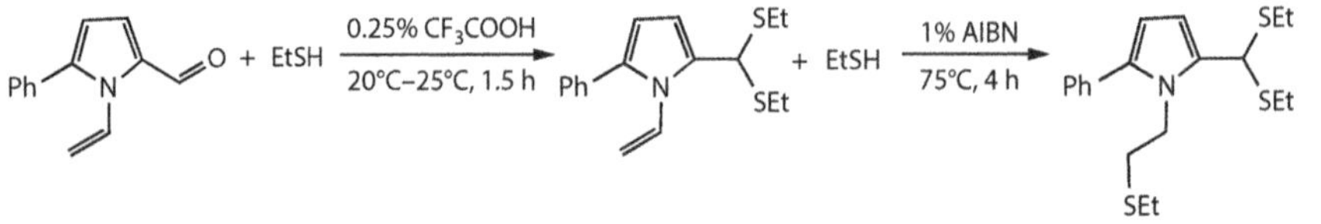

SCHEME 2.159   Two-step exhaustive thiylation of N-vinyl-5-phenylpyrrole-2-carbaldehyde.

place only for N-vinyl-4,5,6,7-tetrahydroindole-2-carbaldehyde, which under the same conditions gives a complex mixture of products gas-liquid chromatography (GLC) owing to higher nucleophilicity of the vinyl group (donating effect of the cyclohexane fragment).

In the presence of the radical initiator (azoisobutyric acid dinitrile [AIBN], benzene), N-vinylpyrrole-2-carbaldehyde reacts with equimolar amount of thiol to regiospecifically form the adducts across the double bond only against the Markovnikov rule, the yields reaching 89% (Scheme 2.158, Table 2.15) [612].

The exhaustive thiylation of 5-phenyl-N-vinylpyrrole-2-carbaldehyde is carried out in two steps, namely, initial formation of mercaptal and its subsequent thiylation across the double bond (Scheme 2.159).

When excess ethanethiol is employed, the reaction leads to the unexpected reduction of the thioacetal function and formation of ethylthiomethyl derivative in up to 80% yield (Scheme 2.160) [612].

Thioacetals, products of N-Vinylpyrrole-2-carbaldehydes thiylation at the carbonyl group, contain reactive vinyl fragment that expands their synthetic potential and possibility of application in synthesis of porphyrins and related pigments.

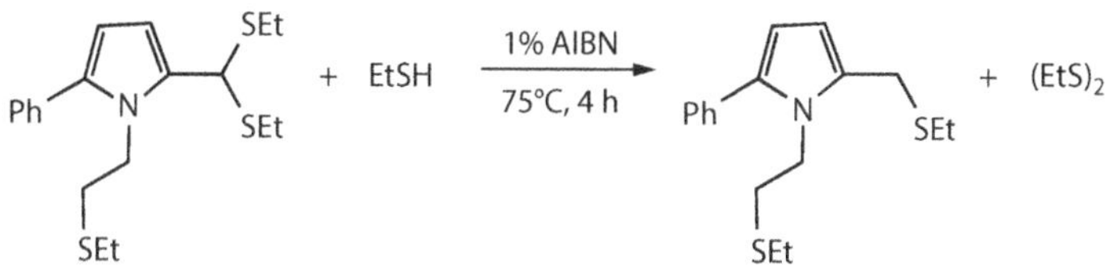

SCHEME 2.160 Reduction of the thioacetal moiety upon free-radical thiylation of N-vinylpyrrolecarbaldehyde thioacetal with EtSH excess.

2.1.8.7 N-Vinylpyrrole-2-carbaldehydes as Precursors of N-Vinylpyrrole-2-carbonitriles

N-Vinylpyrrole-2-carbonitriles are synthesized from N-vinylpyrrole-2-carbaldehyde oximes by two methods: (1) reaction with acetylene (KOH/DMSO, 67% yield, Scheme 2.161) [388] and (2) reaction with acetic anhydride (90°C–100°C, up to 93% yields, Scheme 2.162, Table 2.15) [89].

The experiments have shown that in the first case nitriles are not formed without acetylene. Hence, it follows that they are generated due to the elimination of the vinyl alcohol (acetaldehyde) from the corresponding O-vinyl oxime.

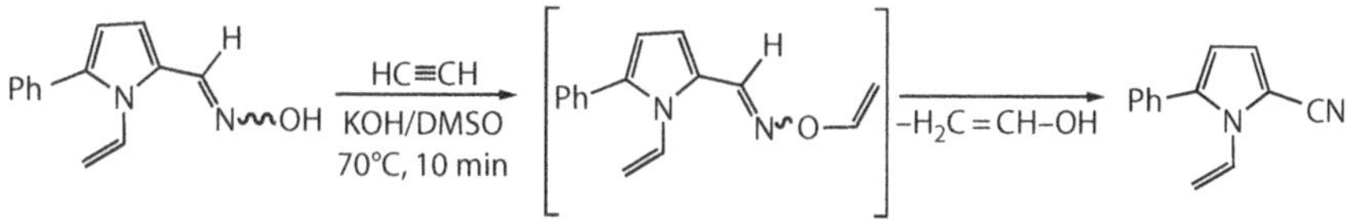

SCHEME 2.161 Synthesis of N-vinylpyrrole-2-carbonitriles via the reaction of N-vinylpyrrole-2-carbaldehyde oximes with acetylene.

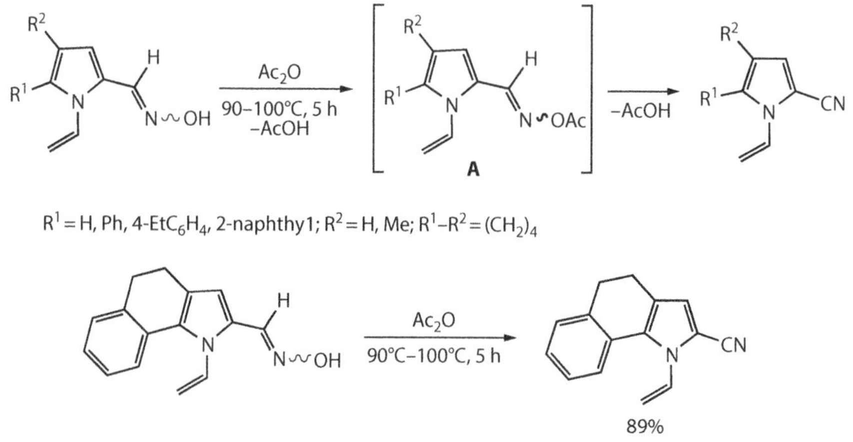

$R^1 = H, Ph, 4-EtC_6H_4, 2-naphthy1; R^2 = H, Me; R^1-R^2 = (CH_2)_4$

SCHEME 2.162 Synthesis of N-vinylpyrrole-2-carbonitriles via acylation of N-vinylpyrrole-2-carbaldehyde oximes.

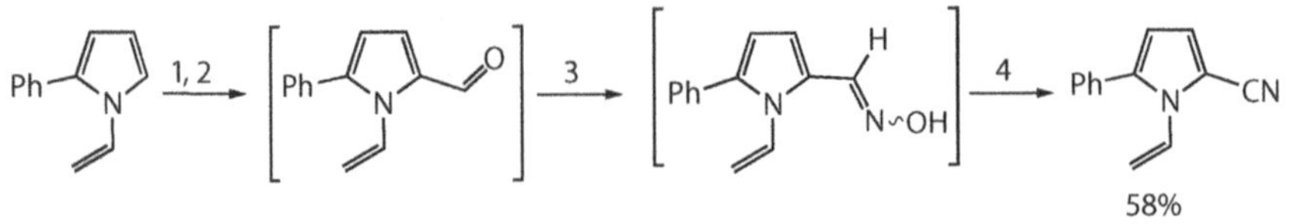

1. Me$_2$NCHO/(COCl)$_2$/CH$_2$Cl$_2$/20°C–25°C/40 min;   2. NaOAc/H$_2$O/20°C–25°C/30 min;
3. NH$_2$OH·HCl/40°C–50°C/10 min;   4. Ac$_2$O/90°C–100°C/5 h

SCHEME 2.163   One-pot synthesis of N-vinylpyrrole-2-carbonitriles from N-vinylpyrroles via their consecutive formylatiom, oximation, and acylation.

In the second case, dehydration apparently proceeds via acylation of oximes to afford acetate **A**, which then eliminates acetic acid.

The nitriles fluoresce intensely (λ_{max} 337–378 nm, Stokes shift 5230–9390 cm^{-1}). The comparison of the UV spectra of N-vinyl-2-phenylpyrrole and its nitrile derivative shows that the nitrile substituent ensures a bathochromic shift (10 nm) and a hypochromic effect (decrease in lg ε from 4.18 to 4.03) due to extension of the conjugation in the molecule. The same effect is observed in UV spectra of other N-vinylpyrrole-2-carbonitriles.

N-Vinylpyrrole-2-carbonitriles can also be synthesized in a one-pot manner via consecutive treatment of N-vinylpyrroles with the complex DMF/(COCl)$_2$, system NH$_2$OH·HCl/NaOAc, and acetic anhydride, the yield of carbonitriles being 58% (Scheme 2.163).

Pyrrole-2-carbonitriles are reactive carriers of the pyrrole nucleus, which can find application in synthesis of rare pyrrole structures. For example, the reaction of pyrrole-2-carbonitrile with hydrazine hydrate leads to 3,6-di(pyrrol-2-yl)-1,2,4,5-tetrazine and 3,5-di(pyrrol-2-yl)-4-amino-1,2,4-triazole (Scheme 2.164, Table 2.15), promising monomers for the design of electroconductive polymers.

2.1.9 TRIFLUOROACETYLATION

The early investigations [613–615] of N-vinylpyrroles trifluoroacetylation have been undertaken in order to assess the relative activity of two most probable nucleophilic sites (α-position of the pyrrole ring and β-position of the N-vinyl moiety) and to develop preparative methods for the synthesis of novel fluorinated pyrrole derivatives [614].

Usually, trifluoroacetyl cation attacks pyrrole ring at the α-position or, when the latter is occupied, at the β-position [616–618]. However, the vinyl group at nitrogen atom, in principle, could significantly affect the reactivity of the pyrrole molecule. The I-effect of the double bond and its competitive conjugation with nitrogen lone electron pair should decrease the aromaticity of the pyrrole cycle.

There is a body of evidences [93] that in N-vinylindoles, electrophile attacks the vinyl group rather than the pyrrole moiety. The same phenomenon is observed in the course of electrophilic addition of water and alcohols to N-vinylpyrroles (see Sections 2.2.1 and 2.2.2). This observation is compatible with the data on ready

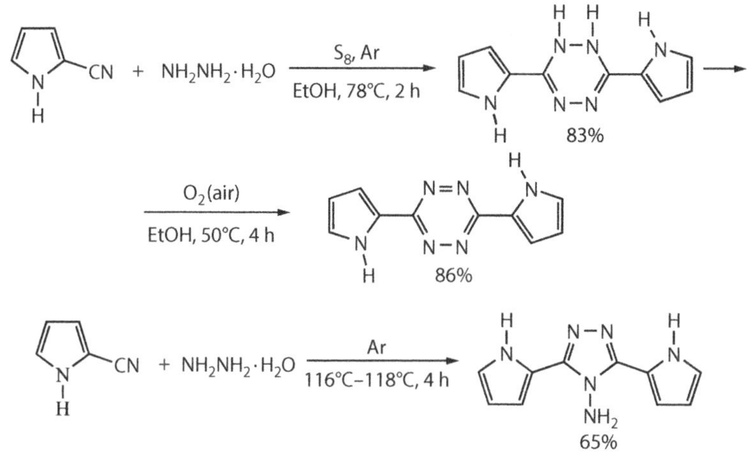

SCHEME 2.164 Synthesis of di(pyrrol-2-yl)-1,2,4,5-tetrazine or 3,5-di(pyrrol-2-yl)-4-amino-1,2,4-triazole from pyrrole-2-carbonitrile and hydrazine hydrate.

acylation of enamines [619] and the electrophilic substitution of hydrogen in the N-vinyl group upon N-vinyl amides trifluoroacetylation [620].

It has been found [613–615] that N-vinylpyrroles are acylated smoothly and regiospecifically with trifluoroacetic anhydride in the presence of pyridine ($-10°C \div 25°C$, CCl_4, CH_2Cl_2, benzene) into the pyrrole ring α-position, yields of trifluoroacylated derivatives reaching 96% (Scheme 2.165, Table 2.16).

For a comparison with N-vinylpyrroles, the trifluoroacetylation of pyrroles (the same conditions) having other substituents such as H, Et, $(CH_2)_2SEt$, and $(CH_2)_2SiEt_3$ has been studied (Scheme 2.166) [615,621]. No noticeable differences in reaction course are revealed. The corresponding α-trifluoroacetylpyrroles are obtained in 77%–98% yields (Table 2.16).

Thus, N-vinylpyrroles do not show significant alienation of nitrogen lone electron pair from heteroaromatic system, and as a consequence, p-π-conjugation of the N-vinyl group is not entirely realized. This leads to qualitative alterations in nature and reactivity of the N-vinyl moiety as compared with vinyl ethers, sulfides, enamines, and N-vinyl amides, which are readily trifluoroacylated under the conditions described earlier at the vinyl group β-position [620]. In other words, of the two

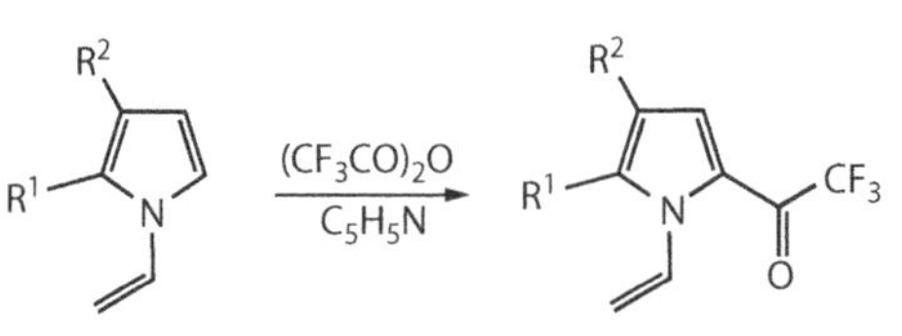

$R^1 = Me, Ph, 4\text{-}EtC_6H_4, 4\text{-}MeOC_6H_4, 4\text{-}BrC_6H_4; R^2 = H, Me; R^1\text{-}R^2 = (CH_2)_4$

SCHEME 2.165 Trifluoroacetylation of N-vinylpyrroles.

TABLE 2.16
Trifluoroacetylpyrroles and Their Derivatives

No.	Structure	Yield (%)	Bp (°C/torr) (Mp, °C)	d_4^{20}	n_D^{20}	References
1		87	(158–159)			[615,621]
2		84	(114–115)			[621]
3		93	(197–199)			[621]
4		92	(169–170)			[621]
5		84	(195–196)			[621]
6		82	(189–191)			[621]
7		83	(193–194)			[621]
8		80	(184–186)			[621]
9		82	73–74/3	1.2590	1.5110	[613–615]
10		76	92–93/4 (37)			[613–615]

(Continued)

TABLE 2.16 (*Continued*)
Trifluoroacetylpyrroles and Their Derivatives

No.	Structure	Yield (%)	Bp (°C/torr) (Mp, °C)	d_4^{20}	n_D^{20}	References
11		73	122–123/3	1.2675	1.5460	[613–615]
12		88	132/1.5	1.2582	1.5883	[614,615,621]
13		96	136/1	1.5810	1.2400	[614,615,621]
14		86	(56–57)			[621]
15		83	(45–47)			[621]
16		87	183/9			[621]
17		87	188/2			[621]
18		90	(62–63)			[621]

(*Continued*)

TABLE 2.16 (*Continued*)

Trifluoroacetylpyrroles and Their Derivatives

No.	Structure	Yield (%)	Bp (°C/torr) (Mp, °C)	d_4^{20}	n_D^{20}	References
19		93	138–140/1	1.2126	1.5732	[615,621]
20		96	158–163/1 (68–70)			[615,621]
21		80	(84–86)			[621]
22		76	(90–93)			[615]
23		98	122/3	1.2572	1.5600	[615]
24		77	164/3–4	1.2419	1.5720	[615]
25		92	170–172/1			[621]
26		87	169–172/ 2–3			[621]
27		83	188–192/1			[621]

(Continued)

TABLE 2.16 (*Continued*)

Trifluoroacetylpyrroles and Their Derivatives

No.	Structure	Yield (%)	Bp (°C/torr) (Mp, °C)	d_4^{20}	n_D^{20}	References
28		82	168–170/1			[621]
29		97	166/1 (44–46)			[615]
30		41	(132–133)			[447]
31		53	(152)			[447]
32		25	124–130/3			[447]
33		71	131/4			[447]
34		38	124–130/3			[447]
35		35	(40–41)			[628]
36		27	148/3–4			[628]

(Continued)

TABLE 2.16 (*Continued*)

Trifluoroacetylpyrroles and Their Derivatives

No.	Structure	Yield (%)	Bp (°C/torr) (Mp, °C)	d_4^{20}	n_D^{20}	References
37		76	132–135/3	1.2709	1.5840	[628]
38		53	120–122/4	1.2664	1.5826	[628]
39		54	130–132/5	1.0790	1.5899	[628]
40		75	(32–33)			[628]
41		60	250 (decomp.)			[621]
42		97	(106–107)			[629,642]
43		84	(90–91)			[642]
44		72	81–82/1 (56)			[7, c. 110]
45		96	(88–89)			[629]
46		72	81–82/1 (56)			[7, c. 110]

(Continued)

TABLE 2.16 (*Continued*)
Trifluoroacetylpyrroles and Their Derivatives

No.	Structure	Yield (%)	Bp (°C/torr) (Mp, °C)	d_4^{20}	n_D^{20}	References
47		95	163–164/3 (70–71)			[629]
48		94	(56–57)			[621]
49		84	(137–165 decomp.)			[630]
50		84	(152–185 decomp.)			[630]
51		79	(150–180 decomp.)			[630]
52		81	(140–180 decomp.)			[630]
53		75	(150–185 decomp.)			[630]
54		83	(142–190 decomp.)			[630]
55		87	(107)			[630]
56		76	(140–170 decomp.)			[630]

(*Continued*)

TABLE 2.16 (*Continued*)

Trifluoroacetylpyrroles and Their Derivatives

No.	Structure	Yield (%)	Bp (°C/torr) (Mp, °C)	d_4^{20}	n_D^{20}	References
57	Ph···COOH	78	(138–173 decomp.)			[630]
58	Me···COOH	82	(165–175 decomp.)			[630]
59	Et···COOH	79	(140–170 decomp.)			[630]
60	MeO···COOH	84	(120–150 decomp.)			[630]
61	Cl···COOH	78	(155–187 decomp.)			[630]
62	Et, Ph···COOH	82	(143–167 decomp.)			[630]
63	n-Am, Ph···COOH	83	(103)			[630]
64	S···COOH	88	(150–175 decomp.)			[630]

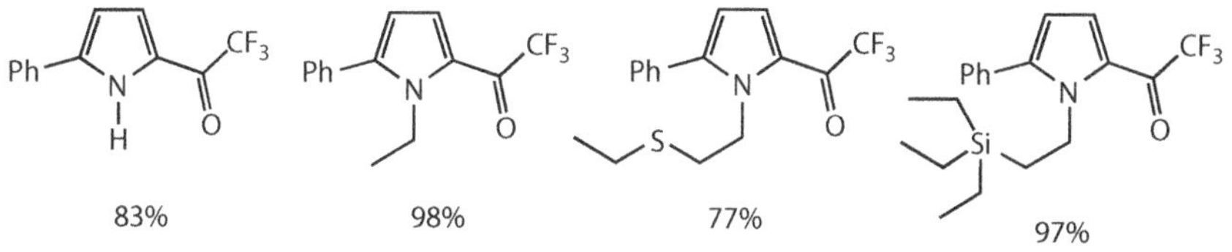

SCHEME 2.166 Representatives of 2-trifluoroacetylpyrroles, obtained by direct reaction of pyrroles with trifluoroacetic anhydride.

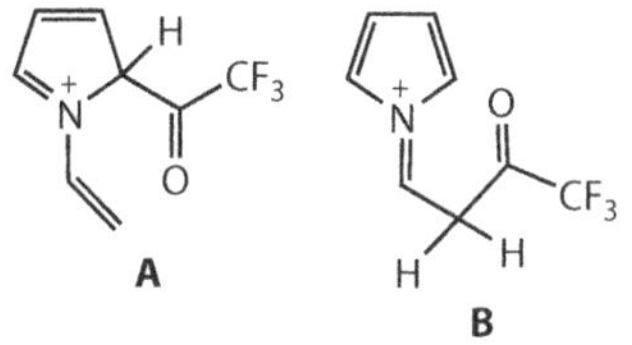

SCHEME 2.167 Intermediate cations in the trifluoroacetylation of N-vinylpyrroles.

alternative cations **A** and **B** (Scheme 2.167), which can be generated during trifluoroacetylation, cation **A** is more stable.

The results obtained indicate certain difficulties related to electrophilic reactions with participation of the N-vinyl group and possibility of synthesis of various α-substituted N-vinylpyrroles via selective electrophilic attack on the pyrrole ring α-position. In view of the fact [93,622] that the double bond of N-vinylindole and especially N-vinyl- and N-alkenylcarbazoles [622] is very prone to electrophilic attack, such a conclusion seems somewhat unexpected.

It is pertinent to emphasize that the trifluoroacetylation of diverse pyrrole systems attracts the attention of researchers for a long time [1,2,623–625]. This reaction has frequently been used to compare the reactivity of pyrroles, furans, and thiophenes in electrophilic substitution processes [617,618,626]. The ratios of the trifluoroacetylation rates of these heterocycles with trifluoroacetylc anhydride are $(5.3 \cdot 10^7)$: $(1.4 \cdot 10^2)$: 1 [627]. Therefore, it is expected that 2-(2-furyl)- and 2-(2-thienyl)pyrroles and their N-vinyl derivatives will be acylated only at position 5 of the pyrrole ring (direction 1) (Scheme 2.168).

It is known [446] that the activity of the pyrrole moiety in the electrophilic substitution reactions, owing to the greater capability of nitrogen atom to delocalize the

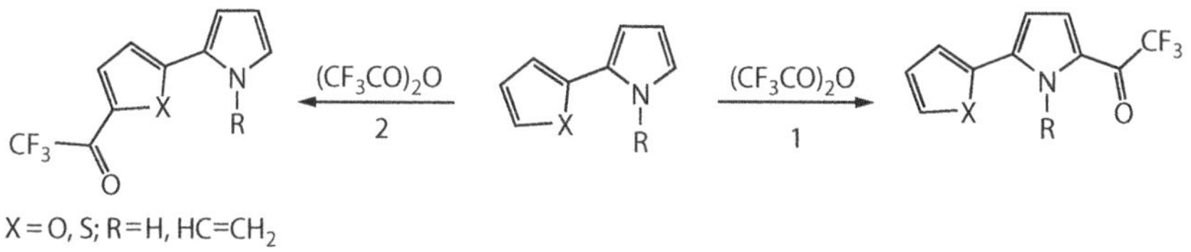

X = O, S; R = H, HC=CH$_2$

SCHEME 2.168 Two directions of trifluoroacetylation of 2-(2-furyl)- and 2-(2-thienyl) pyrroles.

positive charge in cationic σ-complexes in comparison with other heteroatoms, is considerably higher than that of the furan and thiophene rings.

It is shown [447,628] that 2-(2-furyl)- and 2-(2-thienyl)pyrroles are really trifluoroacetylated at the pyrrole ring only.

Unlike 2-phenylpyrrole [615], 2-(2-furyl)- and 2-(2-thienyl)pyrroles are not trifluoroacetylated exhaustively under similar conditions: the conversion is 50%–60% and the yields of the corresponding 5-trifluoroacetyl derivatives range from 40% to 50% (Table 2.16). The use of a twofold excess of the anhydride (relative to 2-(2-furyl)pyrrole) and increase of temperature (50°C) do not change by the reaction direction: only 2-(2-furyl)-5-trifluoroacetylpyrrole is formed, although its yield raises to 75%.

In the course of trifluoroacetylation, 2-(2-furyl)-N-vinylpyrrole behaves in a quite different manner. If 2-(2-thienyl)-N-vinylpyrrole is acylated with both equimolar amount and twofold excess of trifluoroacetyl anhydride only in direction 1, the main direction of 2-(2-furyl)-N-vinylpyrrole trifluoroacetylation is an attack at the α-position of the furan ring (direction 2) (Scheme 2.168). The ratio of the furan and pyrrole trifluoroacetylation products is 9:1 (NMR). Interestingly, 2-(2-thienyl)-N-vinylpyrrole, on trifluoroacetylation, gives better preparative results than its nonvinylated precursor: under comparable conditions, the yields of NH- and N-vinylpyrroles are 53% and 71%, respectively. When twofold excess of the anhydride relative to 2-(2-thienyl)-N-vinylpyrrole is used, the reaction direction does not alter and the yield of the trifluoroacetyl derivative does not increase. At the same time, the overall yield of 2-(2-furyl)-N-vinylpyrrole trifluoroacetylation products is only 50%, and with twofold excess of the anhydride, the yield of isomer acetylated at the pyrrole ring reaches 40% and the total yield slightly rises (63%) [447].

Rationalizing the results obtained, first of all, one should account for the following three factors: (a) significantly higher nucleophilicity of the furan ring as compared to the thiophene cycle, (b) decrease of electronic density in the pyrrole moiety due to competitive conjugation of the vinyl group with nitrogen lone electron pair as well as owing to its negative inductive effect, and (c) the possible spatial hindrance induced by the N-vinyl group for attack at the α-position. The reactivity of the thiophene ring in comparison with that of the furan and especially pyrrole fragments [617,618,626,627] is so low that factors (a) and (c) are probably insufficiently strong to change the reaction direction in the case of 2-(2-thienyl)-N-vinylpyrrole.

The absence of the second acylation products is indicative of very strong deactivating effect of the trifluoroacetyl radical, which is transmitted from the α-position of the pyrrole ring to the α-position of the furan and thiophene functions. Obviously, the mechanism of such long-range interaction (through seven valence bonds) is not only related to the inductive effect but involves formation of the general conjugation system (Scheme 2.169) [447].

Hydrogenation of 2(5)-trifluoroacetyl-N-vinylpyrroles (Raney nickel, 20 wt%, H_2 40–50 atm, 20°C–60°C, ethanol, 5 h) involves reduction of both the vinyl and trifluoroacetyl groups (Scheme 2.170, Table 2.16) [7,621,629].

Free radical addition of mercaptanes to the N-vinyl moiety of 2-trifluoroacetylpyrroles expectedly affords β-adducts (Scheme 2.171, Table 2.16) [621].

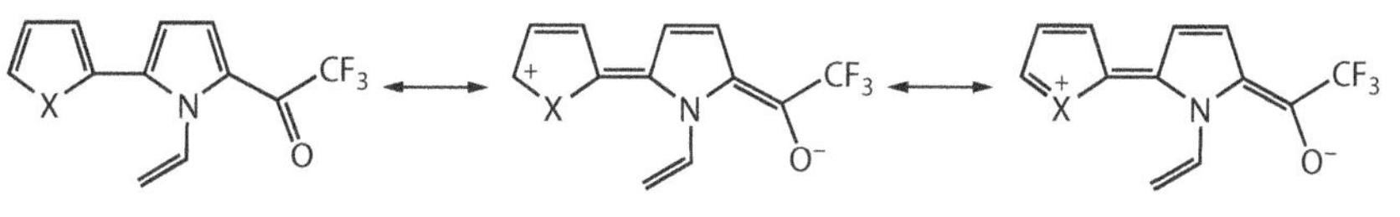
X
N
O
CF3
X
N
O−
CF3
X
N
O−
CF3

SCHEME 2.169 The long-range charge transmission in the intermediate monocations generated during trifluoroacetylation of N-vinyl-2-hetarylpyrroles.

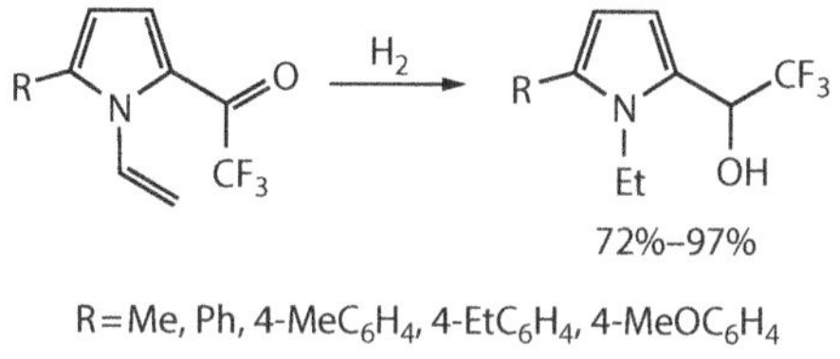
R
N
O
CF3
H2
R
N
CF3
Et
OH
72%–97%
R=Me, Ph, 4-MeC6H4, 4-EtC6H4, 4-MeOC6H4

SCHEME 2.170 Hydrogenation of N-vinyl-2(5)-trifluoroacetylpyrroles.

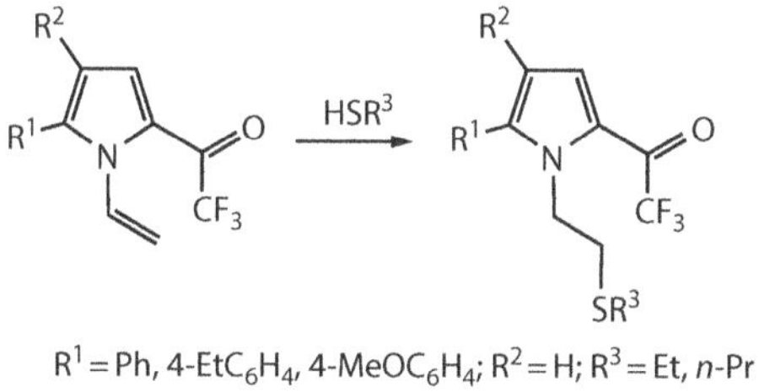
R2
R1
N
O
CF3
HSR3
R2
R1
N
O
CF3
SR3
R1=Ph, 4-EtC6H4, 4-MeOC6H4; R2=H; R3=Et, n-Pr

SCHEME 2.171 Free radical thiolation of N-vinyl-2-trifluoroacetylpyrroles.

Haloform cleavage of 2-trifluoroacetylpyrroles upon boiling their alcohol solutions in the presence of NaOH leads to pyrrolecarboxylic acid salts, which on acidification give the corresponding acids (Scheme 2.172, Table 2.16) [630–632].

2.1.10 Azo Coupling

A series of 2-arylazo-N-vinylpyrroles, new reactive dyes, has been synthesized by a modified azo coupling of readily accessible N-vinylpyrroles with aryldiazonium hydrocarbonates (0°C, aqueous ethanol, 2 h), the yields of the products being 52%–94% (Scheme 2.173, Table 2.17) [88,633].

5-Methylsubstituted 2-arylazo-N-vinylpyrroles **38** [634,635] when refluxed in toluene in air are selectively oxidized at the 5-methyl group retaining the pyrrole, N-vinyl, and azo moieties intact. The major products of the reaction (yield up to 60%) are hydroxymethyl derivatives **39**, along with N-vinylpyrrolylarylazocarbaldehydes **40** (Scheme 2.174, Table 2.17).

Apparently, the oxidation starts with the formation of hydroxyperoxide **A**, which then decomposes to hydroxyl and methoxyl radicals **B** (Scheme 2.175). The latter intercepts hydrogen atom from the next molecule of methylpyrrole to give hydroxymethyl

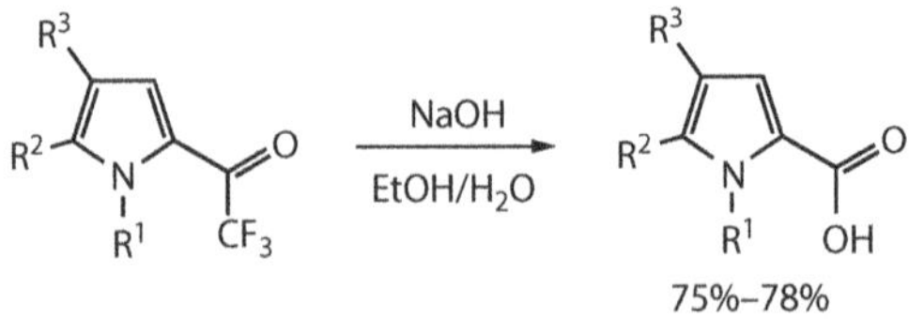

$R^1 = H$, $CH=CH_2$; $R^2 = Ph$, $4\text{-}MeC_6H_4$, $4\text{-}EtC_6H_4$, $4\text{-}MeOC_6H_4$, $4\text{-}ClC_6H_4$,
2-thienyl; $R^3 = H$, Et, n-Am; $R^2\text{-}R^3 = (CH_2)_4$

SCHEME 2.172   Synthesis of pyrrole-2-carboxylic acid from 2-trifluoroacetylpyrroles.

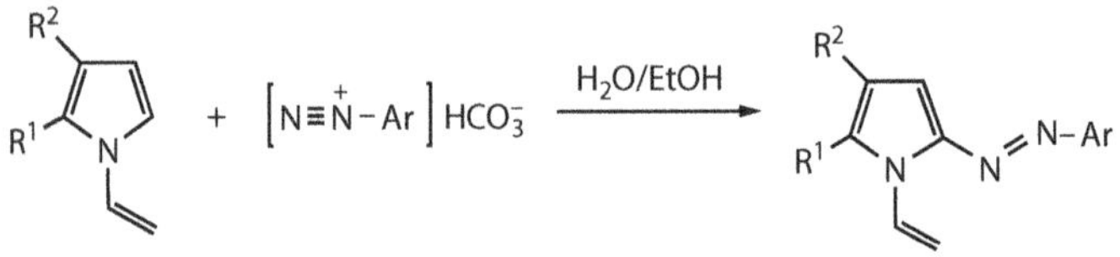

$R^1 = H$, Me, Ph, $4\text{-}MeOC_6H_4$, 2-furyl, 2-thienyl; $R^2 = H$, $R^1\text{-}R^2 = (CH_2)_4$;
Ar = Ph, $4\text{-}EtOC_6H_4$, $4\text{-}O_2NC_6H_4$, $4\text{-}BrC_6H_4$, $4\text{-}N_2C_6H_4C_6H_5$

SCHEME 2.173   Azo-coupling of N-vinylpyrroles with aryldiazonium hydrocarbonates.

derivatives **39** and free radical **C,** which recombines with hydroxyl radical generating another molecule of hydroxymethylpyrrole **39**. The formation of aldehydes **40** is probably due to the decomposition of hydroxyperoxide **A** to release water.

The expected oxidation of the pyrrole cycle, C=C and N=N double bonds, is not observed. The reason is a deep electron density transfer from the N-vinyl fragment to the azo group that decreases susceptibility of the vinyl group and pyrrole ring toward oxidizers. The fact that the stability of the molecule is ensured by the arylazo substituent is confirmed by deep oxidation of 2,3-dimethyl-N-vinylpyrrole in these conditions (NMR spectra of the products show no signals of the vinyl group and pyrrole ring).

2-Phenylazo-N-vinyl-4,5,6,7-tetrahydroindole under similar conditions is oxidized at position 7 of the cyclohexanone ring to deliver 7-hydroxy-2-phenylazo-N-vinyl-4,5,6,7-tetrahydroindole (58% yield, Scheme 2.176, Table 2.17) [635].

This reaction is the first example of easy selective oxidation of the methyl substituent in the pyrrole ring to hydroxymethyl and aldehyde functions. It may be considered as the simple atom-economic synthesis of phenylazopyrrole dyes bearing hydroxymethyl or aldehyde moieties.

2-Arylazo-N-vinylpyrroles are reversibly protonated [88] by excess strong acids (trifluoromethanesulfonic or trifluoroacetic acids, room temperature, several hours) (Scheme 2.177). Unlike alkyl- or aryl- substituted N-vinylpyrroles, their N-vinyl group remains intact.

2-Arylazo-N-vinylpyrroles may act as ligands of palladium complexes exhibiting catalytic activity, for example, in Heck reaction [88]. So, 2-methyl-5-phenylazo-N-vinylpyrrole in combination with $PdCl_2$ catalyzes almost quantitative formation of E-stilbenes from arylbromides and styrene (Scheme 2.178).

TABLE 2.17

2-Arylazo-N-vinylpyrroles and Their Derivatives

No.	Structure	Yield (%)	Mp, °C (n_D^{20})	References
1	2	3	4	5
1		80	72–74	[88,633]
2		52	72–76	[88]
3		84	176–180	[88]
4		80	52–54	[88]
5		76	74–76	[88]
6		85	138–140	[88]
7		76	72–74	[88,633]
8		79	120–122	[88]
9		87	62–66	[88]

(Continued)

TABLE 2.17 (*Continued*)
2-Arylazo-N-vinylpyrroles and Their Derivatives

No.	Structure	Yield (%)	Mp, °C (n_D^{20})	References
1	2	3	4	5
10		80	112–114	[88]
11		94	150–152	[88]
12		82	78–79	[88]
13		81	98–100	[88]
14		53	72–74	[88]
15		56	140–144	[88]
16		57	76–78	[88]
17		56	122–124	[88]
18		53	86–88	[88]

(*Continued*)

TABLE 2.17 (*Continued*)

2-Arylazo-N-vinylpyrroles and Their Derivatives

No.	Structure	Yield (%)	Mp, °C (n_D^{20})	References
1	2	3	4	5
19		53	74–76	[88]
20		79	136–140	[88]
21		23	90–92	[634,635]
22		60	100–102	[635]
23		22	112–114	[635]
24		58	80–82	[635]
25		49	(1.5125)	[636]
26		33	(1.5115)	[636]

(Continued)

TABLE 2.17 (*Continued*)

2-Arylazo-N-vinylpyrroles and Their Derivatives

No.	Structure	Yield (%)	Mp, °C (n_D^{20})	References
1	2	3	4	5
27	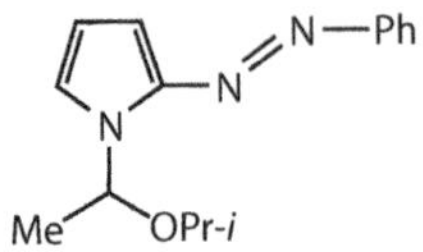 Me OPr-*i* ; N=N—Ph	20	(1.5100)	[636]
28	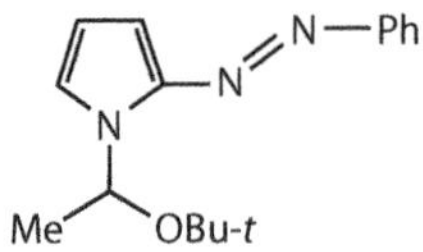 Me OBu-*t* ; N=N—Ph	12	(1.5050)	[636]
29	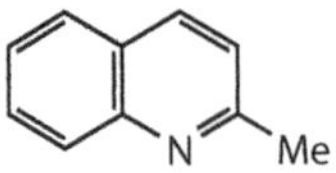 N Me	44	(1.6123)	[637]
30	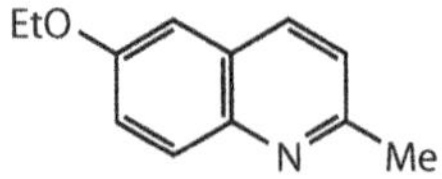 EtO, N Me	54	(1.6090)	[636,637]
31	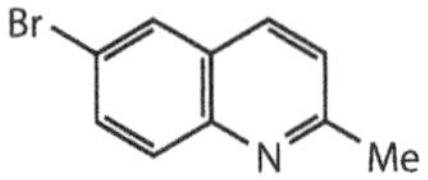 Br, N Me	56	(1.5943)	[637]
32	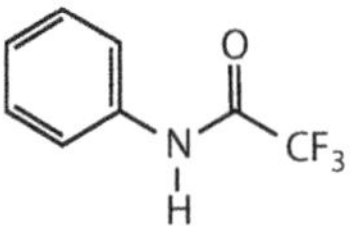 N H C(O) CF₃	32	85–88	[638]
33	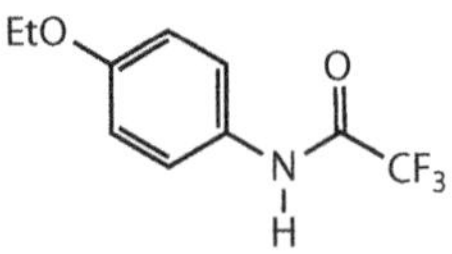 EtO, N H C(O) CF₃	23	140–142	[638]
34	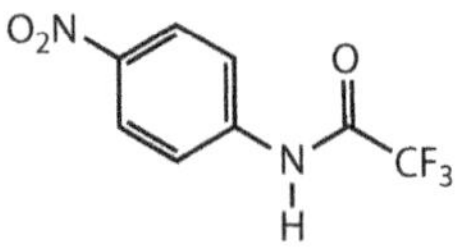 O₂N, N H C(O) CF₃	25	50–52	[638]

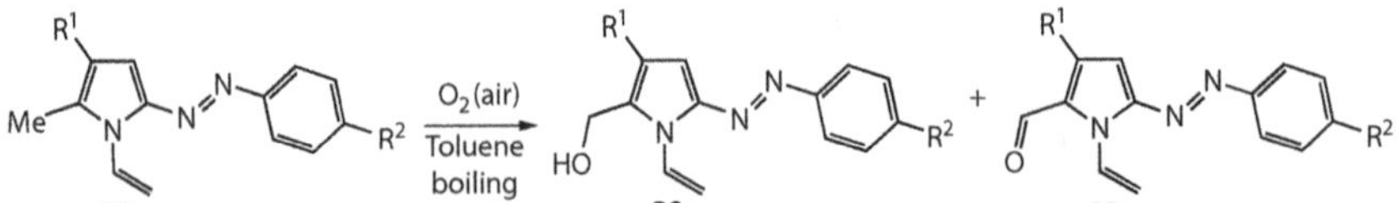

$R^1 = H$, Me; $R^2 = H$, OEt

SCHEME 2.174 Auto-oxidation of methyl group in 5-methyl-2-arylazo-N-vinylpyrroles.

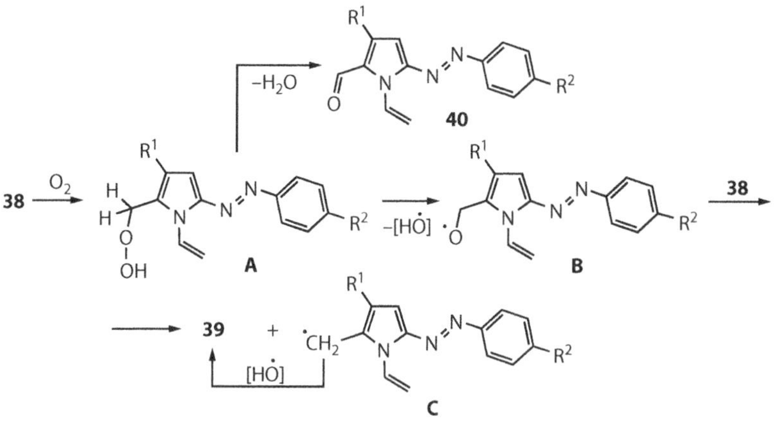

SCHEME 2.175 Plausible steps of auto-oxidation of 5-methyl-2-arylazo-N-vinylpyrroles.

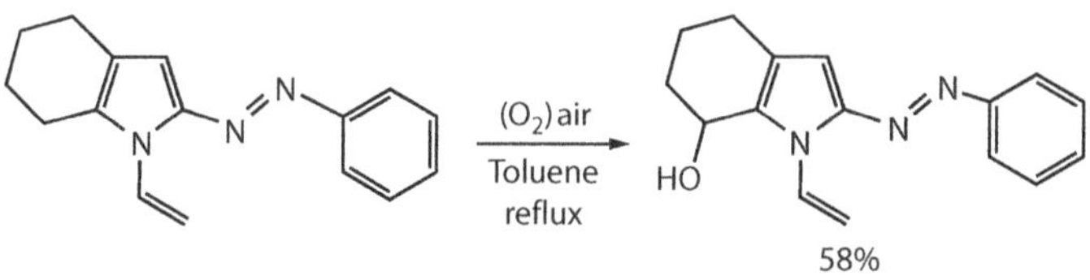

SCHEME 2.176 Auto-oxidation of 2-arylazo-7-methyl-N-vinyl-4,5,6,7-tetrahydroindole.

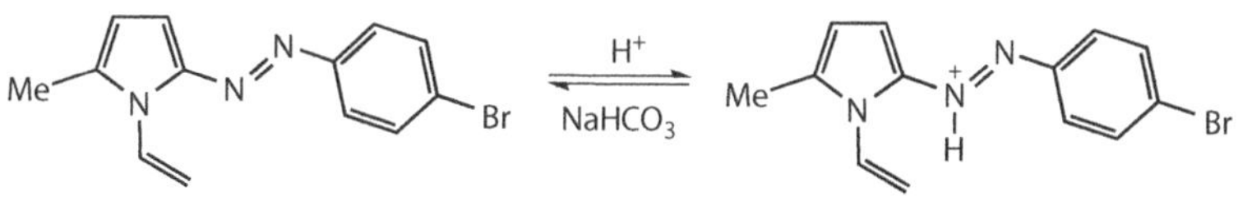

SCHEME 2.177 Reversible protonation of 2-arylazo-N-vinylpyrroles.

R=H, Me, OMe, Ac

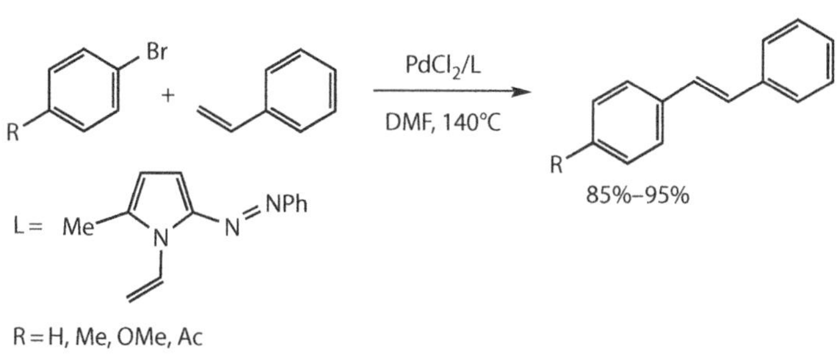

SCHEME 2.178 Pd complexes of 2-arylazo-N-vinylpyrroles as catalyst of the Heck reaction.

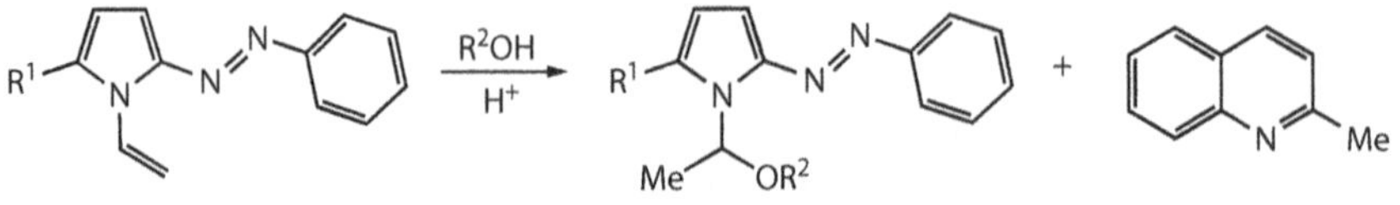

R^1 = H, Me; R^2 = Me, Et, i-Pr, t-Bu

SCHEME 2.179 Addition of alcohols to 2-arylazo-N-vinylpyrroles.

R^1 = H, Me, Ph; R^2 = H, OEt, NO$_2$, Br, OAc, 2,5-Cl$_2$ (polymer)

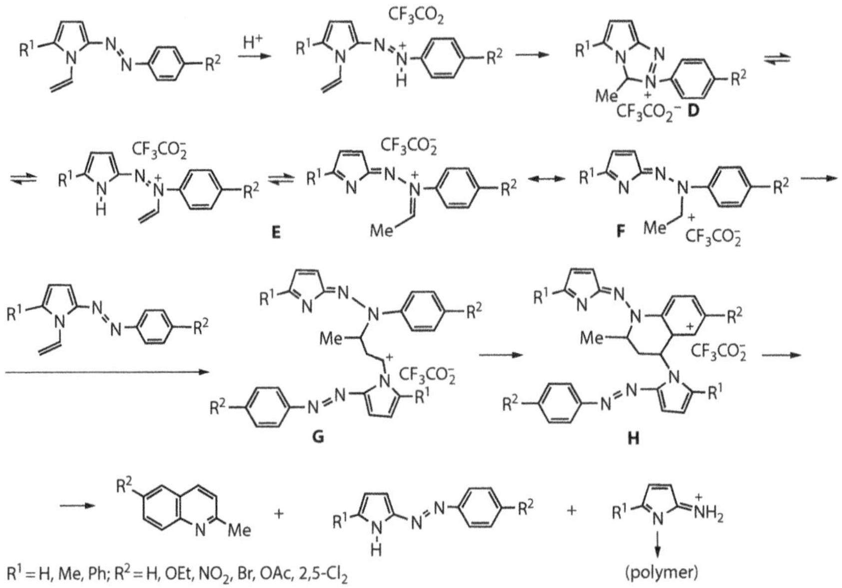

SCHEME 2.180 Tentative steps of the synthesis of 2-methylquinolines from 2-arylazo-N-vinylpyrroles under the action of trifluoroacetic acid.

2-Arylazo-N-vinylpyrroles add alcohols to the vinyl group in the presence of acids or PdCl$_2$ to give N-(1-alkoxyethyl)-2-phenylazopyrroles in up to 50% yield (Scheme 2.179, Table 2.17) [636].

In the presence of TFAA (1%, reflux, 3.5 h), the unexpected formation of 2-methylquinoline (up to 26% yield) is observed. The reaction likely proceeds with participation of the arylazo- and N-vinyl groups (Scheme 2.180).

The yield of adducts expectedly drops with the lower acidity of the alcohol (with the increase in donor effect of the R^2 radical) in keeping with the electrophilic character of the reaction. This corresponds also to the anticipated steric effect of these substituents. The readiness of 2-methylquinoline formation also grows in the same succession (Me → t-Bu).

Later, a series of 2-methylquinolines have been synthesized (up to 56% yield) in the presence of equimolar amount of TFAA (benzene, reflux, 5 h) [637]. NMR monitoring of the reaction indicates that the formation of quinolines starts from protonation of the azo group to give cyclic cation **D**. Further, a transfer of the vinyl

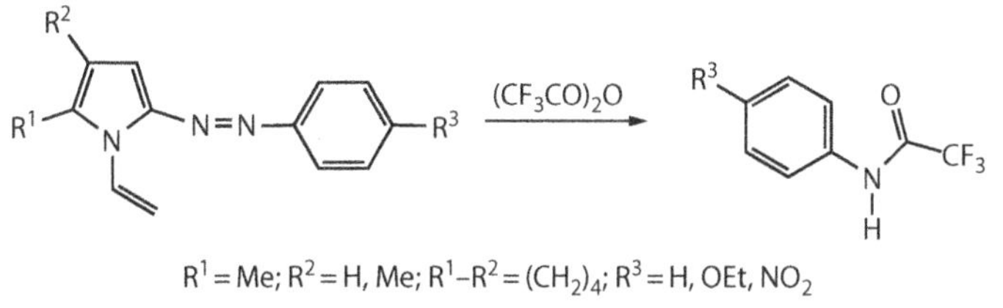

$R^1 = Me; R^2 = H, Me; R^1-R^2 = (CH_2)_4; R^3 = H, OEt, NO_2$

SCHEME 2.181 Formation of N-aryl-2,2,2-trifluoroacetamides from 2-arylazo-N-vinyl-pyrroles under the action of trifluoroacetyl acid anhydride.

group to "aniline" nitrogen atom occurs followed by prototropic rearrangement of cation **E** to iminium cation **F** that adds to the N-vinyl group of arylazopyrrole second molecule to generate cation **G**. Then, the ring closing into tetrahydroquinoline cation **E** follows. The cation **E** is aromatized eliminating 2-arylazopyrrole and protonated 2-imino-2*H*-pyrrole to afford 2-methylquinoline.

2-Arylazo-N-vinylpyrroles under the action of trifluoroacetyl acid anhydride $(CH_2Cl_2$ or benzene, $-5°C$ to $0°C$, 1 h) give N-aryl-2,2,2-trifluoroacetamides (22%–32% yields, Scheme 2.181, Table 2.17) along with conjugated (electroconducting and paramagnetic) polymers [638].

Obviously, the site of the primary attack of trifluoroacetyl cation is nitrogen atom of the azo group adjacent to the benzene ring (Scheme 2.182). In the cation **I**, the positive charge should be mainly concentrated on the pyrrole nitrogen due to electron density transfer from the pyrrole counterpart of the molecule to ionized azo group. One of the tentative pathways of this intermediate fragmentation is elimination of the vinyl cation with subsequent release of proton. Protonation of trifluoroacetamide

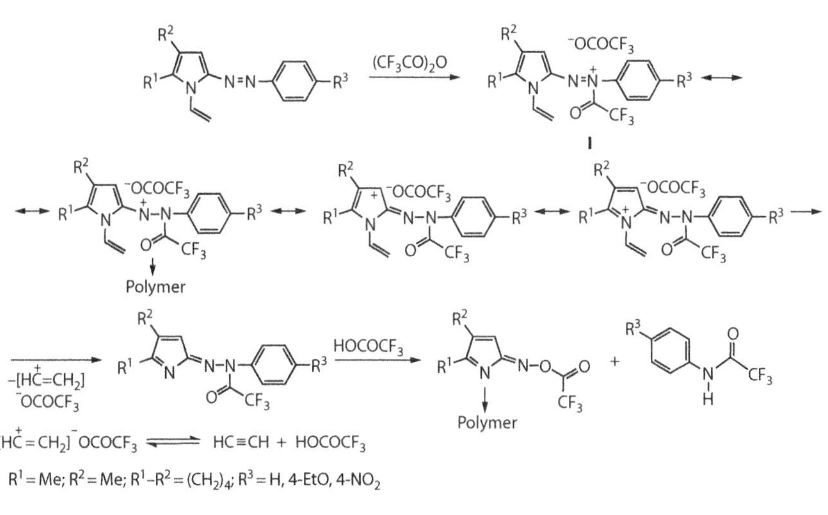

$R^1 = Me; R^2 = Me; R^1-R^2 = (CH_2)_4; R^3 = H, 4-EtO, 4-NO_2$

SCHEME 2.182 Tentative steps of the formation of N-aryl-2,2,2-trifluoroacetamides from 2-arylazo-N-vinylpyrroles under the action of trifluoroacetic anhydride.

moiety with the N–N-bond cleavage affords N-aryl-2,2,2-trifluoroacetamides and unstable 2,2,2-trifluoro-N-(2H-pyrrol-2-ylidene)acetamides.

2.1.11 ASSEMBLY OF 4,4-DIFLUORO-4-BORA-3A,4A-DIAZA-S-INDACENES

2.1.11.1 Via Meso-Aryldipyrromethenes

Pyrroles are widely employed for the synthesis of 4,4-difluoro-4-bora-3a,4a-diaza-s-indacenes (BODIPY) fluorophores [639]. BODIPYs bearing sterically hindered substituents that prevent intermolecular π-electron interaction (π-stacking) attract a special attention.

Recently, 2-mesityl-3-methylpyrrole, synthesized from mesityl ethyl ketone and acetylene via the Trofimov reaction, has been used for the preparation of sterically encumbered fluorophores of BODIPY family (Scheme 2.183) [222]. The fluorophore assembled in a one-pot manner through the following procedure: the pyrrole is condensed with mesityl aldehyde (TFAA, CH$_2$Cl$_2$, 20°C–25°C, 24 h), dipyrromethane is oxidized to dipyrromethene (DDQ, CH$_2$Cl$_2$, 20°C–25°C, 0.5 h), and the latter is treated with BF$_3$·Et$_2$O in the presence of diisopropylethylamine (CH$_2$Cl$_2$, 20°C–25°C, 0.5 h).

In the fluorophore molecule synthesized (Table 2.18), internal rotation of fragments is hindered due to the mesityl groups, and π-stacking is hampered at high concentration [222,640,641] that ensures its intensive fluorescence (λ$_{max}$ 559 nm) and high quantum yield (0.95) in a crystalline state.

The similar reaction sequence is used for the assembly of BODIPY fluorophore from 2-(benzo[b]thiophen-3-yl)pyrrole (Scheme 2.184) [248].

This novel representative of the BODIPY family (Table 2.18) exhibits promising optical properties, superior to the existing analogs. In solutions, it displays an intense red-shifted fluorescence emission (CH$_2$Cl$_2$, λ$_{max}$ 625 nm, quantum yield 0.84) that is fully preserved in the solid state [248].

Sterically hindered BODIPY fluorophores containing [2,2]paracyclophane substituents (Table 2.18) are assembled from 2-([2,2]paracyclophan-4-yl)pyrroles and mesitylaldehyde according to Scheme 2.185 [224].

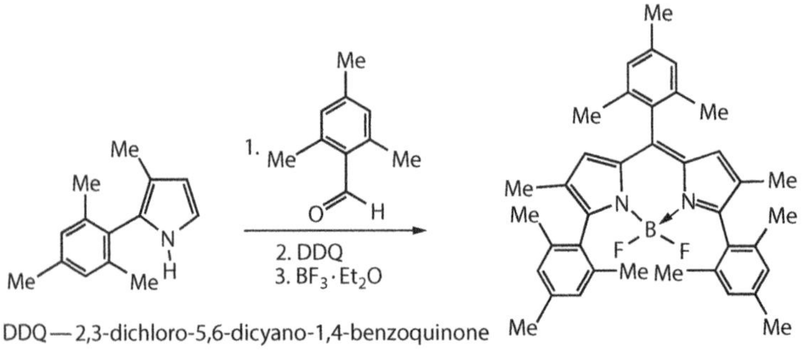

DDQ — 2,3-dichloro-5,6-dicyano-1,4-benzoquinone

SCHEME 2.183 2-Mesityl-3-methylpyrrole-based difluoroboradiazaindacene (BODIPY) fluorophore.

TABLE 2.18
Dipyrromethanes and Their Derivatives

No.	Structure	Yield (%)	Mp (°C)	References
1	2	3	4	5
1		97	114–116	[642]
2		96	136–137	[642]
3		85	144–146	[642]
4		88	Oil	[642]
5		96	68–70	[642]
6		50	156–158	[642]
7		61	136–138	[642]

(Continued)

TABLE 2.18 (*Continued*)

Dipyrromethanes and Their Derivatives

No.	Structure	Yield (%)	Mp (°C)	References
1	2	3	4	5
8		57	150–152	[642]
9		39	192–194	[642]
10		84	162–164	[642]
11		8	>280–300 (decomp.)	[222]
12		30	>250 (decomp.)	[227]

(*Continued*)

TABLE 2.18 (*Contiuned*)

Dipyrromethanes and Their Derivatives

No.	Structure	Yield (%)	Mp (°C)	References
1	2	3	4	5
13		35	>250 (decomp.)	[227]
14		21	>250 (decomp.)	[227]
15		51	>250 (decomp.)	[227]
16		34	>300	[248]

(*Continued*)

TABLE 2.18 (*Contiuned*)

Dipyrromethanes and Their Derivatives

No.	Structure	Yield (%)	Mp (°C)	References
1	2	3	4	5
17		31	>250 (decomp.)	[224]
18		53	>250 (decomp.)	[224]
19		54	>250 (decomp.)	[224]
20		53	>250 (decomp.)	[224]

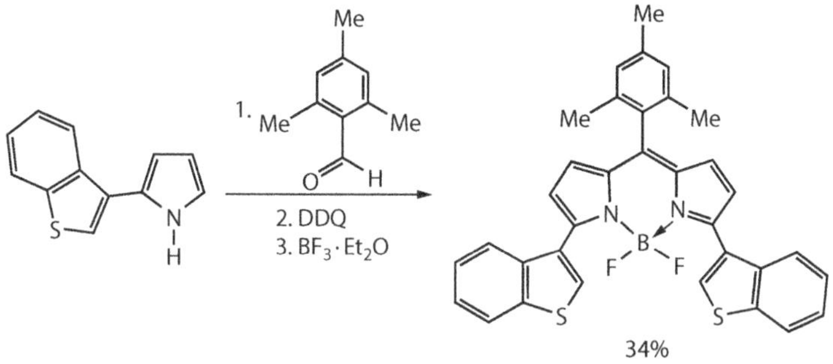

SCHEME 2.184 2-(Benzo[*b*]thiophen-3-yl)pyrrole-based BODIPY.

R = H, Me, Et, *n*-Pr

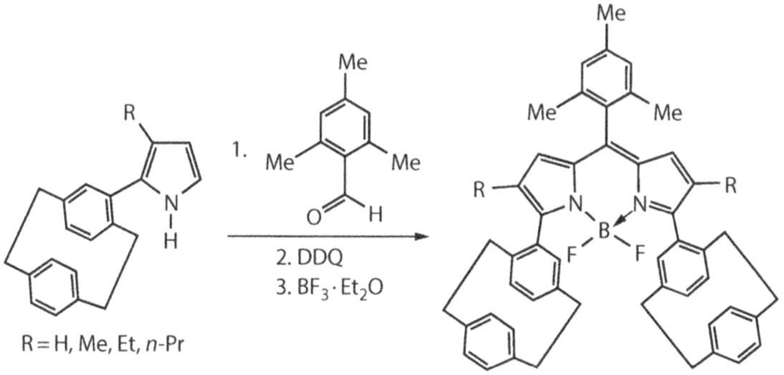

SCHEME 2.185 2-([2,2]Paracyclophan-4-yl)pyrrole-based BODIPYs.

The introduction of the [2.2]paracyclophanyl substituent in the BODIPY skeleton interferes convergence of the molecules and, consequently, π-staking that ensures intense fluorescence of the dyes in solid state.

Another series of earlier inaccessible BODIPY fluorophores (Table 2.18), which intensively fluoresce at 601–606 nm (quantum yield is up to 0.95) [227], has been synthesized from pyrroles bearing polycondensed aromatic substituents in position 2 of the pyrrole ring and mesitylaldehyde (Scheme 2.186).

2.1.11.2 Via Meso-CF₃-Dipyrromethenes

A promising general strategy for the synthesis of symmetric and asymmetric BODIPY fluorophores with a *meso*-CF$_3$-group (Scheme 2.187, Table 2.18) is based on the application of 2,2,2-trifluoro-1-(pyrrol-2-yl)-1-ethanols (Table 2.16), reduction products of available 2-trifluoroacetylpyrroles, as key intermediates [642].

The reaction comprises the following steps: reduction of 2-trifluoroacetylpyrroles, condensation of the alcohols formed with pyrroles to dipyrromethanes (Scheme 2.187, Table 2.18), their oxidation to dipyrromethenes, and complex formation of the latter with boron trifluoride. The last two stages proceed in a one-pot manner.

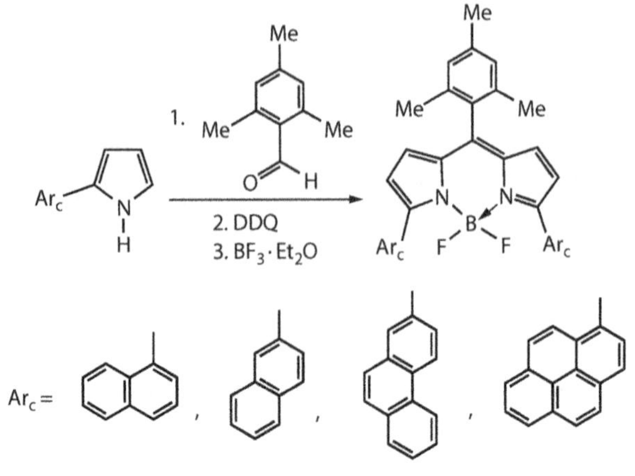

SCHEME 2.186 BODIPYs derived from pyrroles bearing polycondensed aromatic substituents.

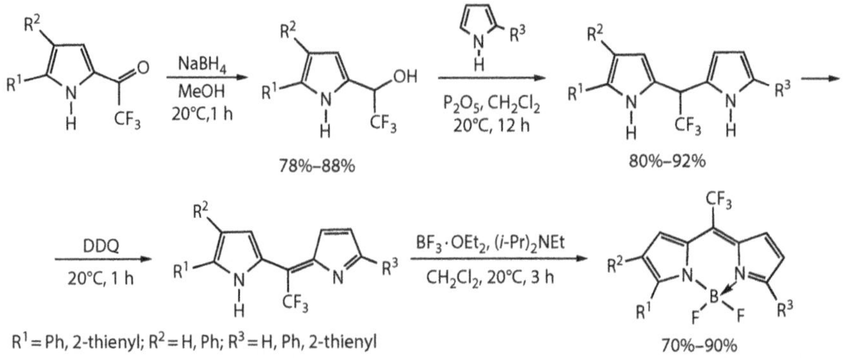

$R^1 = Ph$, 2-thienyl; $R^2 = H$, Ph; $R^3 = H$, Ph, 2-thienyl

SCHEME 2.187 Synthesis of symmetric and asymmetric BODIPY fluorophores with a *meso*-CF$_3$-group.

A key feature and important advantage of this method is a possibility to synthesize the asymmetric representatives of BODIPY (hitherto almost unknown) due to the involvement of α-unsubstituted pyrroles (having other substituents in other positions of the pyrrole ring) in condensation with 2,2,2-trifluoro-1-(pyrrol-2-yl)-1-ethanols. Absorption and emission bands of such fluorophores are red shifted (600–670 nm), the fluorescence quantum yield being close to 1.0 [642].

2.2 REACTIONS WITH PARTICIPATION OF THE VINYL GROUP

2.2.1 HYDROLYSIS

Over many years, the literature contained no information regarding the transformations of N-vinylpyrroles in aqueous acidic media, although without such data,

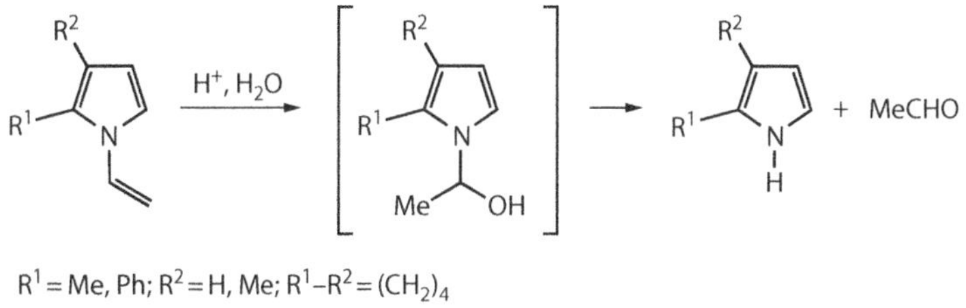

R^1 = Me, Ph; R^2 = H, Me; R^1–R^2 = $(CH_2)_4$

SCHEME 2.188 Expected acid-catalyzed hydrolysis of N-vinylpyrroles.

it is impossible to skillfully use these monomers and intermediates. It was assumed that N-vinylpyrroles would be hydrolyzed in the same way as N-vinylindole [643], N-alkenylcarbazoles [96,643,645], N-vinylphenthiazine [646], and N-vinyllactams (Scheme 2.188) [647].

It has been found [648] that such route of N-vinylpyrroles hydrolysis can be realized only in the dilute solutions if the released acetaldehyde is coupled (e.g., by the reaction with hydroxylamine). Under normal conditions, however, complex acid-catalyzed and oxidative condensation reactions of the initial and formed pyrroles with acetaldehyde as well as between pyrroles themselves take place. Consequently, mixtures of deeply colored oligomers, the composition of which depends on the hydrolysis and isolation conditions, are formed. On the basis of the literature data [1,2,627], spectral (IR, NMR, and ESR) characteristics, and the results of elementary analysis, it has been assumed [648] that these mixtures contain fragments **A–D** (Scheme 2.189).

The possibility of structures **B** formation from pyrroles and acetaldehyde is confirmed by the synthesis of di-(N-vinyl-5-phenylpyrrol-2-yl)methane (46% yield, Scheme 2.190) from 2-phenyl-N-vinylpyrrole and formaldehyde in the dilute acetic acid (dioxane, 70°C–80°C). The intermediate 5-hydroxymethyl-2-phenyl-N-vinyl-pyrrole is isolated in 3% yield [649].

The attack of the intermediate carbimmonium ion (**F**) at the double bond (cationic vinyl polymerization) or at the pyrrole ring with the corresponding intra- and

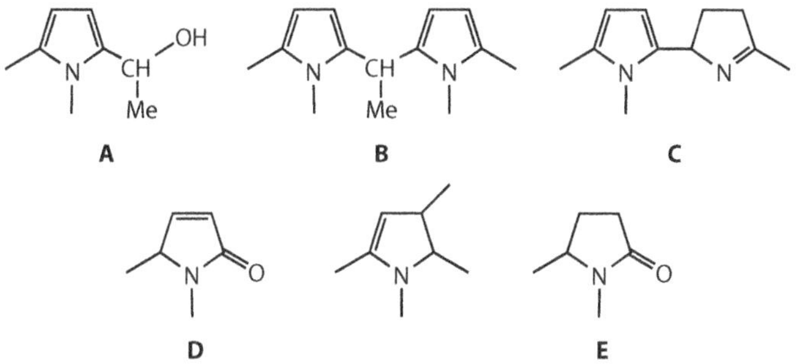

SCHEME 2.189 Plausible fragments of the side products formed in the acid-catalyzed hydrolysis of N-vinylpyrroles.

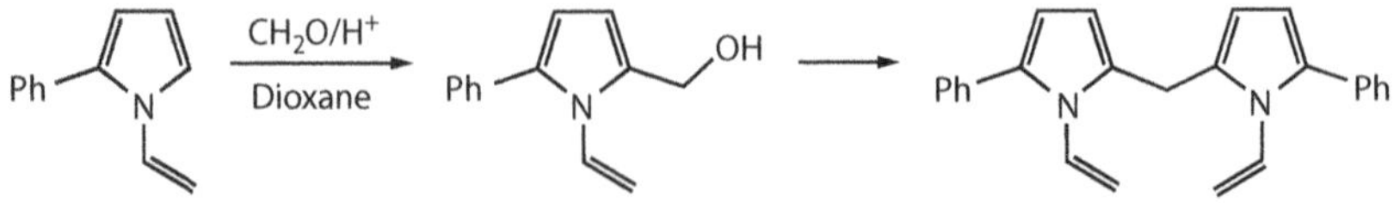

SCHEME 2.190 Reaction of 2-phenyl-N-vinylpyrrole with formaldehyde in the presence of acetic acid.

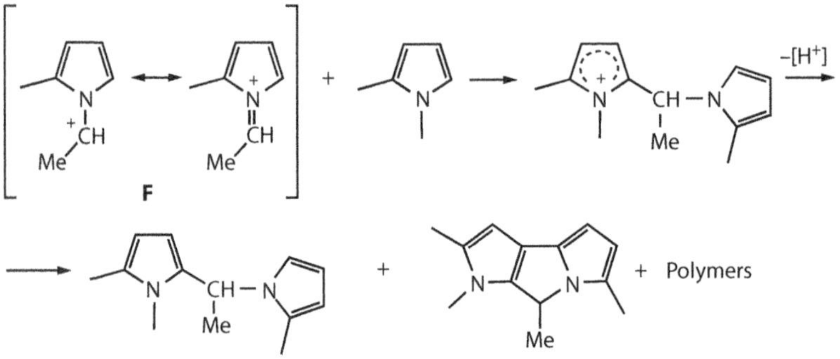

SCHEME 2.191 Plausible dimerization, cyclization, and oligomerization of N-vinylpyrroles during their acid-catalyzed hydrolysis.

intermolecular transformations should also play an appreciable role in the formation of the structure of the products of acidic hydrolysis of N-vinylpyrroles (Scheme 2.191), especially when water is used in insufficient amounts.

As one should expect, N-vinylpyrroles with alkyl substituents in the ring are the least resistant to acidic hydrolysis [648]. They are readily hydrolyzed by dilute acids at room temperature. For example, N-vinyl-4,5,6,7-tetrahydroindole is hydrolyzed by 35% in dilute aqueous solution of HCl (0.6%–1%) at room temperature after 1 h. At 65°C, the same pyrrole is converted completely into an orange-red substance for 2.5 h, whereas at 90°C–96°C, this is achieved for 1.5 h.

The formation of 4,5,6,7-tetrahydroindole in ~5% yield from its vinyl derivative is observed [648] only when hydrolysis is accomplished in the excess acidified $NH_2OH \cdot HCl$. A special experiment has confirmed that 4,5,6,7-tetrahydroindole is unstable under conditions of acid-catalyzed hydrolysis: at 65°C (1% aqueous HCl solution, 2.5 h), it undergoes 66% conversion to a brown-red resin. In the presence of acetaldehyde under the same conditions, the degree of conversion of 4,5,6,7-tetrahydroindole to a polymer is 90%.

As compared with 2- and 2,3-dialkyl-N-vinylpyrroles, 2-phenyl-N-vinylpyrroles are noticeably more resistant to acidic hydrolysis [648]. Without heating, they remain almost unchanged upon prolonged contact (15 h) with a 0.4%–1.3% aqueous solution of HCl. However, they are completely transformed to solid deeply colored (red) resins at 96°C in the same solutions.

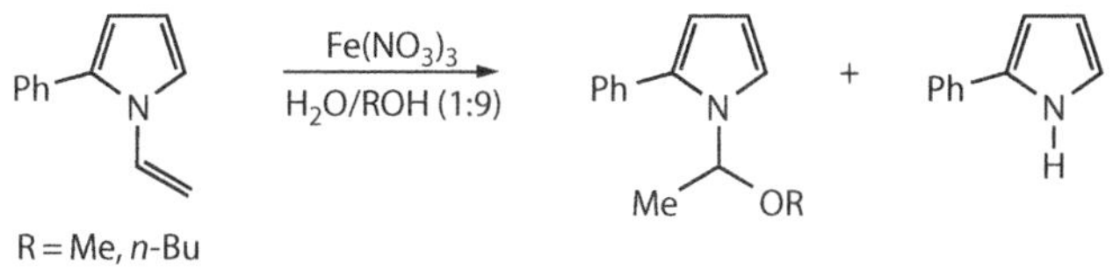

SCHEME 2.192 Electrophilic addition of alcohols to 2-phenyl-N-vinylpyrrole in the $Fe(NO_3)_3/H_2O/ROH$ systems.

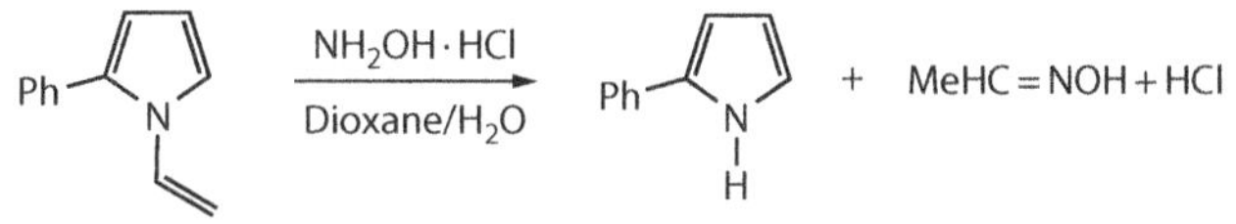

SCHEME 2.193 Hydrolysis of 2-phenyl-N-vinylpyrrole in the presence of $NH_2OH \cdot HCl$.

Ferric nitrate is known [644] to catalyze (room temperature) the hydrolytic decomposition of N-vinylcarbazole in aqueous methanol (1:9) to carbazole and acetaldehyde or (at a higher concentration) causes its dimerization to E-1,2-dicarbazylcyclobutane. However, 2-phenyl-N-vinylpyrrole under these conditions is not almost hydrolyzed (the yield of 2-phenylpyrrole is ~10%) and mainly adds alcohol to furnish N-(1-alkoxyethyl)-2-phenylpyrroles in 76%–85% yields (Scheme 2.192) [648].

2-Phenylpyrrole has been obtained in 52% yield only by hydrolyzing 2-phenyl-N-vinylpyrrole in a dilute aqueous dioxane solution of $NH_2OH \cdot HCl$ (80°C, 2 h), which ties up the acetaldehyde formed (Scheme 2.193) [648].

2.2.2 ELECTROPHILIC ADDITION OF ALCOHOLS AND PHENOLS

Addition of alcohols to N-vinylpyrroles is a route to novel large group of substituted pyrroles and a promising tool of introduction of the pyrrole nucleus into the molecules of complex hydroxyl-containing compounds (polyatomic alcohols, carbohydrates, vitamins, hormones, alkaloids, etc.) to modify their biological action. In the theoretical context, this reaction represents an interesting model for studying the competition of two nucleophilic centers (α-position of the pyrrole ring and β-position of the vinyl group) for addition of electrophile.

It is known that N-vinylindole [650–652] and N-vinylcarbazole [653–655], upon acid-catalyzed interaction with alcohols, give the corresponding N-(α-alkoxyethyl) derivatives (addition according to the Markovnikov rule). In the case of N-vinylindole, the HCl-catalyzed addition competes with the polymerization and considerably reduces the yields of adducts [650]. N-Vinylpyrrolidone adds alcohols and phenols also according to the Markovnikov rule to afford N-[α-alk(ar)oxyethyl] pyrrolidones [647,656]. The benzene nucleus of phenols is reported to be alkylated by N-vinylpyrrolidone, the hydroxyl group remaining intact [647].

N-Vinylpyrroles provide ample opportunities for the synthesis of new families of pyrrole derivatives, in particular, via the reactions with hydroxyl function.

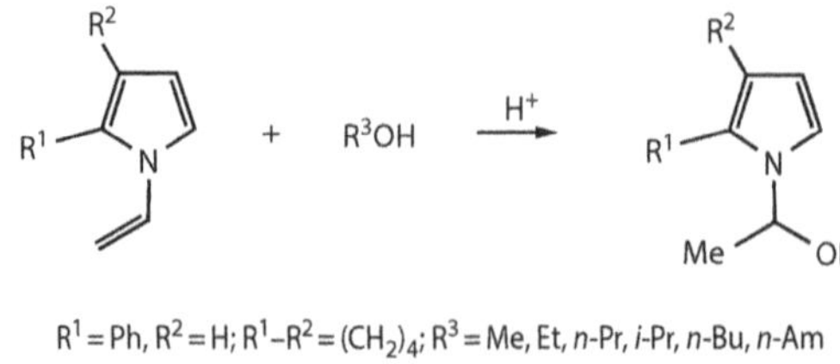

R¹ = Ph, R² = H; R¹–R² = (CH₂)₄; R³ = Me, Et, *n*-Pr, *i*-Pr, *n*-Bu, *n*-Am

SCHEME 2.194 Synthesis of N-(α-alkoxyethyl)pyrroles by electrophilic addition of alcohols to N-vinylpyrroles.

Some regularities of electrophilic addition of alcohols to N-vinylpyrroles have been elucidated on the example of the reaction between 2-phenyl-N-vinylpyrrole and N-vinyl-4,5,6,7-tetrahydroindole resulting in earlier unknown N-(α-alkoxyethyl)pyrroles (Scheme 2.194, Table 2.19) [657].

Protic and aprotic (transition metal salts) acids, either alone or in combination, have been tested as catalysts. Best results are received with $Fe(NO_3)_3$. Noteworthy, Ag, Cr (II), and Co (II) nitrates do not catalyze the reaction. Under comparable conditions, the rate of the addition reaction drops with an increase in the donating power of alcohol radical (R^3), which is in agreement with the electrophilic nature of the process. The character and shape of the kinetic curves obtained at 76°C–78°C indicate reversibility of the reaction: the addition is not complete and content of the adducts in the reaction mixture decreases with time due to polymerization of N-vinylpyrrole, which indirectly influences equilibrium amounts of the adducts. The fact of polymerization is confirmed by a big difference between N-vinylpyrrole conversion and the adducts yielded as well as by a polymer isolation during the reaction mixture treatment. Similar equilibrium has been observed earlier for N-alkenyl- and N-(α-alkoxyalkyl)carbazoles [653]. In contrast to the oxygen and sulfur analogs (acetals and 1-alkoxy-1-alkylthioethanes), N-(α-alkoxyethyl)pyrroles are not prone to disproportionation and alcoholysis: when the reaction is carried out with excess alcohol, neither the corresponding dialkyl acetal nor noticeable amount of NH-pyrrole is detected both in the reaction mixture and distillate. In the aqueous-alcohol medium, along with the addition of alcohol to 2-phenyl-N-vinylpyrrole, a partial hydrolysis to 2-phenylpyrrole takes place (owing to competing electrophilic addition of water to the vinyl group; see the previous section) [657].

The synthesis of N-vinylpyrrole adducts with acetylene alcohols (Scheme 2.195, Table 2.19) is most effectively realized in the presence of perfluorobutyric acid (dioxane, 70°C–96°C, 1%–3% of the catalyst) [658].

Noncatalytic addition of phenols to N-vinyl-4,5,6,7-tetrahydroindole affords N-(1-aroxyethyl)-4,5,6,7-tetrahydroindoles (Scheme 2.196, Table 2.19) [659]. The yields of adducts decrease as a result of N-vinyl-4,5,6,7-tetrahydroindole oligomerization, with an increase in the acidity of phenol and in the presence of CF_3COOH.

Thus, N-vinylpyrroles add alcohols and phenols in the presence of acids or without the catalyst (in the case of phenols possessing higher acidity than alcohols) regioselectively according to the Markovnikov rule and in agreement with the electrophilic nature of the reaction.

TABLE 2.19

Reaction Products of N-Vinylpyrroles with the Participation of Vinyl Group

No.	Structure	Yield (%)	Bp (°C/torr) (Mp, °C)	d_4^{20}	n_D^{20}	References
1	2	3	4	5	6	7
1	Me, N, Me, OEt	50	59–60/5	0.9064	1.4720	[660]
2	Me, Me, N, Me, OEt	60	52–53/1	0.9139	1.4820	[660]
3	N, Me, OBu-n	55	93–94/1	0.9582	1.4960	[657]
4	N, Me, OAm-n	51	110–112/1	0.9568	1.4930	[657]
5	Ph, N, Me, OMe	76	98/1	1.0475	1.5640	[648,657]
6	Ph, N, Me, OEt	63	104/2	1.0302	1.5490	[657,660]
7	Ph, N, Me, OPr-n	70	122–123/3	1.0159	1.5456	[657,660]
8	Ph, N, Me, OPr-i	53	102/1	1.0055	1.5452	[657,660]

(Continued)

TABLE 2.19 (*Continued*)

Reaction Products of N-Vinylpyrroles with the Participation of Vinyl Group

No.	Structure	Yield (%)	Bp (°C/torr) (Mp, °C)	d_4^{20}	n_D^{20}	References
1	2	3	4	5	6	7
9		52	126–130/5	1.0109	1.5550	[648,657]
10		51	104–106/4	1.0296	1.5164	[658]
11		99	(65–66)			[658,660]
12		71	170–175/2	0.9646	1.5084	[658]
13		52	112/3–4	1.0831	1.5446	[628]
14		71	139–141/5	1.1357	1.5870	[628,658]
15		88	142–145/3	0.9717	1.4936	[658]
16		51		0.9957	1.5242	[658]

(*Continued*)

TABLE 2.19 (*Continued*)
Reaction Products of N-Vinylpyrroles with the Participation of Vinyl Group

No.	Structure	Yield (%)	Bp (°C/torr) (Mp, °C)	d_4^{20}	n_D^{20}	References
1	2	3	4	5	6	7
17		77	203–210/1.5	0.9412	1.5917	[658]
18		41	176–178/3	1.0297	1.5189	[658]
19		42	(88–89)			[658]
20		60	98–116/5·10⁻²	1.0798	1.5540	[659]
21		57	124–130/ 4.6·10⁻² (31–34)	1.0588	1.5534	[659]
22		55	124–126/ 2.7·10⁻² (42–45)	1.1435	1.5615	[659]

(*Continued*)

TABLE 2.19 (*Continued*)

Reaction Products of N-Vinylpyrroles with the Participation of Vinyl Group

No.	Structure	Yield (%)	Bp (°C/torr) (Mp, °C)	d_4^{20}	n_D^{20}	References
1	2	3	4	5	6	7
23		42	123–125/ 4.7 · 10⁻² (57–61)	1.1545	1.5665	[659]
24		23	134–136/1	1.1084	1.5480	[659]
25		83	111–112 (1–2)	1.0402	1.5475	[661]
26		82	128–129 (1–2)	1.0313	1.5410	[661]
27		77	116–117 (1–2)	1.0166	1.5390	[661]
28		89	150–151 (1–2)	1.0336	1.5350	[661]
29		75	142–143 (1–2)	1.0321	1.5330	[661]
30		80	125–126 (1–2)	1.0632	1.5375	[661]

(Continued)

TABLE 2.19 (Continued)

Reaction Products of N-Vinylpyrroles with the Participation of Vinyl Group

No.	Structure	Yield (%)	Bp (°C/torr) (Mp, °C)	d_4^{20}	n_D^{20}	References
1	2	3	4	5	6	7
31	Me / SPr-n	82	128–129 (1–2)	0.9991	1.5340	[661]
32	Me / SPr-i	76	110–111 (1–2)	0.9844	1.5335	[661]
33	Me / SBu-n	80	139–140 (1–2)	1.0014	1.5290	[661]
34	Me / SBu-i	73	128–129 (1–2)	1.0128	1.5280	[661]
35	Me, Me / SBu-n	99	103–104 (1–2)	0.9645	1.5153	[7]
36	n-Pr, Me / SBu-n	87	141–142 (1–2)	0.9488	1.5100	[7]
37	n-Am, Me / SBu-n	81	138–139 (1–2)	0.9378	1.5048	[7]
38	Ph / SEt	93	144–145/1–2	1.0766	1.5955	[660,661]

(Continued)

TABLE 2.19 (*Continued*)

Reaction Products of N-Vinylpyrroles with the Participation of Vinyl Group

No.	Structure	Yield (%)	Bp (°C/torr) (Mp, °C)	d_4^{20}	n_D^{20}	References
1	2	3	4	5	6	7
39		94	158–160/1–2	1.0550	1.5826	[661]
40		70 50	166–167/1–2	1.0450	1.5757	[660,661]
41		84	125–130/2–3	1.0976	1.5313	[628]
42		78	155–156/1	1.1185	1.6111	[628]
43		95	186–188/1	1.0556	1.5752	[628]
44		88	175–179/1	1.0614	1.5892	[628]
45		80	167–169/1	1.0876	1.5950	[628]
46		67	177/1	1.4447	1.5440	[631,666, 667]

(Continued)

TABLE 2.19 (*Continued*)

Reaction Products of N-Vinylpyrroles with the Participation of Vinyl Group

No.	Structure	Yield (%)	Bp (°C/torr) (Mp, °C)	d_4^{20}	n_D^{20}	References
1	2	3	4	5	6	7
47		71	165–171/1	1.0699	1.5155	[666,667]
48		53	185/1	1.1688	1.5786	[666,667]
49		58	215–225/1	1.1077	1.5587	[666,667]
50		82	(72–73)			[631,666, 667]
51		79	(58–59)			[631,666, 667]
52		84	174–176/1	1.0961	1.5970	[668]
53		86	198–200/2	1.1571	1.5840	[668]

(*Continued*)

TABLE 2.19 (*Continued*)

Reaction Products of N-Vinylpyrroles with the Participation of Vinyl Group

No.	Structure	Yield (%)	Bp (°C/torr) (Mp, °C)	d_4^{20}	n_D^{20}	References
1	2	3	4	5	6	7
54		81	175–177/1	1.1533	1.6103	[668]
55		88	143–145/1.5	1.0752	1.5868	[668]
56		90	160–162/2	1.1069	1.5750	[668]
57		77	150–152/6	0.8461	1.4873	[671]
58		80	150–152/4	0.8840	1.4824	[671]
59		93	152–153/5	0.9588	1.5103	[671]
60		82	165–166/3.5	0.9418	1.5116	[671]

(Continued)

TABLE 2.19 (*Continued*)

Reaction Products of N-Vinylpyrroles with the Participation of Vinyl Group

No.	Structure	Yield (%)	Bp (°C/torr) (Mp, °C)	d_4^{20}	n_D^{20}	References
1	2	3	4	5	6	7
61	SiMe(Pr-*n*)$_2$	84	196–197/13	0.9264	1.5033	[671]
62	Si(Pr-*n*)$_3$	86	186–187/3	0.9321	1.5026	[671]
63	SiMe(Bu-*n*)$_2$	93	190–191/3	0.9243	1.4987	[671]
64	SiEt(OSiMe$_3$)$_2$	55	160–162/2	0.9622	1.4680	[671]
65	Ph SiMeEt$_2$	76	150–151/1	0.9644	1.5449	[671]
66	Ph SiEt$_3$	79	169–171/1	0.9656	1.5430	[671,672]
67	Ph Si(Pr-*n*)$_3$	68	186–188/4	0.9451	1.5310	[671]
68	Me Ph SiEt$_3$	79	180–182/2	0.9704	1.5365	[671,672]
69	Me Ph Si(Pr-*n*)$_3$	79	188–189/3	0.9499	1.5288	[671]

(*Continued*)

TABLE 2.19 (*Continued*)

Reaction Products of N-Vinylpyrroles with the Participation of Vinyl Group

No.	Structure	Yield (%)	Bp (°C/torr) (Mp, °C)	d_4^{20}	n_D^{20}	References
1	2	3	4	5	6	7
70		35	169–171/1	0.9567	1.5521	[672]
71		7	161/1 (44–46)			[672]
72		40	140–143/1	0.9531	1.5105	[672]
73		70	165–167/4–5	1.0232	1.5525	[672]
74		65	192–196/4	1.0030	1.5420	[672]
75		44	193–196/4	1.0690	1.5400	[672]
76		7	188–191/4 (32–33)			[672]
77		91	Oil			[674]

(*Continued*)

TABLE 2.19 (*Continued*)

Reaction Products of N-Vinylpyrroles with the Participation of Vinyl Group

No.	Structure	Yield (%)	Bp (°C/torr) (Mp, °C)	d_4^{20}	n_D^{20}	References
1	2	3	4	5	6	7
78		91	Oil			[674]
79		91	Oil			[674]
80		90	Oil			[674]
81		100	Oil			[674]
82		93	(30–31)			[674]
83		91	Oil			[677]
84		92	Oil			[677]

(*Continued*)

TABLE 2.19 (*Continued*)

Reaction Products of N-Vinylpyrroles with the Participation of Vinyl Group

No.	Structure	Yield (%)	Bp (°C/torr) (Mp, °C)	d_4^{20}	n_D^{20}	References
1	2	3	4	5	6	7
85		89	Oil			[677]
86		89	Oil			[677]
87		89	Oil			[677]
88		97	(186–188)			[677]
89		97	(110–111)			[677]
90		97	(180–181)			[677]
91		96	Oil			[677]

(Continued)

TABLE 2.19 (*Continued*)
Reaction Products of N-Vinylpyrroles with the Participation of Vinyl Group

No.	Structure	Yield (%)	Bp (°C/torr) (Mp, °C)	d_4^{20}	n_D^{20}	References
1	2	3	4	5	6	7
92		96	Oil			[677]
93		66	(189–190)			[678]
94		37	(142–143)			[678]
95		59	(130–131)			[678]
96		54	(119–121)			[678]
97		56	(146–148)			[678]
98		42	(172–174)			[678]
99		65				[680]

(*Continued*)

TABLE 2.19 (*Continued*)

Reaction Products of N-Vinylpyrroles with the Participation of Vinyl Group

No.	Structure	Yield (%)	Bp (°C/torr) (Mp, °C)	d_4^{20}	n_D^{20}	References
1	2	3	4	5	6	7
100		44	Oil			[680]
101		85				[680]
102		63				[681]
103		77				[681]
104		75	Oil			[680]
105		41	Oil			[680]
106		50	Oil			[680]

(*Continued*)

TABLE 2.19 (*Continued*)

Reaction Products of N-Vinylpyrroles with the Participation of Vinyl Group

No.	Structure	Yield (%)	Bp (°C/torr) (Mp, °C)	d_4^{20}	n_D^{20}	References
1	2	3	4	5	6	7
107		68	Oil			[680]
108		91	(38–39)			[681]
109		85	(81)			[681]
110		94	(75)			[681]
111		95	(128–129)			[681]
112		100	(175)			[681]
113		9	(116–117)			[679]

(*Continued*)

TABLE 2.19 (*Continued*)

Reaction Products of N-Vinylpyrroles with the Participation of Vinyl Group

No.	Structure	Yield (%)	Bp (°C/torr) (Mp, °C)	d_4^{20}	n_D^{20}	References
1	2	3	4	5	6	7
114	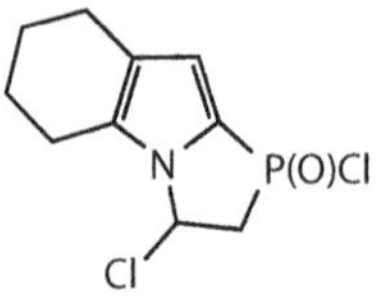	21	(93–94)			[679]

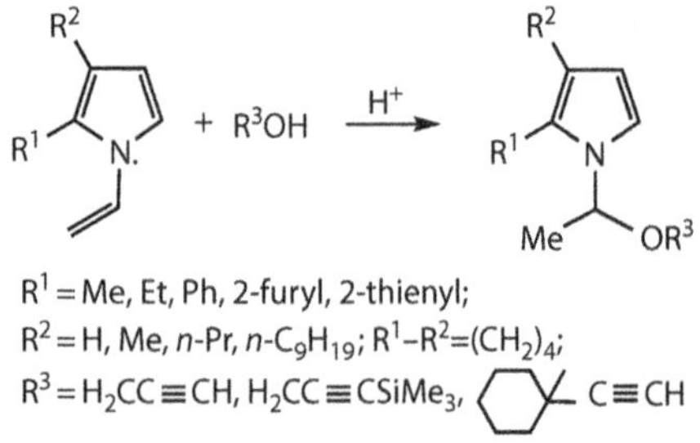

SCHEME 2.195   Electrophilic addition of acetylenic alcohols to N-vinylpyrroles in the presence of perfluorobutyric acid.

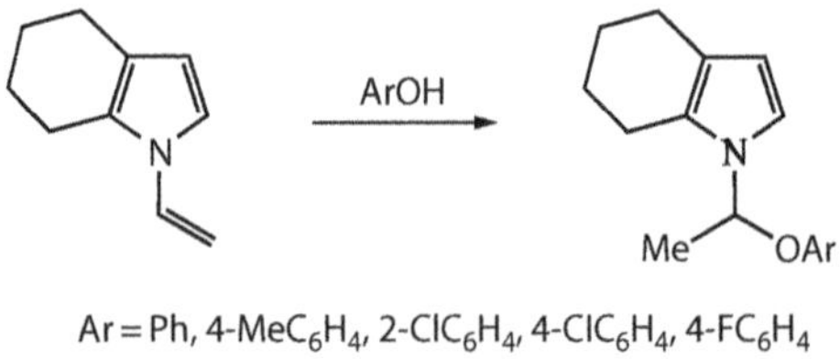

Ar = Ph, 4-MeC₆H₄, 2-ClC₆H₄, 4-ClC₆H₄, 4-FC₆H₄

SCHEME 2.196   Noncatalytic electrophilic addition of phenols to N-vinyl-4,5,6,7-tetrahydroindole.

In the presence of the system AIBN/CCl₄, alcohols add to N-vinylcarbazole to furnish also α-adducts (Scheme 2.197) [653]. This system turns out to be effective for the addition of alcohols to 2- and 2,3-substituted N-vinylpyrroles [660].

At 75°C–80°C, the reaction is completed for 2.5–3 h. Yields of N-(α-alkoxyethyl)pyrroles range from 50% to 70% (Table 2.19).

In the absence of CCl₄, AIBN fails to catalyze the addition of alcohol to N-vinylcarbazole [653] and 2-phenyl-N-vinylpyrrole [660]. Under the conditions studied (75°C–80°C), no catalytic effect is exerted by CCl₄ or AIBN/CHCl₃ system. Thus, the reaction remains electrophilic in nature and is catalyzed by traces of HCl

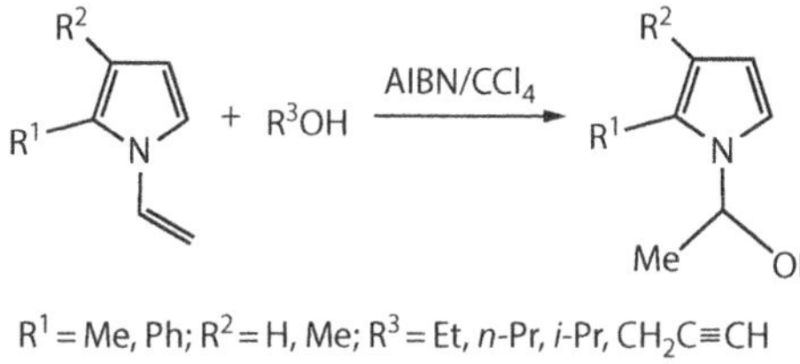

R^1 = Me, Ph; R^2 = H, Me; R^3 = Et, n-Pr, i-Pr, $CH_2C\equiv CH$

SCHEME 2.197 Electrophilic addition of alcohols to N-vinylpyrroles in the presence of AIBN/CCl$_4$ system.

and Cl$_2$, the products of transformations with participation of Cl radical. The latter is generated via AIBN decomposition in CCl$_4$ medium [653].

2.2.3 ADDITION OF THIOLS

The addition of alkanethiols to N-vinylpyrroles with and without free radical initiators has been investigated [661]. The literature contains rather contradictory data on the order of addition of hydrogen sulfide and alkanethiols to the N-vinyl group. It is reported that cyclic enamines [662,663] add alkanethiols in the presence of radical initiators as well as under noncatalytic conditions according to the Markovnikov rule. At the same time, hydrogen sulfide and thiols react with some N-vinyl derivatives containing the pyrrole ring in the anti-Markovnikov manner [93,96,107,109]. For instance, β-addition products (the adducts of thiyl radical to the vinyl group β-carbon) are obtained [664] when hydrogen sulfide is passed through a heated alcoholic alkali solution of N-vinylcarbazole and also when heated N-vinylindole containing potassium metal is treated with hydrogen sulfide. Refluxing N-vinylcarbazole with thiocresol and 2-mercaptoanthraquinone without a catalyst also leads to β-adducts [107,664]. The free radical addition of thiols to N-vinylindole and N-vinylimidazoles under the influence of AIBN proceeds with the formation of a mixture containing α- and β-addition products [665]. According to this work, thiols do not add to N-vinylindole in the absence of initiators.

The paper [661] was aimed at the development of a convenient method for the synthesis of a new group of sulfur-containing pyrroles by the addition of alkanethiols to N-vinylpyrroles as well as at clarification of structural orientation of this reaction. It is established that heating N-vinylpyrroles with alkanethiols both in the presence and absence of a radical initiator produces the β-adducts, N-(β-alkylthioethyl)pyrroles (Scheme 2.198).

Addition proceeds readily at 70°C–80°C and is complete in 18–25 h. When AIBN is used, N-(β-alkylthioethyl)pyrroles are obtained in 70%–94% preparative yields (Table 2.19). Under comparable conditions without an initiator, the yields of adducts drop to 20%–30%.

Analogously, the addition of mercaptoacetic acid O-methyl ester to N-vinylpyrroles delivers β-adducts (53%–71% yields, Scheme 2.199, Table 2.19). The latter react with aqueous ammonia (45°C–50°C, 2 h) and hydrazine hydrate (80°C, 1.5 h) to give the corresponding amides and hydrazides [666].

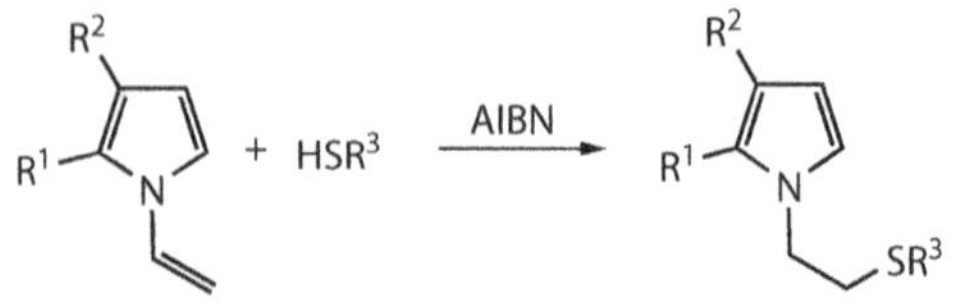

R^1 = Me, Ph; R^2 = H, Me, n-Pr, n-Am; R^1–R^2 = (CH$_2$)$_4$, (CH$_2$)$_3$(CH$_3$)CH;
R^3 = Et, n-Pr, i-Pr, n-Bu, i-Bu

SCHEME 2.198   Synthesis of N-(β-alkylthioethyl)pyrroles by free-radical addition of thiols to N-vinylpyrroles.

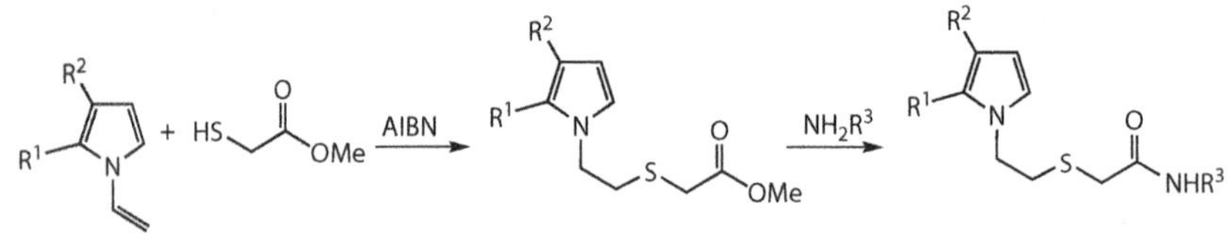

R^1 = Me, Ph; R^2 = H, n-Pr, n-Bu; R^1–R^2 = (CH$_2$)$_4$; R^3 = H, NH$_2$

SCHEME 2.199   Free-radical addition of mercaptoacetic acid O-methyl ester to N-vinylpyrroles.

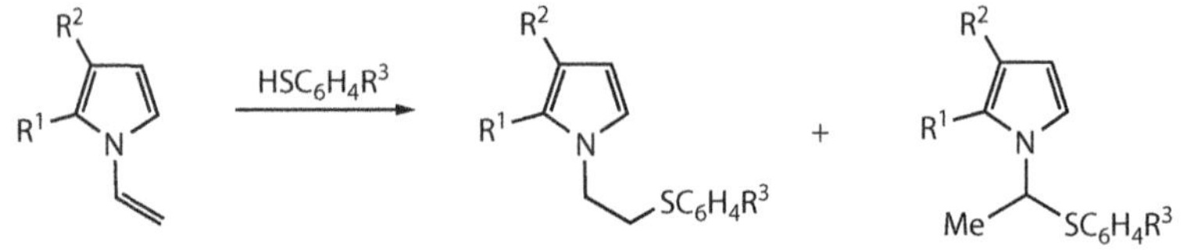

R^1 = Ph; R^2 = H; R^1–R^2 = (CH$_2$)$_4$; R^3 = H, F

SCHEME 2.200   Addition of thiophenols to N-vinylpyrroles.

The system AIBN/CCl$_4$ promotes the radical addition of thiols to N-vinylpyrroles giving rise to β-adducts. For example, from 2-phenyl-N-vinylpyrrole and ethane- and butanethiols, the corresponding N-(β-alkoxyethyl)-2-phenylpyrroles are synthesized in 54% and 50% yields, respectively [660].

N-Vinylpyrroles selectively add thiophenols under the conditions of free radical initiation to form β-adducts, N-(2-arylthioethyl)pyrroles (Scheme 2.200, Table 2.19) [668]. In analogous conditions, the reaction without an initiator leads to a mixture of β- (20%) and a-adduct (80%). Thiylation of a mixture of N-vinyl-4,5,6,7-tetrahydroindole (28.5%), 4,5,6,7-tetrahydroindole (61%), and cyclohexanone oxime both with and without the initiator selectively affords the α-adducts only. Probably, 4,5,6,7-tetrahydroindole and cyclohexanone oxime inhibit the radical addition thus hindering the β-adducts formation, Hence, owing to their increased acidity, thiophenols show a marked tendency to electrophilic addition to produce α-adducts [668].

Thus, diverse N-α- and N-(β-alkyl(aryl)thioethyl)pyrroles can be readily synthesized by the reaction of N-vinylpyrroles with thiols. In most cases, the reactions proceed almost quantitatively, and their orientation can be controlled by the initiation conditions. This opens up interesting prospects of N-vinylpyrroles application for "grafting" the pyrrole rings to enzymes and biologically important amino acids containing SH groups with the purpose of their property modification.

2.2.4 Hydrosilylation

Silicon-containing nitrogen heterocycles are often biologically active and are extensively explored as starting materials for the synthesis of drugs [669]. In this line, the reaction of N-vinylpyrroles hydrosilylation provokes a certain interest. It is known [670] that N-[γ-(trialkoxysilyl)propyl]pyrroles are employed to improve the adhesion of polypyrrole films to *n*-type silicon photoanodes. Such pyrroles are obtained by multistage synthesis (Scheme 2.201).

It is conceivable that the same properties would be found in N-[β-(trialkoxysilyl)-ethyl]pyrroles, which could be prepared by direct addition of trialkoxysilanes to N-vinylpyrroles.

Hydrosilylation of a series of N-vinylpyrroles with trialkylsilanes has been accomplished using $H_2PtCl_6 \cdot 6H_2O$ (0.1 M THF solution) as catalyst (Scheme 2.202) [671].

The reaction is carried out at 80°C–100°C with equimolar ratio of the reactants for 4–10 h, yields of the products being 55%–93% (Table 2.19). High yields are favored by the prolonged reaction time and application of aprotic solvents (tetrahydrofuran, acetone, dioxane).

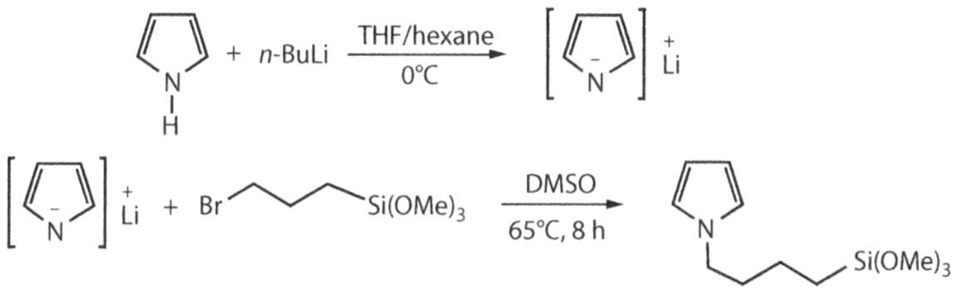
+ n-BuLi
THF/hexane
0°C
Li
+ Br
Si(OMe)3
DMSO
65°C, 8 h
Si(OMe)3

SCHEME 2.201 Step-wise synthesis of N-[γ-(trialkoxysilyl)propyl]pyrroles.

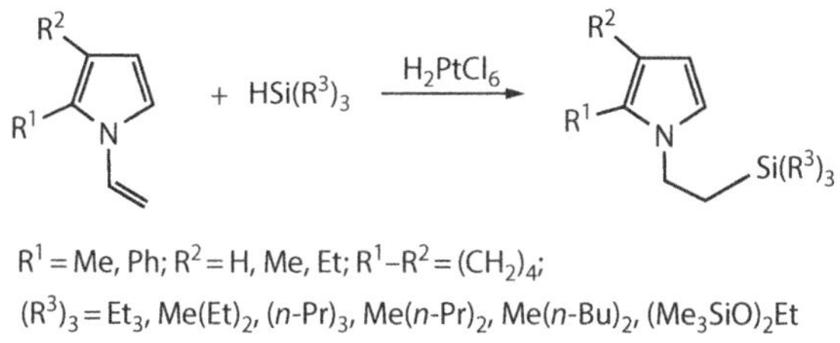
R2
R1
+ HSi(R3)3
H2PtCl6
Si(R3)3
R1 = Me, Ph; R2 = H, Me, Et; R1–R2 = (CH2)4;
(R3)3 = Et3, Me(Et)2, (n-Pr)3, Me(n-Pr)2, Me(n-Bu)2, (Me3SiO)2Et

SCHEME 2.202 Hydrosilylation of N-vinylpyrroles with trialkylsilanes in the presence of H_2PtCl_6.

The yields of adducts depend essentially on the silane structure. For example, an attempt [672] to add trichloro-, alkyldichloro-, and triethoxysilanes to N-vinylpyrroles (H_2PtCl_6) was unsuccessful. It is interesting that the starting N-vinylpyrroles are completely recovered from the reaction, although even a trace of trimethylchlorosilanes readily converts the vinylpyrroles into their dimers and polymers [673]. Only in the reaction of methyldichlorosilane with 2-(2-thienyl)-N-vinylpyrroles, the expected adduct is detected (NMR) in the reaction mixture in insignificant amount (7%) [672].

The rhodium complex $(Ph_3P)RhCl$, which catalyzes the addition of triethylsilane to N-vinylpyrroles (50%–60% yields of adducts) [672], also appears to be inactive with trichloro-, alkyldichloro-, and trialkoxysilanes. No products of N-vinylpyrroles hydrosilylation with the aforementioned silanes are formed when the addition is initiated by UV irradiation [672].

The addition of silanes to the N-vinyl group proceeds regiospecifically to give β-adducts.

The addition of triethylsilane to N-vinylpyrroles bearing aromatic and heteroaromatic substituents in the presence of 0.05 M solution of $H_2PtCl_6 \cdot 6H_2O$ or rhodium complex $(Ph_3P)_3RhCl$ has been studied (Scheme 2.203) [672]. The reaction is carried out in the sealed ampoules at 140°C for 5 h.

Under these conditions, conversion of N-vinylpyrroles is 50%–60%.

Since the catalytic activity of platinum and rhodium complexes usually significantly depends upon the presence of π- and n-donors (potential ligands) in a solution, the assessment of a structure of the starting N-vinylpyrroles effect on the yields of hydrosilylation products in this case is difficult because the reagents themselves contain the groups capable of complex formation and ligand exchange with the catalysts. In other words, the substituents in this series of N-vinylpyrroles should influence the reaction not via the redistribution of electron density in molecules and alteration of their geometry, but through the impact on the catalyzing complexes, that is, changing their ligand environment and, hence, conditions of the reagents coordination. Noteworthy is the effect of the thiophene ring that clearly favors the reaction (yields

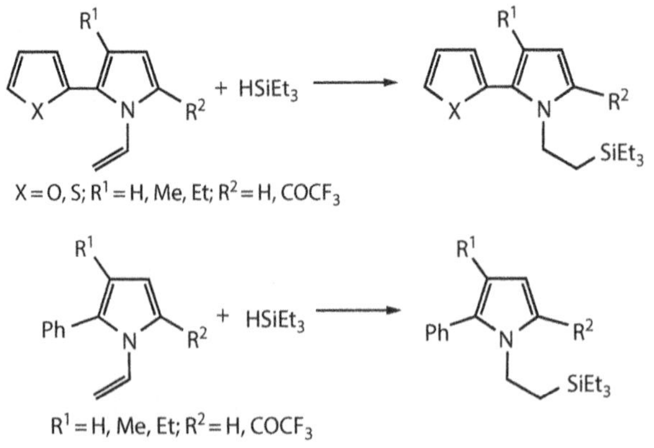

$X = O, S; R^1 = H, Me, Et; R^2 = H, COCF_3$

$R^1 = H, Me, Et; R^2 = H, COCF_3$

SCHEME 2.203 Hydrosilylation of 2-phenyl- and 2-hetaryl-N-vinylpyrroles with triethylsilane in the presence of $H_2PtCl_6 \cdot$ and $Ph_3P)_3RhCl$.

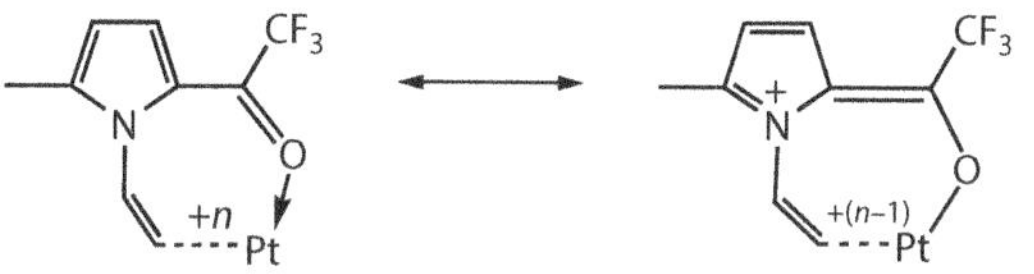

SCHEME 2.204 Chelating of Pt cations by N-vinyl-2-trifluoroacetylpyrroles.

of the products are 65%–70%). Obviously, the thiophene substituent, participating in coordination with platinum cations, forms more active catalytic forms than in the case of other substituents. This is supported by logical drop of the adducts yield (from 70% to 44%) with the increase in volume of the pyrrole substituent R^1 (from H to Et) interfering the coordination. Drastic fall of the products yield (7%) when $R^2 = COCF_3$ is mainly attributable to binding of platinum ions into strong chelates of the following type (Scheme 2.204):

Other sequence of the [β-(trialkylsilyl)]-2-trifluoroacetylpyrroles synthesis seems to be more rational: first hydrosilanes are added and then the adducts obtained are trifluoroacetylated.

2.2.5 ADDITION OF SECONDARY PHOSPHINES

Fundamentally, new methods for the synthesis of phosphorus-containing pyrroles from easily available N-vinylpyrroles are developed. Among them is hydrophosphorylation of the vinyl group with secondary phosphines [674], which are now obtained in a one-stage manner from elemental phosphorus and electrophiles in the superbase systems [675,676].

Under radical initiation (UV irradiation, AIBN), N-vinylpyrroles readily and regioselectively (against the Markovnikov rule) add secondary phosphines to afford β-adducts in almost quantitative yields (88%–91%, Scheme 2.205, Table 2.19). Photochemical initiation turns out to be more effective: UV-promoted reaction proceeds by approximately 20 times faster than in the presence of AIBN.

The addition of secondary phosphines to N-isopropenylpyrroles (AIBN, 65°C) occurs also readily and regiospecifically to deliver the β-adducts only in 89%–92% yields (Scheme 2.206, Table 2.19) [677].

The phosphines thus formed are easily oxidized on air to give phosphine oxides, whereas the reaction with methyl iodide leads to the corresponding phosphonium salts (Scheme 2.207).

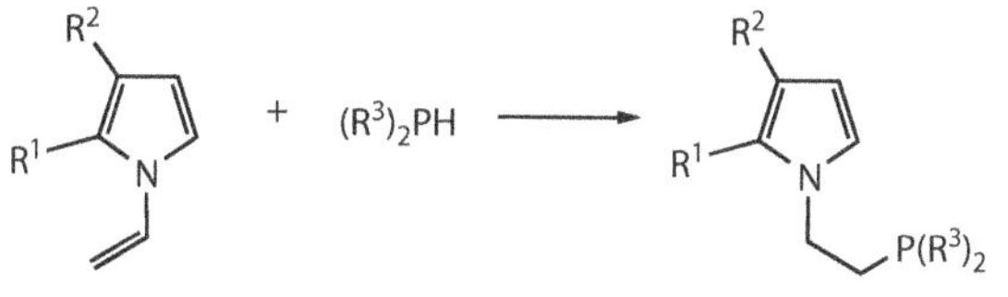

$R^1 = n$-Pr; $R_2 = $ Et; R^1–$R^2 = (CH_2)_4$; $R^3 = n$-Bu, Ph(CH$_2$)$_2$, 2-Py(CH$_2$)$_2$

SCHEME 2.205 Free-radical addition of secondary phosphines to N-vinylpyrroles.

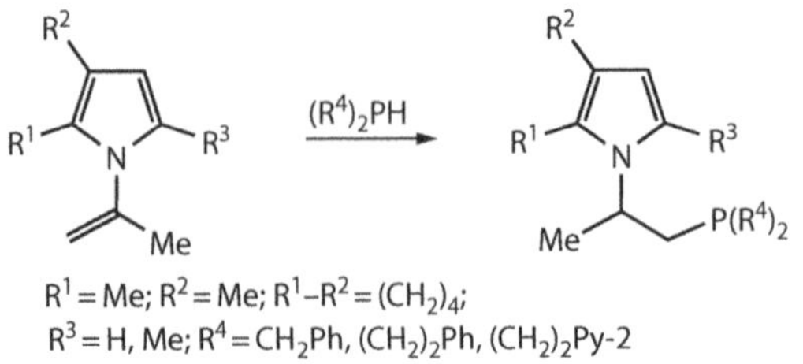

$R^1 = Me; R^2 = Me; R^1–R^2 = (CH_2)_4;$
$R^3 = H, Me; R^4 = CH_2Ph, (CH_2)_2Ph, (CH_2)_2Py-2$

SCHEME 2.206   Free-radical addition of secondary phosphines to N-isopropenylpyrroles.

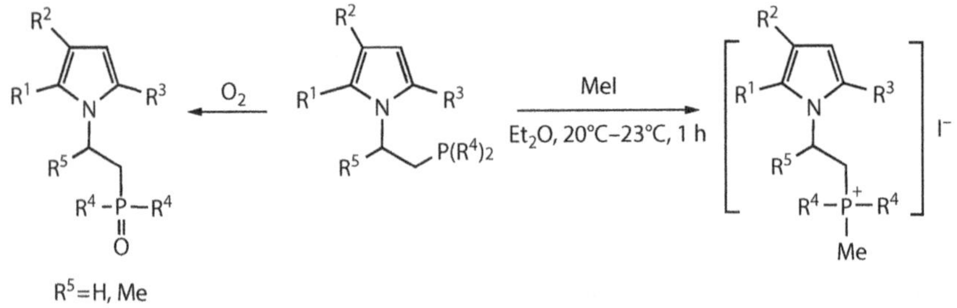

$R^5 = H, Me$

SCHEME 2.207   Formation of phosphine oxides and phosphonium salts from the adducts of secondary phosphines to N-vinylpyrroles.

2.2.6   REACTIONS WITH HALOPHOSPHINES

Phosphorylation of N-vinylpyrroles with phenyldihalophosphines occurs both at the vinyl group and pyrrole ring to furnish unstable phosphine **41**, isolated and characterized (Scheme 2.208, Table 2.19) as phosphine oxide or phosphine sulfide (after treatment with hydrogen peroxide or sulfur, respectively) [678].

2-*Tert*-butyl-N-vinylpyrrole reacts with a mixture of trichloro- and tribromophosphines to produce chlorophosphine, which upon contacting with morpholine is transformed into the corresponding amide. Both phosphines are unstable. Only phosphine sulfide **42** can be isolated as individual compound after treatment of amide with sulfur (Scheme 2.209, Table 2.19) [678].

The reaction of equimolar amounts of N-vinyl-4,5,6,7-tetrahydroindole with PCl_3 can be stopped at the stage of the formation of phosphine **43**. The latter undergoes

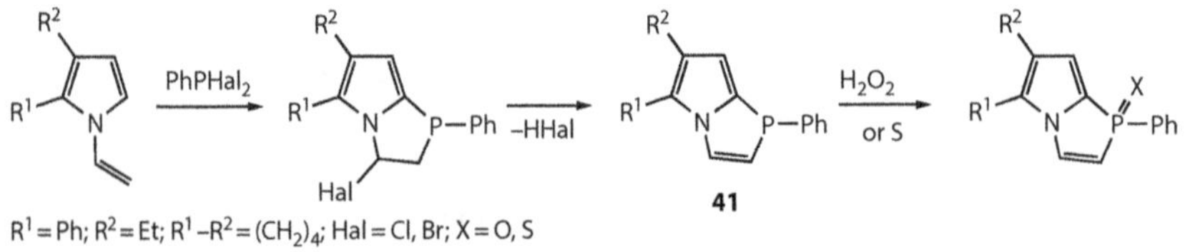

$R^1 = Ph; R^2 = Et; R^1–R^2 = (CH_2)_4; Hal = Cl, Br; X = O, S$

SCHEME 2.208   Phosphorylation of N-vinylpyrroles with phenyldihalophosphines.

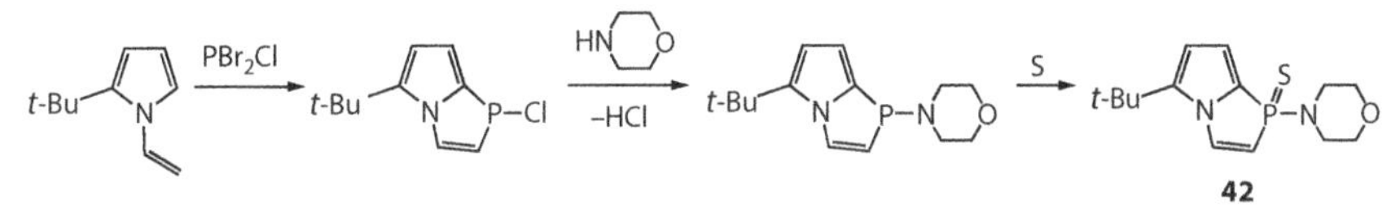

SCHEME 2.209 Phosphorylation of 2-*tert*-butyl-N-vinylpyrrole with the mixture of PCl$_3$ and PBr$_3$.

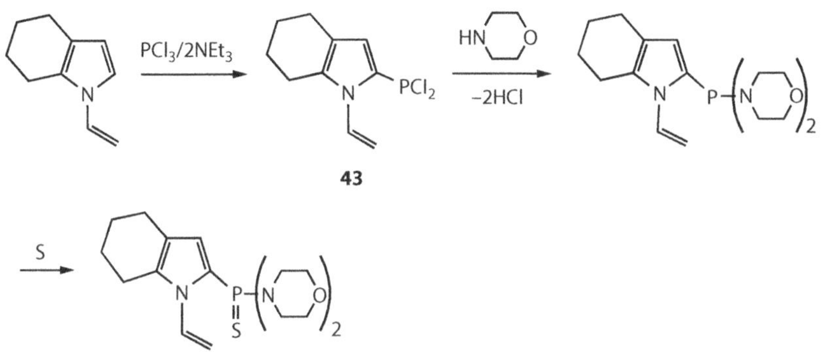

SCHEME 2.210 Phosphorylation of N-vinyl-4,5,6,7-tetrahydroindole with PCl$_3$ in the presence of Et$_3$N and transformations of the product.

some transformations (the reactions with morpholine and sulfur) to yield thiophosphonate **44** (Scheme 2.210, Table 2.19).

Further, compound **44** is treated in pyridine with equimolar amount of phosphorus tribromide to give dibromophosphine **45**, identified (after reactions with morpholine and sulfur) as thiophosphonate **46** (Scheme 2.211, Table 2.19) [678].

2.2.7 REACTIONS WITH PHOSPHORUS PENTACHLORIDE

N-Vinylpyrroles are phosphorylated (benzene, 10°C–15°C) with phosphorus pentachloride, a typical electrophilic reagent, mainly at two nucleophilic centers (α-position of the pyrrole ring and β-position of the vinyl group) to form pyrrolo-trichloro-azaphospholanium hexachlorophosphates **47** (Scheme 2.212, Table 2.19) [679]. In some cases, the products of phosphorylation across the vinyl group only, N-vinylpyrroles **48** (Table 2.19) have been isolated. When treated with sulfur dioxide, hexachlorophosphates **47** and **48** are transformed into chloroanhydrides **49** and **50**, respectively (Table 2.19).

Chloroanhydrides **49** eliminate hydrogen chloride in the presence of triethylamine to form 1-chloro-1-oxo-1*H*-pyrrolo[1,2-*a*][1,3]azaphospholes (Scheme 2.213, Table 2.19) [679].

N-Vinylpyrrole-2-carbaldehydes also react with phosphorus pentachloride by two nucleophilic centers, that is, the N-vinyl group and aldehyde function [680], whereas the third possible object of the electrophilic attack, the pyrrole ring, remains intact.

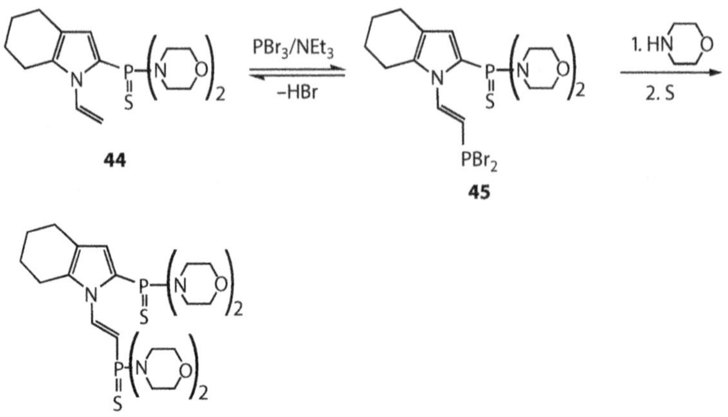

44 45 46

SCHEME 2.211 Phosphorylation of N-vinyl group of thiophosphonate **44** and transformations of the product.

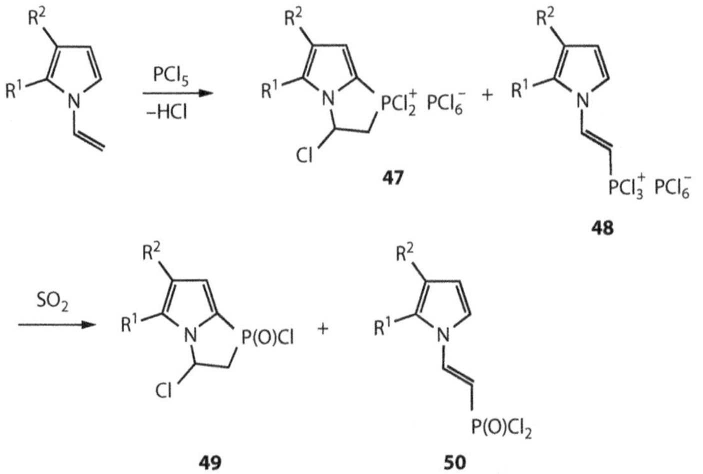

47 48 49 50

R^1 = Me, Ph; R^2 = H, Ph; R^1–R^2 = $(CH_2)_4$

SCHEME 2.212 Phosphorylation of N-vinylpyrroles with PCl_5.

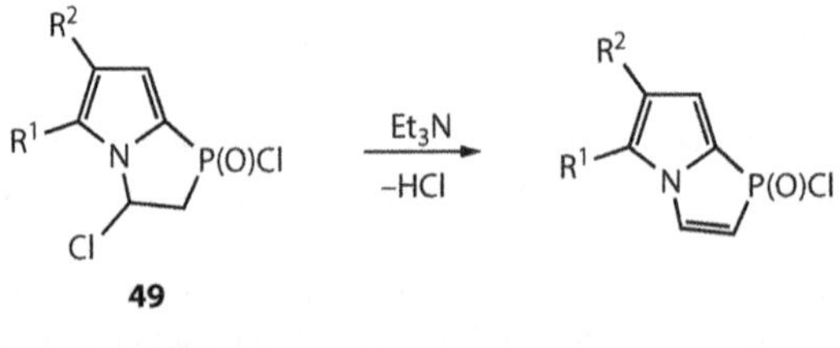

49

SCHEME 2.213 Elimination of HCl from chloroanhydrides **49**.

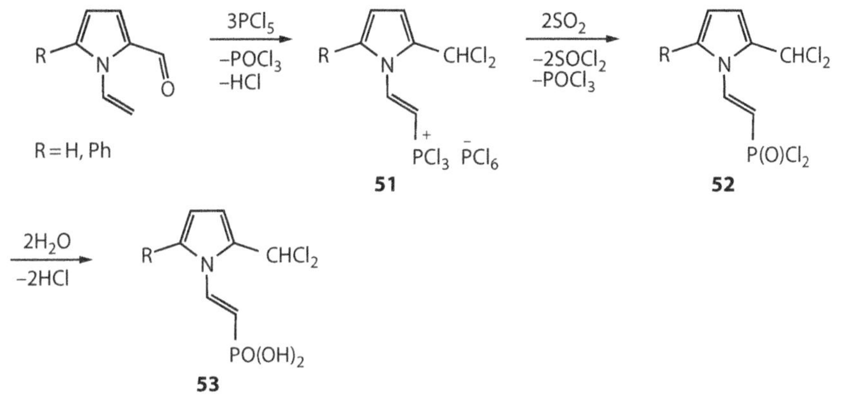

SCHEME 2.214 Reaction of N-vinylpyrrole-2-carbaldehydes with PCl$_5$.

The N-vinyl moiety tolerates the stereospecific reaction of electrophilic substitution to exclusively yield hexachlorophosphate **51** of *E*-configurations (Table 2.19). In parallel, the aldehyde function exchanges oxygen for two chlorine atoms giving rise to the dichloromethyl radical (up to 85% yield). Hexachlorophosphates **51** under the action of sulfur dioxide are converted to chloroanhydrides **52** (75% yield). The latter then undergo hydrolysis to afford the corresponding 2-(α-dichloromethylpyrrolyl) vinylphosphonic acids **53** (Scheme 2.214, Table 2.19).

Similar reactions have been carried out also with N-vinyl-4,5-dihydrobenz[*g*] indole-2-carbaldehyde (Scheme 2.215, Table 2.19) [686].

In the case of unsubstituted N-vinylpyrrole-2-carbaldehyde, a consecutive reaction of intramolecular electrophilic substitution at the α-position of the pyrrole ring involving the P-Cl bond is implemented to form a bicycle, pyrrolo[1,2][1,3]azaphosphole derivative (Scheme 2.216, Table 2.19).

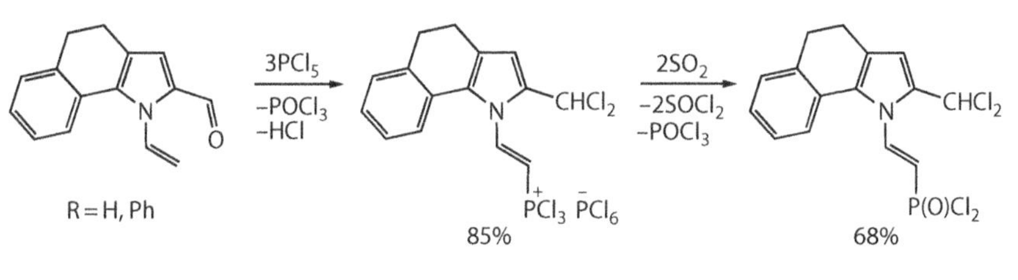

SCHEME 2.215 Phosphorylation of N-vinyl-4,5-dihydrobenz[*g*]indole-2-carbaldehyde with PCl$_5$.

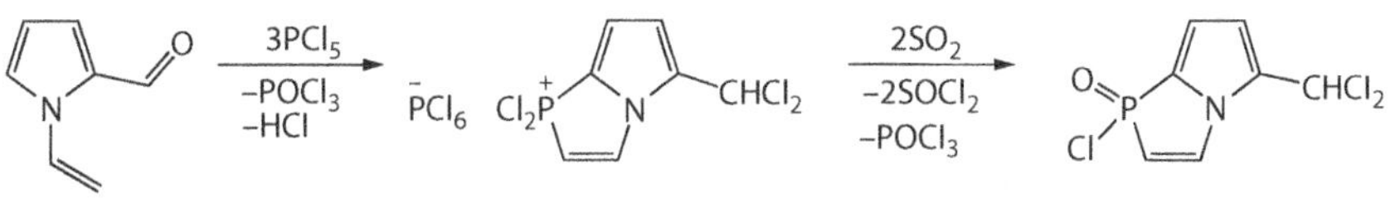

SCHEME 2.216 Phosphorylation of N-vinylpyrrole-2-carbaldehyde with PCl$_5$.

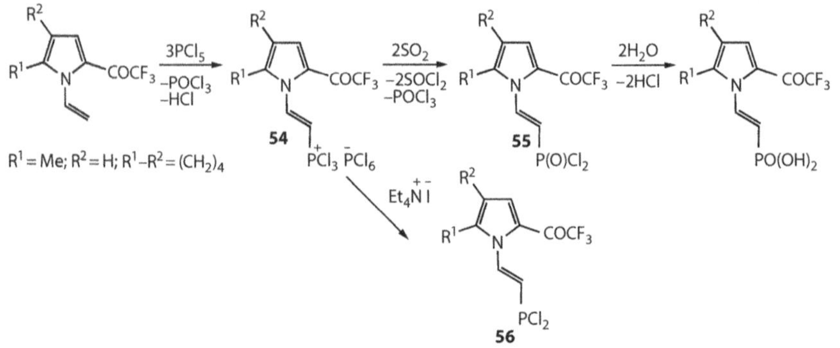

SCHEME 2.217 Phosphorylation of 2-trifluoroacetyl-N-vinylpyrroles with PCl$_5$ and further transformations of hexachlorophosphate **54**.

Thus, under certain conditions, the electrophilic attack of phosphorus pentachloride on N-vinylpyrrole-2-carbaldehyde may proceed involving the third center (the pyrrole ring) [680].

5-Methyl-2-trifluoroacetyl-N-vinyl-pyrroles and 2-trifluoroacetyl-N-vinyl-4,5,6,7-tetrahydroindole are phosphorylated with phosphorus pentachloride selectively at the vinyl group to furnish hexachlorophosphates **54** in high yields, which are further converted under the action of SO$_2$ into phosphonic acid dichloroanhydrides **55** (Scheme 2.217, Table 2.19) [681].

These compounds (stable upon storage) are easily hydrolyzed when treated with water (room temperature, 12 h) into the corresponding phosphonic acids. The reduction of complexes **54** by tetraethylammonium iodide results in the formation of dichlorophosphines **56** (Scheme 2.217, Table 2.19) [681].

To summarize the data discussed in two previous sections, one can ascertain that N-vinylpyrroles represent perspective starting materials for the synthesis of diverse phosphorus-containing compounds of the pyrrole series. They allow synthesizing, depending on the reactant nature, the pyrroles with phosphorus-containing substituents both in the ring and at the vinyl group, as well as products of simultaneous participation of these two nucleophilic centers with phosphorus halides.

2.2.8 CATALYTIC ARYLATION OF THE VINYL GROUP (HECK REACTION)

Direct arylation of N-vinylpyrroles with aryl iodides in the presence of PdCl$_2$ (Heck reaction) leads to hitherto inaccessible N-(α-arylvinyl)pyrroles and pyrrole congeners of stilbenes (Scheme 2.218) [682–684].

The arylation of 2- and 2,3-disubstituted N-vinylpyrroles with various aryl iodides is characterized by high degree of regioselectivity (substitution of α-hydrogen of the vinyl group) and good yields of N-(α-arylethenyl)pyrroles. In the case of unsubstituted N-vinylpyrrole, the reaction proceeds with low regioselectivity to predominantly afford the β-products (α:β = 30:70).

The values of effective charges on α- and β-carbon atoms and character of localization of the vinyl group HOMO and lowest unoccupied molecular orbital (LUMO)

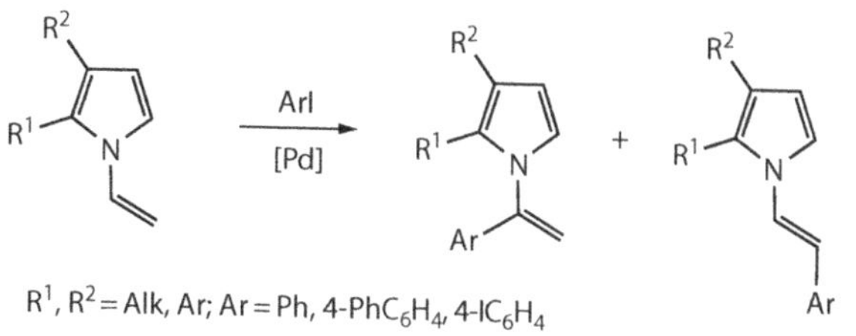

$R^1, R^2 = Alk, Ar; Ar = Ph, 4-PhC_6H_4, 4-IC_6H_4$

SCHEME 2.218 Arylation of N-vinylpyrroles via the Heck reaction.

suggest that neither charge nor orbital control can be considered as reasons of the regioselectivity observed. Quantum-chemical and molecular-mechanical calculations indicate the presence of steric hindrances toward β-arylation, that is, the substituent at the α-position of the pyrrole ring has a determining effect on the reaction regioselectivity.

2.2.9 METALATION OF N-VINYLPYRROLES AND THEIR ANALOGS

2.2.9.1 Metalation of N-Vinylpyrrole

The superbases, n-BuLi, n-BuLi/t-BuOK, and BuLi/tetramethyldiaminoethane, nonselectively deprotonate N-vinylpyrroles (presumably under kinetic conditions) at the α-positions of the vinyl group and the pyrrole ring, carbanions ratio being ~1:1 [685]. Consequently, the treatment of such mixtures with electrophiles (Me$_2$S$_2$, Me$_3$SiCl, pivalaldehyde) leads to the α-substituted products both in the vinyl group and the pyrrole ring, which are also formed in approximately equal ratio (Scheme 2.219, Table 2.20). When diisopropylamine (7–10 mol%) is added to the metalating system n-BuLi/t-BuOK, the reacting electrophiles regioselectively fix a carbanion only with deprotonated α-position of the pyrrole, that is, complete transformation of carbanion with deprotonated α-position of the vinyl group to the α-pyrrole carbanion (a thermodynamic product) takes place.

If n-BuBr is used as an electrophile, the reaction affords the products of double alkylation, simultaneously at the vinyl group and the pyrrole ring.

The double alkylation can be circumvented by adding twofold molar excess of LiBr to the metalating system. Apparently, this shifts equilibrium between lithium and potassium derivatives toward the former, which possess a weaker reactivity, and, therefore, ensures higher selectivity relative to the electrophile attack. Under these conditions, the following electrophiles have been involved into the reactions: methyl

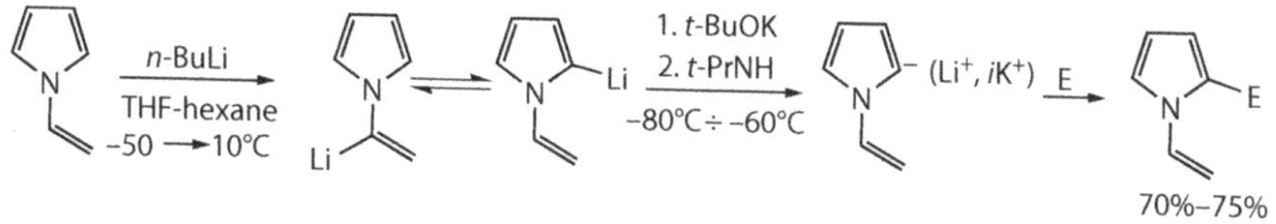

70%–75%

SCHEME 2.219 Metalation of N-vinylpyrrole with the n-BuLi/t-BuOK system.

TABLE 2.20
Metalated N-Vinyl-, N-Allenyl-, and N-Ethynylpyrroles
Derivatives

No.	Structure	Yield (%)	Bp (°C/torr) (Mp, °C)	n_D^{20}	References
1	MeS	70	80/15	1.5766	[685]
2	Me₃Si	75	79/15	1.5085	[685]
3	n-Bu	72	93/15	1.5041	[685]
4	I	90			[685]
5	O	74	85/15	1.5862	[685]
6	HO	65	110/15	1.5525	[685]
7	HO	69	109/15	1.5360	[685]
8	Me, HO	73	130/15	1.5435	[685]
9	Me Me—Me HO	71	120/15	1.5185	[685]
10	Me, Me HO	88	(73–74)		[685]

(Continued)

TABLE 2.20 (*Continued*)
Metalated N-Vinyl-, N-Allenyl-, and N-Ethynylpyrroles Derivatives

No.	Structure	Yield (%)	Bp (°C/torr) (Mp, °C)	n_D^{20}	References
11	Ph, HO	80	(70–71)		[685]
12	HO, cyclohexyl	80	(74–75)		[685]
13	Me₃Si	87	83/1	1.5237	[686,687]
14	Me₃Si, SiMe₃, SiMe₃	16		1.5077	[687]
15	n-Bu	86	57–61/1	1.5247	[686,687]
16	MeS	60			[686,687]
17	Me, Me, OH	88	(52–53)		[687]
18	HO, cyclohexyl	76	(59–60)		[686,687]

(Continued)

TABLE 2.20 (*Continued*)
Metalated N-Vinyl-, N-Allenyl-, and N-Ethynylpyrroles
Derivatives

No.	Structure	Yield (%)	Bp (°C/torr) (Mp, °C)	n_D^{20}	References
19		70			[687]
20		78	114/15	1.4930	[688]
21		62	78/15	1.4960	[688]
22		33			[688]
23		49	(26–27)		[688]
24		52	(49–50)		[688]
25		45	(43–44)		[688]

iodide, butyl bromide, dimethyl disulfide, trimethylchlorosilane, oxirane, paraform, acetaldehyde, pivalaldehyde, benzaldehyde, acetone, cyclohexanone, dimethylformamide, and iodine. As a result, the corresponding 2-functionalized pyrroles have been synthesized (Scheme 2.220, Table 2.20) [685].

2.2.9.2 Metalation of N-Allenylpyrrole

Unlike N-vinylpyrrole, N-allenylpyrrole is selectively metalated with one equivalent of *n*-BuLi (THF/hexane, 10°C, 40 min) at the α-position of the allene

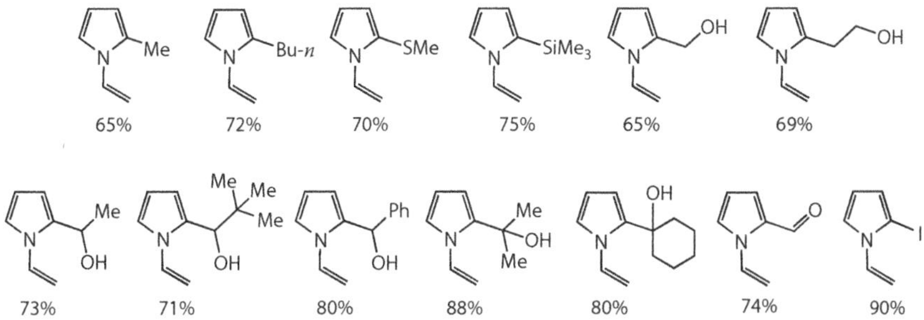

SCHEME 2.220 2-Functionalized N-vinylpyrroles synthesized via metalation of N-vinylpyrroles and subsequent reactions of α-pyrrole carbanion with electrophilies.

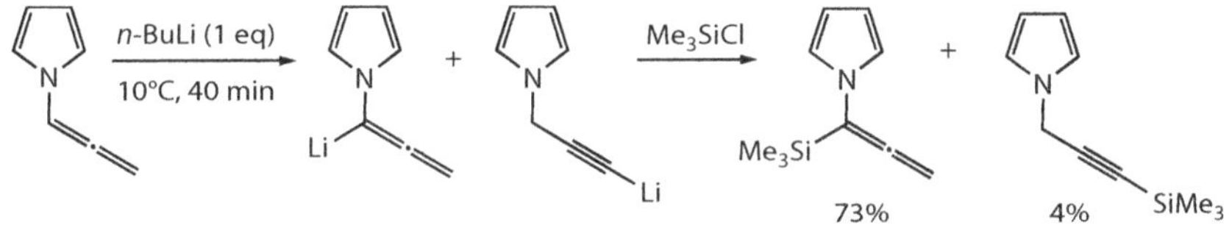

SCHEME 2.221 Lithiation of N-allenylpyrrole with n-BuLi.

group [686,687]. The treatment of the metalation product with trimethylchlorosilane affords N-(1-trimethylsilylpropadienyl)pyrrole (83%, Scheme 2.221, Table 2.20). The isomer of the latter, N-trimethylsilylpropargylpyrrole, is also formed in small amount (4%). Hence, it follows that the monodeprotonated N-allenylpyrrole contains minor (~4%) quantities of isomeric propargylic carbanions.

When 3.5-fold excess of N-butyllithium is employed, the double and exhaustive metalation of the allene group takes place, the pyrrole ring remaining intact. In this case, the treatment of the reaction mixture with trimethylchlorosilane gives a mixture of all four possible products (the content of these compounds is shown in Scheme 2.222). The product of the exhaustive silylation of the allene group has been isolated from this mixture in 13% yield.

Apart from lithium derivatives of N-allenylpyrrole, potassium, copper (I), and zinc derivatives have been tested in the reaction with trimethylchlorosilane. Copper and zinc compounds are easily prepared from organolithium ones by the exchange with the corresponding salts in THF. Potassium derivatives (apparently,

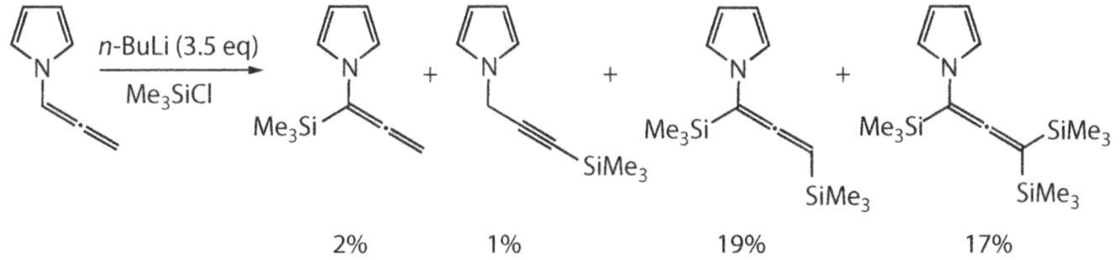

SCHEME 2.222 Lithiation of N-allenylpyrrole with excess of n-BuLi.

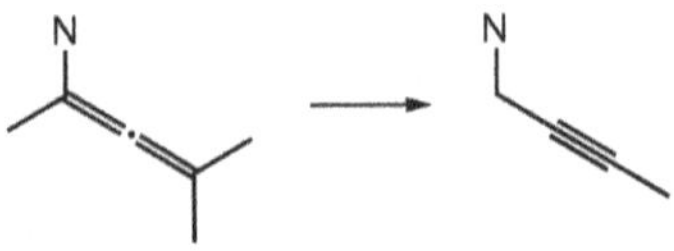

SCHEME 2.223 Isomerization of the allenyl group to the propargylic moiety in the presence of *n*-BuLi/*t*-BuOK system.

in equilibrium mixture with lithium compounds) are obtained by metalation of N-allenylpyrrole with the *n*-BuLi/*t*-BuOK system (THF/hexane, 2:1). Noteworthy, even with equimolar ratio of the reagents, the metalation of N-allenylpyrrole does not stop at the stage of monometal derivative formation, but continues up to exhaustive metalation of the allene group. The yields of trisilylated derivatives increase with the augmentation of the reaction temperature: at −80°C, the contents and yield of the product are 13% and 7%, respectively, whereas at −50°C ÷ −30°C, these values change to 34% and 16%. The unexpected conclusion follows therefrom that trimetalated allenylpyrrole (with potassium counterion, probably in a combination with lithium counterion) is kinetically and thermodynamically more stable than monometalated allenylpyrrole.

Additionally, in this case, the ratio of prototropic isomerization of the allenyl group to the propargylic moiety dramatically increases and the yields of propargylic derivatives grow with decrease of the reaction temperature (37% at −80°C) (Scheme 2.223).

Trimethylchlorosilane, dimethyldisulfide, butyl bromide, butyl iodide, paraform, acetic and pivalic aldehydes, benzaldehyde, acetone, and cyclohexanone are used as electrophilic reagents relative to N-allenylpyrrole metalated with butyllithium. The conditions of the second stage of the reaction, electrophilic substitution, or addition (for aldehydes and ketones) depend upon the electrophile employed. In all the cases, α-allenic derivatives are formed as a rule (Table 2.20).

However, the reaction with aldehydes and ketones affords, along with allenic alcohols, their N-ethynyl isomers, the content of which in the products mixture can reach 60% (in the case of the reaction with paraform) (Scheme 2.224). The mixture composition is defined to a large extent by the reaction conditions and the electrophile nature.

In the case of ketones, the content of acetylene isomers in the crude product does not exceed 5%.

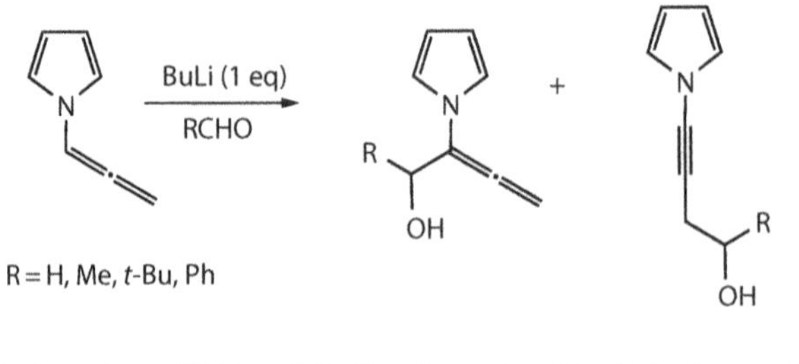

R = H, Me, *t*-Bu, Ph

SCHEME 2.224 Reaction of lithiated N-allenylpyrrole with aldehydes and ketones.

The formation of N-ethynyl derivatives testifies that under these conditions, the allenyl carbanion undergoes prototropic isomerization to N-propargylic (C_{sp3}-centered) carbanion.

Copper and zinc derivatives of N-allenylpyrrole are much less active in the reaction with electrophiles than the corresponding lithium and potassium compounds [687].

2.2.9.3 Metalation of N-Ethynylpyrrole

The treatment of N-1,2-dichlorovinylpyrrole with two equivalents of n-BuLi leads to N-lithioethynylpyrrole [688]. With three equivalents of n-BuLi (THF, $-40°C \div -35°C$), a further lithiation occurs at the α-position of the pyrrole ring to produce 2-lithio-1-(lithioethynyl)pyrrole. The reaction of the latter with trimethylchlorosilane, butyl iodide, and dimethylformamide results in the corresponding ethynylpyrroles (Scheme 2.225, Table 2.20). The same dilithium derivative upon contacting with elemental sulfur, selenium, and tellurium furnishes (t-butanol/DMF, 0°C–35°C) pyrrolo[2,1-b][1,3]thiazole, pyrrolo[2,1-b][1,3]selenazole, and pyrrolo[2,1-b][1,3]tellurazole, respectively (Table 2.20) [688].

2.2.10 DEVINYLATION

N-Vinylpyrroles are considered [689] as protected pyrroles, since the vinyl group can be deprotected by hydrolysis (see Sections 2.1.8 and 2.2.1) or under the action of some other reagents.

For example, N-vinylpyrroles are acetoxymercurated [690] under mild neutral conditions (with the subsequent treatment of the reaction mixture with sodium borohydride) to readily give NH-pyrroles (Scheme 2.226).

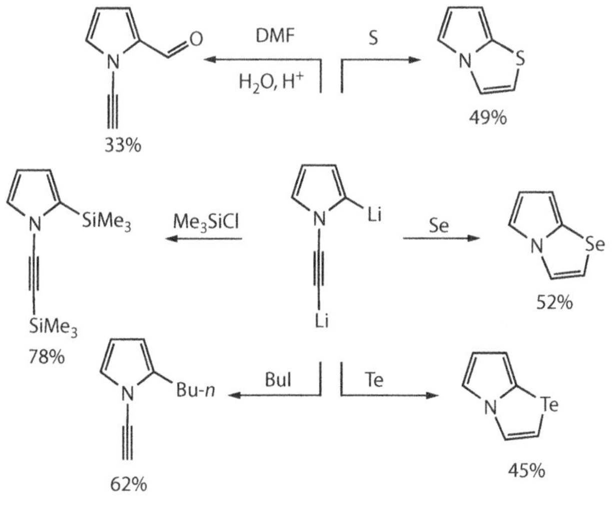

SCHEME 2.225 Lithiation of N-ethynylpyrrole and synthetic application of the dianion formed.

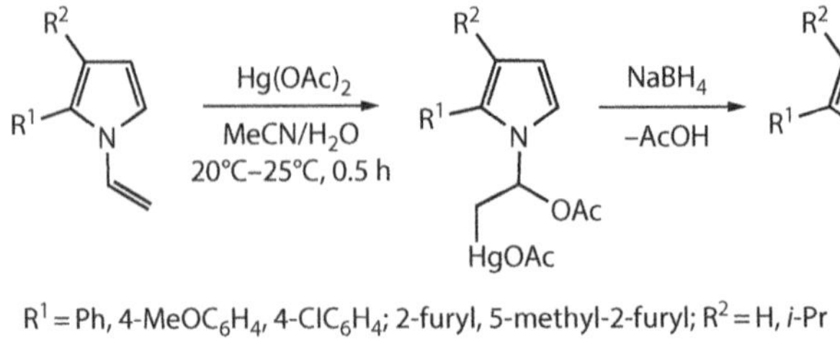

R^1 = Ph, 4-MeOC$_6$H$_4$, 4-ClC$_6$H$_4$; 2-furyl, 5-methyl-2-furyl; R^2 = H, i-Pr

SCHEME 2.226 Devinylation of N-vinylpyrroles under the subsequent action Hg(OAc)$_2$ and NaBH$_4$.

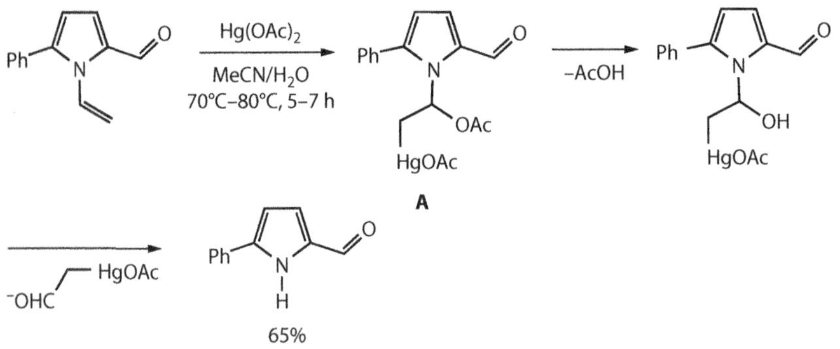

SCHEME 2.227 Devinylation of 2-phenyl-N-vinylpyrrole-2-carbaldehyde in the presence of Hg(OAc)$_2$.

Despite the complete consumption of N-vinylpyrroles at the mercuration stage, they are formed again in some amounts after reduction with sodium borohydride, apparently, due to reversibility of acetoxymercuration reaction.

Acetoxymercuration of N-vinylpyrroles is a typical reaction of electrophilic addition to the N-vinyl group. Its key stage involves an attack of the mercury acetoxy cation on the double bond to generate adduct **A** (Scheme 2.227). The latter, upon hydrolysis, can exchange acetoxy group for the hydroxy moiety, and unstable hemiaminal should further decompose to give free pyrrole and mercury acetaldehyde. Thus, the removal of vinyl protection can be achieved without application of sodium borohydride that is confirmed by the example of 5-phenylpyrrole-2-carbaldehyde [691], which reduces its aldehyde group under usual conditions (with NaBH$_4$).

From the preparative viewpoint, the most attractive is oxidative elimination of the N-vinyl group by potassium permanganate, which has been reported for di- and tribromo-N-vinylpyrazoles [692]. Therefore, it seems important to elucidate whether devinylation of N-vinylpyrroles with KMnO$_4$ is possible, taking into account that the known ability of pyrrole ring to undergo oxidation [2] could give rise to undesirable side processes.

It has been found that some N-vinylpyrroles are capable of losing the N-vinyl group with conservation of the pyrrole ring upon treatment with a dilute (0.4%–2.0%) aqueous solution of KMnO$_4$ for few minutes (Scheme 2.228) [693].

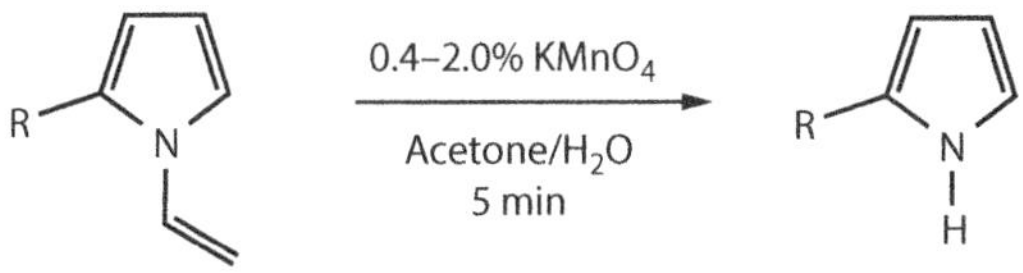

R = Ph, 2-naphthyl

SCHEME 2.228 Oxidative devinylation of N-vinylpyrroles in the presence of KMnO$_4$.

The reactions proceed at −20°C to 20°C, and the unoptimized yields of NH-pyrroles are 22%–24%.

The efficacy of the process depends on the structure of the initial N-vinylpyrrole. For example, only traces of the corresponding NH derivative are obtained from N-vinyl-4,5,6,7-tetrahydroindole (^{1}H NMR data), though the substrate is consumed completely. Obviously, in this case, successful devinylation requires even milder conditions (lower temperature, shorter time, and smaller oxidant concentration).

Nevertheless, further optimization of this reaction could lead to the development of a convenient express procedure for devinylation of *N*-vinylpyrroles.

2.3 CONCLUSIONS

The chemistry of pyrroles continues to stand at the forefront of modern research related to the synthesis and practical application of heterocyclic compounds. The interest in this field of science is steadily growing. Open-access publications [694], fundamental monographs [695], and chapters in books on chemistry of natural compounds [696] are dedicated to this hemisphere of knowledge.

Over the last decades, a novel branch of the pyrrole chemistry, based on the reaction of ketones (ketoximes) with acetylenes in the superbase catalytic systems (the Trofimov reaction), has emerged and keeps progressing rapidly. New facile and highly efficient approaches to the diverse compounds of the pyrrole series have been developed. The investigations in this line are gathering momentum. Not only the discoverers and developers of novel pyrrole synthesis (the authors of this monograph) but many other research teams both in Russia and abroad are engaged in these studies.

It is essentially important that the new reaction allows not only NH-pyrroles but also almost hitherto unknown their N-vinyl derivatives to be readily synthesized. Thus, absolutely original stratum of the pyrrole chemistry with a promising and unexplored synthetic potential has been unveiled.

Indeed, the pyrrole ring and the double bond, linked through the nitrogen atom, represent a poorly studied combination of chemically highly reactive fragments that impart novel, but not always explicit, properties to the molecule.

As shown in this chapter, the participation of nitrogen atom, and hence the whole pyrrole system, in stabilization of active intermediate particles, cations, and radicals is a peculiar feature of electrophilic and radical addition to N-vinylpyrroles. The character of electron redistributions and rearrangements in such particles and their chemical consequences can be discussed only at qualitative level. The subsequent kinetic, spectral, and quantum-chemical investigations should facilitate the searches for novel synthetic opportunities in the field. It is not excluded that in some cases and at certain structural architecture, cationoid, cation radical, and radical intermediates generated by N-vinylpyrroles will be quite stable and, therefore, suitable for their detailed experimental studying and preparative application.

Owing to a high sensitivity of the pyrrole ring to electrophilic attack, an unexpected direction of cationic oligomerization of N-vinylpyrroles becomes possible. The key stage here is electrophilic substitution of the pyrrole hydrogen in one molecule by the immonium cation formed in the other molecule [7,18,697–700].

The same sequence of reactions, which are carried out intramolecularly, can lead to the new fused polycyclic and macrocyclic pyrrole systems. However, these opportunities arising from the synthesis of N-vinylpyrrole dimers [7,18,701–708] have not been realized still. Here, there is a plenty of room for research creativity of experts in the field of heterocyclic compounds synthesis and polymerization.

The investigations into radical polymerization are also gathering force [709–728].

The cyclization reactions, which can be originated from functionalization of the pyrrole ring or vinyl group of the N-vinylpyrroles, wait for their researchers.

Indeed, the phosphorylation of N-vinylpyrroles provides evidences that the reactions of electrophilic substitution can tolerate not only the pyrrole ring but also the vinyl group of N-vinylpyrroles. As it has been supposed, the target products turn out to be capable of cyclization involving the pyrrole moiety. The synthesis of pyrrole-phosphazoles is a brilliant illustration of preparative power of this direction.

Among new pages of the pyrrole chemistry are the results on addition of diverse compounds (alcohols, thiols, carboxylic acids, hydrogen halides, hydrosilanes, phosphines, and phosphine halides) to the N-vinyl substituent. In combination with the ability of the pyrrole ring to oxidative polymerization and cycle opening under the action of acids (in the case of bi- and polyfunctional addends), this creates an approach to the curable film-forming oligomer, which upon structuring can give electroconductive materials, polypyrroles of new generation.

Functionalized pyrroles and N-vinylpyrroles, the syntheses of which have been covered in Chapter 1 of the present monograph, represent highly reactive building blocks for the direct preparation of new drugs, biologically active compounds, and materials for advanced technologies.

For 40 years since its discovery, the reaction of ketones (ketoximes) with acetylene has recommended itself as a reliable preparative tool for wide application in chemistry of pyrroles and their N-vinyl derivatives. Systematic development of novel syntheses of pyrroles, and especially N-vinylpyrroles via the Trofimov reaction, is still progressing. There is reason to hope that this will lead not only to the further enhancement of the preparative importance of this synthesis but also to the discovery of its new variants and analogs. The increased availability of N-vinylpyrroles stimulates synthetic, physical–chemical [507,729–755], and theoretical [756–759] investigations in the field and also works on polymerization and practical application of these compounds [697–700,709–728].

To conclude, this monograph not only sums up development of the new direction in chemistry of pyrroles but also introduces a novel branch of synthesis of important heterocyclic compounds, where a lot of things lie ahead.

References

1. Gossauer, A. 1974. *Die Chemie der Pyrrole*. Berlin, Germany: Springer-Verlag.
2. Jones, R.A. and G.P. Bean. 1977. *The Chemistry of Pyrroles*. London, U.K.: Academic Press.
3. Trofimov, B.A., S.V. Amosova, A.I. Mikhaleva et al. 1977. Reactions of acetylene in superbase media. In *Fundamental Investigations. Chemical Sciences*, ed. G.K. Boreskov, pp. 174–178. Novosibirsk, Russia: Nauka.
4. Trofimov, B.A. and A.I. Mikhaleva. 1980. Reactions of ketoximes with acetylene: A novel general method for the synthesis of pyrroles. *Khim Geterocicl* 10:1299–1312.
5. Trofimov, B.A. 1981. *Heteroatom Derivatives of Acetylene. Novel Polyfunctional Monomers, Reagents and Semi-Products*. Moscow, Russia: Nauka.
6. Trofimov, B.A. 1981. Reactions of acetylene in superbasic media. *Usp Khim* 50 (2):248–272.
7. Trofimov, B.A. and A.I. Mikhaleva. 1984. *N-Vinylpyrroles*. Novosibirsk, Russia: Nauka.
8. Trofimov, B.A., A.I. Mikhaleva, and L.V. Morozova. 1985. Polymerization of N-vinylpyrroles. *Usp Khim* 54 (6):1034–1050.
9. Trofimov, B.A. 1986. New intermediates for organic synthesis based on acetylene. *Z Chem* 26 (2):41–49.
10. Trofimov, B.A. 1986. Superbasic media in acetylene chemistry. *Zh Org Khim* 22 (9):1991–2006.
11. Trofimov, B.A. and A.I. Mikhaleva. 1987. Azirines in reactions of pyrrole ring formation. *Khim Geterocicl* 10:1299–1312.
12. Trofimov, B.A., L.N. Sobenina, and A.I. Mikhaleva. 1987. *Advances in Pyrrole Synthesis*. Moscow, Russia: VINITI.
13. Sobenina, L.N., A.I. Mikhaleva, and B.A. Trofimov. 1989. Synthesis of pyrroles from aliphatic compounds. *Usp Khim* 58 (2):275–333.
14. Trofimov, B.A. 1989. Perspectives of pyrrole chemistry. *Usp Khim* 58 (10):1703–1720.
15. Sobenina, L.N., A.I. Mikhaleva, and B.A. Trofimov. 1989. Synthesis of pyrroles from heterocyclic compounds. *Khim Geterocicl* 3:291–308.
16. Trofimov, B.A. 1990. Preparation of pyrroles from ketoximes and acetylenes. *Adv Heterocycl Chem* 51:177–301.
17. Bean, G.P. 1990. The synthesis of 1*H*-pyrroles. In *The Chemistry of Heterocyclic Compounds*, vol. 48, Pt. 1, ed. E.C. Taylor, pp. 105–294. New York: Wiley.
18. Trofimov, B.A. 1992. Vinylpyrroles. In *The Chemistry of Heterocyclic Compounds*, vol. 48, Pt. 2, ed. R.A. Jones, pp. 131–298. New York: Wiley.
19. Trofimov, B.A. and A.I. Mikhaleva. 1994. Further development of the ketoxime-based pyrrole synthesis. *Heterocycles* 37 (2):1193–1232.
20. Trofimov, B.A. 1994. Sulfur-containing pyrroles. *Phosphorus Sulfur* 95–96:145–163.
21. Sigalov, M.V. and B.A. Trofimov. 1995. 1-Vinylpyrrolium ions. *Zh Org Khim* 31 (6):801–826.
22. Trofimov, B.A. 1995. Some aspects of acetylene chemistry. *Zh Org Khim* 31 (9):1368–1387.
23. Trofimov, B.A. and A.I. Mikhaleva. 1996. From ketones to pyrroles in two stages. *Zh Org Khim* 32 (8):1127–1141.
24. Trofimov, B.A. and M.V. Sigalov. 1998. N-Vinylpyrrolium, furanium and thiophenium ions. *Main Group Chem News* 6 (2–3):30–41.

25. Korostova, S.E., A.I. Mikhaleva, A.M. Vasil'tsov, and B.A. Trofimov. 1998. Arylpyrroles: Development of classical and modern synthetic methods. Part I. *Zh Org Khim* 34 (7): 967–1000.

26. Korostova, S.E., A.I. Mikhaleva, A.M. Vasil'tsov, and B.A. Trofimov. 1998. Arylpyrroles: Development of classical and modern synthetic methods. Part II. *Zh Org Khim* 34 (12): 1767–1785.

27. Korostova, S.E., A.I. Mikhaleva, and B.A. Trofimov. 1999. Bipyrroles, furyl- and thienylpyrroles. *Usp Khim* 68 (6):506–530.

28. Sobenina, L.N., A.P. Demenev, A.I. Mikhaleva, and B.A. Trofimov 2002. Synthesis of C-vinylpyrroles. *Usp Khim* 71 (7):641–671.

29. Mikhaleva, A.I. and E.Yu. Schmidt. 2002. Two-step synthesis of pyrroles from ketones and acetylenes through the Trofimov reaction. In *Selected Methods for Synthesis and Modification of Heterocycles*, vol. 1, ed. V.G. Kartsev, pp. 334–352. Moscow, Russia: IBS Press.

30. Trofimov, B.A. 2003. Pyrrolecarbodithioates: Synthesis and application in design of complex heterocyclic system. In: *Chemistry and Biological Activity of Natural Compounds. Oxygen- and Sulfur-Containing Compounds*, vol. 1, ed. V.G. Kartsev, pp. 119–135. Moscow, Russia: IBS Press.

31. Trofimov, B.A., L.N. Sobenina, A.P. Demenev, and A.I. Mikhaleva. 2004. C-Vinylpyrroles as pyrrole building blocks. *Chem Rev* 104 (5):2481–2506.

32. Trofimov, B.A. 2004. Superbase catalysts and reagents: Concept, application, perspectives. *Modern Problem of Organic Chemistry* 14:131–175.

33. Mikhaleva, A.I., A.B. Zaitsev, and B.A. Trofimov. 2006. Oximes as reagents. *Usp Khim* 75 (9):884–912.

34. Trofimov, B.A. and N.K. Gusarova. 2007. Acetylene: New possibilities of classical reactions. *Usp Khim* 76 (6):550–570.

35. Trofimov, B.A. 2008. From chemistry of acetylene to chemistry of pyrrole. *Khimiya v interesakh ustoichivogo razvitiya* 16 (1):105–118.

36. Trofimov, B.A. and N.A. Nedolya. 2008. Pyrroles and their benzo derivatives: Reactivity. In *Comprehensive heterocyclic chemistry III. A Review of the Literature 1995–2007*, vol. 3, eds. A.R. Katritzky, C.A. Ramsden, E.F.V. Scriven, and R.J.K. Taylor, pp. 45–268. Amsterdam, the Netherlands: Elsevier.

37. Nedolya, N.A. 2008. Chemistry of heterocyclic compounds at the A.E. Favorsky Irkutsk institute of chemistry over 50 years. *Khim Geterocicl* 10:1443–1502.

38. Trofimov, B.A. and L.N. Sobenina. 2009. Ethynylation of pyrrole nucleus with haloacetylenes on active surfaces. In *Targets in Heterocyclic Systems: Chemistry and Properties*, vol. 12, eds. O.A. Attanasi and D. Spinelli, pp. 92–119. Rome, Italy: Societa Chemica Italiana.

39. Trofimov, B.A., A.I. Mikhaleva, E.Yu. Schmidt, and L.N. Sobenina. 2010. Pyrroles and N-vinylpyrroles from ketones and acetylenes: Recent strides, Chapter 7. *Adv Heterocycl Chem* 99:209–254.

40. Pinna, G.A., M. Sechi, G. Paglietti, and M.A. Pirisi. 2003. Addition reactions of acetylenic esters to 6,7-dihydrobenzo[*b*]furan-4(5*H*)-one, 6,7-dihydroindol-4(5*H*)-one, 5,6-dihydrobenzo[b]furan-7(6*H*)-one and 5,6-dihydroindol-7(6*H*)-one ketoximes. Formation of reduced furo[*g*]- and pyrrolo[*g*]-indoles. *J Chem Res* 3:362–380.

41. Galenko, A.V., S.I. Selivanov, P.S. Lobanov, and A.A. Potekhin. 2007. Rearrangement of O-vinyl-α-(aminocarbonyl)acetamidoximes to 2-aminopyrroles and 2-aminopyrrolinones. *Chem Heterocycl Compd* 43 (9): 1124–1130.

42. Ngwerume, S. and J.E. Camp. 2010. Synthesis of highly substituted pyrroles via nucleophilic catalysis. *J Org Chem* 75:6271–6274.

43. Wang, H.-Y., D.S. Mueller, R.M. Sachwani et al. 2010. Carbon–carbon bond formation and pyrrole synthesis via the [3,3] sigmatropic rearrangement of O-vinyl oxime ethers. *Org Lett* 12:2290–2293.

44. Estévez, V., M. Villacampa, and J.C. Menéndez. 2010. Multicomponent reactions for the synthesis of pyrroles. *Chem Soc Rev* 39:4402–4421.

45. Wang, H.-Y., D.S. Mueller, R.M. Sachwani et al. 2011. Regioselective synthesis of 2,3,4- or 2,3,5-trisubstituted pyrroles via [3,3] or [1,3] rearrangements of O-vinyl oximes. *J Org Chem* 76:3203–3221.

46. Ngwerume, S. and J. Camp. 2011. Gold-catalysed rearrangement of O-vinyl oximes for the synthesis of highly substituted pyrroles. *Chem Commun* 47:1857–1859.

47. Attanasi, O.A., S. Berretta, L. De Crescentini et al. 2011. Zinc(II) triflate-catalyzed divergent synthesis of polyfunctionalized pyrroles. *Adv Synth Catal* 353:595–605.

48. Ng, E.P.J., Y.-F. Wang, and S. Chiba. 2011. Manganese(III)-catalyzed formal [3+2] annulation of vinyl azides and β-keto acids for synthesis of pyrroles. *Synlett* 6:783–786.

49. Camp, J. 2012. Synthesis of heterocycles via gold multifaceted catalysis. *Chem Today Catal Appl* 30:6–9.

50. Madabhushi, S., V.S. Vangipuram, K.K. Mallu et al. 2012. Europium(III) triflate-catalyzed Trofimov synthesis of polyfunctionalized pyrroles. *Adv Synth Catal* 354:1413–1416.

51. Ngwerume, S., W. Lewis, and J.E. Camp. 2013. Development of a gold-multifaceted catalysis approach to the synthesis of highly substituted pyrroles: Mechanistic insights via Huisgen cycloaddition studies. *J Org Chem* 78:920–934.

52. Reddy, M.V.R., D.J. Faulkner, Y. Venkateswarlu, and M.R. Rao. 1997. New lamellarin alkaloids from an unidentified ascidian from the Arabian Sea. *Tetrahedron* 53 (10):3457–3466.

53. Reddy, R.M.V., M.R. Rao, D. Rhodes et al. 1999. Lamellarin α 20-sulfate, an inhibitor of HIV-1 integrase active against HIV-1 virus in cell culture. *J Med Chem* 42 (11):1901–1907.

54. Dayan, F.A. and E.A. Dayan. 2011. A class of pigment molecules binds King George III, vampires and herbicides. *Am Sci* 99:236–243.

55. Faulkner, D. 2000. Marine natural products. *J Nat Prod Rep* 17:7–55.

56. Bailly, C. 2004. Lamellarins, from A to Z: A family of anticancer marine pyrrole alkaloids. *Curr Med Chem Anticancer Agents* 4 (4):363–378.

57. Kaneda, M., S. Akamura, N. Ezaki, and Y. Iitaka. 1981. Structure of pyrrolomycin B, a chlorinated nitro-pyrrole antibiotic. *J Antibiot* 34 (10):1366–1368.

58. Liu, W.-Y., W.-D. Zhang, H.-S. Chen et al. 2003. Pyrrole alkaloids from bolbostemma paniculatum. *J Asian Nat Prod Res* 5 (3):159–163.

59. Takahata, H. and T. Momose. 1993. Simple indolizidine alkaloids. In *The Akaloids*, vol. 44, ed. G.A.V. Cordel, pp. 189–256. San Diego, CA: Academic Press.

60. Dembitski, V.M. and G.A. Tolstikov. 2003. *Natural Halogenated Organic Compounds*. Novosibirsk, Russia: SO RAN.

61. Raggatt, M.E., T.J. Simpson, and S.K. Wrigley. 1999. Biosynthesis of XR587 (strepto-pyrrole) in *Streptomyces rimosus* involves a novel carbon-to-nitrogen rearrangement of a proline-derived unit. *Chem Commun* 11:1039–1040.

62. Breinholt, J., H. Gurtler, A. Kjar et al. 1998. Streptopyrrole: An antimicrobial metabolite from *Streptomyces armeniacus*. *Acta Chem Scand* 52 (8):1040–1044.

63. Furstner, A., H. Weintritt, and A. Hupperts. 1995. A new, titanium-mediated approach to pyrroles: First synthesis of Lukianol A and Lamellarin O dimethyl ether. *J Org Chem* 60 (20):6637–6641.

64. Gupton, G.T., K.E. Krumpe, B.S. Burnham et al. 1998. The application of disubstituted vinylogous iminium salts and related synthons to the regiocontrolled preparation of unsymmetrical 2,3,4-trisubstituted pyrroles. *Tetrahedron* 54 (20):5075–5088.

65. Khanna, I.K., R.M. Weier, Y. Yu et al. 1997. 1,2-Diarylpyrroles as potent and selective inhibitors of cyclooxygenase-2. *J Med Chem* 40 (11):1619–1633.

66. Thurkauf, A., J. Yuan, X. Chen et al. 1995. 1-Phenyl-3-(aminomethyl)pyrroles as potential antipsychotic agents. Synthesis and dopamine receptor binding. *J Med Chem* 38 (25):4950–4952.

67. Carson, J.R., R.J. Carmosin, P.M. Pitis et al. 1997. Aroyl(aminoacyl)pyrroles, a new class of anticonvulsant agents. *J Med Chem* 40 (11):1578–1584.

68. Anderson, H.J. and C.E. Loader. 1985. The synthesis of 3-substituted pyrroles from pyrrole. *Synthesis* 4:353–364.

69. Lainton, J.A.H., J.W. Huffman, B.R. Martin, and D.R. Compton. 1995. 1-Alkyl-3-(1-naphthoyl)pyrroles: A new class of cannabinoid. *Tetrahedron Lett* 36 (9):1401–1404.

70. DeLeon, C.Y. and B. Ganem. 1997. A new approach to porphobilinogen and its analogs. *Tetrahedron* 53 (23):7731–7752.

71. Jacobi, P.A., L.D. Coutts, J.S. Gou et al. 2000. New strategies for the synthesis of biologically important tetrapyrroles. The "B,C + D + A" approach to linear tetrapyrroles. *J Org Chem* 65 (1):205–213.

72. Portevin, B., C. Tordjman, P. Pastoureau et al. 2000. 1,3-Diaryl-4,5,6,7-tetrahydro-2*H*-isoindole derivatives: A new series of potent and selective COX-2 inhibitors in which a sulfonyl group is not a structural requisite. *J Med Chem* 43 (24):4582–4593.

73. Magnus, P. and Y.-S. Or. 1983. Initial studies on the synthesis of the antitumor agent CC-1065: 3,4-Disubstituted pyrroles and 3,3′-bipyrroles. *J Chem Soc Chem Commun* 1:26–27.

74. Skladanowski, A., M. Koba, and L. Konopa. 2001. Does the antitumor cyclopropyl-pyrroloindole antibiotic CC-1065 cross-link DNA in tumor cells? *Biochem Pharmacol* 61:67–72.

75. Oldfield, E. 2010. Targeting isoprenoid biosynthesis for drug discovery: Bench to bedside. *Acc Chem Res* 43:1216–1226.

76. Lindsley, C.W. 2010. The top prescription drugs of 2009 in the US: CNS therapeutics rank among highest grossing. *ACS Chem Neurosci* 1:407–408.

77. Vernitskaya, T.V. and O.N. Efimov. 1997. Polypyrrole as conducting polymer, its synthesis, properties, applications. *Usp Khim* 66 (5):489–505.

78. Gimenez, I.F. and O.L. Alves. 1999. Formation of a novel polypyrrole/porous phosphate class ceramic nanocomposite. *J Braz Chem Soc* 10 (2):167–168.

79. Nizuski-Mann, R.E. and M.P. Cava. 1992. Synthesis of mixed thiophene-pyrrole heterocycles. *Heterocycles* 34 (10):2003–2007.

80. Silva, C.D. and D.A. Walker. 1998. Acid-base catalysis in the synthesis of arylmethylene and alkylmethine pyrroles. *J Org Chem* 63 (19):6715–6718.

81. Chou, S.-S.P. and Y.-H. Yeh. 2001. Synthesis of novel sulfonyl-substituted pyrrole chromophores for second-order nonlinear optics. *Tetrahedron Lett* 42 (7):1309–1312.

82. Costello, B.P.J., P. Evans, B.P.J. de Lacy et al. 2000. The synthesis of a number of 3-akyl and 3-carboxy substituted pyrroles; their chemical polymerisation onto poly(vinylidene fluoride) membranes, and their use as gas sensitive resistors. *Synth Met* 114 (2):181–188.

83. Ezaki, O., T. Shomura, M. Koyama et al. 1981. New chlorinated nitro-pyrrole antibiotics, pyrroiomycin A and B (SF-2080 A and B). *J Antibiot* 34 (10):1363–1365.

84. Ackrell, J., F. Franco, R. Greenhouse et al. 1980. New heterocyclic systems. Benzo[4,5]cyclohepta[1,2-*b*]pyrrole and benzo[5,6]cyclohepta[1,2-f]pyrrolizidine derivatives. *J Heterocycl Chem* 17:1081–1086.

85. Carmona, O., R. Greenhouse, R. Landeros, and J.M. Muchowski. 1980. Synthesis and rearrangement of 2-(arylsulfinyl)- and 2-(alkylsulfinyl)pyrroles. *J Org Chem* 45 (26): 5336–5339.

86. Rokach, J., P. Hamel, and M. Kakushima. 1981. A simple and efficient route to β-substituted pyrroles. *Tetrahedron Lett* 22:4901–4904.

87. Mikhaleva, A.I., A.B. Zaitsev, A.V. Ivanov et al. 2006. Expedient synthesis of 1-vinylpyrroles-2-carbaldehydes. *Tetrahedron Lett* 47 (22):3693–3696.

88. Trofimov, B.A., E.Yu. Schmidt, A.I. Mikhaleva et al. 2006. 2-Arylazo-1-vinylpyrroles: A novel promising family of reactive dyes. *Eur J Org Chem* 17:4021–4033.

89. Trofimov, B.A., A.M. Vasil'tsov, A.I. Mikhaleva et al. 2009. Synthesis of 1-vinylpyrrole-2-carbonitriles. *Tetrahedron Lett* 50 (1): 97–100.

90. Shafi, S., K. Kędziorek, and K. Grela. 2011. Cross metathesis of n-allylamines and α,β-unsaturated carbonyl compounds: A one-pot synthesis of substituted pyrroles. *Synlett* 1:124–128.

91. Sims, V.A. 1962. Vinylation of aromatic heterocyclic nitrogen compounds. US Patent 3,047,583.

92. Nomori, H., M. Hatano, and S. Kambara. 1966. Polymerization of N-vinylindole and N-vinylpyrrole with organic electron acceptors. *Polym Lett* 4:623–628.

93. Shostakovsky, M.F., G.G. Skvortsova, and E.S. Domnina. 1969. N-Vinyl compounds of pyrrole series. *Usp Khim* 38 (5):892–916.

94. Cooper, G., W.J. Irwin, and D.L. Wleeler. 1971. 1-Vinylpyrroles. *Tetrahedron Lett* 45:4321–4324.

95. Reppe, W. 1956. Vinylierung. *Lieb Ann* 601:128–138.

96. Vatsulik, P. 1960. *Chemistry of Monomers*. Moscow, Russia: IL.

97. Gossauer, A. 1994. Pyrrole. In *Methoden der Organischen Chemie*, vol. E6a, part 1, ed. R.R. Kreher, pp. 556–798. Stuttgart, Germany: Thieme.

98. Tedeschi, R.J. 2001. Acetylene. In *Encyclopedia of Physical Science and Technology*, 3rd edn., vol. 1, ed. R.A. Meyers, pp. 55–89. San Diego, CA: Academic Press.

99. Pozharsky, A.F., V.A. Anisimova, and E.B. Tsupak. 1988. *Practical Works on Chemistry of Heterocycles*. Rostov, Russia: Rostov University.

100. Kim, D.G. 2000. *Organic Chemistry in Reactions. A Study Guide*. Chelyabinsk, Russia: Chelyabinsk State University.

101. Li, J.J. 2010. *Name Reactions for Carbocyclic Ring Formations*. Hoboken: John Wiley & Sons.

102. Li, J.J., and E.J. Corey. 2011. *Name Reactions in Heterocyclic Chemistry II*. Hoboken: John Wiley & Sons.

103. Wang, Z. 2010. *Comprehensive Organic Name Reactions and Reagents*, part 3, pp. 2793–2796. London, U.K.: Wiley.

104. Trost, B.M., J.-P. Lumb, and J.M. Azzarelli. 2011. An atom-economic synthesis of nitrogen heterocycles from alkynes. *J Am Chem Soc* 133:740–743.

105. Trofimov, B.A. 2002. Acetylene and its derivatives in reactions with nucleophiles: Recent advances and current trends. *Curr Org Chem* 6 (13):1121–1162.

106. Trofimov, B.A. 2010. Ideas of academician A.E. Favorsky in modern chemistry of acetylene. In *Modern Problems of Organic Chemistry*, ed. M.A. Kuzhetsov, pp. 23–48. Saint-Petersburg, Russia: LEMA.

107. Reppe, W. 1949. *Neue Entwicklungen auf dem Gebiet der Chemie des Acetylen und Kohlenoxyds*. Berlin, Germany: Springer.

108. Copenhaver, J.W. and M.H. Bigelow. 1949. *Acetylene and Carbon Monoxide Chemistry*. New York: Reinhold Publishing Corporation.

109. Miller, S.A. 1966. *Acetylene. Its Properties, Manufacture and Uses*, vol. 2. London, U.K.: Ernest Benn Limited.

110. Shostakovsky, M.F. 1952. *Vinyl Ethers*. Moscow, Russia: AN SSSR.
111. Tedeschi, R.J. 1982. *Acetylene-Based Chemicals from Coal and Other Natural Resources*. New York: Marcel Dekker.
112. Boguslawski, Z. 1970. Tylko etylen—czy acetylen i ethylen? *Przem Chem* 49:709–713.
113. Wilks, P.H. 1976. The current state and future directions of industrial plasma chemistry. *Pure Appl Chem* 48:195–197.
114. Tedeschi, R.J. 1973. Applications of pressure reactions to acetylene chemistry. *Ann NY Acad Sci* 214:40–61.
115. Kucherov, V.F., M.V. Mavrov, and A.R. Derzhinsky. 1972. *Natural Polyacetylenic Compounds*. Moscow, Russia: Nauka.
116. Aylward. N. and N. Bofinger. 2006. A plausible prebiotic synthesis of pyridoxal phosphate: Vitamin B6—A computational study. *Biophys Chem* 123 (2–3):113–121.
117. Volkov, A.N. and A.N. Nikolskaya. 1977. α-Cyanoacetylenes. *Usp Khim* 46 (4):712–739.
118. Miller, S.L. and H.J. Cleaves. 2006. Prebiotic chemistry on the primitive earth. In *System Biology, Genomics*, vol. 1, eds. I. Rigoutsos and G. Stephanopoulos, pp. 3–56. Oxford, U.K.: Oxford University Press.
119. Sanchez, R.A., J.P. Ferris, and L.E. Orgel. 1966. Cyanoacetylene in prebiotic synthesis. *Science* 154:784–785.
120. Ungerechts, H., C.M. Walmsley, and G. Winnewisser. 1980. Ammonia and cyanoacetylene observations of the high density core of L 183 (L 134 B). *Astron Astrophys* 88:259–266.
121. Simonetta, M. and A. Gavezzotti. 1978. General and theoretical aspects of the acetylenic compounds. In *The Chemistry of the Carbon–Carbon Triple Bond*, part 1, ed. S. Patai, pp. 1–50. Chichester, U.K.: Wiley & Sons.
122. Hencher, J.L. 1978. The structural chemistry of the C≡C bond. In *The Chemistry of the Carbon–Carbon Triple Bond*, part 1, ed. S. Patai, pp. 57–74. Chichester, U.K.: Wiley & Sons.
123. Leprince, P. 1964. Acetylene. 1. Basic data. *Rev Just Fr Petrole* 19:613–626.
124. Kirkpatrick, D.M. 1976. Acetylene from calcium carbide is an alternate feedstock route. *Oil Gas J* 74:133–134.
125. Key chemicals. Ethylene. Capacity is ample. Demand is rising steadily. Business is good. 1976. *Chem Eng News* 54 (20):10–14.
126. Favorsky, A.E. 1961. *Selected Works*. Moscow, Russia: AN SSSR.
127. Patai, S. (ed.) 1978. *The Chemistry of the Carbon–Carbon Triple Bond*, parts 1 and 2. Chichester, U.K.: Wiley & Sons.
128. Stang, P.J. and F. Diederich (eds.). 1995. *Modern Acetylene Chemistry*. Weinheim, Germany: VCH.
129. Diederich, F., P.J. Stang, and R.R. Tykwinsky. 2005. *Acetylene Chemistry*. Weinheim, U.K.: Wiley-VCH.
130. Trofimov, B.A., E.Yu. Schmidt, N.V. Zorina et al. 2008. A short-cut from 1-acetyl adamantane to 2-(1-adamantyl)pyrroles. *Tetrahedron Lett* 49:4362–4365.
131. Puciova, M., P. Ertl, and S. Toma. 1994. Synthesis of ferrocenyl-substituted heterocycles: The beneficial effect of the microwave irradiation. *Collect Czech Chem Commun* 59 (1):175–185.
132. Schmidt, E.Yu., N.V. Zorina, A.B. Zaitsev et al. 2004. A selective synthesis of 2-([2,2] paracyclophan-5-yl)pyrrole from 5-acetyl[2,2]paracyclophane via the Trofimov reaction. *Tetrahedron Lett* 45:5489–5491.
133. Vasil'tsov, A.M., E.Yu. Schmidt, A.I. Mikhaleva et al. 2005. Synthesis and electrochemical characterization of dipyrroles separated by diphenyleneoxide and diphenylenesulfide spacers via the Trofimov reaction. *Tetrahedron* 61:7756–7762.
134. Trofimov, B.A., A.M. Vasil'tsov, E.Yu. Schmidt et al. 2005. Synthesis, structure and spectral properties of di(2-pyrrolyl)pyridines. *Eur J Org Chem* 20:4338–4345.

135. Trofimov, B.A., A.I. Mikhaleva, E.Yu. Schmidt et al. Method for the preparation of 4,5,6,7-tetrahydroindole. 2007. Russian Federation Patent 2,297,410.

136. Trofimov, B.A., A.I. Mikhaleva, E.Yu. Schmidt et al. 2007. Method for the preparation of indole. Russian Federation Patent 2,307,830.

137. Trofimov, B.A., A.I. Mikhaleva, E.Yu. Schmidt et al. 2009. Catalyst of dehydrating 4,5,6,7-tetrahydroindole into indole and method of its obtaining. Russian Federation Patent 2,345,066.

138. Gordon, J. 1979. *Organic Chemistry of Electrolyte Solutions*. Moscow, Russia: Mir.

139. Stewart, R., J. Donnell, D. Cram D, and B. Rickborn. 1962. Rate-equilibrium correlation for dissociation of a carbon acid. *Tetrahedron* 18:917–922.

140. Dolman, D. and R. Stewart. 1967. Strongly basic systems. VIII. The H-function for dimethyl sulfoxide–water–tetramethyl ammonium hydroxide. *Can J Chem* 45:911–924.

141. Kruus, P. and B.E. Poppe. 1979. Modeling of interactions in solutions: Alkali halides in DMSO. *Can J Chem* 57:538–551.

142. Hiller, F. and J.H. Krueger. 1967. The rate and mechanism of the iodine-formate reaction in dimethyl sulfoxide–water solvents. *Inorg Chem* 6:528–533.

143. Poum, T.M. and P.A. Zuman. 1976. Acidity function for solutions of sodium hydroxide in water–ethanol and water–dimethyl sulfoxide mixtures. *J Org Chem* 41:1614–1619.

144. Makosza, M. 1977. Reactions of carbanions and halocarbenes in two-phase systems. *Usp Khim* 46 (12):2174–2202.

145. Dehmlow, E.V. 1977. New synthetic methods. 20. Advances in phase-transfer catalysis. *Angew Chem* 89:521–533.

146. Weber, V. and G. Gockel. 1980. *Phase Transfer Catalysis in Organic Synthesis*. Moscow, Russia: Mir.

147. Yakubov, R.D., I.N. Azerbaev, B.A. Trofimov et al. 1963. Hydration of acetylene via vinyl ethers of ethylene- and diethylene glycols. *Vestn AN Kaz SSR* 7:21–31.

148. Shostakovsky, M.F., A.S. Atavin, and B.A. Trofimov. 1964. Vinyl ethers of di- and tri-ethylene glycols. *Zh Org Khim* 34:2112–2116.

149. Gasson, E.J., D.C. Quin, and F.E. Salt. 1949. Manufacture of vinyl ethers. Great Britain Patent 616,197.

150. Miller, S.A. and G.A. Weeks. 1954. Improvements in or relating to the preparation of vinyl ethers. Great Britain Patent 717,051.

151. Antonov, V.N., I.P. Azerbaev, M.F. Shostakovsky et al. 1996. Method for preparation of diethylene glycol divinyl ether. USSR Author's Certificate 180,579.

152. Otsuka, S., Y. Matsui, and S. Muraliaslii S. 1959. Studies of vinylation reaction. XIII. Vinylation of alcohols. Part 2. Mechanism of vinylation of β-ethoxyethane. XIV. Vinylation of alcohols. Part 3. Effects of cations, solvents and inhibitors. *Nippon kagaku zasshi J Chem Soc Jap Pure Chem Sec* 80:1153–1160.

153. Caubere, P. 1978. Complex bases and complex reducing agents: Now tools in organic synthesis. In *Topics in Current Chemistry, Organic Chemistry*, vol. 73, ed. F.L. Boschke, pp. 50–103. Berlin, Germany: Springer-Verlag.

154. Moskovskaya, T.E., N.M. Vitkovskaya, and B.A. Trofimov. 1982. Ab initio study of acetylene complexes with cations H^+, Li^+, Na^+. *B Acad Sci USSR Ch* 4:891–895.

155. Moskovskaya, T.E., N.M. Vitkovskaya, V.G. Bernshtein, and B.A. Trofimov. 1982. Ab initio study of complexes with alkali metal cations. 2. Specifics of complex with cation K^+. *B Acad Sci USSR Ch* 7:1474–1476.

156. Vitkovskaya, N.M., T.E. Moskovskaya, and B.A. Trofimov. 1982. Ab initio study of complexes with alkali metal cations. 3. Complexes with two or three molecules of acetylene. *B Acad Sci USSR Ch* 7:1477–1479.

157. Cox, B.G., P. De Maria, and A. Fini. 1970. Dissociation constants of benzenethiols in mixture of water and dimethyl sulfoxide. *Gazz Chim Ital* 106:817–821.

158. Buncel, E. and H. Wilson. 1977. Physical organic chemistry of reactions in dimethyl sulfoxide. *Adv Phys Org Chem* 14:133–202.

159. Mikhaleva, A.I., A.N. Vasiliev, and B.A. Trofimov. 1981. Pyrroles from ketoximes and acetylene. 17. A study of the reaction of cyclohexanone oxime with acetylene in super-base media. *Zh Org Khim* 17 (9):1977–1980.

160. Trofimov, B.A., S.E. Korostova, L.N. Balabanova, and A.I. Mikhaleva. 1978. Pyrroles from ketoximes and acetylene. 6. A study of the reaction of aceto- and propiophenox-imes with acetylene. *Zh Org Khim* 14 (8):1733–1736.

161. Mikhaleva, A.I., B.A. Trofimov, and A.N. Vasiliev. 1979. Pyrroles from ketoximes and acetylene. 8. Synthesis of 4,5,6,7-tetrahydroindole and its 1-vinyl derivative. *Khim Geterocicl* 2:197–199.

162. Trzhtsinskaya, B.V., L.F. Teterina, V.K. Voronov, and G.G. Skvortsova. 1976. Synthesis and properties of vinyl derivative of 4,5-diphenyl-2-mercaptoimidazole. *Khim Geterocicl* 4:516–520.

163. Trofimov, B.A., S.E. Korostova, L.N. Balabanova, and A.I. Mikhaleva. 1978. Pyrroles from ketoximes and acetylene. 3. Synthesis of 2-aryl- and 1-vinyl-2-arylpyrroles. *Khim Geterocicl* 4:489–491.

164. Trofimov, B.A., S.E. Korostova, L.N. Balabanova, and A.I. Mikhaleva. 1978. Pyrroles from ketoximes and acetylene. 7. 3-Alkyl(phenyl)-2-phenylpyrroles and their N-vinyl derivatives. *Zh Org Khim* 14 (10):2182–2184.

165. Trofimov, B.A., S.E. Korostova, A.I. Mikhaleva et al. Method for the preparation of α-phenyl- or α,β-diphenylpyrroles. 1978. USSR Author's Certificate 601,282.

166. Shostakovsky, M.F., B.A. Trofimov, A.S. Atavin, and V.I. Lavrov. 1968. Methods for synthesis of vinyl ethers containing functional groups and heteroatoms. *Usp Khim* 37 (11):2070–2093.

167. Kotlyarevsky, I.L., M.S. Shvartsberg, and L.B. Fisher. 1967. *Reactions of Acetylenic Compounds*. Novosibirsk, Russia: Nauka.

168. Trofimov, B.A. and S.V. Amosova. 1983. *Divinyl Sulfide and its Derivatives*. Novosibirsk, Russia: Nauka.

169. Skvortsov, Yu.M., A.G. Mal'kina, B.A. Trofimov et al. 1980. Cyanoacetylene and its derivatives. 3. Base-catalyzed autoheterocyclization of cyanoacetylenic carbanions. *B Acad Sci USSR Ch* 6:1349–1353.

170. Khan, M.M.T. and A.E. Martel. 1974. *Homogenous Catalysis by Metal Complexes. Activation of Alkenes and Alkynes*. New York: Academic Press.

171. Eisenstein, O., G. Procter, and J.D. Dunitz. 1978. Nucleophilic addition to a triple bond: Preliminary ab initio study. *Helv Chim Acta* 61:2538–2541.

172. Miller, S.A. 1969. *Acetylene, Its Properties, Manufacture and Application*, vol. 1. Leningrad, Russia: Khimiya.

173. Temnikova, T.I. 1971. Modern notions on organic solvents and their role in substitution processes in aliphatic series. In *Reactivity and Reaction Mechanisms of Organic Compounds*, eds. T.A. Favorskaya and I.A. Favorskaya, pp. 26–41. Leningrad, Russia: LGU.

174. Trofimov, B.A., A.I. Mikhaleva, A.S. Atavin, and E.G. Chebotareva. 1975. New direction of acetone oxime condensation with acetylene. *Khim Geterocicl* 10:1427.

175. Trofimov, B.A., A.I. Mikhaleva, and A.N. Vasiliev. 1976. Cyclohexanone oxime as "carbonyl component" in the Favorsky reaction. *Zh Org Khim* 12 (6):1180–1183.

176. Wolford, R.K. 1964. Kinetics of the acid-catalyzed hydrolysis of acetal in dimethyl sulfoxide–water solvents at 15, 25, and 35. *J Phys Chem* 68:3392–3398.

177. Werblan, L. and J. Lesinski. 1978. Structure and selected properties of water–dimethyl sulfoxide mixtures. *Pol J Chem* 52:1211–1219.

178. Strizhevsky, I.I. and V.F. Zakaznov. 1968. Study of explosive and detonation decomposition of acetylene relating to explosion prevention of syntheses using acetylene under pressure. In *Chemistry of Acetylene*, ed. M.F. Shostakovsky, pp. 472–476. Moscow, Russia: Nauka.

179. Mikhaleva, A.I., B.A. Trofimov, A.N. Vasiliev et al. 1981. Pyrroles from ketoximes and acetylene. 16. Synthesis of N-vinyltetrahydroindole on pilot plant. *B Acad Sci USSR Ch* 2 (1):150–152.

180. Trofimov, B.A., A.I. Mikhaleva, E.Yu, Schmidt et al. 2010. New technology for synthesis of 4,5,6,7-tetrahydroindole. *Dokl Chem* 435 (1):60–63.

181. Schmidt, E.Yu., A.I. Mikhaleva, A.M. Vasil'tsov et al. 2005. A straightforward synthesis of pyrroles from ketones and acetylene: A one-pot version of the Trofimov reaction. *Arkivoc* vii:11–17.

182. Mikhaleva, A.I., E.Yu. Schmidt, A.V. Ivanov et al. 2007. Selective synthesis of 1-vinyl-pyrroles directly from ketones and acetylene: Modification of the Trofimov reaction. *Zh Org Khim* 43 (2):236–238.

183. Vasil'tsov, A.M., A.V. Ivanov, A.I. Mikhaleva, and B.A. Trofimov. 2010. A three-component domino reaction of 2-tetralone, hydroxylamine and acetylene: A one-pot highly regioselective synthesis of 4,5-dihydrobenz[*e*]indoles. *Tetrahedron Lett* 51 (13):1690–1692.

184. Trofimov, B.A., A.V. Ivanov, E.Yu. Schmidt, and A.I. Mikhaleva. 2010. One-pot selective synthesis of N-vinyl-4,5-dehydrobenz[*g*]indole from 1-tetralone and acetylene in the system NH₂OH·HCl-KOH-DMSO. *Khim Geterocicl* 6:941–943.

185. Trofimov, B.A., A.S. Atavin, A.I. Mikhaleva et al. 1975. Method for the preparation of N-vinyl-2,3-dialkylpyrroles. USSR Author's Certificate 463,666.

186. Mikhaleva, A.I., B.A. Trofimov, and A.N. Vasiliev. 1979. Pyrroles from ketoximes and acetylene. IX. Synthesis of 2-alkyl- and 2,3-dialkylpyrroles and their *N*-vinyl derivatives. *Zh Org Khim* 15 (3):602–609.

187. Trofimov, B.A., A.I. Mikhaleva, S.E. Korostova et al. 1979. Method for the preparation of 2- or 2,3-substituted pyrroles. USSR Author's Certificate 694,504.

188. Trofimov, B.A., A.I. Mikhaleva, A.N. Vasiliev, and M.V. Sigalov. 1978. Pyrroles from ketoximes and acetylene. I. Structural orientation of the reaction with unsymmetrical ketoximes. NMR spectra of 1-vinylpyrroles and effects of alkyl substituents. *Khim Geterocicl* 1:54–59.

189. Korostova, S.E., I.N. Domnin, N.B. Viktorov et al. 1996. Pyrroles from ketoximes and acetylene. 48. Synthesis of 2-cyclopropylpyrroles. *Zh Org Khim* 32 (8): 1219–1224.

190. Trofimov, B.A., A.I. Mikhaleva, G.A. Kalabin et al. 1975. Synthesis of 1-vinyl- and 1-vinyl-7-methyl-4,5,6,7-tetrahydroindoles, their spectral properties and transformations. Presented at IVth All-Union colloquium on chemistry and pharmacology of indole compounds, pp. 24–25. Kishinev, Moldova: Shttintsa.

191. Trofimov, B.A., A.S. Atavin, A.I. Mikhaleva et al. 1977. An indole derivative. Great Britain Patent 1,463,228.

192. Trofimov, B.A., A.S. Atavin, A.I. Mikhaleva et al. 1977. 1-Vinyl-4,5,6,7-tetrahydroindol und Verfahren zu dessen Herstellung. West Germany Patent 2,543,850.

193. Trofimov, B.A., A.S. Atavin, A.I. Mikhaleva et al. 1978. 1-Vinyl-4,5,6,7-tetrahydroindole and preparation thereof. US Patent 4,077,975.

194. Trofimov, B.A., A.S. Atavin, A.I. Mikhaleva et al. 1977. 1-Vinyl-4,5,6,7-tetrahydroindole. Japan Patent 1,090,993.

195. Trofimov, B.A., A.S. Atavin, A.I. Mikhaleva et al. 1973. Condensation of cyclohexanone oxime with acetylene. *Zh Org Khim* 9 (10):2205.

196. Trofimov, B.A., S.E. Korostova, and A.I. Mikhaleva. 1978. Method for the preparation of 4,5,6,7-tetrahydroindole. USSR Author's Certificate 620,486.

197. Mikhaleva, A.I., E.Yu. Schmidt, N.I. Protsuk et al. 2008. Effect of alkali metal cations on synthesis of 4,5,6,7-tetrahydroindole and its vinyl derivative from cyclohexanone oxime and acetylene in the systems MOH-DMSO (M = Li, Na, K, Cs). *Dokl Chem* 423 (1):66–68.

198. Gonzalez, F., J.F. Sanz-Cervera, and R.M. Williams. 1999. Synthetic studies on Asperparaline A. Synthesis of the spirosuccinimide ring system. *Tetrahedron Lett* 40:4519–4522.

199. Trofimov, B.A., A.I. Mikhaleva, and R.N. Nesterenko. 1978. 4,5-Dihydrobenz[*g*]indole from oxime of *α*-tetralone and acetylene. *Zh Org Khim* 14 (5):1119.

200. Chen, J., A. Burghart, A. Derecskei-Kovacs, and K. Burges. 2000. 4,4-Difluoro-4-bora-3a,4a-diaza-s-indacene (BODIPY) dyes modified for extended conjugation and restricted bond rotations. *J Org Chem* 65 (10):2900–2906.

201. Trofimov, B.A., A.I. Mikhaleva, A.S. Atavin et al. 1976. Method for the preparation of pyrroles. USSR Author's Certificate 518,493.

202. Vasil'tsov, A.M., E.A. Polubentsov, A.I. Mikhaleva, and B.A. Trofimov. 1990. Cycloalka[*b*]pyrroles from ketoximes and acetylene: Synthesis and kinetic study. *B Acad Sci USSR Ch* 4:864–867.

203. Trofimov, B.A., A.M. Vasil'tsov, A.I. Mikhaleva et al. 1991. Stereochemical aspects of pyrroles formation from substituted piperidin-4-one oximes and acetylene. *Khim Geterocicl* 10:1365–1370.

204. Varlamov, A.V., L.G. Voskresensky, T.N. Borisova et al. 1999. Heterocyclization of tropinone and 3-methyl-3-azabicyclo[3.3.1]nonan-9-one oximes with acetylene in superbasic medium. *Khim Geterocicl* 5:683–687.

205. Trofimov, B.A., A.B. Shapiro, R.N. Nesterenko et al. 1998. Pyrroles from ketoximes and acetylene. XXXVI. Synthesis of 4,4,6,6-tetramethyl-4,5,6,7-tetrahydro-5-azaindole, its nitroxylic, vinyl derivatives and spin-labeled copolymer. *Khim Geterocicl* 3:350–355.

206. Voskresensky, L.G., T.N. Borisova, and A.V. Varlamov. 2004. Heterocyclization of 3,5-dimethyl(1,3,5-trimethyl)-2,6-diphenylpiperidin-4-one and N-benzylpyrrolidin-3-one oximes with acetylene in superbasic medium. *Khim Geterocicl* 3:401–409.

207. Mikhaleva, A.I., M.V. Sigalov, and G.A. Kalabin. 1982. An example of a novel synthetic approach to alkenylpyrroles. *Tetrahedron Lett* 23:5063–5066.

208. Vasil'tsov, A.M., A.I. Mikhaleva, R.N. Nesterenko, and M.V. Sigalov. 1991. Alkenylpyrroles from 5-hexen-2-one oxime: Prototropic isomerization under conditions of the Trofimov reaction. *Khim Geterocicl* 4:477–480.

209. Trofimov, B.A., A.M. Vasil'tsov, E.Yu. Schmidt et al. 1994. 2-Methyl-3-alkenylpyrroles from oximes of terpenoid ketones and acetylene. *Zh Org Khim* 30 (4):576–580.

210. Vasil'tsov, A.M., B.A. Trofimov, and S.V. Amosova. 1987. Mathematical model of superbase system alkali-DMSO in the region of low concentrations of water. *B Acad Sci USSR Ch* 8:1785–1791.

211. Albuquerque, E.X., J.W. Daly, and B. Witkop. 1971. Batrachotoxin: Chemistry and pharmacology. *Science* 172:995–1002.

212. Vasil'tsov, A.M., E.Yu. Schmidt, A.I. Mikhaleva et al. 2001. First example of construction of pyrrole ring bonded to a steroid system by the Trofimov reaction. *Khim Geterocicl* 12:1641–1645.

213. Vasil'tsov, A.M., A.B. Zaitsev, A.I. Mikhaleva et al. 2002. Annelation of pyrrole ring to a stereoid skeleton by the Trofimov reaction. *Khim Geterocicl* 1:66–70.

214. Zhang, X. and Z. Sui. 2003. An efficient synthesis of novel estrieno[2.3-*b*]- and [3.4-*c*] pyrroles. *Tetrahedron Lett* 44:3071–3073.

215. Trost, B.M. and E. Keinan. 1980. Pyrrole annulation onto aldehydes and ketones *via* palladium-catalyzed reactions. *J Org Chem* 45 (14):2741–2746.

216. Orlov, A.V., G.R. Khazipova, N.G. Komissarova et al. 2011. Trofimov synthesis of betulin derivatives with 2,3-annelated pyrrole. *Chem Nat Compd* 46 (6):906–909.

217. Zaitsev, A.B., A.M. Vasil'tsov, E.Yu. Schmidt et al. 2003. Oximes of ketosteroids in the Trofimov reaction: Steroid-pyrrole structures. *Zh Org Khim* 39 (10):1479–1483.

218. Petersen, Q.R. and E.E. Sowers. 1964. 4-Cholesten-3-one ethylene ketal. *J Org Chem* 29 (6):1627–1629.

219. Trofimov, B.A., A.I. Mikhaleva, O.V. Petrova, and L.N. Sobenina. 2010. Method for the preparation of 2-phenylpyrrole. Russian Federation Patent 2,397,974.

220. Trofimov, B.A., A.I. Mikhaleva, O.V. Petrova, and L.N. Sobenina. 2010. Method for the preparation of 1-vinyl-2-phenylpyrrole. Russian Federation Patent 2,399,615.

221. Korostova, S.E., S.G. Shevchenko, E.Yu. Schmidt et al. 1992. Pyrroles from ketoximes and acetylene. 42. Peculiarities of synthesis of novel 2-arylpyrroles. *Khim Geterocicl* 5:609–613.

222. Zaitsev, A.B., R. Méallet-Renault, E.Yu. Schmidt et al. 2005. Synthesis of 2-mesityl-3-methylpyrrole *via* the Trofimov reaction for a new BODIPY with hindered internal rotation. *Tetrahedron* 61:2683–2688.

223. Varlamov, A.V., T.N. Borisova, N. Bonifas et al. 2004. Synthesis and some transformations of 2-([2,2]paracyclophan-5-yl)pyrrole. *Khim Geterocicl* 2:201–211.

224. Vu, T.T., S. Badre, C. Dumas-Verdes et al. 2009. New hindered BODIPY derivatives: Solution and amorphous state fluorescence properties. *J Phys Chem C* 113:11844–11855.

225. Korostova, S.E., B.A. Trofimov, L.N. Sobenina et al. 1982. Pyrroles from ketoximes and acetylene. XXIII. 2-(1-Naphthyl)- and 2-(2-naphthyl)pyrroles and their 1-vinyl derivatives. *Khim Geterocicl* 10:1351–1353.

226. Markova, M.V., L.V. Morozova, I.V. Tatarinova et al. 2010. Selective synthesis and radical polymerization of N-vinyl-2-(1-naphthyl)- and N-vinyl-2-(2-naphthyl)pyrroles. *Dokl Chem* 434 (1):65–69.

227. Schmidt, E.Yu., N.V. Zorina, M.Yu. Dvorko et al. 2011. A general synthetic strategy for the design of new BODIPY fluorophores based on pyrroles with polycondensed aromatic and metallocene substituents. *Chem Eur J* 17:3069–3073.

228. Korostova, S.E., S.G. Shevchenko, M.V. Sigalov, and N.I. Golovanova. 1991. Pyrroles from ketoximes and acetylene. 45. New route to the synthesis of 2,2-dipyrroles. *Khim Geterocicl* 4:460–463.

229. Yurovskaya, M.A., V.V. Druzhinina, M.A. Tyurekhodzhaeva, and Yu.G. Bundel. 1984. Synthesis of *O*-vinyl ethers of 3-acetylindole oximes and their heterocyclization to pyrrolylindoles. *Khim Geterocicl* 1:69–72.

230. Trofimov, B.A., E.Yu. Schmidt, A.I. Mikhaleva et al. 2007. Unexpected formation of 4-methyl-1-vinyl-δ-carboline in the reaction of 3-acetylindole oxime with acetylene. *Mendeleev Commun* 17 (1):40–42.

231. Trofimov, B.A., A.I. Mikhaleva, R.I. Polovnikova et al. 1978. N-Vinyl-2-(2-furyl) pyrroles from acylfurans and acetylene. *Zh Org Khim* 14 (12):2628.

232. Trofimov, B.A., V.K. Voronov, A.I. Mikhaleva et al. 1979. Remote spin–spin interaction $H^1–H^1$ in 2-(2′-furyl)pyrroles. *B Acad Sci USSR Ch* 10:2372–2376.

233. Trofimov, B.A., A.I. Mikhaleva, R.I. Polovnikova et al. 1981. Pyrroles from ketoximes and acetylene. XVIII. 2-(2-Furyl)pyrroles and their 1-vinyl derivatives. *Khim Geterocicl* 8:1058–1061.

234. Zaitsev, A.B., E.Yu. Schmidt, A.M. Vasil'tsov et al. 2005. Unusually facile pyrrolization of 2-acetylcoumarone oxime with acetylene. *Khim Geterocicl* 4:524–529.

235. Engel, N. and W. Steglich. 1978. Einfache Synthese von 2-Aryl- und 2-Heteroaryl-pyrrolen aus N-Allylcarbonsäureamiden. *Angew Chem* 90 (9):719–720.

236. Boukou-Poba, J.P., M. Farnier, and R. Guilard. 1979. A general method for the synthesis of 2-arylpyrroles. *Tetrahedron Lett* 20 (19):1717–1720.

237. Ghabrial, S.S., I. Thomsen, and K.B.G. Torssell. 1987. Synthesis of biheteroaromatic compounds via the isoxazoline route. *Acta Chem Scand* 41:426–434.

238. Reeves, J.T., J.J. Song, Z. Tan et al. 2007. A general synthesis of substituted formylpyrroles from ketones and 4-formyloxazole. *Org Lett* 9 (10):1875–1878.

239. Maurizio, D'A., D.L. Eliana, M. Giacomo et al. 1977. Photochemical substitution of halogenopyrrole derivatives. *J Chem Soc Perkin Trans 1* 16:2369–2373.

240. Steter, H. and W. Haese. 1984. Addition von Aldehyden an Aktivierte Doppelbindungen, XXXIV1. Addition von Aldehyden an Cyclische α-Methylenketone. *Chem Ber* 117 (2):682–693.

241. Hansford, K.A., G.S.A. Perez, W.G. Skene, and W.D. Lubell. 2005. Bis(pyrrol-2-yl) arylenes from the tandem bidirectional addition of vinyl Grignard reagent to aryl diesters. *J Org Chem* 70 (20):7996–8000.

242. Raposo, M.M., A.M.R.C. Sousa, A.M.C. Fonseca, and G. Kirsch. 2006. Synthesis of formyl-thienylpyrroles: Versatile building blocks for NLO materials. *Tetrahedron* 62 (15):3493–3501.

243. Raposo, M.M., A.M.B.A. Sampaio, and G. Kirsch. 2005. Synthesis and characterization of new thienylpyrrolyl-benzothiazoles as efficient and thermally stable nonlinear optical chromophores. *Synthesis* 2:199–210.

244. Trofimov, B.A., A.I. Mikhaleva, R.N. Nesterenko et al. 1977. One-pot synthesis of 2,2′-thienylpyrroles from methyl-2-thienyl ketoxime and acetylene. *Khim Geterocicl* 8:1136–1137.

245. Korostova, S.E., A.I. Mikhaleva, R.N. Nesterenko et al. 1985. Pyrroles from ketoximes and acetylene. XVII. 2-(2-Thienyl)-3-alkylpyrroles and their 1-vinyl derivatives. *Zh Org Khim* 21 (2):406–411.

246. Pozo-Gonzalo, C., J.A. Pomposo, J.A. Alduncin et al. 2007. Orange to black electrochromic behaviour in poly(2-(2-thienyl)-1*H*-pyrrole) thin films. *Electrochim Acta* 52 (14):4784–4791.

247. Pozo-Gonzalo, C., M. Salsamendi, J.A. Pomposo et al. 2008. Influence of the introduction of short alkyl chains in poly(2-(2-thienyl)-1*H*-pyrrole) on its electrochromic behavior. *Macromolecules* 41 (19):6886–6894.

248. Schmidt, E.Yu., B.A. Trofimov, A.I. Mikhaleva et al. 2009. Synthesis and optical properties of 2-(benzo[b]thiophene-3-yl)pyrroles and a new BODIPY fluorophore (BODIPY=4,4-difluoro-4-bora-3a,4a-diaza-s-indacene). *Chem Eur J* 15 (23):5823–5830.

249. Mikhaleva, A.I., R.N. Nesterenko, A.M. Vasil'tsov et al. 1992. Synthesis of 2-(2-selenienyl)pyrrole from methyl-2-selenienylketoxime and acetylene. *Khim Geterocicl* 5:708–710.

250. Trofimov, B.A., E.Yu. Schmidt, A.I. Mikhaleva et al. 2009. Synthesis of 2-(selenophen-2-yl)pyrroles and their electropolymerization to electrochromic nanofilms. *Chem Eur J* 15 (26):6435–6445.

251. Petrova, O.V., A.I. Mikhaleva, L.N. Sobenina et al. 1997. Synthesis of 1H- and 1-vinyl-2-(2-pyridyl)pyrroles by the Trofimov reaction. *Zh Org Khim* 33 (7):1078–1080.

252. Petrova, O.V., A.I. Mikhaleva, L.N. Sobenina et al. 1997. Synthesis of 1*H*- and 1-vinyl-2-pyridylpyrroles by the Trofimov reaction. *Mendeleev Commun* 4:162–163.

253. Trofimov, B.A., E.B. Oleinikova, M.V. Sigalov et al. 1980. Pyrroles from ketoximes and acetylene. XII. Abnormal reactions of α- and β-oxyketoximes. *Zh Org Khim* 16 (2):410–415.

254. Trofimov, B.A., A.G. Kalabin, A.S. Atavin et al. 1975. ¹H and ¹³C NMR spectra of N-vinylpyrroles. *Khim Geterocicl* 3:360–363.

255. Trofimov, B.A., E.B. Oleinikova, Yu.M. Skvortsov, and A.I. Mikhaleva. 1978. Abnormal case of alkylpyrroles formation from β-oxyketone oximes and acetylene. *B Acad Sci USSR Ch* 10:2426.

256. Rapoport, H. and K.G. Holden. 1962. The synthesis of prodigiosin. *J Am Chem Soc* 84 (4):635–642.

257. Williams, R.P. and W.R. Hearn. 1967. Prodigiosin. *Antibiotics* 2:410–432.

258. Boger, D.L. and M. Patel. 1987. Total synthesis of prodigiosin. *Tetrahedron Lett* 28 (22):2499–2502.

259. Boger, D.L. and M. Patel. 1988. Total synthesis of prodigiosin, prodigiosene, and desmethoxyprodigiosin: Diels-Alder reactions of heterocyclic azadienes and development of an effective palladium(ii)-promoted 2,2′-bipyrrolle coupling procedure. *J Org Chem* 53:1405–1415.

260. Furstner, A., H. Szillat, B. Gabor, and R. Mynott. 1998. Platinum- and acid-catalyzed enyne metathesis reactions: Mechanistic studies and applications to the syntheses of steptorubin B and metacycloprodigiosin. *J Am Chem Soc* 120 (33):8305–8314.

261. Carte, B. and D.J. Faulkner. 1983. Defensive metabolites from three nembrothid nudibranchs. *J Org Chem* 48 (14):2314–2318.

262. Kojiri, K., S. Nakajima, H. Suzuku et al. 1993. A new antitumor substance, BE-18591, produced by a streptomycete. I. Fermentation, isolation, physico-chemical and biological properties. *J Antibiot* 46 (12):1799–1803.

263. Eldo, J., A. Arunkumar, and A. Ajayaghosh. 2000. Fluorescent bispyrroles. New building blocks for novel π-conjugated polymers. *Tetrahedron Lett* 41 (32):6241–6244.

264. Ajayaghosh, A. and J. Eldo. 2001. A novel approach toward low optical band gap polysquaraines. *Org Lett* 3 (16):2595–2598.

265. Aldissi, M. (ed.) 1993. *Intrinsically Conducting Polymers: Emerging Technology*. Dordrecht, the Netherlands: Kluwer Academic.

266. Soloducho, J. 1999. Convenient synthesis of polybispyrrole system. *Synth Met* 99:181–189.

267. Soloducho, J., J. Doskocz, J. Cabaj, and S. Roszak. 2003. Practical synthesis of bis-substituted tetrazines with two pendant 2-pyrrolyl or 2-thienyl groups, precursors of new conjugated polymers. *Tetrahedron* 59 (26):4761–4766.

268. Rabias, I., I. Hamerton, B.J. Howline, and P.-J.S. Foot. 1998. Theoretical studies of conducting polymers based on substituted polypyrroles. *Comp Theor Polym S* 3–4:265–271.

269. Vasil'tsov, A.M., A.B. Zaitsev, E.Yu. Schmidt et al. 2001. Unexpected formation of 1-vinyl-2-[2′-(6′-methylpyridyl)]pyrrole from dimethylglyoxime and acetylene in the Trofimov reaction. *Mendeleev Commun* 2:74–75.

270. Zaitsev, A.B., E.Yu. Schmidt, A.M. Vasil'tsov et al. 2006. 1,2-Dioximes in the Trofimov reaction. *Khim Geterocicl* 1:39–46.

271. Zaitsev, A.B., E.Yu. Schmidt, A.M. Vasil'tsov et al. 2005. Dioximes of 1,3-diketones in the Trofimov reaction: Novel 3-substituted pyrroles. *Khim Geterocicl* 6:839–847.

272. Trofimov, B.A., A.B. Zaitsev, E.Yu. Schmidt et al. 2004. From 1,4-diketones to N-vinyl derivatives of 3,3′-bipyrroles and 4,8-dihydropyrrolo[2,3-*f*]indole in just two preparative steps. *Tetrahedron Lett* 45:3789–3791.

273. Vashchenko, A.V., A.M. Vasil'tsov, E.Yu. Schmidt et al. 2006. Structure of dihydropyrroloindoles: Quantum-chemical estimation of conjugation in cyclohexadiene systems. *Zh Strukt Khim* 47 (4):636–641.

274. Korostova, S.E. and A.I. Mikhaleva. 1982. 1,4-Bis(1-vinyl-2-pyrrolyl)benzene. *Zh Org Khim* 18 (12):2620–2621.

275. Schmidt, E.Yu., A.I. Mikhaleva, N.V. Zorina et al. 2008. Unexpected behaviour of 4,4′-diacetyldiphenyl dioxime in the reaction with acetylene in the systems MOH-DMSO (M = Li, K, Cs). *Dokl Chem* 421 (6):779–782.

276. Karabatsos, G.I. and R.A. Taller. 1968. Structural studies by nuclear magnetic resonance-XV: Conformations and configurations of oximes. *Tetrahedron* 24 (8):3347–3360.

277. Potapov, V.M. 1976. *Stereochemistry*. Moscow, Russia: Khimiya.

278. Spencer, T.A. and C.W. Long. 1975. Regioselectivity in α-proton abstraction from ketone methoximes. *Tetrahedron Lett* 16 (45):3889–3892.

279. Cuvigny, T., J.F. Leborgne, M. Larchevegner, and H. Normant. 1976. Hyperbasic media I. Metallation of hydrazones. A direct synthesis of nitriles. *Synthesis* 4:237–238.

280. Le Borgne, J.F., T. Cuvigny, M. Larchevegner, and H. Normant. 1976. Hyperbasic media. II. Metallation of hydrazones. Synthesis of substituted or functional nitriles, 2-iminotetrahydrofurans and γ-butyrolactones. *Synthesis* 4:238–240.

281. Jung, M.E., P.A. Blaier, and J.A. Lowe. 1976. Reactions of oxime dianions: Stereospecificity in alkylation. *Tetrahedron Lett* 17 (18):1439–1442.

282. Trofimov, B.A., S.E. Korostova, L.N. Sobenina et al. 1982. Pyrroles from ketoximes and acetylene. XIX. Regioselectivity of the reaction of alkyl benzyl ketoximes with acetylene. *Khim Geterocicl* 2:193–198.

283. Trofimov, B.A., A.I. Mikhaleva, A.N. Vasiliev et al. 1981. Stereospecific participation of Z-isomer of 1-methyl-2-acetylbenzimidazole oxime in the synthesis of 1-methyl-2-(2-pyrrolyl)benzimidazole by the reaction with acetylene. *Khim Geterocicl* 10:1422.

284. Trofimov, B.A., O.A. Tarasova, A.I. Mikhaleva et al. 2000. A novel facile synthesis of 2,5-di- and 2,3,5-trisubstituted pyrroles. *Synthesis* 11:1585–1590.

285. Miller, S.I. and R. Tanaka. 1970. Nucleophilic additions to acetylenes. In *Selective Organic Transformations*, vol. 1, ed. B.S. Thyagarajan, pp. 143–238. New York: Wiley-Interscience.

286. Trofimov, B.A., A.I. Mikhaleva, S.E. Korostova, and G.A. Kalabin. 1977. Reaction of ketoximes with phenylacetylene as a route to α-phenylpyrroles. *Khim Geterocicl* 2:994.

287. Yurovskaya, M.A., A.Z. Afanasiev, and Yu.G. Bundel. 1984. Synthesis of isomeric (O-phenylvinyl)acetophenoximes and their rearrangement to pyrroles. *Khim Geterocicl* 8:1077–1079.

288. Petrova, O.V., L.N. Sobenina, I.A. Ushakov et al. 2009. Reaction of acetophenone and benzylphenylketone oximes with phenylacetylene: Synthesis of di- and triphenylpyrroles. *Arkivoc* iv:14–20.

289. Korostova, S.E., A.I. Mikhaleva, B.A. Trofimov et al. 1992. Condensation of ketoximes with phenylacetylene. *Khim Geterocicl* 4:485–488.

290. Trofimov, B.A., T.E. Glotova, M.Yu. Dvorko et al. 2010. Triphenylphosphine as an effective catalyst for ketoximes addition to acylacetylenes: Regio- and stereospecific synthesis of (E)-(O)-2-(acyl)vinylketoximes. *Tetrahedron* 66:7527–7532.

291. Glotova, T.E., E.Yu. Schmidt, M.Yu. Dvorko et al. 2010. Synthesis of O-2-(acyl)vinyl-ketoximes and their unusual rearrangements into 2- and 3-acyl-substituted pyrroles. *Tetrahedron Lett* 51:6189–6191.

292. Sobenina, L.N., A.I. Mikhaleva, S.E. Korostova, and M.V. Sigalov. 1990. Interaction of vinylacetylene with ketoximes under the conditions of the Trofimov reaction. *Zh Org Khim* 26 (1):53–56.

293. Sokolyanskaya, L.V., A.N. Volkov, and B.A. Trofimov. 1976. Condensation of diacetylene with ketoximes. *Zh Org Khim* 12 (4):905.

294. Trofimov, B.A., L.V. Sokolyanskaya, A.N. Volkov, and A.I. Mikhaleva. 1980. Addition of ketoximes to diacetylene. *B Acad Sci USSR Ch* 12:2803–2805.

295. Vasil'tsov, A.M., A.I. Mikhaleva, R.N. Nesterenko, and G.A. Kalabin. 1989. Unexpected transformation of phenylthioacetylene to 1,2-di(phenylthio)ethene in the system KOH-DMSO-cyclohexanone oxime. *B Acad Sci USSR Ch* 7:1702.

296. Morita, K., H. Hashimoto, and K. Mutsumura. 1971. Method for the preparation of 3-aminoisoxazole. US Patent 3,598,834.

297. Trofimov, B.A., A.I. Mikhaleva, R.N. Nesterenko et al. 1988. Isomerization during addition of ketoximes to propargylic alcohols. *Zh Org Khim* 24 (12):2618–2619.

298. Trofimov, B.A. and A.I. Mikhaleva. 1980. Reaction of ketoximes with vinylhalides as a novel route to pyrroles and N-vinylpyrroles. *Zh Org Khim* 16 (3):672.

299. Trofimov, B.A. and A.I. Mikhaleva. 1979. Synthesis of pyrroles and N-vinylpyrroles from ketoximes and dihaloethanes. *B Acad Sci USSR Ch* 12:2840.

300. Trofimov, B.A., A.I. Mikhaleva, A.N. Vasiliev et al. 1981. Method for the preparation of 4,5,6,7-tetrahydroindole. USSR Author's Certificate 840,038.

301. Mikhaleva, A.I., I.A. Aliev, R.N. Nesterenko, and G.A. Kalabin. 1982. One-pot synthesis of 4,5-dihydrobenz[g]indole and its 1-vinyl derivative from α-tetralone oxime and vinyl chloride. *Zh Org Khim* 18 (10):2229–2230.

302. Mikhaleva, A.I., B.A. Trofimov, A.N. Vasiliev, and G.A. Komarova. Method for the preparation of 2- or 2,3-substituted pyrroles. 1982. USSR Author's Certificate 979,337.

303. Mikhaleva, A.I., B.A. Trofimov, A.N. Vasiliev et al. 1982. Pyrroles from ketoximes and acetylene. XXII. Dihaloethanes instead of acetylene in the reaction with cyclohexanone oxime. *Khim Geterocicl* 9:1202–1204.

304. Sheradsky, T. 1970. The rearrangement of O-vinyloximes. A new synthesis of substituted pyrroles. *Tetrahedron Lett* 11 (1):25–26.

305. Freeman, J.P. 1973. Less familiar reactions of oximes. *Chem Rev* 73 (4):283–292.

306. Winterfeldt, E. and W. Krohn. 1967. Cycloaddition mit Ketoximen. *Angew Chem* 79:722–723.

307. Winterfeldt, E. and W. Krohn. 1969. Additionen an die Dreifachbindung. XII. Darstellung des N-Oxides eines nicht basischen vinylogen Urethans. *Chem Ber* 102:2336–2345.

308. Winterfeldt, E. and W. Krohn. 1969. Additionen on die Dreifachbindung. XIII. Nitron-Addukte mit Acetylendicarbonsauredimethylester. *Chem Ber* 102:2346–2361.

309. Ochiai, M., M. Obayashi, and K. Morita. 1967. A new 1,3-dipolar cycloaddition reaction. Synthesis of some isoxazolidine derivatives. *Tetrahedron* 23:2641–2648.

310. Yakimovich, S.I. and K.G. Golodova. 1974. In *Reactivity and Mechanisms of Reactions of Organic Compounds*, part 2, ed. T.I. Temnikova, pp. 186–213. Leningrad, Russia: LGU.

311. Trofimov, B.A., A.I. Mikhaleva, A.N. Vasiliev, and M.V. Sigalov. 1979. O-Vinylacetoxime. *B Acad Sci USSR Ch* 3:695–696.

312. Tarasova, O.A., S.E. Korostova, A.I. Mikhaleva et al. 1994. Nucleophilic addition to alkynes in superbase catalytic systems. V. Vinylation of ketoximes. *Zh Org Khim* 30 (6):810–816.

313. Trofimov, B.A., A.I. Mikhaleva, A.M. Vasil'tsov et al. 2000. Synthesis and thermal stability of O-vinylketoximes. *Synthesis* 8:1125–1132.

314. Zaitsev, A.B., A.M. Vasil'tsov, E.Yu. Schmidt et al. 2002. O-vinyldiaryl- and O-vinylaryl(hetaryl)ketoximes: A breakthrough in O-vinyloxime chemistry. *Tetrahedron* 58 (50):10043–10046.

315. Trofimov, B.A., S.E. Korostova, A.I. Mikhaleva et al. 1983. Transformation of O-vinylalkylaryl(hetaryl)ketoximes to pyrroles in the system KOH-DMSO. *Khim Geterocicl* 2:273.

316. Mikhaleva, A.I., O.V. Petrova, V.M. Bzhezovsky, and N.A. Kalinina. 1987. Unexpected transformation of O-allyl ether of cyclohexanone oxime in superbase media. *B Acad Sci USSR Ch* 7:1677–1678.

317. Trofimov, B.A., A.I. Mikhaleva, and O.V. Petrova. 1991. Synthesis of 5,6,7,8-tetrahydroquinoline from O-allyl ether of cyclohexanone oxime in superbase media. *Zh Org Khim* 27 (9):1941–1946.

318. Koyama, J., Y. Suzuta, H. Irie, and I. Katayama. 1979. A new method for constructing pyridine ring: Thermolysis of oxime O-allyl ether. *Heterocycles* 12:157.

319. Ranganathan, S., D. Ranganathan, R.S. Sidhu, and A.K. Mehrotra. 1973. The novel [2,3[-sigmatropic rearrengement of oxime-O-allyl ethers. *Tetrahedron Lett* 37:3577–3578.

320. Eckersley, A. and N.A.J. Rogers. 1974. Thermolysis of oxime O-allyl ethers. *Tetrahedron Lett* 18:1661–1664.

321. Pinna, G.A., M.A. Pirisi, and G. Paglietti. 1990. Addition reactions of acetylenic esters upon 6-substituted-α-tetralone ketoximes and conversion of the adducts into 4,5-dihydro-1H-benz[g]indoles. *J Chem Res Synopses* 11:360–361.

322. Trofimov, B.A., S.E. Korostova, A.I. Mikhaleva et al. 1983. 4H-2-Oxy-2,3-dihydropyrroles as intermediates in the formation of pyrroles from ketoximes and acetylene in the system KOH-DMSO. *Khim Geterocicl* 2:276.

323. Trofimov, B.A., S.G. Shevchenko, S.E. Korostova et al. 1985. Novel route to 3*H*-pyrroles. *Khim Geterocicl* 11:1573–1574.

324. Korostova, S.E., S.G. Shevchenko, and M.V. Sigalov. 1991. Novel synthesis of 3*H*-pyrroles. *Khim Geterocicl* 10:1371–1374.

325. Trofimov, B.A., S.G. Shevchenko, S.E. Korostova et al. 1989. Diene autocondensation of 3,3-dimethyl-2-phenyl- and 3,3-dimethyl-2-(2-thienyl)-3*H*-pyrroles. *Khim Geterocicl* 11:1566–1567.

326. Pivnenko, V.P. 1970. Synthesis of vinylacetoxime ester. *Zh Org Khim* 6 (10):2146.

327. Shmidt, E.Yu., N.V. Zorina, I.A. Ushakov et al. 2009. Unexpected formation of 2,4,6-triphenylpyridine in the synthesis of 2,5-diphenylpyrrole from acetophenone oxime and phenylacetylene. *Khim Geterocicl* 7:1089–1091.

328. Irie, H., I. Katayama, Y. Mizuno et al. 1979. Thermolysis of oxime O-allyl ethers: A new method for synthesis of pyridine derivatives. *Heterocycles* 12 (6):771–773.

329. Jones, R.R. and R.G. Bergman. 1972. p-Benzyne. Generation as an intermediate in a thermal isomerization reaction and trapping evidence for the 1,4-benzenediyl structure. *J Am Chem Soc* 94 (2):660–661.

330. Maretina, I.A. and B.A. Trofimov. 2006. Endiyne antibiotics and their models: New possibilities of acetylene chemistry. *Usp Khim* 75 (9):913–935.

331. Knunyanz, I.L. and B.P. Fabrichny. 1954. In *Reactions and Methods of Investigations of Organic Compounds*, vol. 3, Beckmann rearrangement, eds. V.M. Rodionov, B.A. Kazansky, N.N. Mel'nikov, pp. 137–251. Moscow, Russia: Goskhimizdat.

332. Petrova, O.V., A.I. Mikhaleva, L.N. Sobenina, and B.A. Trofimov. 2010. *napa*-Terphenyl as unexpected side product of 2,5-diphenylpyrrole synthesis from acetophenone oxime and phenylacetylene. *Zh Org Khim* 46 (3):456–458.

333. Trofimov, B.A., A.I. Mikhaleva, O.V. Petrova, and M.V. Sigalov. 1985. Rearrangement in reductive methylenation of 2-methylcyclohexanone with the system KOH-DMSO. *Zh Org Khim* 21 (12):2613–2614.

334. Trofimov, B.A., A.M. Vasil'tsov, O.V. Petrova, and A.I. Mikhaleva. 1988. Unexpected transformation of cyclohexanone to methyl(1-cyclohexenyl)methylsulfoxide in the system KOH-DMSO. *Zh Org Khim* 24 (9):2002–2003.

335. Trofimov, B.A., A.I. Mikhaleva, O.V. Petrova, and M.V. Sigalov. 1988. Reductive methylenation of ketones with a suspension KOH-dimethylsulfoxide. *Zh Org Khim* 24 (10):2095–2101.

336. Trofimov, B.A., A.M. Vasil'tsov, A.I. Mikhaleva et al. 1989. Mechanoactivation of the superbase system KOH-DMSO: Efficiency of application in the reaction with cyclohexanone. *B Acad Sci USSR Ch* 12:2879–2880.

337. Trofimov, B.A., O.V. Petrova, A.M. Vasil'tsov, and A.I. Mikhaleva. 1990. The reaction of cyclic ketones with the system KOH-DMSO. *B Acad Sci USSR Ch* 7:1601–1605.

338. Vasiltsov, A.M., O.V. Petrova, A.I. Mikhaleva, and B.A. Trofimov. 1991. Study of the base-catalysed condensation of DMSO with cyclohexanone under mechanoactivation. *Sulfur Lett* 13 (4):171–174.

339. Petrova, O.V., A.I. Mikhaleva, B.A. Trofimov, and A.M. Vasil'tsov. 1992. One-pot synthesis of 1-methyl-2-methylenecyclohexanol from 2-methylcyclohexanone. *Zh Org Khim* 28 (1):92–94.

340. Petrova, O.V., A.M. Vasil'tsov, A.I. Mikhaleva et al. 1992. Reaction of acetophenone with the system KOH/DMSO under mechanoactivation. *Dokl Chem* 326 (2):279–281.

341. Schmidt, E.Yu., N.V. Zorina, A.I. Mikhaleva et al. 2010. 1,3,5-Triphenylbenzene as unexpected side product of 2,5-diphenylpyrrole synthesis from acetophenone oxime and phenylacetylene. *Zh Org Khim* 46 (3):461–462.

342. Petrova, O.V., I.A. Ushakov, L.N. Sobenina et al. 2010. 2,3,6-Triphenylpyridine – an unusual side product of the synthesis of N-vinyl-2,3-diphenylpyrrole from benzylphenyl ketoxime and acetylene in the system KOH-DMSO. *Khim Geterocicl* 6:938–940.

343. Petrova, O.V., I.A. Ushakov, L.N. Sobenina et al. 2010. Formation of N-benzylbenzamide from benzylphenylketone oxime in the system C_2H_2-KOH-DMSO: Key role of acetylene. *Zh Org Khim* 46 (9):1412–1413.

344. Hoch, J. 1934. The action of organomagnesium compounds on ketoximes. *C R Hebd Seances Acad Sci* 198:1865–1868.

345. Campbell, K.N. and J.F. McKenna. 1939. The action of Grignard reagents on oximes. 1. The action of phenylmagnesium bromide on mixed ketoximes. *J Org Chem* 4:198–205.

346. Campbell, K.N., B.K. Campbell, and E.P. Chaput. 1943. The reaction of Grignard reagents with oximes. II. The action of aryl Grignard reagents with mixed ketoximes. *J Org Chem* 8:99–102.

347. Campbell, K.N., B.K. Campbell, J.F. McKenna, and E.P. Chaput. 1943. The reaction of Grignard reagents with oximes. III. The mechanism of the action of arylmagnesium halides on mixed ketoximes. A new synthesis of ethylenimines. *J Org Chem* 8:103–109.

348. Campbell, K.N., B.K. Campbell, L.G. Hess, and I.J. Schaffner. 1944. The action of Grignard reagents on oximes. IV. Aliphatic Grignard reagents and mixed ketoximes. The action of Grignard reagents on oximes. *J Org Chem* 9:184–186.

349. Bowie, J.H. and B. Nussey. 1973. Thermal rearrangements of 2,3-diphenyl-2H-azirine. *J Chem Soc Perkin Trans 1* 1:1693–1696.

350. Petrova, O.V., I.A. Ushakov, L.N. Sobenina et al. 2011. 2,4,5-Triphenylimidazole as indicator of intermediate formation of 1,2-diphenyl-2*H*-azirine in the synthesis of 1-vinyl-2,3-diphenylpyrrole from benzylphenyl ketoxime and acetylene. *Dokl Chem* 436 (6):785–787.

351. Korostova, S.E., A.I. Mikhaleva, S.G. Shevchenko et al. 1986. A peculiar methylthiylation in the course of the Trofimov reaction. *Sulfur Lett* 5 (2):39–46.

352. Korostova, S.E., A.I. Mikhaleva, S.G. Shevchenko et al. 1988. Dipyrrolylethanes from ketoximes and acetylene. *Zh Org Khim* 24 (8):1789–1790.

353. Korostova, S.E. and S.G. Shevchenko. 1991. Unexpected formation of α-ethylpyrroles in the Trofimov reaction. *Zh Org Khim* 27 (9):2027.

354. Arzel, E., P. Grelier, M. Labaeid et al. 2001. New synthesis of benzo-δ-carbolines, cryptolepines, and their salts: In vitro cytotoxic, antiplasmodial, and antitrypanosomal activities of δ-carbolines, benzo-δ-carbolines, and cryptolepines. *J Med Chem* 44 (6): 949–960.

355. Anders, J.T. and P. Langer. 2004. Domino cyclization/electrocyclization/elimination reactions of arylacetonitriles with N,N'-bis(1-naphthyl)oxaldiimidoyl dichlorides: Efficient synthesis of fluorescent 15H-benzo[*h*]benzo[6,7]indolo[3,2-*b*]quinolines. *Eur J Org Chem* 5020–5026.

356. Bonjean, K., M.C. De Pauw-Gillet, M.P. Defresne et al. 1998. The DNA intercalating alkaloid cryptolepine interferes with topoisomerase II and inhibits primarily DNA synthesis in B16 melanoma cells. *Biochemistry* 37:5136–5146.

357. Tripathi, R.P., R.C. Mishra, N. Dwivedi, N. Tewari, and S.S. Verma. 2005. Current status of malaria control. *Curr Med Chem* 12 (22):2643–2659.

358. Castagnoli, N., Jr., J. Rimoldi, J. Bloomquist, and K.P. Castagnoli. 1997. Potential metabolic bioactivation pathways involving cyclic tertiary amines and azaarenes. *Chem Res Toxicol* 10:924–940.

359. Sevodin, V.P., V.S. Velezheva, and N.N. Suvorov. 1981. Indole derivatives. 122. Novel synthesis of δ-carbolines from 1-acetylindolinone-3. *Khim Geterocicl* 3:368–371.

360. Rocca, P., F. Marsais, A. Codard, and G. Queguiner. 1993. Connection between metalation and cross-coupling strategies. A new convergent route to azacarbazoles. *Tetrahedron* 49:49–64.

361. Trofimov, B.A., E.Yu. Schmidt, A.I. Mikhaleva et al. 2009. One-pot assembly of 4-methylene-3-oxa-1-azabicyclo[3.1.0]hexanes from alkyl aryl(hetaryl) ketoximes, acetylene, and aliphatic ketones: A new three-component reaction. *Tetrahedron Lett* 50:3314–3317.

362. Hili, R. and A.K. Yudin. 2006. Readily available unprotected amino aldehydes. *J Am Chem Soc* 128:14772–14773.

363. He, Z. and A.K. Yudin. 2010. A versatile synthetic platform based on strained propargyl amines. *Angew Chem Int Edit* 49:1607–1610.

364. Bergmeier, S.C. and S.J. Katz. 2002. A method for the parallel synthesis of multiply substituted oxazolidinones. *J Comb Chem* 4:162–166.

365. Kakeya, H., M. Morishita, H. Koshino et al. 1999. Cytoxazone: A novel cytokine modulator containing a 2-oxazolidinone ring produced by streptomyces sp. *J Org Chem* 64:1052–1053.

366. Kearney, J.A., K. Barbadora, E.O. Mason et al. 1999. In vitro activities of the oxazolidinone compounds linezolid (PNU-100766) and eperzolid (PNU-100592) against middle ear isolates of streptococcus pneumoniae. *Int J Antimicrob Ag* 12:141–144.

367. Sauerberg, P., J. Chen, E. WoldeMussie, and H. Rapoport. 1989. Cyclic carbamate analogs of pilocarpine. *J Med Chem* 32:1322–1326.

368. Muller, G., J. Napier, M. Balestra et al. 2000. (–)-Spiro[1-azabicyclo[2.2.2]octane-3,5′-oxazolidin-2′-one], a conformationally restricted analogue of acetylcholine, is a highly selective full agonist at the α7 nicotinic acetylcholine receptor. *J Med Chem* 43:4045–4050.

369. Foley, P.J., Jr. 1969. A novel method of converting aldehydes into nitriles under mild conditions. The reaction of dialkyl hydrogen phosphonates with oximes. *J Org Chem* 34:2805–2806.

370. Clive, D.L.Y. 1970. A new method for conversion of aldoximes into nitriles: Use of chlorothionoformates. *Chem Commun* 16:1014–1015.

371. Rosini, G., C. Baccolini, and S. Cacchi. 1973. Nitriles from aldoximes. A new reaction of phosphonitrilic chloride. *J Org Chem* 38:1060–1061.

372. Suzuki, H., T. Fuchito, A. Iwaso et al. 1978. Diphosphorus tetraiodide as a reagent for converting epoxides into olefins and aldoximes into nitriles under mild conditions. *Synthesis* 12:905–908.

373. Rogic, M.M., J.F. Van Peppen, K.P. Klein et al. 1974. New facile method for conversion of oximes to nitriles. Preparation and acid-catalyzed transformation of aldehyde oxime ortho esters. *J Org Chem* 39:3424–3426.

374. Sosnovsky, G. and J.A. Krogh. 1979. The utilization of sulfur, sulfenyl, selenenyl, and seleninyl chlorides in the conversion of aldoximes to nitriles. *Z Naturforsch B* 34:511–515.

375. Olah, G.A. and Y.D. Vankar. 1978. Synthetic methods and reactions. 52. Preparation of nitriles from aldoximes *via* dehydration with trimethylamine/sulfur dioxide complex. *Synthesis* 9:702–703.

376. Sosnovsky, G. and J.A. Krogh. 1978. A versatile method for the conversion of aldoximes to nitriles using selenium dioxide. *Synthesis* 9:703–705.

377. Krause, J.G. and S. Shaikh. 1975. Nitriles from aldoximes *via* oxime-O-sulfonates. *Synthesis* 8:502.

378. Carotti, A., F. Compagna, and R. Ballini. 1979. An easy, high-yield conversion of aldoximes to nitriles. *Synthesis* 1:56–58.

379. Hendrickson, J.B., K.W. Bair, and P.M. Keehn. 1976. An efficient synthesis of nitriles from aldoximes. *Tetrahedron Lett* 17 (8):603–604.

380. Vesterager, N.O., E.B. Pedersen, and S.O. Laweson. 1974. Studies on organophosphorus compounds. VIII. The reaction of oximes with HMPA at elevated temperature. *Tetrahedron* 30:2509–2514.

381. Chaabouni, R. and A. Laurent. 1973. Action d'organomagnesien sur l'oxime du camphre: Formation de nitril. *Tetrahedron Lett* 14 (13):1061–1064.

382. Kukhar, V.P. and V.I. Pasternak. 1974. Nitriles from aldoximes and N,N-dimethyldichloromethaniminium chloride. *Synthesis* 8:503.

383. Sakamoto, T. and K. Ohsawa. 1999. Palladium-catalyzed cyanation of aryl and heteroaryl iodides with copper(I) cyanide. *J Chem Soc Perkin Trans 1* 16:2323–2326.

384. Hughes, T.V. and M.P. Cava. 1999. Electrophilic cyanations using 1-cyanobenzotriazole: Sp2 and sp carbanions. *J Org Chem* 64 (1):313–315.

385. Sardarian, A.R., Z. Shahsavari-Fard, H.R. Shahsavari, and Z. Ebrahimi. 2007. Efficient Beckmann rearrangement and dehydration of oximes *via* phosphonate intermediates. *Tetrahedron* 48 (14):2639–2643.

386. Dohi, T., K. Morimoto, N. Takenaga et al. 2007. Direct cyanation of heteroaromatic compounds mediated by hypervalent iodine(III) reagents: In situ generation of PhI(III)–CN species and their cyano transfer. *J Org Chem* 72 (1):109–116.

387. Trofimov, B.A., A.I. Mikhaleva, S.E. Korostova et al. 1976. Base-catalyzed dehydration of aldoximes to DMSO. *B Acad Sci USSR Ch* 3:690–691.

388. Schmidt, E.Yu., A.I. Mikhaleva, E.Yu. Senotrusova et al. 2008. Formation of 1-vinyl-5-phenylpyrrol-2-carbonitrile during vinylation of 1-vinyl-5-phenylpyrrol-2-carbaldehyde oxime with acetylene. *Zh Org Khim* 44 (9):1412–1413.

389. Hantzsch, A. 1891. Über Oxime von Aldehyden und α-Ketonsaure. *Chem Ber* 24:36–51.

390. Hantzsch, A. and A. Lucas. 1895. Über die oxime des symmetrischen trimethylbenzaldehyds. *Chem Ber* 28:744–753.

391. Reissert, A. 1908. Über einige oxime der *o*-nitrotoluol-reihe und ihre umwandlungen. *Chem Ber* 41:3810–3816.

392. Jordan, E. and C.R. Hauser. 1936. Conversion of aldoximes to carboxylic acids by means of hot alkali. The elimination of water from aldoximes. *J Am Chem Soc* 58:1227–1232.

393. Tigchelaar-Lutjeboer, H.D., H. Bootsma, and J.F. Arens. 1960. Chemistry of acetylenic ethers. XLIII. Reactions of oximes with ethoxyethylene. *Rec Trav Chem* 79:888–894.

394. Bernhart, C. and C.G. Wermuth. 1977. Benzonitriles from benzaldoximes and enamine. *Synthesis* 5:338–339.

395. Heaney, F., B.M. Kelly, S. Bourke et al. 1999. Site selectivity in the addition of ketoximes to activated allenes and alkynes; N- versus O-alkylation. *J Chem Soc Perkin Trans 1* 2:143–148.

396. Trofimov, B.A., E.Yu. Schmidt, A.I. Mikhaleva et al. 2000. An unusually fast nucleophilic addition of amidoximes to acetylene. *Mendeleev Commun* 1:29–30.

397. Trofimov, B.A., E.Yu. Schmidt, A.M. Vasil'tsov et al. 2001. Synthesis and properties of O-vinylamidoximes. *Synthesis* 16:2427–2430.

398. Heindel, N.D. and V.C. Chun. 1971. Imidazole carboxylates by a Claisen-type rearrangement of amidoxime-propiolate adducts. *Tetrahedron Lett* 12 (18):1439–1440.

399. Elay, F. and R. Lenaers. 1962. The chemistry of amidoximes and related compounds. *Chem Rev* 62:155–183.

400. Volkov, A.N., L.V. Sokolyanskaya, and B.A. Trofimov. 1976. Condensation of diacetylene with amidoximes. *B Acad Sci USSR Ch* 6:1430.

401. Park, C.A., C.F. Beam, E.M. Keiser et al. 1976. Preparation of 2-isoxazolines from C(α), O-dilithiooximes and aldehydes and ketones. *J Heterocyclic Chem* 13:449–453.

402. Sandifer, R.M., L.M. Shaffer, W.M. Hollinger et al. 1976. An easy method for the preparation of 4-acylisoxazoles from C(α), O-dianions of oximes. *J Heterocycl Chem* 13:607.

403. Epiotis, N.D. 1973. Attractive nonbonded interactions in organic molecules. *J Am Chem Soc* 95:3087–3096.

404. Mastroianni, M., M. Pikal, and S. Lindenbaum. 1972. Effect of dimethyl sulfoxide, urea, guanidine hydrochloride, and sodium chloride on hydrophobic interactions. Heats of dilution of tetrabuthylammonium bromide and lithium bromide in mixed aqueous solvent systems. *J Phys Chem* 76:3050.

405. Caubere, P. 1978. In *Topics in Current Chemistry, Organic Chemistry*, vol. 73, pp. 50–103.

406. Sanders, J.F., K. Hovins, and J.B. Engberts. 1974. Nucleophilic addition of aliphatic hydroxyl amines to p-tolylsulfonylacetylenes. Competitive nitrogen and oxygen attack. *J Org Chem* 39:2641–2643.

407. Smith, P.A. 1966. In *The Chemistry of Open-Chain Organic Nitrogen Compounds.* vol. VII, p. 10. New York: N.A. Benjamin.

408. Aurich, H.G. and K. Hahn. 1979. Bildung von Nitronen und anderen Folgeprodukten bei der Umsetzung von N-Alkyl- und N-Arylhydroxylaminen mit verschiedenen Acetylenderivaten. ed. H. Zahn, *Chem Ber* 112:2769–2775.

409. Sheradsky, T. 1966. Application of the Fischer indole synthesis to the preparation of benzofurans. *Tetrahedron Lett* 7 (43):5225–5227.

410. Sheradsky, T. 1966. O-(2,4-Dinitrophenyl)oximes. Synthesis and cyclization to 5,7-dinitrobenzofurans. *J Heterocycl Chem* 4:413–414.

411. Mooradian, A. 1967. Rearrangement of substituted O-aryl oximes to 5- and 7-substituted benzofurans. *Tetrahedron Lett* 8 (5):407–408.

412. Kaminsky, D., J. Shavel, and R.I. Meltzer. 1967. Fischer indole-like synthesis. An approach to the preparation of benzofurans and benzothiophenes. *Tetrahedron Lett* 8 (10):859–861.

413. Mooradian, A. and P.E. Dupont. 1967. The rearrangement of O-aryl oximes. *Tetrahedron Lett* 8 (30):2867–2870.

414. Unterhalt, B. 1974. In *Methodicum Chimicum*, ed. F. Korte, Bd. 6, pp. 413–447. New York: Academic Press.

415. Cattanach, C.J. and R.G. Rees. 1971. Preparation of 4a-alkoxy-1,2,3,4,4a,9b-hexahydro- and -1,2,3,4-tetrahydro-benzofuro[3,2-c]pyridines. *J Chem Org Soc C:* 1:53–60.

416. Sheradsky, T. and G. Salemnick G. 1971. Thermal rearrangement of O-(2-pyridyl) oximes. *J Org Chem* 36:1061–1063.

417. Sigalov, M.V., E.Yu. Schmidt, and B.A. Trofimov. 1987. Peculiarities of N-vinylpyrroles protonation. *B Acad Sci USSR Ch* 5:1146–1149.

418. Sigalov, M.V., E.Yu. Schmidt, and B.A. Trofimov. 1988. Protonated forms of N-vinylpyrroles. *Khim Geterocicl* 3:334–338.

419. Sigalov, M.V., E.Yu. Schmidt, A.I. Mikhaleva et al. 1993. NMR spectra and structure of protonated 1-vinylpyrroles. *Khim Geterocicl* 1:48–57.

420. Sigalov, M.V., G.A. Kalabin, A.I. Mikhaleva et al. 1980. Pyrroles from ketoximes and acetylene. 11. Conformation of 1-vinylpyrroles: ^{1}H NMR study. *Khim Geterocicl* 3:328–330.

421. Sigalov, M.V., B.A. Trofimov, A.I. Mikhaleva, and G.A. Kalabin. 1981. ^{1}H and ^{13}C NMR study conformational and electronic structure of 1-vinylpyrroles. *Tetrahedron* 37:3051–3059.

422. Sigalov, M.V., S. Toyota, M. Oki, and B.A. Trofimov. 1994. Dynamic NMR as a non-destructive method for determination of rates of dissociation. XXI. Dissociation in 1-(1-haloethyl)pyrrolium cations. *B Chem Soc Jpn* 67:1161–1169.

423. McDaniel, D.H. and R.E. Vallee. 1963. Strong hydrogen bonds. I. The halide-hydrogen halide systems. *Inorg Chem* 2:996–1001.

424. Sigalov, M.V., S. Toyota, M. Oki M, and B.A. Trofimov. 1994. Addition of hydrogen halides to 1,2-dialkylpyrrolium cations. *B Chem Soc Jpn* 67 (7):1872–1878.

425. Sigalov, M.V., E.Yu. Schmidt, S.E. Korostova, and B.A. Trofimov. 1991. Protonation of 1,4-bis(1-vinylpyrrolyl)benzene. *Khim Geterocicl* 8:1041–1045.

426. Sigalov, M.V., E.Yu. Schmidt, and B.A. Trofimov. 1987. A case of successful competition during protonation of the furan and pyrrole ring. *B Acad Sci USSR Ch* 9:2136–2137.

427. Sigalov, M.V., E.Yu. Schmidt, A.B. Trofimov, and B.A. Trofimov. 1989. Protonated forms of 2-(2-furyl)pyrroles and their interconversions. *Khim Geterocicl* 10:1343–1355.

428. Sigalov, M.V., A.B. Trofimov, E.Yu. Schmidt, and B.A. Trofimov. 1993. Protonated forms of 2-(2-thienyl)pyrroles. ^{1}H NMR and MNDO studies. *Khim Geterocicl* 6:825–833.

429. Sigalov, M.V., E.Yu. Schmidt, A.B. Trofimov, and B.A. Trofimov. 1992. Protonated forms of 2-(2-furyl)pyrroles and their interconversion: ^{1}H NMR and quantum-chemical (MNDO) study. *J Org Chem* 57:3934–3938.

430. Sigalov, M.V., A.B. Trofimov, E.Yu. Schmidt, and B.A. Trofimov. 1993. Protonation of 2-(2-thienyl)pyrrole and 2-(2-thienyl)-1-vinylpyrroles. *J Phys Org Chem* 6:471–477.

431. Trofimov, A.B., B.A. Trofimov, N.M. Vitkovskaya, and M.V. Sigalov. 1991. Quantum-chemical study of protonated forms of 2-(2-furyl)pyrrole. *Khim Geterocicl* 6:746–753.

432. Trofimov, B.A., S.E. Korostova, A.I. Mikhaleva et al. 1977. Selective hydrogenation of the vinyl group of N-vinylpyrroles. *Khim Geterocicl* 2:215–216.

433. Trofimov, B.A., S.E. Korostova, L.N. Balabanova et al. 1978. Pyrroles from ketoximes and acetylene. 2. Exhaustive hydrogenation of 4,5,6,7-tetrahydroindole and its N-vinyl derivative. *Khim Geterocicl* 3:347–349.

434. Rice, R.G. and E.J. Kohn. 1955. Raney nickel catalyzed N-alkylation of aniline and benzidine with alcohols. *J Am Chem Soc* 77:4052–4054.

435. Rice, R.G., E.J. Kohn, and L.W. Daasch. 1958. Alkylation of amines with alcohols catalyzed by Raney nickel. II. Aliphatic amines. *J Org Chem* 23:1352–1354.

436. Booth, H. and F.E. King. 1958. Synthetic and stereochemical investigation of reduced cyclic bases. VI. Synthesis of *trans*-octahydroindole and reinvestigation of its Hofmann degradation products. *J Chem Soc* 2688–2693.

437. Kojima, M. and Y. Tomioka. 1975. Aminoethylation. II. Reactions of meso-*cis*-cyclohexenimine and its derivative with diethylmalonate. Synthesis of *trans*-octahydroindole. *Yakugaku Zasshi* 95:889–892.

438. Kroy, L.R. and N.G. Reinecke. 1967. A convenient preparation of pyrrolizidine by reductive cyclization. *J Org Chem* 32:225–227.

439. Trofimov, B.A., A.I. Mikhaleva, A.M. Vasil'tsov et al. 2010. Selective dehydrogenation of 4,5,6,7-tetrahydroindole to indole in nickel-sulfide catalyst. *Dokl Chem* 434 (5): 636–638.

440. Trofimov, B.A., A.I. Mikhaleva, E.Yu. Schmidt et al. 2011. Catalyst of selective dehydrogenation of 4,5,6,7-tetrahydroindole to indole. *Dokl Chem* 437 (4):504–506.

441. Lim, S., I. Jabin, and G. Revial. 1999. Reaction of cyclohexanone imines with substituted nitroolefins. New synthesis of tetrahydroindole derivatives. *Tetrahedron Lett* 40:4177–4180.

442. Trofimov, B.A., A.M. Vasil'tsov, I.A. Ushakov et al. 2007. A peculiar selective rearrangement during NiS-catalyzed dehydrogenation of 1*H*-4,5-dihydrobenz[g]indole. *Mendeleev Commun* 17 (5):296–298.

443. Sobenina, L.N., L.A. Es'kova, A.I. Mikhaleva et al. 1999. Nucleophilic addition of pyrroles to vinyl sulfones. *Zh Org Khim* 35 (8):1226–1231.

444. Vokin, A.I., T.I. Vakul'skaya, and N.M. Murzina. 1999. Spectral monitoring of tricyanovinylation of 2-phenylpyrrole and 4,5,6,7-tetrahydroindole. Self-association of the target products. *Zh Org Khim* 36 (10):1539–1544.

445. Trofimov, B.A., L.N. Sobenina, A.P. Demenev et al. 2001. Tricyanovinylpyrroles: Synthesis, conformational structure, photosensitizing properties and electrical conductivity. *Arkivoc* ix:37–48.

446. Belen'kii, L.I. 1980. Activity and selectivity in the electrophilic substitution of five-membered heterocycles. *Khim Geterocicl* 12:1587–1605.

447. Trofimov, B.A., S.E. Korostova, A.I. Mikhaleva et al. 1982. Pyrroles from ketoximes and acetylene. XXI. Trifluoroacetylation of 2-(2-furyl)- and 2-(2-thienyl)pyrroles and their 1-vinyl derivatives. *Zh Org Khim* 18 (4):894–899.

448. Drichkov, V.N., L.N. Sobenina, T.I. Vakul'skaya et al. 2008. Copper (I)-promoted selective 2-tricyanoethenylation of N-methylpyrroles. *Synthesis* 16:2631–2635.

449. Sobenina, L.N., V.N. Drichkov, O.V. Petrova et al. 2008. Unprecedented easy migration of functionalized enol substituent in the pyrrole ring. *Zh Org Khim* 44 (2):246–255.

450. Gorshkov, A.G., V.K. Turchaninov, G.N. Kurov, and G.G. Skvortsova. 1979. On interaction of 10-vinylphenothiazine with tetracyanoethylene. *Zh Org Khim* 15 (6):767–770.

451. Gorshkov, A.G., E.S. Domnina, V.K. Turchaninov et al. 1983. Reaction of tetracyanoethylene with 1-vinylindole. *Khim Geterocicl* (7):951–954.

452. Gorshkov, A.G., E.S. Domnina, A.I. Mikhaleva, and G.G. Skvortsova. 1985. Unexpected transformation of 1-(2,2,3,3-tetracyanoethyleno-1-cyclobutyl)pyrroles in methanol. *Khim Geterocicl* 6:848.

453. Trofimov, B.A., L.N. Sobenina, V.N. Drichkov et al. 2009. Tricyanovinylation of 2-aryl-1-vinylpyrroles: Solvent- and substituent controlled chemo- and regioselectivity. *Tetrahedron* 65:4326–4331.

454. Reppe, W. 1954. *Polyvinylpyrrolidon*. Weinheim/Bergstr:VCH, Germany.

455. Musaev, U.N., S.Kh. Nasiroba, T.A. Azimov et al. 1974. Synthesis of pharmacologically active copolymers of N-vinylpyrrolidone and their properties. *Khim Farm Zh* 8:36–38.

456. Ikeda, M., H. Sato, E. Torii et al. 1974. Electrophotographic element. Great Britain Patent 1,348,437.

457. Matsumoto, S. and T. Kaieko. 1974. Electrophotographic material. Japan Patent 4,902,8455.

458. Irwin, W. J. and D.L. Wleeler. 1972. The reaction of methyl pyrrole-2-carboxylate with epoxides. *Tetrahedron* 28:1113–1121.

459. Otsuki, H., I. Okano, and T. Takeda. 1946. Synthesis of vinylcarbazole. *J Soc Chem Ind Jn* 49:169–170.

460. Otsuki, H. 1949. Polyvinylcarbazole from 9-ethylolcarbazole. Japan Patent 174,356.

461. Miller, H.F. and R.G. Flowers. 1947. 9-Vinylcarbazole. US Patent 2,426,465

462. Hisao, O., M. Tadashi, and A. Kiichi. 1978. Studies of heterocyclic compounds. XIV. Synthesis of 1-vinylpyrazoles by dehydrohalogenation of 1-(2-haloethyl)pyrazoles. *J Pharm Soc Jn* 98:165–171.

463. Kolesnikov, S.G. 1960. *Synthesis of Vinyl Derivatives of Aromatic and Heteroaromatic Compounds*. Moscow, Russia: AN SSSR.

464. Johnson, W. 1937. Vinylpyrroles. Great Britain Patent 470,077.

465. Reppe, W., E. Keyssner, and F. Nicolai. 1937. Vinyl compounds. Germany Patent 646,995.

466. Lebedev, A.Y., V.V. Izner, D.N. Kazyul'kin et al. 2002. Palladium-catalyzed stereocontrolled vinylation of azoles and phenothiazine. *Org Lett* 4:623–626.

467. Settambolo, R., M. Mariani, and A. Caiazzo. 1998. Synthesis of 1,2- and 1,3-divinylpyrrole. *J Org Chem* 63:10022–10026.

468. Abele, E., O. Dzenitis, K. Rubina, and E. Lukevics. 2002. Synthesis of N- and S-vinyl derivatives of heteroaromatic compounds under phase-transfer catalysis. *Khim Geterocicl* 6:776–779.

469. Di Santo, R., A. Tafi, R. Costi et al. 2005. Antifungal agents. 11. N-Substituted derivatives of 1-[(aryl)(4-aryl-1*H*-pyrrol-3-yl)methyl]-1*H*-imidazole: Synthesis, anti-*candida* activity, and QSAR studies. *J Med Chem* 48:5140–5153.

470. Filimonov, V.D., E.E. Sirotkina, I.L. Gabel et al. 1974. New convenient method for the synthesis of 9-vinylcarbazole. *Zh Org Khim* 10 (8):1790.

471. Filimonov, V.D., S.G. Gorbachev, and E.E. Sirotkina. 1980. 9-Alkenylcarbazoles. 6. Synthesis and structure of *cis*-9-propenylcarbazoles. *Khim Geterocicl* 3:340–343.

472. Filimonov, V.D., V.A. Anfinogenov, and E.E. Sirotkina. 1978. 9-Alkenylcarbazoles. 3. Stereospecific synthesis and NMR spectra. *Zh Org Khim* 14 (12):2607–2611.

473. Domnina, E.S., G.G. Skvortsova, N.P. Glazkova et al. 1966. Synthesis and polymerization of N-vinylindole. *Khim Geterocicl* 3:390–394.

474. Shostakovsky, M.F., G.G. Skvortsova, E.S. Domnina et al. 1966. Method for the preparation of vinylindole. USSR Author's Certificate 181,116.

475. Nedolya, N.A., O.A. Tarasova, A.I. Albanov et al. 2010. First example of the synthesis of N-vinylpyrrole from methoxyallene and methoxyethyl isothiocyanate: Unprecedented elimination of methanol from β-substituted methyl ethyl ether. *Khim Geterocicl* 1:72–76.

476. Tarasova, O.A., N.A. Nedolya, A.I. Albanov, and B.A. Trofimov. 2010. New route to N-vinylpyrroles. *Zh Org Khim* 46 (5):777–780.

477. Nedolya, N.A., O.A. Tarasova, A.I. Albanov, and B.A. Trofimov. 2010. A one-pot synthesis and mild cleavage of 2-[2- or 5-(alkylsulfanyl)pyrrol-1-yl]ethyl vinyl ethers by *t*-BuOK/DMSO: A novel and facile approach to N-vinylpyrroles. *Tetrahedron Lett* 51 (40):5316–5318.

478. Neidlein, R. and G. Jeromin. 1982. Synthese neuer N-Vinylpyrrole. *Chem Ber* 115:714–721.

479. Movassaghi, M. and A.E. Ondrus. 2005. Palladium-catalyzed synthesis of N-vinylpyrroles and indoles. *J Org Chem* 70:8638–8641.

480. Reppe, W. and E. Keyssner. 1935. N-Vinyl compounds. Germany Patent 618,120.

481. Natsuo, S., N. Sigeru, Ya. Masakhiro et al. 1974. Method for the preparation of N-vinylcarbazole on the basis of acetylene. Japan Patent 4,900,9467.

482. Wolf, W. 1937. N-Vinyl compounds. Germany Patent 651,734.

483. Matsui, E. 1942. Vinylcarbazole and its polymer. *J Soc Chem Ind Jn* 45:1192–1193.

484. Trofimov, B.A., A.I. Mikhaleva, S.E. Korostova et al. 1977. Vinylation of pyrroles in dimethylsulfoxide. *Khim Geterocicl* 2:213–214.

485. Mikhaleva, A.I., B.A. Trofimov, S.E. Korostova et al. 1979. Vinylation of pyrroles under atmospheric pressure. *B Acad Sci USSR Ch* 4 (2):105–106.

486. Trofimov, B.A., R.N. Nesterenko, A.I. Mikhaleva et al. 1986. New examples of NH-heterocycles vinylation with acetylene under atmospheric pressure in the system KOH-DMSO. *Khim Geterocicl* 4:481–485.

487. Trofimov, B.A., S.E. Korostova, S.E. Shevchenko et al. 1990. Effect of pyrroles structure on rate of their vinylation. *Zh Org Khim* 26 (5):1110–1113.

488. Tarasova, O.A., A.G. Mal'kina, A.I. Mikhaleva et al. 1994. An efficient procedure for the N-vinylation of pyrrole. *Synth Commun* 24 (14):2035–2037.

489. Filimonov, V.D. 1981. 9-Alkenylcarbazoles. 7. Regio- and stereospecific addition of carbazole and indole to phenylacetylene. Structure and some properties of 9-(2-phenyl-vinyl)carbazole. *Khim Geterocicl* 2:207–210.

490. *N-Vinylpyrroles*. 1977. Booklet. Moscow, Russia: AN SSSR.

491. Trofimov, B.A., A.I. Mikhaleva, R.N. Nesterenko et al. 1996. Method for the preparation of N-vinylcarbazole. USSR Author's Certificate 1,262,907.

492. Trofimov, B.A., A.I. Mikhaleva, R.N. Nesterenko et al. 1994. Method for the preparation of N-vinylcarbazole. Russian Federation Patent 1,536,761.

493. Mikhaleva, A.I., A.M. Vasil'tsov, L.N. Sobenina et al. 2003. Technology of azoles vinylation. *Nauka Proizvodstvu* 6 (62):27.

494. Trofimov, B.A., S.E. Korostova, S.E. Shevchenko, and A.I. Mikhaleva. 1990. Features of one-electron transfer in vinylation of pyrroles with acetylene. *Zh Org Khim* 26 (5): 940–943.

495. Trofimov, B.A., T.I. Vakul'skaya, S.E. Korostova et al. 1990. One-electron transfer in vinylation of 4,5,6,7-tetrahydroindole with acetylene in the system KOH-DMSO. *B Acad Sci USSR Ch* 1:142–144.

496. Vakul'skaya, T.I., B.A. Trofimov, A.I. Mikhaleva et al. 1992. Formation of free radicals in vinylation of 2-substituted pyrroles with acetylenes in the system KOH-DMSO. *Khim Geterocicl* 8:1056–1062.

497. Trofimov, B.A., T.I. Vakul'skaya, T.V. Leshina et al. 1998. The ESR study of radical particles formed in the reaction of pyrroles with cyanoacetylenes in the system KOH-DMSO. *Zh Org Khim* 34 (11):1738–1740.

498. Trofimov, B.A., Yu.D. Tsvetkov, T.V. Leshina et al. 1998. Spin chemistry of elementoorganic compounds. In *Integration Programs of Fundamental Research*, ed. V.D. Ermikov, pp. 468–477. Novosibirsk, Russia: SO RAN.

499. Trofimov, B.A., L.N. Sobenina, A.I. Mikhaleva et al. 2003. N- and C-vinylation of pyrroles with disubstituted activated acetylenes. *Synthesis* 8:1272–1277.

500. Sobenina, L.N., A.I. Mikhaleva, I.A. Ushakov et al. 2003. First example of C-vinylation of pyrroles with activated acetylene under conditions of base catalysis. *Zh Org Khim* 39 (8):1266–1267.

501. Sobenina, L.N., A.I. Mikhaleva, I.A. Ushakov et al. 2003. Unexpected 1:2 annelation of 4,5,6,7-tetrahydroindole with 1-benzoyl-2-phenylacetylene. *Khim Geterocicl* 8:1269–1270.

502. Vakul'skaya, T.I., Sobenina, A.I. Mikhaleva et al. 2003. Radical intermediates of nucleophilic addition of pyrroles to disubstituted activated acetylenes. *Dokl. Chem.* 390 (4):484–487.

503. Trofimov, B.A., Z.V. Stepanova, L.N. Sobenina et al. 1998. An example of the facile C-vinylation of pyrroles. *Mendeleev Commun* 3:119–120.

504. Trofimov, B.A., Z.V. Stepanova, L.N. Sobenina et al. 1999. Pyrroles as C-nucleophiles in the reaction with acylacetylenes. *Izv Akad Nauk Ser Khim* 8:1562–1567.

505. Trofimov, B.A., Z.V. Stepanova, L.N. Sobenina et al. 1999. Reaction of 2-phenylpyrrole with 2-acyl-1-phenylacetylenes on silicon oxide. *Khim Geterocicl* 9:1253–1254.

506. Chipanina, N.N., Z.V. Stepanova, G.A. Gavrilova et al. 2000. Synthesis and IR spectra of 2-(2-trichloroacetyl-1-phenylethenyl)-5-phenylpyrrol. *Izv Akad Nauk Ser Khim* 11:1945–1947.

507. Chipanina, N.N., V.K. Turchaninov, I.I. Vorontsov et al. 2002. Strong intramolecular hydrogen bonding N–H···O in 2-(2-acyl-1-phenylethenyl)-5-phenylpyrroles. *Izv Akad Nauk Ser Khim* 1:107–111.

508. Turchaninov, V.K., N.N. Chipanina, Z.V. Stepanova et al. 2003. Solvatochromism of heteroaromatic compounds. XVIII. Reversible effects of the medium on UV spectrum of 2-(2-benzoyl-1-phenylethenyl)-5-phenylpyrrol. *Zh Obshch Khim* 73 (3): 471–480.

509. Ushakov, I.A., A.B. Afonin, V.K. Voronov et al. 2003. ^{1}H and ^{13}C NMR study of steric and electronic structure of 2-(2-acylethenyl)pyrroles. *Zh Org Khim* 39 (9): 1391–1397.

510. Stepanova, Z.V., L.N. Sobenina, A.I. Mikhaleva et al. 2003. Synthesis and x-ray study of 2-(2-acyl-1-phenylethenyl)pyrroles. *Zh Org Khim* 39 (11):1705–1712.

511. Trofimov, B.A., Z.V. Stepanova, L.N. Sobenina et al. 2001. C-Vinylation of 1-vinylpyrroles with benzoylacetylene on silica gel. *Synthesis* 12:1878–1882.

512. Ushakov, I.A., A.B. Afonin, V.K. Voronov et al. 2002. ^{1}H and ^{13}C NMR study of steric and electronic structure of 2-(2-acylethenyl)-1-vinylpyrroles. *Zh Org Khim* 38 (12): 1836–1842.

513. Sobenina, L.N., S.G. Dyachkova, Z.V. Stepanova et al. 1999. Nucleophilic substitution of chlorine in alkylthiochloroacetylenes with pyrrolide-anions. *Zh Org Khim* 35 (6): 941–945.

514. Trofimov, B.A., Z.V. Stepanova, L.N. Sobenina et al. 2004. Ethynylation of pyrroles with 1-acyl-2-bromoacetylenes on alumina: A formally "inverse Sonogashira coupling". *Tetrahedron Lett* 34:6513–6516.

515. Trofimov, B.A., Z.V. Stepanova, L.N. Sobenina et al. 2005. 2-(2-Benzoylethynyl)-5-phenylpyrrole: Fixation of *cis*- and *trans*-rotamers in crystal state. *Mendeleev Commun* 6:229–232.

516. Trofimov, B.A., L.N. Sobenina, Z.V. Stepanova et al. 2006. Synthesis of 2-benzoylethynylpyrroles by cross-coupling of 2-arylpyrroles with benzoylbromoacetylene over aluminum oxide. *Zh Org Khim* 42 (9):1366–1372.

517. Trofimov, B.A., L.N. Sobenina, A.P. Demenev et al. 2007. A palladium- and copper-free cross-coupling of ethyl 3-halo-2-propynoates with 4,5,6,7-tetrahydroindoles on alumina. *Tetrahedron Lett* 48 (27):4661–4664.

518. Trofimov, B.A., L.N. Sobenina, Z.V. Stepanova et al. 2007. Regioselective cross-coupling of 1-vinylpyrroles with acylbromoacetylenes on Al_2O_3: A synthesis of 1-vinyl-2-(2-acylethynyl)pyrroles. *Synthesis* 3:447–451.

519. Trofimov, B.A., L.N. Sobenina, Z.V. Stepanova et al. 2008. Reactions of 2-phenylpyrrole with bromobenzoylacetylene on metal oxides active surfaces. *Tetrahedron* 64:5541–5544.

520. Stepanova, Z.V., L.N. Sobenina, A.I. Mikhaleva et al. 2004. Silica-assisted reactions of pyrroles with 1-acyl-2-bromacetylenes. *Synthesis* 16:2736–2742.

521. Sobenina, L.N., D.N. Tomilin, O.V. Petrova et al. 2010. Cross-coupling of 4,5,6,7-tetrahydroindole with functionalized haloacetylenes on active surfaces of metal oxides and salts. *Zh Org Khim* 46 (9):1371–1375.

522. Trofimov, B.A., L.N. Sobenina, Z.V. Stepanova et al. 2008. Chemo- and regioselective ethynylation of 4,5,6,7-tetrahydroindoles with ethyl 3-halo-2-propynoates. *Tetrahedron Lett* 49:3946–3949.

523. Sobenina, L.N., A.P. Demenev, A.I. Mikhaleva et al. 2006. Ethynylation of indoles with 1-benzoyl-2-bromoacetylene on Al_2O_3. *Tetrahedron Lett* 47:7139–7141.

524. Petrova, O.V., L.N. Sobenina, I.A. Ushakov, and A.I. Mikhaleva. 2008. Reaction of indoles with ethylbromopropynoate on Al_2O_3 surface. *Zh Org Khim* 44 (10):1534–1538.

525. Passarela, D., G. Lesma, M. Deleo et al. 1999. Convenient synthesis of methyl indol-2-ylpropiolate. *J Chem Soc Perkin Trans 1* 2669–2670.

526. Bharate, S.B. 2006. 2,3-Dichloro-5,6-dicyano-1,4-benzoquinone (DDQ). *Synlett* 3:496–497.

527. Cavdar, H. and N. Saracoglu. 2006. Synthesis of new 2-vinylation products of indole *via* a Michael-type addition reaction with dimethyl acetylenedicarboxylate and their Diels-Alder reactivity as precursors of new carbazoles. *J Org Chem* 71:7793–7799.

528. Trofimov, B.A., L.N. Sobenina, Z.V. Stepanova et al. 2010. Facile [2+2]-cycloaddition of DDQ to an alkyne: Synthesis of pyrrolyl- and indolylbicyclo[4.2.0]octadienes from C-ethynylpyrroles or C-ethynylindoles. *Synthesis* 470–477.

529. Trofimov, B.A., L.N. Sobenina, Z.V. Stepanova et al. 2010. Peculiar rearrangement of the [2+2]-cycloadducts of DDQ and 2-ethynylpyrroles. *Tetrahedron Lett* 51:5028–5031.

530. Pappas, S.P., B.S. Pappas, and N.A. Portnoy. 1969. Alkyne-quinone photoaddition. Formation and solvolytic rearrangement of 1-methoxybicyclo[4.2.0]octa-3,7-diene-2,5-diones. *J Org Chem* 34:520–525.

531. Sobenina, L.N., Z.V. Stepanova, I.A. Ushakov et al. 2011. From 4,5,6,7-tetrahydroindole to functionalized furan-2-one-4,5,6,7-tetrahydroindole-cyclobutene sequence in two steps. *Tetrahedron* 67:4832–4837.

532. Sobenina, L.N., D.N. Tomilin, O.V. Petrova et al. 2010. Hydroamination of 2-ethynyl-4,5,6,7-tetrahydroindoles: Towards 2-substituted aminoderivatives of indole. *Synthesis* 14:2468–2474.

533. Katritzky, A.R., Ch.V. Marson, and H. Faid-Allah. 1987. Heterocyclic N-dithiocarboxylic acids. *Heterocycles* 26:1657–1670.

534. Oddo, B. and C. Alberti. 1938. The pyrrole-indole group. Series II. Note XXIII. Derivatives of N-pyrrole dithiocarboxylic acids and α-pyrroledithiocarboxylic acid. *Gazz Chim Ital* 68:204–214.

535. El A'mma, A.G. and R.S. Drago. 1977. Unusual oxidation-state stabilization of iron complexes by the pyrrole-N-carbodithioate ligand. *Inorg Chem* 16:2975–2977.

536. Bereman, R.D. and D. Nalewajek. 1977. Preparation and characterization of pyrrole-N-carbodithioate complexes of selected transition elements. *Inorg Chem* 16:2687–2691.

537. Trofimov, B.A., L.N. Sobenina, A.I. Mikhaleva et al. 1991. Convenient synthesis of 2-pyrroledithiocarboxylic acids. *Khim Geterocicl* 7:996–997.
538. Trofimov, B.A., L.N. Sobenina, A.I. Mikhaleva et al. 1992. Reaction of pyrroles with carbon disulfide in the system KOH-DMSO. *Khim Geterocicl* 9:1176–1181.
539. Trofimov, B.A., L.N. Sobenina, A.I. Mikhaleva et al. 1992. Reaction of pyrroles with carbon disulfide in KOH/DMSO system. *Sulfur Lett* 15 (5):219–226.
540. Sobenina, L.N., L.E. Protasova, M.P. Sergeeva et al. 1995. Synthesis and redox properties of pyrroledithiocarboxylic acid esters. *Khim Geterocicl* 1:47–54.
541. Sobenina, L.N., A.I. Mikhaleva, and B.A. Trofimov. 2005. Pyrrolecarbodithioates: Synthesis and reactions. *Russ Khim Zh* 6:97–108.
542. Trofimov, B.A., L.N. Sobenina, A.I. Mikhaleva et al. 2000. Reaction of pyrrole anions with carbon disulfide. Synthesis of pyrrole-3-carbodithioates. *Tetrahedron* 56:7325–7329.
543. Trofimov, B.A., N.M. Vitkovskaya, V.B. Kobychev et al. 2001. Addition of pyrrole anions to carbon disulfide. Theoretical analysis. *Sulfur Lett* 24 (4):181–190.
544. Kobychev, V.B., N.M. Vitkovskaya, I.L. Zaitseva et al. 2001. Ab initio quantum chemical study of the reaction of pyrrole anions with carbon disulfide. *Zh Strukt Khim* 42 (4):645–653.
545. Kobychev, V.B., N.M. Vitkovskaya, I.L. Zaitseva et al. 2002. Theoretical analysis of pyrrole anions addition to carbon disulfide and carbon dioxide. *Int J Quant Chem* 88:542–548.
546. Kobychev, V.B., N.M. Vitkovskaya, I.L. Zaitseva, and B.A. Trofimov. 2004. Quantum chemical study of the profiles of reactions that form pyrrole anion N-adducts with CS_2 and CO_2. *Zh Strukt Khim* 45 (6):990–993.
547. Kobychev, V.B., N.M. Vitkovskaya, I.L. Zaytseva, and B.A. Trofimov. 2004. Pyrrole anion addition to carbon disulfide: An ab initio study. *Int J Quant Chem* 100 (4):360–366.
548. Sobenina, L.N., A.I. Mikhaleva, O.V. Petrova et al. 1999. Addition of pyrrole-1- and pyrrole-2-carbodithioates to acryl systems. *Zh Org Khim* 35 (10):1534–1537.
549. Sobenina, L.N., A.P. Demenev, A.I. Mikhaleva et al. 2001. The addition reactions of pyrrolecarbodithioates to activated alkenes and alkynes. *Synthesis* 2:293–299.
550. Drozd, V.N., M.L. Petrov, N.Ya. Kuz'mina, and A.S. Vyazgin. 1988. Additions of dithioacids to unsaturated compounds. *Ush Khim* 57:94–113.
551. Drozd, V.N., O.A. Popova, A.S. Vyazgin, and D.B. Dmitriev. 1983. Anionic [3+2]-cycloaddition of dithiobenzoate-anion across the activate multiple bonds. *Zh Org Khim* 19 (4):847–853.
552. Kuz'mina, N.Ya., M.L. Petrov, and A.A. Petrov. 1984. The effects of substituents in dithioate-anions on the course of the reaction with acetylenedicarboxylic acid dimethyl ester. *Zh Org Khim* 20 (12):2511–2517.
553. Kuz'mina, N.Ya. 1985. Interaction of dithioic acids with activated acetylenes. PhD dissertation, Leningrad Technological Institute, Leningrad, Russia.
554. Sobenina, L.N., A.P. Demenev, A.I. Mikhaleva et al. 2001. Interaction of pyrrole-1- and pyrrole-2-carbodithioates with activated acetylenes. *Zh Org Khim* 37 (4):582–586.
555. Trofimov, B.A., L.N. Sobenina, A.I. Mikhaleva et al. 1992. Synthesis of 2-(1-alkylthio-2-cyanoethenyl)pyrroles. *Zh Org Khim* 28 (8):1766–1767.
556. Sobenina, L.N., A.I. Mikhaleva, M.P. Sergeeva et al. 1995. Pyrrole-2-dithiocarboxylates: Synthesis of 2-(1-alkylthio-2-cyanoethenyl)pyrroles. *Tetrahedron* 51:4223–4230.
557. Murzina, N.M., A.I. Vokin, L.N. Sobenina, and V.K. Turchaninov. 2002. Solvatochromism of heteroaromatic compounds. XVI. Theoretic study of the effect of nonspecific salvation on the rotational isomerism and spectral characteristics of 2-(1-methylthio-2, 2-dicyanovinyl)-5-methylpyrrole. *Zh Obshch Khim* 72 (5):848–854.

558. Murzina, N.M., A.I. Vokin, S.V. Fedorov et al. 2002. Solvatochromism of heteroaromatic compounds. XVII. Effect of aprotic inert solvents on the structure of 2-(1-methylthio-2,2-dicyanovinyl)-5-methylpyrrole. *Zh Obshch Khim* 72 (6):1011–1014.

559. Demenev, A.P., L.N. Sobenina, A.I. Mikhaleva, and B.A. Trofimov. 2003. Reaction of pyrrole-2-carbodithioates with CH-acids: Stereospecific synthesis of new functional 2-vinylpyrroles. *Sulfur Lett* 26:95–100.

560. Trofimov, B.A., A.P. Demenev, L.N. Sobenina et al. 2003. The first chemoselective synthesis of functionalized 3-vinylpyrroles. *Tetrahedron Lett* 44:3501–3503.

561. Sobenina, L.N., A.P. Demenev, A.I. Mikhaleva et al. 2002. Functionally substituted 1,3-diethenyl[1,2-*c*][1,3]pyrrolothiazoles from pyrrole-2-carbodithioates. *Sulfur Lett* 25 (3):87–93.

562. Sobenina, L.N., A.P. Demenev, A.I. Mikhaleva et al. 2002. Synthesis of functionally substituted pyrrolothiazolidines from pyrrole-2-carbodithioates, CH-acids and haloacetylenes. *Khim Geterocicl* 1:95–104.

563. Trofimov, B.A., L.N. Sobenina, A.I. Mikhaleva et al. 1992. Synthesis of 1-alkylthio-3*H*-indolizin-3-ones. *Khim Geterocicl* 7:998–999.

564. Sobenina, L.N., A.I. Mikhaleva, M.P. Sergeeva et al. 1996. 3*H*-Pyrrolizin-3-ones. *Khim Geterocicl* 7:919–924.

565. Tominaga, Y., Y. Matsuoka, S. Kohra, and A. Hosomi. 1987. A novel preparation of polarized ethylenes by the reaction of thioamides or dithiocarboxylates with tetracyanoethylene oxide. Synthesis of pyrazoles and pyrimidines. *Heterocycles* 26 (3):613–616.

566. Tominaga, Y., Y. Matsuoka, and A. Hosomi. 1988. A new synthesis of 5-aza[2.2.3]cyclazines by [8+2]cycloaddition of 3-imino-3*H*-pyrrolizines with dimethyl acetylenedicarboxylate. *Heterocycles* 27 (12):2791–2793.

567. Tominaga, Y. 1989. Synthesis of heterocyclic compounds using carbon disulfide and their products. *J Heterocycl Chem* 26:1167–1204.

568. Trofimov, B.A., O.V. Petrova, L.N. Sobenina et al. 2006. Vinylic nucleophilic substitution in functionalized 2-vinylpyrroles: A route to a new family of stable enols. *Tetrahedron* 17:4146–4152.

569. Trofimov, B.A., O.V. Petrova, L.N. Sobenina et al. 2006. Easy α- to β-migration of the enol moiety over the pyrrole ring. *Tetrahedron Lett* 47:3645–3648.

570. Choi, D.-S., S. Huang, M. Huang, T.S. Barnard, R.D. Adams, J.M. Seminario, and J.M. Tour. 1998. Revised structures of N-substituted dibrominated pyrrole derivatives and their polymeric products. Termaleimide models with low optical band gaps. *J Org Chem* 63:2646–2655.

571. Sobenina, L.N., A.I. Mikhaleva, D.-S.D. Toryashinova et al. 1997. Exchange of the ethylthio group for an amino group upon cyclization of 2-(2-cyano-1-ethylthioethenyl)pyrroles to 4,5,6,7-tetrahydropyrrolo[1,2-*a*]indoles. *Sulfur Lett* 20:205–212.

572. Khutsishvili, S.S., Yu.Yu. Rusakov, L.B. Krivdin et al. 2008. ^{13}C–^{13}C spin–spin coupling constants in structural studies. XLIII. Stereochemical study of functionalized 3-iminopyrrolizines. *Zh Org Khim* 44 (9):1354–1360.

573. Sobenina, L.N., A.I. Mikhaleva, and B.A. Trofimov. 1995. Cyclization of 2-(1-alkylthio-2,2-dicyanoethenyl)pyrroles. *Khim Geterocicl* 3:418–419.

574. Sobenina, L.N., A.I. Mikhaleva, D.-S.D. Toryashinova, and B.A. Trofimov. 1996. Exchange of the ethylthio group in 1-ethylthio-2-cyano-3-imino-4,5,6,7-tetrahydrocyclohexa[*c*]-3*H*-pyrrolizine for an amino group. *Sulfur Lett* 20:9–14.

575. Tominaga, Y., Y. Matsuoka, Y. Oniyama et al. 1990. Polarized ethylenes. IV. Synthesis of polarized ethylene using thioamides and methyl dithiocarboxylates and their application to syntheses of pyrazoles, pyrimidines, pyrazolo[3,4-*d*]pyrimidines and 5-aza[2.2.3] cyclazines. *J Heterocycl Chem* 27 (3):647–660.

576. Trofimov, B.A., L.N. Sobenina, A.I. Mikhaleva et al. 2007. Facile coupling of 2-(1-ethylthioethenyl)pyrroles with amines: A route to 2-(1-aminoethenyl)pyrroles and 1-amino-3-iminopyrrolizines. *J Heterocycl Chem* 44 (3):505–514.

577. Sobenina, L.N., A.I. Mikhaleva, O.V. Petrova et al. 1999. Synthesis of 5-amino-3-(2-pyrrolyl)pyrazoles. *Zh Org Khim* 35 (8):1241–1245.

578. Sobenina, L.N., A.P. Demenev, A.I. Mikhaleva et al. 2000. Synthesis and reactions with carbon disulfide of 5-amino-3-(2-pyrrolyl)pyrazoles. *Sulfur Lett* 24 (1):1–12.

579. Yang, Z., K. Zhang, F. Gong et al. 2001. A highly selective fluorescent sensor for fluoride anion based on pyrazole derivative: Naked eye "no-yes" detection. *J Photochem Photobiol A* 217:29–34.

580. Petrova, O.V., L.N. Sobenina, A.P. Demenev, and A.I. Mikhaleva. 2003. Synthesis of functionalized 2-(pyrrol-2-yl)pyrazolo[1,5-a]pyrimidines. *Zh Org Khim* 39 (10): 1540–1545.

581. Baikalova, L.V., L.N. Sobenina, A.I. Mikhaleva et al. 2001. Assembling of complex heterocyclic ensembles, Schiff bases, from 5-amino-3[2-(4.5.6.7-tetrahydroindolyl)] pyrazoles and 1-vinyl(ethyl)-2-formylimidazoles. *Zh Org Khim* 37 (12):1817–1821.

582. Sobenina, L.N., V.N. Drichkov, A.I. Mikhaleva et al. 2005. Synthesis of 3- and 5-amino-5-(3)-(pyrrol-2-yl)isoxazoles. *Tetrahedron* 61 (20):4841–4849.

583. Kashima, C., N. Yoshiwara, and Y. Omote. 1982. Alkylation of aminohydroxy anion, dissociated species of hydroxylamine. *Tetrahedron Lett* 23:2955–2956.

584. Sobenina, L.N., V.N. Drichkov, A.I. Mikhaleva et al. 2005. The reaction of 1-ethyl-thio-3-iminopyrrolizines with hydroxylamine. A new synthesis of 3-aminoisoxazoles. *Arkivoc* vii:28–35.

585. Yang, Z., K. Zhang, F. Gong et al. 2011. A new fluorescent chemosensor for fluoride anion based on a pyrrole–isoxazole derivative. *Beilstein J Org Chem* 7:46–52.

586. Tarasova, O.A., E.Yu. Schmidt, A.I. Albanov et al. 1999. Synthesis and prototropic isomerization of 1-(2-propenyl)pyrrole in the system KOH-DMSO. *Zh Org Khim* 35 (10):1530–1533.

587. Van Eyk, S.J., H. Naarmann, and N.P.C. Walker. 1992. 1,6-Dipyrrol-1-yl-2,4-hexadiyne. Preparation and characterization. *Synthetic Met* 48 (3):295–300.

588. Paley, M.S., D.O. Frazier, H. Abeledeyem et al. 1992. Synthesis. Vapor growth, polymerization, and characterization of thin films of novel diacetylene derivatives of pyrrole. The use of computer modeling to predict chemical and optical properties of these diacetylenes and poly(diacetylenes). *J Am Chem Soc* 114 (9):3247–3251.

589. Brandsma, L., O.A. Tarasova, N.A. Kalinina et al. 2002. Efficient synthesis of 1-propargylpyrrole. *Zh Org Khim* 38 (7):1115–1117.

590. Tarasova, O.A., B.A. Trofimov, A.I. Mikhaleva et al. 1991. Direct allenylation of pyrroles with propargyl chloride. *Zh Org Khim* 27 (8):1798.

591. Tarasova, O.A., L. Brandsma, and B.A. Trofimov. 1993. Facile one-pot synthesis of 1-allenylpyrroles. *Synthesis* 6:571–572.

592. Tarasova, O.A., E.Yu. Schmidt, L.V. Klyba et al. 1998. N-Allenylation of pyrrole, di- and triazoles with proparhyl bromide in the system KOH-DMSO. *Zh Org Khim* 34 (5):730–734.

593. Dumont, M.J.-L. 1985. Propargylation des amines aromatiques. Préparation d'ynamines et de diynamines. *C R Acad Sc Paris* 261 (7):1710–1712.

594. Hubert, A.J. and H. Reimlinger. 1968. Base-catalysed prototropic isomerisations. Part II. The isomerisation of N-prop-2-ynyl heterocycles into N-substituted allenes and acetylenes. *J Chem Soc C* 5:606–608.

595. Tarasova, O.A., A.I. Mikhaleva, T.L. Markova, and B.A. Trofimov. 1992. N-Allenylation of pyrroles with 2,3-dichloro-1-propene. *Khim Geterocicl* 1:125.

596. Tarasova, O.A., T.L. Markova, and B.A. Trofimov. 1991. N-Allenylation of pyrroles with 1,2,3-trichloropropane. *B Acad Sci USSR Ch* 9:2161–2162.

597. Trofimov, B.A., O.A. Tarasova, and L. Brandsma. 1994. Facile synthesis of N-ethynylpyrrole. *Zh Org Khim* 30 (2):314.

598. Brandsma, L., A.G. Malkina, and B.A. Trofimov. 1994. An improved procedure for N-ethynylpyrrole. *Synthetic Commun* 24 (19):2721–2724.

599. Markova M.V., L.V. Morozova, E.Yu. Schmidt et al. 2009. Preparation of 4,5,6,7-tetrahydroindole alkaline salts and their reaction with chloromethyloxirane. *Arkivoc* iv:57–63.

600. Trofimov, B.A., M.V. Markova, L.V. Morozova et al. 2001. Protected bis(hydroxyorganyl) polysulfides as modifiers of Li/S battery electrolyte. *Electrochim Acta* 56:2458–2463.

601. Mikhaleva, A.I., A.V. Ivanov, E.V. Skital'tseva et al. 2009. An efficient route to 1-vinyl-pyrrole-2-carbaldehydes. *Synthesis* 4:587–590.

602. Pozo-Gonzalo, C., J.A. Pomposo, J. Rodriguez et al. 2007. Synthesis and electrochemical study of narrow band gap conducting polymers based on 2,2'-dipyrroles linked with conjugated aza-spacers. *Synthetic Met* 157 (1):60–65.

603. Trofimov, B.A., A.V. Ivanov, E.V. Skital'tseva et al. 2009. A straightforward synthesis of 2-(1-vinyl-1*H*-pyrrol-2-yl)-1*H*-benzimidazoles from 1-vinyl-1*H*-pyrrole-2-carbaldehydes and *o*-phenylenediamine. *Synthesis* 21:3603–3610.

604. Vasil'tsov, A.M., K. Zhang, A.V. Ivanov et al. 2009. 1-Vinylpyrrole-2-carbaldehyde oximes: Synthesis, isomerisation and spectral properties. *Monatsh Chem* 140 (12): 1475–1480.

605. Trofimov, B.A., A.V. Ivanov, I.A. Ushakov et al. 2011. Stereospecific protonation of pyrrole-2-carboxaldehyde Z-oximes as a result of through-space cation stabilization with oxime hydroxyl. *Mendeleev Commun* 21:103–105.

606. Mikhaleva, A.I., A.M. Vasil'tsov, A.V. Ivanov et al. 2008. Direct synthesis of semicarbazones, thiosemicarbazones and guanylhydrazones of 1-vinylpyrrole-2-carbaldehydes. *Khim Geterocicl* 9:1384–1390.

607. Ivanov, A.V., I.A. Ushakov, K.B. Petrushenko et al. 2010. Chemo-, regio- and stereospecific synthesis of novel unnatural fluorescent amino acids via condensation of L-lysine and 1-vinylpyrrole-2-carbaldehydes. *Eur J Org Chem* 24:4554–4558.

608. Belyaeva, K.V., L.V. Andriyankova, L.P. Nikitina et al. 2011. Synthesis of 1-vinylpyrrole-imidazole alkaloids. *Synthesis* 17:2843–2847.

609. Fattorusso, E. and O. Taglialatela-Scafati. (eds.) 2008. *Modern Alkaloids: Structure, Isolation, Synthesis and Biology*. Weinheim, Germany: Wiley-VCH.

610. Chenoweth, D.M. and P.B. Dervan. 2010. Structural basis for cyclic Py-Im polyamide allosteric inhibition of nuclear receptor binding. *J Am Chem Soc* 132 (41):14521–14529.

611. Jacobi, N. and T. Lindel. 2010. Assembly of the bis(imidazolyl)propene core of nagelamides C and S by double Grignard reaction. *Eur J Org Chem* 28:5415–5425.

612. Vasil'tsov, A.M., A.V. Ivanov, I.A. Ushakov et al. 2007. Selective thiylation of 1-vinyl-pyrrole-2-carbaldehydes: Synthesis of 2-[bis(ethylsulfanyl)methyl]-1-vinylpyrroles and 1-(2-ethylthioethyl)pyrrole-2-carbaldehydes – novel pyrrole synthons. *Synthesis* 3:452–456.

613. Trofimov, B.A., A.I. Mikhaleva, G.A. Kalabin et al. 1977. N-Vinyl-α-trifluoroacetylpyrroles. *B Acad Sci USSR Ch* 11:2639–2640.

614. Trofimov, B.A., A.I. Mikhaleva, S.E. 1979. Korostova et al. Method for the preparation of 1-vinyl-2(5)-trifluoroacetylpyrroles. USSR Author's Certificate 698,981.

615. Trofimov, B.A., A.I. Mikhaleva, S.E. Korostova et al. 1979. Pyrroles from ketoximes and acetylene. 10. Trifluoroacetylation of 1-vinylpyrroles. *Zh Org Khim* 15 (10):2042–2046.

616. Cooper, W.D. 1958. Synthesis of 2-trifluoroacetylpyrrole. *J Org Chem* 23:1382.

617. Clementi, S. and G. Marino. 1972. Dependence of the ρ-values of electrophilic substitutions upon the nature of the aromatic substrate. Trifluoroacetylation of substituted furans, thiophenes and pyrroles. *J Chem Soc Perkin Trans II* 1:71–73.

618. Clementi, S. and G. Marino. 1969. Electrophilic substitution in five-membered hetero-cyclic systems. VIII. Relative rates of trifluoroacetylation. *Tetrahedron* 25:4599–4603.

619. Shmushkovich, J. 1966. Enamines. In *Advances in Organic Chemistry*, vol. 4, ed. I.L. Knunyanz, pp. 5–123. Moscow, Russia: Mir.

620. Hojo, M., R. Masuda, G. Kokuryo et al. 1976. Electrophilic substitutions of olefinic hydrogens. II. Acylation of vinyl ethers and N-vinyl amides. *Chem. Lett* 5:499–502.

621. Korostova, S.E., A.I. Mikhaleva, L.N. Sobenina et al. 1989. Pyrroles from ketoximes and acetylene. 38. New representatives of trifluoroacetylpyrroles. Synthesis and transformations. *Khim Geterocicl* 1:48–52.

622. Gorbachev, S.G. 1978. Synthesis of 9-alkenylcarbazoles and some aspects of their reactivity in cationic polymerization. PhD dissertation, Tomsk.

623. Genkina, N.K., V.N. Eraksina, and N.N. Suvorov. 1980. Two directions in benzindoles acylation. *Zh Org Khim* 16 (10):2154–2156.

624. Kimie, J. and Terukijo H. 1976. Trifluoroacetylation at the 3 position of 1-(2-aminophenyl)-2,5-dimethylpyrrole in trifluoroacetic acid. *Bull Chem Soc Jpn* 49:1363–1365.

625. Mackie, R.K., S. Mhatre, and J.M. Tedder. 1977. C-Trifluoroacetylation. A convenient route to carboxylic acids. *J Fluor Chem* 10:437–445.

626. Marino, J. 1973. Quantitative effect of electrophilic substitution in furan, thiophene, pyrrole and other five-membered heteroaromatic systems. *Khim Geterocicl* 5:579–589.

627. Joule, J. and G. Smith. 1975. *Heterocyclic Chemistry*. Moscow, Russia: Mir.

628. Korostova, S.E., R.N. Nesterenko, A.I. Mikhaleva et al. 1991. Pyrroles from ketoximes and acetylene. 41. Some transformations of 1-vinyl-2-(2-furyl)- and -2-(2-thienyl)pyrroles. *Khim Geterocicl* 3:337–342.

629. Trofimov, B.A., A.I. Mikhaleva, A.I. Belyaevsky et al. 1981. Synthesis and antimicrobial activity of some pyrrole derivatives. *Khim Farm Zh* 3:25–29.

630. Sobenina, L.N., M.P. Sergeeva, A.I. Mikhaleva et al. 1990. Synthesis of pyrrole-2-carboxylic acids and their N-vinyl derivatives. *Khim Geterocicl* 5:612–616.

631. Korobchenko, L.V., G.V. Vladyko, E.I. Boreko et al. 1992. Synthesis and antiviral activity of pyrrolecarboxylic acids and their derivatives. *Khim Farm Zh* 11–12:57–59.

632. Golovanova, N.I., D.-S.D. Toryashinova, L.N. Sobenina et al. 1992. Transmission of influence of substituent in 2-arylpyrrole-2-carboxylic acids. *Khim Geterocicl* 9:1182–1186.

633. Rusakov, Yu.Yu., L.B. Krivdin, E.Yu. Senotrusova et al. 2007. Conformational study of 2-arylazo-1-vinylpyrroles. *Magn Reson Chem* 45:142–151.

634. Trofimov, B.A., E.Yu. Schmidt, E.Yu. Senotrusova et al. 2009. Selective thermooxidation of 1-vinyl-2,3-dimethyl-5-(phenyldiazenyl)pyrrole to the corresponding pyrrole-2-carbaldehyde: Stabilization of the pyrrole nucleus with phenylazo substituent. *Zh Org Khim* 45 (10):1579–1580.

635. Schmidt, E.Yu., E.Yu. Senotrusova, I.A. Ushakov et al. 2010. First example of auto-oxidation of methyl and cyclohexano groups attached to the pyrrole ring: Stabilization effect of phenyldiazenyl substituents. *Arkivoc* ii:352–359.

636. Schmidt, E.Yu., E.Yu. Senotrusova, I.A. Ushakov et al. 2007. Electrophilic addition of alcohols to 1-vinyl-2-phenylazopyrroles and unexpected formation of 2-methylquino-line. *Zh Org Khim* 43 (10):1502–1508.

637. Schmidt, E.Yu., E.Yu. Senotrusova, I.A. Ushakov et al. 2009. The peculiar reaction of 2-arylazo-1-vinylpyrroles with trifluoroacetic acid: A novel synthesis of 2-methylquino-lines. *Tetrahedron* 65:4855–4858.

638. Schmidt, E.Yu., E.Yu. Senotrusova, N.V. Zorina et al. 2011. The reaction of 2-arylazo-1-vinylpyrroles with trifluoroacetic anhydride: Unexpected formation of *N*-aryl-2,2,2-trifluoroacetamides and conjugated polymers. *Mendeleev Commun* 21:36–37.

639. Loudet, A. and K. Burgess. 2007. BODIPY dyes and their derivatives: Syntheses and spectroscopic properties. *Chem Rev* 107:4891–4932.

640. Badré, S., V. Monnier, R. Méallet-Renault et al. 2006. Fluorescence of molecular micro- and nanocrystals prepared with BODIPY derivatives. *J Photochem Photobiol A* 183:238–246.

641. Méallet-Renault, R., G. Clavier, C. Dumas-Verdes et al. 2008. Synthesis of novel BODIPY from sterically hindered pyrroles and study of their photophysical behaviour in solution, polystyrene nanoparticles and solid phase. *Russ Khim Zh* 52 (1):91–99.

642. Sobenina, L.N., A.M. Vasil'tsov, O.V. Petrova et al. 2011. A general route to symmetric and asymmetric *meso*-CF$_3$-3(5)-aryl(hetaryl)- and 3,5-diaryl(dihetaryl)-BODIPY dyes. *Org Lett* 13 (10):2524–2527.

643. Shostakovsky, M.F., G.G. Skvortsova, E.S. Domnina et al. 1965. Hydrolytic cleavage of vinylindole. *J Appl Chem USSR* 38:2602–2604.

644. McKinley, S., J.V. Crawford, and C.-H. Wang. 1966. The formation of a dimer of N-vinylcarbazole. *J Org Chem* 31:1963–1964.

645. Filimonov, V.D., V.A. Anfinogenov, and E.E. Sirotkina. 1978. 9-Alkenylcarbazoles. IV. Mechanism of acidic hydrolysis. *Zh Org Khim* 14 (12):2550–2555.

646. Trotsenko, L.I., G.N. Kurov, V.K. Turchaninov et al. 1979. Kinetics of acidic hydrolysis of N-vinyl derivatives of phenthiazine and carbazole. *Zh Obshch Khim* 49 (4): 904–909.

647. Sidel'kovskaya, F.P. 1970. *Chemistry of N-Vinylpyrrolidone and Its Polymers*. Moscow, Russia: Nauka.

648. Trofimov, B.A., S.E. Korostova, A.I. Mikhaleva et al. 1982. Pyrroles from ketoximes and acetylene. 24. Acidic hydrolysis of 1-vinylpyrroles. *Khim Geterocicl* 12:1631–1639.

649. Korostova, S.E., A.I. Mikhaleva, and M.V. Sigalov. 1987. Synthesis of di(1-vinyl-5-phenyl-2-pyrrolyl)methane. *Zh Org Khim* 23 (2):448.

650. Skvortsova, G.G., E.S. Domnina, and N.P. Glazkova. 1969. Synthesis of alkoxyethylindoles. *Khim Geterocicl* 2:255–257.

651. Skvortsova, G.G., B.V. Trzhtsinskaya, and L.F. Teterina. 1973. Method for the preparation of N-(α-alkoxy)ethylindoles. USSR Author's Certificate 367,096.

652. Skvortsova, G.G., B.V. Trzhtsinskaya, L.A. Usov et al. 1975. Synthesis and some pharmacological properties of N-(α-alkoxyethyl)indoles. *Khim Farm Zh* 9:16–18.

653. Filimonov, V.D., V.A. Anfinogenov, and E.E. Sirotkina. 1979. 9-Alkenylcarbazoles. 5. A facile method of synthesis and electrophilic addition of alcohols under the action of CCl$_4$. *Khim Geterocicl* 4:497–502.

654. Filimonov, V.D., E.E. Sirotkina, and N.A. Tsekhanovskaya. 1979. One-stage synthesis of 9-vinylcarbazoles by vinylation of carbazoles with vinyl ethers. *Zh Org Khim* 15 (1): 174–177.

655. Lopatinsky, V.P., Yu.P. Shekhirev, and E.E. Sirotkina. 1966. Investigations in the field of carbazole derivatives chemistry. 31. Addition of carbazole to vinyl ethers. *Khim Geterocicl* 3:398–402.

656. Smith, R.F. Preparation of ether adducts of N-vinyl-2-pyrrolidone. 1974. US Patent 3,823,160.

657. Trofimov, B.A., S.E. Korostova, L.N. Sobenina et al. 1980. Pyrroles from ketoximes and acetylene. 15. Electrophilic addition of alcohols to 1-vinylpyrroles. *Zh Org Khim* 16 (9): 1964–1968.

658. Korostova, S.E., A.I. Mikhaleva, and S.G. Shevchenko. 1986. Pyrroles from ketoximes and acetylene. 33. Addition of acetylenic alcohols to 1-vinylpyrroles. *Zh Org Khim* 22 (12):2489–2496.

659. Markova, M.V., A.I. Mikhaleva, M.V. Sigalov et al. 1989. Electrophilic addition of phenols to 1-vinyl-4,5,6,7-tetrahydroindole. *Khim Geterocicl* 5:604–606.

660. Korostova, S.E., A.I. Mikhaleva, B.A. Trofimov et al. 1982. Pyrroles from ketoximes and acetylene. 20. Addition of alcohols to 1-vinylpyrroles in the presence of the system azoisobutyric acid dinitrile—CCl$_4$. *Zh Org Khim* 18 (3):525–528.

661. Mikhaleva, A.I., S.E. Korostova, A.N. Vasil'ev et al. 1977. Free-radical addition of alkane thiols to 1-vinylpyrroles. *Khim Geterocicl* 12:1636–1639.
662. Leonard, N.J. and A.S. Hay. 1956. Unsaturated amines. V. The attack of ternary iminium compounds by nucleophilic reagents. *J Am Chem Soc* 78:1984–1987.
663. Lawesson, S.O., E.H. Larsen, and H.J. Jakobsen. 1964. Enamine chemistry. II. Addition of thiophenols to enamines. *Rec Trav Chim* 83:461–463.
664. Reppe, W. and F. Nicolai. Sulfurized condensation products. 1936. Germany Patent 624,622.
665. Skvortsova, G.G., N.P. Glazkova, E.S. Domnina et al. 1970. Addition of mercaptanes to N-vinyl derivatives of indole and imidazole. *Khim Geterocicl* 2:167–172.
666. Sobenina, L.N., A.I. Mikhaleva, M.P. Sergeeva et al. 1992. Radical addition of methyl mercaptoacetate to N-vinylpyrroles. *Sulfur Lett* 15 (5):227–232.
667. Trofimov, B.A., L.N. Sobenina, A.I. Mikhaleva et al. 1992. Addition of methylmercaptoacetate to N-vinylpyrroles. *Khim Geterocicl* 4:481–484.
668. Aliev, I.A., A.I. Mikhaleva, and B.R. Gasanov. 1980. Addition of thiophenols to N-vinylpyrroles. *Khim Geterocicl* 6:750–752.
669. Voronkov, M.G., G.I. Zelchan, and E.Ya. Luketitz. 1978. *Silicon and Life*. Riga, Latvia: Zinatne.
670. Simon, R.A., A.J. Ricco, M.S. Wrighton. 1982. Synthesis and characterization of a new surface derivatizing reagent to promote the adhesion of polypyrrole films to *n*-type silicon photoanodes: N-[3-(trimethoxysilyl)propyl]pyrrole. *J Am Chem Soc* 104:2031–2034.
671. Pukhnarevich, V.B., L.I. Kopylova, S.E. Korostova et al. 1979. Hydrosilylation of 1-vinylpyrroles. *Zh Obshch Khim* 49 (1):116–119.
672. Kopylova, L.I., S.E. Korostova, L.N. Sobenina et al. 1981. Hydrosilylation of 2- and 2,3-substituted 1-vinylpyrroles. *Zh Obshch Khim* 51 (8):1778–1781.
673. Trofimov, B.A., A.I. Mikhaleva, L.V. Morozova et al. 1983. Dimerization of 1-vinyl-4,5,6,7-tetrahydroindole under the action of HCl. *Khim Geterocicl* 2:269–270.
674. Trofimov, B.A., S.F. Malysheva, B.G. Sukhov et al. 2003. Addition of secondary phosphines to N-vinylpyrroles. *Tetrahedron Lett* 44 (13):2629–2632.
675. Trofimov, B.A., N.K. Gusarova, and L. Brandsma. 1996. The system elemental phosphorus—Strong bases as synthetic reagents. *Main Group Chem News* 4:18–24.
676. Trofimov, B.A., S.N. Arbuzova, and N.K. Gusarova. 1999. Phosphine in the synthesis of organophosphorus compounds. *Usp Khim* 68 (3):215–227.
677. Trofimov, B.A., N.K. Gusarova, and B.G. Sukhov et al. 2005. Atom-economic, solvent-free, high yield synthesis of 2-(pyrrol-1-yl)propyl diorganyl phosphines. *Synthesis* 6:965–970.
678. Tolmachev, A.A., S.I. Dovgopoly, A.N. Kostyuk et al. 1997. Phosphorus-containing heterocyclic compounds derived from N-vinylpyrroles. *Heteroatom Chem* 8 (6):495–499.
679. Rozinov, V.G., G.A. Pensionerova, V.I. Donskikh et al. 1986. Phosphorus-containing enamines. IV. Reaction of alkyl- and phenylsubstituted N-vinylpyrroles with phosphorus pentachloride. *Zh Obshch Khim* 56 (4):790–804.
680. Dmitrichenko, M.Yu., A.V. Ivanov, I.A. Bidusenko et al. 2011. Reaction of 1-vinylpyrrole-2-carbaldehydes with phosphorus pentachloride: A stereoselective synthesis of *E*-2-(2-dichloromethylpyrrol-1-yl)vinylphosphonyl dichlorides. *Tetrahedron Lett* 52 (12):1317–1319.
681. Rozinov, V.G., G.A. Pensionerova, V.I. Donskikh et al. 1984. Phosphorus-containing enamines. III. Phosphorylation of N-vinylsubstituted trifluoroacetylpyrroles. *Zh Obshch Khim* 51 (10):2241–2246.
682. Schmidt, A.F., T.A. Vladimirova, E.Yu. Schmidt, and T.V. Dmitrieva. 1995. Regioselective α-arylation of N-vinylpyrroles by Heck reaction. *Izv Akad Nauk Ser Khim* 4:786.

683. Schmidt, A.F., T.A. Vladimirova, and E.Yu. Schmidt. 1997. Regioselectivity of the olefin introduction stage across the Pd–C bond in Heck reaction. *Kinet Catal* 38 (2): 268–273.

684. Schmidt, A.F., A. Khalaika, L.O. Nindakova, and E.Yu. Schmidt. 1998. Mechanism of alkene introduction across the Pd–Ar bond in Heck reaction. *Kinet Catal* 39 (2):216–222.

685. Mal'kina, A.G., O.A. Tarasova, H.D. Verkruijsse et al. 1995. Metallation and functionalization of N-vinylpyrrole. *Recl Trav Chim Pays-Bas* 114 (1):18–21.

686. Tarasova, O.A., F. Taherirastgar, H.D. Verkruijsse et al. 1996. Metallation and functionalization of 1-allenylpyrrole. *Recl Trav Chim Pays-Bas* 115 (2):145–147.

687. Tarasova, O.A., L. Brandsma, H.D. Verkruijsse et al. 1996. Metallation of N-allenylpyrrole with superbase reagents. Synthesis of new N-substituted pyrroles. *Zh Org Khim* 32 (8):1208–1218.

688. Mal'kina, A.G., R. den Besten, A.C.H.T.M. van der Kerk et al. 1995. Dimetallation of N-ethynylpyrrole and subsequent regiospecific derivatization. *J Organomet Chem* 493:271–273.

689. Ganzalez, C., R. Greenhouse, R. Tallabs, and J.M. Muchowski. 1983. Protecting groups for the pyrrole nitrogen atom. The 2-chloroethyl, 2-phenylsulfonylethyl, and related moieties. *Can J Chem* 61 (8):1697–1702.

690. Trofimov, B.A., S.E. Korostova, S.G. Shevchenko et al. 1996. N-Vinylpyrroles as protected pyrroles. *Zh Org Khim* 32 (9):897–899.

691. Schmidt, E.Yu., A.M. Vasil'tsov, N.V. Zorina et al. 2011. Devinylation of N-vinylpyrroles with mercury (II) acetate. *Khim Geterocicl* 10:1570–1573.

692. Iddon, B., J.E. Tonde, M. Hosseini, and M. Begtrup. 2007. The N-vinyl group as a protection group of the preparation of 3(5)-substituted pyrazoles via bromine–lithium exchange. *Tetrahedron* 63:56–61.

693. Trofimov, B.A., E.Yu. Schmidt, A.I. Mikhaleva et al. 2008. Oxidative devinylation of N-vinylpyrroles. *Zh Org Khim* 44 (8):1258–1259.

694. Surhone, L.M., M.T. Timpledon, and S.F. Marseken. 2010. *Pyrrole*. Saarbrücken, Germany: VDM Verlag.

695. Paulson, M. 2010. *Novel Methods for Pyrrole Synthesis*. Saarbrücken, Germany: Lambert Academic Publishing.

696. Mal, D., B. Shome, and B.K. Dinda. 2011. Pyrrole and its derivatives. In *Heterocycles in Natural Products Synthesis*, eds. K.C. Majumdar and S.K. Chattopadhyay, pp. 187–220. Weinheim, Germany: Wiley-VCH Verlag.

697. Trofimov, B.A., L.V. Morozova, M.V. Sigalov et al. 1987. An unexpected mode of cationic oligomerization of 1-vinyl-4,5,6,7-tetrahydroindole. *Macromol Chem* 188:2251–2257.

698. Trofimov, B.A., L.V. Morozova, A.I. Mikhaleva et al. 1989. Oligomerization of N-vinylpyrroles under the action of metal sodium. *Khim Geterocicl* 10:1420–1430.

699. Trofimov, B.A., L.V. Morozova, E.I. Brodskaya et al. 1989. Oligomerization of 1-vinyl-4,5,6,7-tetrahydroindole under the action of metal sodium. *Vysokomol Soed* 31 (12):897–902.

700. Annenkov, V.V., I. Alsarsur, E.N. Danilovtseva et al. 1999. Interaction of copolymers of 1-vinyl-4,5,6,7-tetrahydroindole and maleinic acid with transition metal ions. *Vysokomol Soed A* 41 (9):1404–1408.

701. Trofimov, B.A., A.I. Mikhaleva, L.V. Morozova et al. 2008. Method for the preparation of 1-vinyl-2-[(1–4,5,6,7-tetrahydroindolyl)-ethyl-]-4,5,6,7-tetrahydroindole. USSR Author's Certificate 1,115,430.

702. Morozova, L.V., A.I. Mikhaleva, and M.V. Sigalov. 1986. Dimerization of N-vinylpyrroles in the presence of Bronsted and Lewis acids. *B Acad Sci USSR Ch* 8 (3):128.

703. Morozova, L.V. and A.I. Mikhaleva. 1987. Dimerization of N-vinyl-4,5,6,7-tetrahydroindole in the presence of acids. *Khim Geterocicl* 4:479–480.

704. Trofimov, B.A., A.I. Mikhaleva, L.V. Morozova et al. 2008. Method for the preparation of N-vinyl-2-[1-(1–4,5,6,7- tetrahydroindolyl)-ethyl-]-4,5,6,7-tetrahydroindole. USSR Author's Certificate 1,529,678.
705. Morozova, L.V., A.I. Mikhaleva, M.V. Markova et al. 1996. Dimerization of 1-vinyl-4,5,6,7-tetrahydroindole in the presence of acids. *Izv Akad Nauk Ser Khim* 2:423–425.
706. Markova, M.V. 1999. New aspects of N-vinylpyrroles polymerization. PhD dissertation, Irkutsk State University, Irkutsk, Russia.
707. Trofimov, B.A., M.V. Markova, L.V. Morozova et al. 2001. Synthesis of 5-[1-(1H-pyrrol-1-yl)ethyl]-1-vinyl-1H-pyrroles. *Arkivoc* (ix):24–30.
708. Trofimov, B.A., M.V. Markova, L.V. Morozova et al. 2010. Cationic and radical polymerization of N-vinyl-2-phenylpyrrole: Synthesis of electroconducting, paramagnetic and fluorescent oligomers. *Synthetic Met* 160:1539–1543.
709. Trofimov, B.A., L.V. Morozova, I.L. Filimonova et al. 2008. Photosensitive layer of electrophotographic material. USSR Author's Certificate 997,540.
710. Morozova, L.V., A.I. Mikhaleva, and G.F. Myachina. 1987. Copolymerization of N-vinyl-4,5,6,7-tetrahydroindole with ethanol amine vinyl ether. *J Prikl Khim* 5:1193–1196.
711. Morozova, L.V., M.V. Markova, A.I. Mikhaleva et al. 1990. Synthesis and study of absorption capacity of N-vinyl-4,5,6,7-tetrahydroindole copolymers. *J Prikl Khim* 63 (9):2022–2025.
712. Danilovtseva, E.N., V.V. Annenkov, E.S. Domnina et al. 1996. Copolymerization of 1-vinyl-4,5,6,7-tetrahydroindole with styrene. *Vysokomol Soed B* 38 (11):1925–1927.
713. Morozova, L.V., O.A. Tarasova, A.I. Mikhaleva et al. 1997. Polymerization of 1-allenyl-pyrroles. *Izv Akad Nauk Ser Khim* 11:1958–1961.
714. Shaglaeva, N.S., A.I. Mikhaleva, G.I. Sarapulova et al. 1997. Dehydrochlorination in the course of copolymerization of N-vinyl-4,5,6,7-tetrahydroindole with vinyl chloride or vinylidene dichloride. *Izv Akad Nauk Ser Khim* 12:2267–2268.
715. Danilovtseva, E.N., V.V. Annenkov, and A.I. Mikhaleva. 1998. Synthesis and properties of copolymers of 1-vinyl-4,5,6,7-tetrahydroindole with maleinic acid. *Vysokomol Soed B* 40 (2):366–368.
716. Morozova, L.V., M.V. Markova, O.A. Tarasova et al. 1998. Radical polymerization and copolymerization of 1-allenylpyrrole. *Vysokomol Soed B* 40 (10):1687–1690.
717. Lebedeva, O.V., N.S. Shaglaeva, L.V. Kanitskaya et al. 2001. Coplymerization of N-vinyl-4,5,6,7-tetrahydroindole with acrylamide. *J Prikl Khim* 74 (2):345–346.
718. Annenkov, V.V., O.V. Lebedeva, E.N. Danilovtseva et al. 2001. Synthesis and polyelectrolyte properties of carboxyl-containing copolymers of 1-vinyl-4,5,6,7-tetrahydroindole. *Vysokomol Soed B* 43 (9):1560–1564.
719. Shaglaeva, N.S., O.V. Lebedeva, A.I. Mikhaleva et al. 2002. Coplymerization of 1-vinyl-4,5,6,7-tetrahydroindole with vinyl chloride in the presence of a radical initiator. *J Prikl Khim* 75 (9):1494–1496.
720. Morozova, L.V., I.V. Tatarinova, M.V. Markova et al. 2007. Synthesis and modification of polymers of 1-vinyl-2-pyrrolecarbaldehydes. *Izv Akad Nauk Ser Khim* 11:2134–2138.
721. Trofimov, B.A., M.V. Markova, L.V. Morozova et al. 2007. 2-Arylazo-1-vinylpyrroles: radical polymerization and copolymerization. *Vysokomol Soed B* 49 (12):2200–2205.
722. Petrova, O.V., M.V. Markova, L.N. Sobenina et al. 2009. Synthesis and polymerization of sterically hindered 1-vinyl-2,5-diphenyl- and 1-vinyl-2,3,5-triphenylpyrroles. *Izv Akad Nauk Ser Khim* 1:115–118.
723. Markova, M.V., L.V. Morozova, Yu.B. Monakov et al. 2007. Cholesterol vinyl ether in the reactions of radical homo- and copolymerization. *Vysokomol Soed B* 49 (3):553–558.
724. Shaglaeva, N.S., O.V. Lebedeva, Yu.N. Pozhidaev et al. 2008. Copolymerization of vinyl chloride with 1-vinyl-4,5,6,7-tetrahydroindole and 2-methyl-5-vinylpyridine. *Vysokomol Soed B* 50 (11):2035–2041.

725. Trofimov, B.A., M.V. Markova, I.V. Tatarinova et al. 2010. I$_2$-Doped and pyrrole ring-iodinated semi-conducting oligomers of N-vinyl-3-alkyl-2-phenylpyrroles. *Synthetic Met* 160:2573–2580.

726. Trofimov, B.A., M.V. Markova, L.V. Morozova et al. 2010. Polymerization of 1,4-bis[2-(N-vinyl)pyrrolyl]benzene. *Vysokomol Soed B* 52 (4):662–666.

727. Tatarinova, I.V., L.V. Morozova, M.V. Markova et al. 2011. Copolymerization of N-vinylpyrrole-2-carbaldehydes with styrene, N-vinylpyrrolidone and ethylene glycol vinyl glycidyl ether. *Vysokomol Soed B* 53 (3):475–481.

728. Trofimov, B.A., L.V. Morozova, A.I. Mikhaleva et al. 1991. Polybutylvinyl ether as macromonomer in the process of radical copolymerization. *J Prikl Khim* 9:1959–1963.

729. Krivdin, L.B., D.F. Kushnarev, G.A. Kalabin et al. 1984. Direct ^{13}C–^{13}C spin–spin coupling constants in the vinyl group of monosubstituted ethylene. *Zh Org Khim* 20 (5):949–951.

730. Afonin, A.V., V.K. Voronov, A.I. Mikhaleva et al. 1987. Structure of N-vinyl-2(2-furyl) pyrroles: NMR study. *B Acad Sci USSR Ch* 1:184–186.

731. Afonin, A.V., M.V. Sigalov, V.K. Voronov et al. 1987. Stereospecificity of direct ^{13}C–^{1}H spin–spin coupling constants in the vinyl group of N-vinylpyrroles. *B Acad Sci USSR Ch* 6:1418–1421.

732. Afonin, A.V., M.V. Sigalov, S.E. Korostova et al. 1988. Study of structural effects of substituents in N-vinyl-2-arylpyrroles. *B Acad Sci USSR Ch* 12:2765–2769.

733. Turchaninov, V.K., A.F. Ermikov, S.E. Korostova et al. 1989. Photoelectron spectra and conformational structure of 2-substituted 1-vinylpyrroles. *Zh Obshch Khim* 59 (4):791–796.

734. Afonin, A.V., M.V. Sigalov, S.E. Korostova et al. 1990. Intramolecular interaction in N-vinyl-2-arylpyrroles. *Magn Reson Chem* 28 (7):580–586.

735. Afonin, A.V., M.V. Sigalov, B.A. Trofimov et al. 1991. Conformational study of N-vinyl-2-cyclopropylpyrroles by NMR technique. *B Acad Sci USSR Ch* 5:1031–1038.

736. Strashnikova, N.V., M.V. Sigalov, S.E. Korostova et al. 1993. Electron absorption spectra of protonated 2-(2-furyl)pyrroles, 2-(2-thienyl)pyrroles and 2-arylpyrroles. *B Acad Sci USSR Ch* 6:1060–1062.

737. Golovanova, N.I., L.N. Sobenina, A.I. Mikhaleva, and B.A. Trofimov. 1993. The effects of substituents on frequencies of stretching vibrations in NH-pyrroles. *Zh Org Khim* 29 (7):1319–1324.

738. Afonin, A.V., M.V. Sigalov, and B.A. Trofimov. 1998. Steric effects in ^{1}H and ^{13}C NMR spectra of the annelated cycle in 2,3-annelated 1-vinylpyrroles. *Zh Org Khim* 34 (9):1394–1399.

739. Afonin, A.V., I.A. Ushakov, O.V. Petrova et al. 2000. ^{1}H and ^{13}C NMR study of spatial and electronic structure of a series of 2-(pyridyl)pyrroles. *Zh Org Khim* 36 (7):1074–1080.

740. Turchaninov, V.K., N.M. Murzina, A.I. Vokin et al. 2001. Solvatochromism of hetero-aromatic compounds. VII. Bifurcational hydrogen bonding in solvate complexes of 2-(1,2,2-tricyanovinyl)pyrrole. *Zh Obshch Khim* 71 (2):261–270.

741. Turchaninov, V.K., N.M. Murzina, A.I. Vokin, and B.A. Trofimov. 2001. Solvatochromism of heteroaromatic compounds. VIII. 2-(1,2,2-Tricyanovinyl)-5-phenylpyrrole. *Zh Obshch Khim* 71 (11):1795–1801.

742. Afonin, A.V., S.Yu. Kuznetsova, I.A. Ushakov et al. 2002. Spatial and electronic structure of 2-(2-furyl)- and 2-(2-thienyl)pyrroles according to ^{1}H and ^{13}C NMR data. *Zh Org Khim* 38 (11):1712–1717.

743. Afonin, A.V., I.A. Ushakov, S.Yu. Kuznetsova et al. 2002. C–H⋯X (X = N, O, S) intramolecular interaction in 1-vinyl-2-(2′-heteroaryl)pyrroles as monitored by ^{1}H and ^{13}C NMR spectroscopy. *Magn Res Chem* 40 (7):114–122.

744. Khulugurov, V.M., L.I. Bryukvina, K.B. Petrushenko et al. 2005. Non-linear optical properties of α-arylpyrroles. *Dokl Chem* 402 (3):353–354.

745. Ushakov, I.A. 2006. Structural peculiarities of C2-substituted pyrroles according to ^{1}H, ^{13}C and ^{15}N NMR spectral data. PhD dissertation, Irkutsk State University, Irkutsk, Russia.

746. Rusakov, Yu.Yu., L.B. Krivdin, E.Yu. Schmidt et al. 2007. ^{13}C–^{13}C Spin–spin coupling constants in structural studies. XL. Conformational analysis of N-vinylpyrroles. *Zh Org Khim* 43 (6):882–889.

747. Krivdin, L.B., Yu.Yu. Rusakov, E.Yu. Schmidt et al. 2007. Stereochemical study of 2-substituted N-vinylpyrroles. *Aust J Chem* 60 (8):583–589.

748. Afonin, A.V., I.A. Ushakov, D.E. Simonenko et al. 2007. Comparison of electronic and steric structures of 1-vinyl- and 1-propenylpyrroles according to ^{1}H and ^{13}C NMR data. *Zh Org Khim* 43 (3):398–406.

749. Kobychev, V.B., N.M. Vitkovskaya, E.Yu. Schmidt et al. 2007. Ab initio quantum-chemical study of vinylation of pyrrole and 2-phenylazopyrrole with acetylene in the system KOH/DMSO. *Zh Strukt Khim* 48:S107–S116.

750. Petrushenko, I.K., V.I. Smirnov, K.B. Petrushenko et al. 2007. Fluorescence quenching and laser photolysis of dipyrrolylbenzenes in the presence of chloromethanes. *Zh Obshch Khim* 77 (8):1307–1316.

751. Afonin, A.V., I.A. Ushakov, D.E. Simonenko et al. 2008. Stereospecificity of shielding constants of carbon-13 nuclei in ^{13}C NMR spectra of oximes with heterocyclic substituents. *Khim Geterocicl* 10:1523–1531.

752. Petrushenko, I.K., K.B. Petrushenko, V.I. Smirnov et al. 2009. Deprotonation of short-living radicals of pyrrolylbenzenes. *Khim Geterocicl* 5:705–710.

753. Petrushenko, I.K. 2009. Excited states and photoinduced reactions of pyrrole-based tricyclic conjugated systems. PhD dissertation, Irkutsk State University, Irkutsk, Russia.

754. Petrushenko, K.B., I.K. Petrushenko, V.I. Smirnov et al. 2010. The nature of transitions in electronic absorption spectra of radical cations of dipyrroles with phenylene bridging groups. *Izv Akad Nauk Ser Khim* 4:763–767.

755. Galangau, O., C. Dumas-Verdes, E.Yu. Schmidt et al. 2011. N-Vinyl ferrocenophane pyrrole: Synthesis and physical and chemical properties. *Organometallics* 30:6476–6481.

756. Rusakov, Yu.Yu., L.B. Krivdin, E.Yu. Schmidt et al. 2006. Nonempirical calculations of NMR indirect spin–spin coupling constants. Part 15: Pyrrolylpyridines. *Magn Reson Chem* 44:692–697.

757. Shcherbina, N.A., N.V. Istomina, L.B. Krivdin et al. 2007. ^{13}C–^{13}C spin–spin coupling constants in structural studies. Ab initio calculations: Heteroaromatic oximes. *Zh Org Khim* 43 (6):874–881.

758. Rusakov, Yu.Yu. 2007. Stereochemical investigations of pyrrole derivatives using NMR technique and quantum-chemical calculations. PhD dissertation, Irkutsk Institute of Chemistry, Irkutsk, Russia.

759. Afonin, A.V., A.V. Vashchenko, I.A. Ushakov et al. 2008. Comparative analysis of hydrogen bonding with participation of the nitrogen, oxygen and sulfur atoms in the 2(2'-heteroaryl)pyrroles and their trifluoroacetyl derivatives based on the ^{1}H, ^{13}C, ^{15}N spectroscopy and DFT calculations. *Magn Reson Chem* 46 (5):441–447.

Index